Inhomogeneous
Superconductors

Inhomogeneous Superconductors

Granular and Quantum Effects

EUGEN ŠIMÁNEK

Professor of Physics
University of California
Riverside

New York   Oxford
OXFORD UNIVERSITY PRESS
1994

o545492x
PHYSICS

Oxford University Press

Oxford   New York   Toronto
Delhi   Bombay   Calcutta   Madras   Karachi
Kuala Lumpur   Singapore   Hong Kong   Tokyo
Nairobi   Dar es Salaam   Cape Town
Melbourne   Auckland   Madrid

and associated companies in
Berlin   Ibadan

Copyright © 1994 by Oxford University Press, Inc.

Published by Oxford University Press, Inc.,
200 Madison Avenue, New York, New York 10016

Oxford is a registered trademark of Oxford University Press

Library of Congress Cataloging-in-Publication Data
Šimánek, Eugen.
Inhomogeneous superconductors :
granular and quantum effects /
Eugen Šimánek.
p. cm.
Includes bibliographical references and index.
ISBN 0-19-507828-4
1. Superconductors.
2. Inhomogeneous materials.
3. Fluctuations (Physics)
I. Title.
QC612.S8S57 1994 537.6'236 — dc20 93-13074

2 4 6 8 9 7 5 3 1

Printed in the United States of America
on acid-free paper

PREFACE

The objective of this book is to provide an up-to-date introduction to the theory of quantum fluctuations in inhomogeneous superconductors. The book has evolved from a set of lecture notes for a graduate course ("Troisième Cycle") organized by the Universities of French Switzerland in the summer of 1981. Since many new developments have occurred in the intervening years, a considerable amount of revising, updating, and expanding of the original notes was necessary. My own involvement with the subject began in the late 1970s when I became interested in the possibility of a superconducting analog of the metal-insulator transition in normal systems. Early investigations, based on mean-field approximations, have shown that a superconductor-insulator transition can be induced in a granular superconductor by the charging energy competing with the Josephson coupling between the grains. In a sense the original "Troisième Cycle" notes were written around this theme. A significant point in further development of the subject was the realization that quantum fluctuations of the phase difference across the Josephson junction can be mitigated by the presence of dissipation. A large portion of this book deals with the various ramifications of this idea. The main emphasis is on phase transitions in resistively shunted Josephson arrays, but a thorough discussion is also given of the dynamics and quantum tunneling of vortices in both granular and bulk superconductors.

As far as possible, I have tried to present the material in a unified and deductive way. The effects of dissipation on quantum fluctuations in superconductors are conveniently described in terms of the path-integral methods. I have tried to present these methods in the forms that are actually needed in relevant applications. All derivations are done explicitly, which at times may appear pedantic to a sophisticated reader. But, I believe that a graduate student, attempting to move rapidly to the forefront of this research field, may find this useful. To reduce cluttering of the main text, lengthy derivations are deferred to numerous appendices. A basic knowledge of solid state physics and quantum mechanics is assumed. A certain acquaintance with the phenomenology of superconductors (Ginzburg-Landau approach) is also required. The excellent books *Superconductivity in Metals and Alloys* by de Gennes (1966) and

Introduction to Superconductivity by Tinkham (1975) are recommended for this purpose. Microscopic derivations of superconducting tunneling, in the present book, are done in terms of the thermal Green's function method. A particularly lucid presentation of this method can be found in *Quantum Theory of Many-Particle Systems* by Fetter and Walecka (1971).

I am pleased to thank many friends with whom I have discussed various parts of topics of this book. Although it is impossible to list all here, I should like to acknowledge explicitly collaborations or communications over the years with E. Abrahams, H. Beck, R. Brown, H. Cooper, F. Cummings, P. G. de Gennes, G. Deutscher, J. M. Duan, P. Fazekas, M. P. A. Fisher, B. Giovannini, S. Gregory, A. F. Hebard, J. V. José, H. M. Jaeger, S. Kobayashi, P. Lindenberg, D. E. MacLaughlin, P. Martinoli, A. C. Mota, and K. A. Müller. It was the late W. L. McLean who had initiated my interest in the subject matter of this book. It would not have been possible to complete this book without the collaboration and essential assistance of Richard Brown. The speedy typing, done with perfection by Shelly Couch, provided an incentive to continued progress.

Riverside, Calif.   E.Š.
May, 1993

CONTENTS

Inhomogeneous
Superconductors

1

INTRODUCTION

Inhomogeneous superconductors are often classified as composites, with one component superconducting and the other one normal. The latter may be an insulator, semiconductor, or a normal metal. The structure of an inhomogeneous superconductor may involve layers, superconducting filaments in a normal metal matrix, and superconducting grains embedded in a normal matrix. This book is mostly concerned with the properties of granular superconductors involving an insulating matrix. In such systems the movement of a Cooper pair takes place via Josephson tunneling, which is also responsible for establishing a global phase-coherent state. Our aim is to study the deleterious effects of macroscopic quantum fluctuations on the phase coherence in these and other materials, such as dirty films or high-T_c oxides. A milestone in the development of the subject was the suggestion of Abeles (1977) that long-range phase coherence in granular arrays can be destroyed by the charging energy needed to transfer the Cooper pair across a low-capacitance intergrain junction. When this charging energy exceeds a certain critical value (proportional to the Josephson coupling energy), the granular superconductor undergoes a transition into an insulating state, which can be viewed as a superconductor analog of the metal–insulator transition in the Hubbard model (Hubbard, 1964). Since then there has been extensive literature on the subject of phase ordering in granular superconductors in the presence of quantum-phase fluctuations (see Šimánek, 1985, and references therein).

The subject received its next major impetus from the ideas of Caldeira and Leggett (1981), concerning the role of dissipation in macroscopic quantum tunneling. The extension of these ideas to granular arrays with charging energy led to the experiments of the Minnesota group (Jaeger et al., 1989), which showed that the onset of global-phase coherence in thin films at low temperatures is mostly controlled by the strength of the dissipation. On the theoretical side, there has been a surge of activity focused on the understanding of the observed *universal value of the normal sheet resistance* of the films, at which phase coherence sets in (see Chakravarty et al. 1987, and references therein). The difficulty with these theories derives from the vexed question about the origin of the dissipation. Specifically, the conductance channel due to quasiparticle tunneling is presumably blocked at low temperatures owing to the presence of a superconducting energy gap. Subsequent efforts took into account the renormalization of intergrain capacitance by virtual quasiparticle-tunneling processes, to derive a criterion for the onset of phase coherence at low temperatures (Chakravarty et al., 1987; Ferrell and Mirhashem, 1988). Though these theories bypass the difficulty caused by the absence of quasiparticle dissipation, they derive a criterion for the critical resistance that agrees only qualitatively with the experiments. Moreover, drastic assumptions about the capacitance matrix of the arrays are made, in order to obtain a simplified, near-universal formulation of the problem. In our view, the understanding of the critical sheet resistance is far from complete and requires further work. In this context, it is interesting to invoke the idea of Fisher et al. (1990) suggesting that, at the superconductor–insulator transition, the Cooper pairs are capable of ordinary diffusion, thus furnishing a conductance channel that persists even at $T = 0$.

At finite temperatures, the description of the onset of superconductivity in two-dimensional systems requires a different approach (from the $T = 0$ case) because of the diverging thermal-phase fluctuations, which preclude the possibility of broken gauge symmetry. Berezinskii (1971) and Kosterlitz and Thouless (1973) developed a model of the superfluid–normal transition, involving proliferation of thermally induced vortices. Below the critical temperature, the vortices and antivortices remain *bound*, and the film is *superfluid*. Above this temperature, vortex unbinding takes place and the free vortices lead to a destruction of superfluidity. For this mechanism to work, it is essential that the vortex–antivortex interaction exhibits a *logarithmic dependence* on the distance. While this condition is satisfied in a neutral superfluid, screening by superelectrons in a superconducting film causes the interaction at large intervortex distances to decay faster than logarithmically, preventing the vortex binding at any finite temperature. Beasley et al. (1979) have pointed out that in a thin and dirty film, the effective screening length may exceed the size of the sample so that the logarithmic-interaction law holds approximately within the sample. Thus a vortex-unbinding transition should be present as in other finite-size systems. A large number of experimental studies on superconducting films been published since (Mooij, 1984). Of particular significance for further theoretical development of the subject have been a study of resistive transitions in granular lead films by Hebard (1979) and Hebard and Vandenberg (1980). Their data show an abrupt drop of the vortex-unbinding temperature T_{KT} when the sheet resistance increases beyond 13 kΩ. This is in contrast with the Berezinskii–Kosterlitz–Thouless theory, which predicts a smooth dependence (T_{KT} being inversely proportional to the sheet resistance). In an attempt to resolve this discrepancy, we have considered the effect of quantum fluctuations on vortex unbinding in a granular superconductor with charging energy (Šimánek, 1980b). As the sheet resistance increases, the Josephson coupling energy decreases and so does the intergrain capacitance. This leads to an increase of the quantum-phase fluctuations until the point of instability, at which the renormalized superconducting sheet density drops to zero. Several approaches have been developed to calculate the vortex-unbinding temperature renormalized by these zero-point fluctuations. The self-consistent harmonic approximation (SCHA) predicts a discontinuous drop of the renormalized superfluid density at the critical value of the sheet resistance (Šimánek, 1980b). In the mean-field approximation (MFA), developed by Maekawa et al. (1981), the renormalized vortex-unbinding temperature is proportional to $\langle \cos \theta \rangle_{\mathrm{MFA}}^2$, where θ is the local-phase variable on a grain. Using the method, previously developed for calculating the MFA phase diagram in three-dimensional arrays (Šimánek, 1979), the authors find $\langle \cos \theta \rangle_{\mathrm{MFA}}^2$ to be a nonmonotonic function of the temperature. This leads to a *reentrant* transition of the Kosterlitz–Thouless variety. Specifically, this means that as temperature is lowered the vortices first bind, but on further cooling an unbinding (caused by quantum-phase fluctuations) takes place. The origin of this reentrance is the same as that in the theory of the superconductor–insulator transition in three-dimensional arrays: By allowing the phase variable to vary over the interval $(-\infty, \infty)$, the mean-field Hamiltonian describes a fictitious particle moving in a $\cos \theta$ potential in an *unrestricted* θ space. By virtue of the Floquet theorem, the periodic eigenfunctions are not all 2π periodic; for instance, there is a set of low-energy 4π periodic solutions that are directly responsible for the reentrance. One unusual aspect of these solutions, which has led to considerable controversy, is that they are associated with *odd* eigenvalues of the electron number operator. Clearly, such a mechanism of reentrance would not apply to a lattice-boson model of a granular superconductor. On the other hand, an array of Josephson junctions involving ohmic dissipation may be a good candidate for this mechanism (Kampf and Schön, 1987).

Another interesting realization of macroscopic quantum effects in superconductors can be seen in the dynamics of vortices. For a two-dimensional incompressible fluid, the classical equations of motion for the coordinates of the vortex center are very different from the Newtonian dynamics (Onsager, 1949). Because of the absence of any kinetic energy term in the Hamiltonian, the force acting on the "particle" determines its velocity rather than its acceleration. As a result, the x and y coordinates of the vortex center play the role of canonically conjugate variables. In a canonical quantization procedure, these variables are replaced by a pair of noncommuting operators, yielding an uncertainty relation associated with the zero-point fluctuations of the vortex center (Fetter et al., 1966).

More recently it has been realized that the dynamics of a vortex in a granular superconductor can be *endowed with a Newtonian character*, owing to the fact that a *kinetic energy* term appears in the Hamiltonian of a moving vortex, which has its origin in the *charging energy* (Šimánek, 1983). Like an ordinary quantum particle with *nonzero* inertial mass, such a vortex is capable of *quantum-mechanical tunneling*. Since the vortex motion is usually accompanied by viscous dissipation, the theory of dissipative quantum tunneling of Caldeira and Leggett (1981) can be applied to it (Glazman and Fogel, 1984).

Several experimental studies have been published recently that have focused on measuring possible manifestations of quantum vortex tunneling. Mota et al. (1987, 1988) have observed anomalously high decay rates of the magnetization in high-T_c superconductors at millikelvin temperatures, which they attribute to flux motion by quantum tunneling. Two other relevant experiments study the temperature dependence of the resistance in two-dimensional superconductors. One of them is carried out on an artificially constructed underdamped Josephson junction array (van der Zant et al., 1991). The other investigates disordered ultrathin films (Liu et al., 1992). The results of these studies suggest that vortex tunneling may be responsible for the resistance that has been measured at low temperatures. However, detailed comparison with theory is hindered by our poor knowledge of the detailed vortex structure in the core region.

The book starts with a general introduction of theoretical concepts needed to describe these phenomena. It turns out that most of the essential physics (concerned with quantum fluctuations in superconductors) can be derived from the time-dependent Ginzburg–Landau model, usually known as the "relativistic superconductor." Chapter 2 sets out a classical field theory based on this model. We derive the Hamilton equations of motion for the superfluid density (playing the role of the canonical momentum) and the conjugate variable, the phase of the order parameter. By considering a simple example of an isolated small superfluid, we show that quantum-phase fluctuations do not have to be associated exclusively with the electrostatic charging energy, as traditionally assumed in the field of granular superconductivity. In fact, *the Hamiltonian of a small superfluid particle already* describes the kinetic energy of a planar rotator (rotating in the complex plane of the order parameter). The moment of inertia is proportional to the number of Cooper pairs in the particle and inversely proportional to the Fermi energy. Though this result first emerges from the framework based on the phenomenological time-dependent Ginzburg–Landau equation, it can also be verified by a microscopic calculation using the Green's function method (see Chapter 5). In a Josephson junction with charging energy, the quantum-phase fluctuations are then described by a Hamiltonian in which the charging energy (per pair) is *augmented* by an energy proportional to the electron-level separation, near the Fermi energy (Kubo, 1962). For that reason, we refer to the source of this term as the δ effect. The physical origin of this term can be traced to the compressibility energy associated with the density fluctuation created by adding or removing a Cooper pair from the grain. The δ effect turns out to have ramifications throughout the rest of the book. Experimentally, it is expected to play a role, whenever the charging energy is exceeded by the compressibility energy. For instance, small superconducting grains imbedded in a matrix, made of a normal metal or an insulator with a very high dielectric constant, may be good candidates for the δ effect. For a small particle of ^{3}He superfluid, the δ effect was first reported by Efetof and Salomaa (1981). Chapter 2 also contains a derivation of the mean-square fluctuations of the phase variable in a $(1 + 1)$-dimensional neutral superfluid. It demonstrates the absence of symmetry breaking due to the logarithmically diverging zero-point phase fluctuations (Ferrell, 1964). These results also bring out another important aspect of formal nature. Namely, the phase fluctuations in a one-dimensional system at $T = 0$ are the same as those calculated from classical statistical mechanics in a two-dimensional system. This kind of mapping is generalized to a one-dimensional system in the last section of Chapter 2.

Chapter 3 describes various models of ordered arrays of Josephson grains. After a brief review of superconductivity in an isolated grain, we consider the xy model for a three-dimensional array without the charging energy. The phase-locking transition in this model is described using the MFA. This is followed by a matrix description of the Coulomb energies in regular arrays. Particular attention is given to the geometric capacitance matrix and its relation to the range of the Coulomb interactions. In Section 3.5 we review the calculation of the MFA phase diagram of a three-dimensional array with a short-range form of the Coulomb energy (self-charging model) (Šimánek, 1979). The same type of junction array is described, in Section 3.6, in terms of the $S = 1$ pseudospin model (de Gennes, 1978). Though the MFA results of Sections 3.5 and 3.6 agree, as far as the onset of phase ordering at $T = 0$ is concerned, there is a difference at finite temperatures in that the pseudospin model exhibits no reentrance. This is not surprising, since the pseudospin model admits only integer changes in the number of Cooper pairs on a grain. The MFA phase diagram of an array, involving the off-diagonal Coulomb interactions is discussed in Section 3.7. We describe in detail the method of the phase correlator, developed by Efetof (1980). The last two sections of Chapter 3 consider again the self-charging model of Section 3.5 but with the use of different approaches. In Section 3.8 we introduce, for the first time, the Feynman path integral to study the MFA phase diagram of the self-charging model. There is a remarkable agreement with the ordinary quantum-mechanical approach, substantiating our understanding of the reentrance as a direct consequence of the infinite interval for the phase variable. Section 3.9 describes a variational method, in which the Josephson cosine term (of the self-charging model) is replaced by a trial quadratic term. Solving the relevant self-consistent equation, the amplitude of this term is found to exhibit a discontinuous drop as the ratio of the Josephson coupling constant to the charging energy exceeds a certain critical value. This corresponds to the same instability of global coherence as seen in the MFA approach; however, there are important distinctions. First, in the MFA method, the superfluid density goes to zero continuously at the superconductor–insulator boundary. The precipitous drop found in the SCHA method is believed to be an artifact of the approximation. Moreover, the SCHA method is not capable of capturing the reentrant features, which are inevitably linked to the periodic potential due to the Josephson term in the Hamiltonian.

The effects of dissipation on quantum mechanics of the phase variable, in a resistively shunted tunnel junction, are discussed in Chapter 4. We describe the model of Caldeira and Leggett (1981), in which dissipation is generated by the interaction between the fictitious particle (phase) and a set of environmental oscillators. The effective action of the particle, resulting from integrating out the oscillator degrees of freedom, is derived in detail. As an example of the suppression of quantum effects by dissipation, we present the evaluation of the mean-square displacement of a harmonic oscillator with ohmic dissipation. We also show how the *quantum tunneling* of a particle, coupled to a reservoir, can be formulated in terms of the *functional-integral method*. Explicit expressions for the tunneling rate are derived for the case of weak and strong dissipation. Both limits find their application in Chapter 8, where we discuss quantum tunneling of vortices.

Ambegaokar et al. (1982) have derived the effective action of a Josephson junction with quasiparticle dissipation starting from a microscopic model. The resulting dissipative part of the action differs in an important way from the case of ohmic dissipation, first discussed by Caldeira and Leggett (1981), in that it is a *periodic* functional of the phase difference across the junction. This reflects the discrete nature of the charge of the tunneling quasiparticles. This periodicity has implications for the phase ordering in dissipative junction arrays, which are discussed in Chapter 6. Since the method of Ambegaokar et al. (1982) involves a series of formally involved steps, we divide the problem, for pedagogical reasons, into several parts. We start with the derivation of the partition function of a neutral superfluid. This gives us an opportunity to introduce the method of the Stratonovich (1958) transformation (to treat the electron–electron interaction). As a simple application of this theory, we derive the effective action of a small superfluid particle. The result confirms our previous prediction of the δ effect, based on the time-dependent Ginzburg–Landau equation (see Chapter 2). In Section 5.4 we consider a Josephson tunnel junction, formed by two *identical neutral* superfluid particles. The effective action then contains, in addition to the phase inertial terms (δ effect), two new contributions produced by the tunneling: the Josephson coupling term and the quasiparticle-dissipation term. In the absence of any pair-breaking effects, real quasiparticle-tunneling processes are blocked, as $T \rightarrow 0$, owing to the nonzero superconducting gap. However, virtual transfers of quasiparticles are still present and leading to an effective phase inertia, which, in the case of charged carriers, causes a renormalization of the junction capacitance. This tunneling-induced phase inertia is discussed in Section 5.4.3. Josephson junction with charging energy is studied in Section 5.5. The Coulomb interaction is treated by means of a Stratonovich transformation, which introduces an extra path integration over the *random electrostatic potential* across the junction. The partition function (in the form of a path integral over this potential and the phase variable) is derived, which agrees with the work of Ambegaokar et al. (1982). These authors *eliminate the potential* from the problem by assuming that the time derivative of the phase is pinned to the potential. In Section 5.5.1 we use direct Gaussian path integration to eliminate the potential. The resulting effective action is different from that derived by using the pinning constraint, in that it includes the contribution of the δ effect. Specifically, the effective charging action is modified so that it involves an effective capacitance, corresponding to a *series arrangement* of the *geometric* junction capacitance and the *fictitious* capacitance. The latter capacitance is inversely proportional to the δ energy in the grains. For large-size grains, the δ energy is small and the effective capacitance of the series combination reduces to the geometric capacitance. In this case, the effective action of the junction takes the *standard form*, which agrees with the work of Ambegaokar et al. (1982). In the opposite limit of very small grains, the fictitious capacitance may be smaller than the geometric one. The *extreme* example is the Josephson junction formed by *neutral* superfluid particles. An important role in the contribution of the δ effect plays the fact that the Cooper pairs, transferred by Josephson tunneling, reside in a *thin layer* on the surface of the grains. This is caused by Debye shielding with a screening length, which is of the order of the inverse Fermi wave vector. This effectively *reduces the volume* of the superfluid, thus producing a *larger energy-level splitting* at the Fermi level. Numerical estimates indicate that the δ effect starts playing an important role for junctions of aluminum particles of diameter about 100 Å or less.

In Chapter 6 we discuss Josephson junction arrays with charging energy and dissipation. By extending the Stratonovich transformation of Chapter 5 to an array, the effective action of the system is derived in a form that automatically incorporates the δ effect into the phase inertial term. Derivation of the MFA phase diagram of a dissipative self-charging model is reviewed, following the work by Šimánek and Brown (1986). Then we discuss the work by Chakravarty et al. (1986), which derives the phase diagram of the same model with use of the SCHA. This is followed by the already mentioned MFA theories taking into account the *renormalization of the intergrain capacitance* (Chakravarty et al., 1987; Ferrell and Mirhashem, 1988). These authors assume from the outset that the geometric intergrain capacitance can be neglected compared to the quasiparticle-induced capacitance. In Section 6.7.2 we analyze the criterion for the onset of global phase coherence, taking into account a nonzero geometric intergrain capacitance and using the effective action with the δ effect incorporated. We find that the nonuniversality of the critical sheet resistance, produced by the presence of the geometric intergrain resistance, can be mitigated by the δ effect. This may explain why a universal sheet resistance is usually observed on films involving *smaller* metallic grains (Yamada et al., 1992). In Section 6.8 we derive, following the method of Doniach (1981), the mapping of the two-dimensional Josephson array (with charging energy and at $T = 0$), to a $(2 + 1)$-dimensional Ginzburg–Landau functional. From the latter functional, we determine the criterion for the onset of global coherence at $T = 0$. Within the Gaussian approximation, this criterion agrees precisely with the MFA theory of Efetof (1980). Taking into account the renormalization of the phase fluctuation propagator, due to the quartic term of the functional, we predict that the MFA theory tends to overestimate the critical resistance. Chapter 6 is concluded by a short section on the correlation length in a two-dimensional array with the charging energy.

Chapter 7 is devoted to problems involved in the description of the onset of superconductivity in two-dimensional superconductors at *finite temperatures*. In Section 7.2 we show that, in spite of the *absence* of *conventional ordering*, a two-dimensional Meissner effect exists, as long as the *phase stiffness is nonzero*. As a prelude to the vortex-unbinding theories of the Berezinskii–Kosterlitz–Thouless variety, we present a simple SCHA calculation of the renormalized stiffness constant (originally due to Pokrovskii and Uimin, 1973). The interesting feature of this calculation is that it predicts a precipitous *drop* of the superfluid density at a temperature, *close* to the *vortex-unbinding temperature*. In Section 7.3 we consider vortices in a two-dimensional neutral superfluid, modeled as a continuum version of the xy model. We derive the energy of a single vortex as well as the Hamiltonian of the interacting system of vortices. The vortex-unbinding transition is discussed in Section 7.4 using the dielectric's screening model of the two-dimensional neutral plasma of logarithmically interacting vortex charges. Section 7.5 discusses the modifications of the vortex interactions due to the screening by charged Cooper pairs in superconducting films. The formula for vortex-unbinding temperature as a function of the sheet resistance is derived, following the work by Beasley et al. (1979). The effects of quantum-phase fluctuations on the vortex unbinding are considered in Section 7.6.

Chapter 8 describes the dynamics of superfluid and superconducting vortices. Starting with the classical two-dimensional incompressible fluid, we present a derivation of the basic hydrodynamic theorem, according to which the vortex center moves at the local fluid velocity. With the help of this theorem, we obtain the Hamilton equations of motion for the rectangular coordinates of the vortex. Next, we consider vortex quantization, which leads to a two-dimensional electron analogy first pointed out by Haldane and Wu (1985). In Section 8.4 we pursue this analogy further by evaluating the phase change of the superfluid wave function, as the vortex is transported around a closed a path. In accord with the vortex-electron analogy, this geometric phase is equal to 2π times the mean number of superfluid particles within the contour. By comparing this phase with the Aharonov and Bohm (1959) phase factor for an electron transported around a magnetic flux, we find that to every superfluid particle there is attached a flux line of pseudomagnetic field. In a translationally invariant ideal fluid, this field produces a pseudo-Lorentz force acting on a moving vortex, which exactly cancels the Magnus force. In Section 8.5 we consider the dynamics of vortices in two-dimensional superconductors. We argue that, as long as the material is homogeneous, the pseudo-Lorentz force is present, and the equation of motion for the vortex is of the same form as that in a neutral superfluid. Only if the Galilean invariance is spoiled by inhomogeneities can the pseudo-Lorentz force become small, leading to a small Hall resistance, as usually observed (Tinkham, 1975).

The remaining parts of Chapter 8 involve studies of vortices endowed with a nonzero inertial mass. Starting from the time-dependent Ginzburg–Landau equation of Chapter 2, we discuss the core- and electromagnetic-mass contributions first derived by Suhl (1965). By assuming a perfect electrostatic screening, he sets the charge density on the vortex line equal to zero. This is tantamount to forcing a perfect pinning of the phase variable to the electrostatic potential. Following the analogy with the δ effect in a granular superconductor, we eliminate the potential by a direct Gaussian integration, thus relaxing the pinning constraint. This leads to a deviation of the electromagnetic mass from the prediction of Suhl (1965). It also helps us resolve difficulties with mass infinities resulting in the limit of negligible charging energy. A similar approach is used to deal with the massive vortices in granular superconductors. In Section 8.7.1 we consider a vortex in a two-dimensional array with vanishing intergrain geometric capacitance. This property ensures that the *inverse* capacitance matrix is of *short range*. The resulting effective action contains two kinds of phase inertial terms:

1. The *site-diagonal* terms, which are proportional to an effective capacitance corresponding to a *series combination* of the diagonal geometric capacitance and the fictitious "δ effect capacitance."

2. The *site off-diagonal* terms, proportional to the intergrain capacitance of the *quasiparticle* origin.

A special situation arises when the intergrain capacitance vanishes, so that only the site-diagonal terms are left in the effective action. This leads to the presence of acoustic spin waves, which couple strongly to the moving vortex. In Section 8.7.2 we derive, following Eckern and Schmid (1989), the effective vortex action, describing a dissipation due to this coupling. In Section 8.7.4 we derive the vortex dissipative action for a vortex in a granular array with ohmic dissipation. The action has the form predicted by Caldeira and Leggett (1981) for a particle coupled to a set of environmental oscillators. Moreover, the dissipative coefficient agrees with the vortex

viscosity previously derived by Bardeen and Stephen (1965) for bulk superconductors. The equation of motion for a massive vortex, in a granular
two-dimensional array, is considered in Section 8.7.5. In Section 8.8.1 we
use this equation to discuss the experimental data on underdamped arrays
due to van der Zant et al. (1991). Section 8.8.2 is devoted to a discussion of
the unusual T dependence of the resistivity in disordered superconducting
films, recently measured by Liu et al. (1992). These authors assume that
the vortices contributing to the resistance are strongly damped by an ohmic
dissipation. We argue that such a strong damping would lead to a dissipative localization of vortices, similar to the one recently suggested for regular
arrays by Mooij and Schön (1992). The same section also contains a short
review of the vortex transport via the variable-range hopping (Fisher et al.,
1991) and its possible modifications due to the dissipative localization of
vortices. Section 8.8.4 describes the theory of quantum-flux creep in bulk
superconductors proposed by Blatter et al. (1991) to explain the magnetization decay rates in high-T_c superconductors. We conclude this chapter
by a brief discussion of problems caused by nonadiabatic transitions among
fermion bound states in a core of a *moving* vortex.

Since a host of physical effects have been obtained from a particular framework (based on the time-dependent Ginzburg–Landau model of
Chapter 2), it is legitimate to ask to what degree are these results model
dependent. Let us therefore consider an alternative model, in which the
superconductor represented by a *nonrelativistic* fluid of charged bosons
(Feynman, 1972). Eliminating the density fluctuations from the partition
function, an effective action is obtained that describes phase modes propagating as acoustic phonons at a speed proportional to the square root of
the boson-interaction potential. As shown in Chapter 2, a formally similar phase phonon mode is obtained from the *relativistic* time-dependent
Ginzburg–Landau equation. Consequently, the phase fluctuations, resulting in the fermion case from the δ effect, have a *bosonic* analog. For instance,
the phase variable in a small droplet of ^{4}He tends to diffuse quantum mechanically with a characteristic time proportional to the droplet volume
and inversely proportional to the short-range boson-interaction potential.
Recently, the model of interacting bosons has been applied to granular
superconductors to describe the superconductor–insulator transition (Cha
et al., 1991, and references therein). In many respects, the boson picture
is equivalent to the standard model of Josephson coupled arrays with the
charging energy. The distinction appears mainly at finite temperatures,
owing to the thermal breaking of Cooper pairs. For instance, the boson
model does not yield the reentrant features due to the 4π periodic states.
On the other hand, an important possibility of Cooper pair conduction at
the superconductor–insulator transition has been derived from the boson
model (Fisher, 1990). Moreover, recent fertile ideas of charge vortex duality
have been formulated in terms of this model (Fisher, 1990).

2

DYNAMICS AND FLUCTUATIONS OF THE ORDER PARAMETER

2.1. Fluctuations in classical Ginzburg–Landau model

To start talking about quantum fluctuations of the superconducting order parameter, one must first establish the laws of the classical dynamics of the condensate wave function. Conceptually, it seems reasonable to abandon the microscopic approach and proceed directly from the phenomenological Ginzburg–Landau model. The static version of this model has been used with great success to interpret the thermodynamic and magnetic properties of both the first and second kind of superconductors. A well-written review of these topics can be found in the monographs by de Gennes (1966) and Tinkham (1975). In the static Ginzburg–Landau model, the quantum-mechanical state of a macroscopic number of Cooper pairs is described by a complex order parameter $\psi(\mathbf{r}) = |\psi(\mathbf{r})| \exp[i\theta(\mathbf{r})]$, which is a function of the position $\mathbf{r}$. The function ψ, being identified with wave function of the superconducting component of the system, is often called the *superconducting wave function*. To establish our own notation, we follow Tinkham (1975) and write the Ginzburg–Landau (GL) free-energy density in the superconducting state in the form

$$f = f_{\mathrm{no}} + \alpha|\psi|^2 + \frac{\beta}{2}|\psi|^4 + \frac{1}{2m^*}\left|\left(\frac{\hbar}{i}\boldsymbol{\nabla} - \frac{e^*}{c}\mathbf{A}\right)\psi\right|^2 + \frac{\mathbf{H}^2}{8\pi} \qquad (2.1)$$

where the last term is the magnetic field energy in vacuum, f_{no} is the free-energy density in the normal state, and m^* and e^* are the effective mass and charge of the Cooper pair, respectively. The parameter $\beta(T)$ is regular at the temperature $T = T_c$, whereas the parameter $\alpha(T)$ changes its sign from positive to negative at T_c. Hence, below T_c, a minimum of the free energy (in the absence of fields and gradients of ψ) is established when the order parameter ψ has a magnitude:

$$|\psi_0| = \left(\frac{|\alpha|}{\beta}\right)^{\frac{1}{2}} \qquad (2.2)$$

An infinitesimally small "order parameter field," proportional to $\cos\theta$, can be introduced to fix the phase to $\theta = 0$. Thus, the classical condensate wave

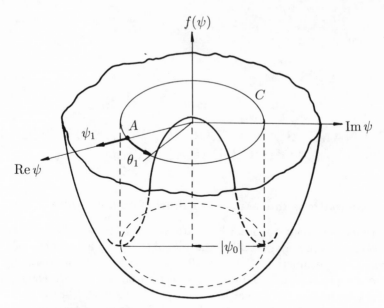

FIG. 2.1.   Plot of the free-energy density $f(\psi) = \alpha|\psi|^2 + \frac{1}{2}\beta|\psi|^4$ as a function of the real and imaginary parts of the complex order parameter ψ. Circle C is the locus of minima of $f(\psi)$. The amplitude and phase fluctuations about the broken-symmetry state (point A) are labeled ψ_1 and θ_1, respectively.

function ψ_{class} exhibits a spontaneously broken gauge symmetry, implying $\langle\psi\rangle = \psi_{\text{class}}$. In the classical Ginzburg–Landau model, gauge symmetry can be restored only by thermal fluctuations of the order parameter. The nature of these fluctuations can be discussed in terms of the plot of $f(\psi)$ in the complex ψ plane. For simplicity, we disregard the gauge fields and assume $T < T_c$. In this case, the plot takes the well-known form of the "Mexican hat" shown in Fig. 2.1 The minima of $f(\psi)$ consist of points on a circle of radius $|\psi_0|$ given by Eq. (2.2). The fluctuations of ψ about the broken-symmetry state (shown as point A in Fig. 2.1) have two components: one along the tangent to the circle at $\theta = 0$, which corresponds to a pure phase fluctuation $\theta_1(\mathbf{r})$, and the other one, which is an amplitude fluctuation ψ_1. Rice (1965) has studied the correlation function $\langle\psi(\mathbf{r})\psi^*(\mathbf{r}')\rangle$ within the quadratic approximation in which the phase and the amplitude fluctuations contribute separately to the free-energy functional. Due to the gauge invariance of Eq. (2.1), the phase fluctuation is a Goldstone boson. This implies that the phase variable enters the free energy as a gradient term proportional to $|\psi_0|^2(\boldsymbol{\nabla}\theta_1)^2$. As a result, the correlator $\langle\psi(\mathbf{r})\psi^*(\mathbf{r}')\rangle$ receives a significant contribution from the phase fluctuation. Specifically, $\langle\psi(\mathbf{r})\psi^*(\mathbf{r}')\rangle$ is proportional to the phase correlator, which can be found by a simple Gaussian path integration and is given by the expression (the derivation of which is given in Appendix A9):

$$I_\theta = \langle \exp\{i[\theta_1(\mathbf{r}) - \theta_1(\mathbf{r}')]\} \rangle$$

$$= \exp\left(-\frac{k_B T m^*}{\Omega |\psi|^2 \hbar^2} \sum_k \frac{1}{k^2}[1 - \cos \mathbf{k} \cdot (\mathbf{r} - \mathbf{r}')]\right) \quad (2.3)$$

For a one-dimensional superconductor, the k sum in Eq. (2.3) yields a factor proportional to $|x - x'|$ so that $I_\theta^{(1d)} \sim \exp(-|x - x'|/\eta)$, where η is a correlation length given by

$$\eta = \frac{2\hbar^2 |\psi|^2}{m^* k_B T} \quad (2.4)$$

The condition for the presence of the off-diagonal long-range order (ODLRO) is (Yang, 1962)

$$\langle \psi(\mathbf{r})\psi^*(\mathbf{r}') \rangle \xrightarrow[|\mathbf{r}-\mathbf{r}'|\to\infty]{} \text{finite constant} \quad (2.5)$$

Then we see from Eqs. (2.3)–(2.5) that the ODLRO is absent in a one-dimensional superconductor, as long as the temperature is finite. In two dimensions, the k summation in Eq. (2.3) yields a term proportional to $\ln|\mathbf{r} - \mathbf{r}'|$, which implies a power law for the phase correlator

$$I_\theta^{(2d)} \sim |\mathbf{r} - \mathbf{r}'|^{-\frac{k_B T m^*}{2\pi\hbar^2 |\psi_0|^2}} \quad (2.6)$$

Recalling Eq. (2.5), this again implies the absence of the ODLRO in the two-dimensional superconductor. In spite of this, the system may retain the superconductivity property as a result of topological order, which keeps the topological defects (vortices) bound below the Kosterlitz–Thouless transition temperature (Kosterlitz and Thouless, 1973). We will revisit this type of ordering in the chapter on two-dimensional superconductors. In three dimensions, the k sum in Eq. (2.3) yields a constant term (independent of $|\mathbf{r} - \mathbf{r}'|$). This assures that the condition (2.5) holds and ODLRO is not destroyed by thermal fluctuations. At $T = 0$, the expression (2.4) suggests that ODLRO is possible even in one dimension. However, it turns out that quantum fluctuations of the phase variable are capable of destroying the ODLRO even at $T = 0$. This has been first shown by Ferrell (1964) and later rediscovered by Coleman (1973), as the absence of spontaneous symmetry breaking in $(1 + 1)$ dimensions (see Section 2.8).

2.2. Time-dependent Ginzburg–Landau model

There are two possible frameworks in which to study the dynamics of the superconducting order parameter. In one of them, the superconducting condensate is formed by a bosonic charged fluid. It satisfies a nonlinear Schrödinger equation of a form similar to the Gross–Pitaevskii equation used previously for superfluid helium (see Fetter and Walecka, 1971; Feynman, 1972). The other prototype model is the $T = 0$ time-dependent Ginzburg–Landau model (TDGL) described by the following Lagrangian density

$$
\mathcal{L} = \frac{3c^2}{2m^* v_f^2} \left| \left(\frac{i\hbar}{c} \frac{\partial}{\partial t} - \frac{e^*}{c} A_0 \right) \psi \right|^2 - \frac{1}{2m^*} \left| \left(\frac{\hbar}{i} \nabla - \frac{e^*}{c} \mathbf{A} \right) \psi \right|^2
$$

$$
- \alpha |\psi|^2 - \frac{\beta}{2} |\psi|^4 + \frac{1}{8\pi} \left(\mathbf{E}^2 - \mathbf{H}^2 \right)
$$

$$
= \mathcal{L}_m + \mathcal{L}_{EM} \tag{2.7}
$$

where $\mathcal{L}_m$ and $\mathcal{L}_{EM}$ denote the matter and the electromagnetic field Lagrangians, respectively. A microscopic derivation of Eq. (2.7) was first attempted by Stephen and Suhl (1964). They considered the problem near T_c and obtained a wavelike TDGL equation for ψ by expanding the Green's functions in small variations of ψ. Subsequently, Abrahams and Tsuneto (1966) have shown that, near T_c, the wavelike variation is dominated by a diffusive term that has been omitted in the work of Stephen and Suhl (1964). Furthermore, Abrahams and Tsuneto (1966) have been able to derive a wavelike TDGL equation at $T = 0$, but of a form somewhat different from Eq. (2.7). Nevertheless, the Lagrangian density (2.7) is a proper gauge-invariant generalization of the static model, which leads to correct results for the quantum phase fluctuations. In Chapter 5, we derive, from a microscopic theory, a phase fluctuation Lagrangian that agrees with that derived from Eq. (2.7). The difference between this equation and the microscopic approach of Abrahams and Tsuneto (1966), as far as the magnitude of the order parameter is concerned, has recently been discussed by Duan and Leggett (1992).

2.2.1. *Noether's theorem and charge conservation*

According to Noether's theorem (see Ramond, 1981), there is associated with the gauge invariance a constant of motion—the electric charge. To show this, consider a global gauge transformation of the complex condensate wave function

$$
\psi(\mathbf{r}, t) \to \psi'(\mathbf{r}, t) = e^{i\alpha} \psi(\mathbf{r}, t)
$$

$$
\psi^*(\mathbf{r}, t) \to \psi'^*(\mathbf{r}, t) = e^{-i\alpha} \psi^*(\mathbf{r}, t) \tag{2.8}
$$

where α is a constant. For α small, Eqs. (2.8) can be written as

$$
\psi \to \psi' = \psi + \delta\psi = \psi + i\alpha\psi
$$

$$
\psi^* \to \psi'^* = \psi^* + \delta\psi^* = \psi^* - i\alpha\psi^* \tag{2.9}
$$

We now vary ψ and ψ^* independently and obtain, for arbitrary $\delta\psi$ and $\delta\psi^*$, a small change in $\mathcal{L}$ that is equal to

$$\delta\mathcal{L} = \frac{\partial\mathcal{L}}{\partial\psi}\delta\psi + \frac{\partial\mathcal{L}}{\partial\psi^*}\delta\psi^* + \frac{\partial\mathcal{L}}{\partial\psi_\mu}\delta\psi_\mu + \frac{\partial\mathcal{L}}{\partial\psi_\mu^*}\delta\psi_\mu^* \qquad (2.10)$$

where $\psi_\mu = \partial\psi/\partial x^\mu$ and $x^\mu = (t, \mathbf{r})$. Using the identity

$$\frac{\partial}{\partial x^\mu}\left(\frac{\partial\mathcal{L}}{\partial\psi_\mu}\delta\psi\right) = \frac{\partial\mathcal{L}}{\partial\psi_\mu}\delta\psi_\mu + \left[\frac{\partial}{\partial x^\mu}\left(\frac{\partial\mathcal{L}}{\partial\psi_\mu}\right)\right]\delta\psi \qquad (2.11)$$

in Eq. (2.10), we obtain

$$\delta\mathcal{L} = \left[\frac{\partial\mathcal{L}}{\partial\psi} - \frac{\partial}{\partial x^\mu}\left(\frac{\partial\mathcal{L}}{\partial\psi_\mu}\right)\right]\delta\psi + \left[\frac{\partial\mathcal{L}}{\partial\psi^*} - \frac{\partial}{\partial x^\mu}\left(\frac{\partial\mathcal{L}}{\partial\psi_\mu^*}\right)\right]\delta\psi^*$$
$$+ \frac{\partial}{\partial x^\mu}\left(\frac{\partial\mathcal{L}}{\partial\psi_\mu}\delta\psi + \frac{\partial\mathcal{L}}{\partial\psi_\mu^*}\delta\psi^*\right) \qquad (2.12)$$

According to the principle of minimum action, we have for the $(d+1)$-dimensional space–time

$$\int_\Omega \delta\mathcal{L}\, d^d\mathbf{r}\, dt = 0 \qquad (2.13)$$

with the boundary condition $\delta\psi = \delta\psi^* = 0$ at the boundary of the volume Ω. We now insert into Eq. (2.13) the variation (2.12). Transforming the volume integral of the last term, by a Gauss theorem, into a surface integral that vanishes on account of the boundary condition, we obtain

$$\frac{\partial\mathcal{L}}{\partial\psi} - \frac{\partial}{\partial x^\mu}\left(\frac{\partial\mathcal{L}}{\partial\psi_\mu}\right) = 0 \qquad (2.14)$$

and a complex conjugate of Eq. (2.14). Assuming that ψ and ψ^* satisfy these Lagrange equations, we go back to Eq. (2.12) and obtain for the variation of $\mathcal{L}$

$$\delta\mathcal{L} = \frac{\partial}{\partial x^\mu}\left(\frac{\partial\mathcal{L}}{\partial\psi_\mu}\delta\psi + \frac{\partial\mathcal{L}}{\partial\psi_\mu^*}\delta\psi^*\right) \qquad (2.15)$$

For the special choice of $\delta\psi$ and $\delta\psi^*$, given by the *global gauge* transformation (2.9), the gauge invariance of $\mathcal{L}$ implies that $\delta\mathcal{L} = 0$, so that Eq. (2.15) yields the equation of continuity

$$i\alpha\frac{\partial}{\partial x^\mu}\left(\frac{\partial\mathcal{L}}{\partial\psi_\mu}\psi - \frac{\partial\mathcal{L}}{\partial\psi_\mu^*}\psi^*\right) = 0 \qquad (2.16)$$

which implies the conservation of the total electric charge. Using the Lagrangian density of Eq. (2.7), we have

$$\frac{\partial\mathcal{L}}{\partial\psi_0} = \frac{\gamma i\hbar}{2m^*c}\left(\frac{\hbar}{ic}\frac{\partial\psi^*}{\partial t} - \frac{e^*}{c}A_0\psi^*\right)$$
$$\frac{\partial\mathcal{L}}{\partial\psi_\mu} = \frac{\hbar}{2m^*i}\left(\frac{\hbar}{i}\frac{\partial\psi^*}{\partial x^\mu} + \frac{e^*}{c}A_\mu\psi^*\right) \qquad (2.17)$$

where $\gamma = 3c^2/v_f^2$ and $\mu = 1, 2, 3$. Introducing the expressions (2.17) into Eq. (2.16), we obtain

$$\left(\frac{i\alpha\hbar}{2m^*}\right) \left\{ \frac{\partial}{\partial t} \left[\frac{\gamma}{c} \left(\frac{\hbar}{c} \frac{\partial \psi^*}{\partial t} - \frac{ie^*}{c} A_0 \psi^* \right) \psi - \text{c.c.} \right] \right.$$
$$\left. - \frac{\partial}{\partial x^\mu} \left[\left(\hbar \frac{\partial \psi^*}{\partial x^\mu} + \frac{ie^*}{c} A_\mu \psi^* \right) \psi - \text{c.c.} \right] \right\} = 0 \qquad (2.18)$$

where $\mu = 1, 2, 3$. Comparing Eq. (2.18) with the standard equation of continuity, the charge and current density of the time-dependent Ginzburg–Landau model are found in the form

$$\rho = \frac{ie^*\gamma\hbar}{2m^*c^2} \left(\psi^* \frac{\partial \psi}{\partial t} - \psi \frac{\partial \psi^*}{\partial t} \right) - \frac{\gamma e^{*2}}{m^*c^2} \psi^* \psi A_0 \qquad (2.19)$$

$$\mathbf{j} = \frac{e^*\hbar}{2m^*i} (\psi^* \boldsymbol{\nabla} \psi - \psi \boldsymbol{\nabla} \psi^*) - \frac{e^{*2}}{m^*c} \psi^* \psi \mathbf{A} \qquad (2.20)$$

where the overall factor, multiplying both the expressions (2.19) and (2.20), is adjusted so that $\mathbf{j}$ agrees with the current density of the time-independent Ginzburg–Landau model (see Tinkham, 1975). We note that these expressions also follow directly from the relations of classical electrodynamics

$$\rho = -\frac{\partial \mathcal{L}}{\partial A_0} \qquad (2.21)$$

$$\mathbf{j} = c \frac{\partial \mathcal{L}}{\partial \mathbf{A}} \qquad (2.22)$$

2.3.   Dynamics of phase and density

The phase variable $\theta(\mathbf{r}, t)$ can be introduced into the Lagrangian density (2.7) by making the substitution

$$\psi(\mathbf{r}, t) = \sqrt{n_0(\mathbf{r}, t)} e^{i\theta(\mathbf{r}, t)} \qquad (2.23)$$

which yields the following form for the matter part of Eq. (2.7)

$$\mathcal{L}_m = \frac{\gamma}{2m^*} \left| \frac{i\hbar}{c} \left(\frac{\partial \sqrt{n_0}}{\partial t} + i\sqrt{n_0} \frac{\partial \theta}{\partial t} \right) - \frac{e^*}{c} \sqrt{n_0} A_0 \right|^2$$
$$- \frac{1}{2m^*} \left| \frac{\hbar}{i} (\boldsymbol{\nabla} \sqrt{n_0} + i\sqrt{n_0} \boldsymbol{\nabla} \theta) - \frac{e^*}{c} \sqrt{n_0} \mathbf{A} \right|^2$$
$$- \alpha n_0 - \frac{\beta}{2} n_0^2$$
$$= \frac{\gamma}{2m^*} \left[\left(\frac{\hbar}{c} \frac{\partial \theta}{\partial t} + \frac{e^*}{c} A_0 \right)^2 n_0 + \frac{\hbar^2}{c^2} \left(\frac{\partial \sqrt{n_0}}{\partial t} \right)^2 \right]$$
$$- \frac{1}{2m^*} \left[\left(\hbar \boldsymbol{\nabla} \theta - \frac{e^*}{c} \mathbf{A} \right)^2 n_0 + \hbar^2 (\boldsymbol{\nabla} \sqrt{n_0})^2 \right]$$

$$- \alpha n_0 - \frac{\beta}{2} n_0^2 \tag{2.24}$$

Inserting this expression into Eq. (2.21), we have (writing $\partial \theta / \partial t = \dot{\theta}$)

$$\rho = -\frac{\partial \mathcal{L}}{\partial A_0} = -\frac{e^*}{\hbar} \frac{\partial \mathcal{L}_m}{\partial \dot{\theta}} \tag{2.25}$$

so that the number density ρ_v (equal to the number of Cooper pairs in a unit volume) satisfies

$$\rho_v = \frac{\rho}{e^*} = -\frac{\partial \mathcal{L}_m}{\partial (\hbar \dot{\theta})} = -\frac{\delta L}{\delta (\hbar \dot{\theta})} \tag{2.26}$$

where L is the volume integral of the Lagrangian density $\mathcal{L}$. Equation (2.26) shows that ρ_v and $-\hbar \theta$ are *canonically conjugate field variables*. This fact constitutes the starting point for the quantization of these fields. It should be pointed out that this result is quite general, as it follows from the *gauge invariance* of the Lagrangian density (2.7). In fact, the relation (2.26) can also be derived from other time-dependent models, such as bosonic charged superfluid.

The conjugation of the number density and $\hbar \theta$ holds also for the neutral versions of the models discussed. Introducing $e^* = 0$ into Eq. (2.24), we have

$$\mathcal{L}_m^{\text{neutral}} = \frac{\gamma}{2m^*} \left(\frac{\hbar}{c} \right)^2 \left[\left(\frac{\partial \theta}{\partial t} \right)^2 n_0 + \left(\frac{\partial \sqrt{n_0}}{\partial t} \right)^2 \right]$$
$$- \frac{\hbar^2}{2m^*} \left[(\nabla \theta)^2 n_0 + (\nabla \sqrt{n_0})^2 \right]$$
$$- \alpha n_0 - \frac{\beta}{2} n_0^2 \tag{2.27}$$

which implies

$$\frac{\partial \mathcal{L}^{\text{neutral}}}{\partial \dot{\theta}} = \frac{\gamma \hbar^2 n_0}{m^* c^2} \dot{\theta} \tag{2.28}$$

The number density $\rho^{\text{neutral}} = \rho / e^*$ is obtained by dividing Eq. (2.19) by e^* and setting $e^* = 0$ afterwards, yielding

$$\rho^{\text{neutral}} = \frac{i \gamma \hbar}{2 m^* c^2} \left(\psi^* \frac{\partial \psi}{\partial t} - \psi \frac{\partial \psi^*}{\partial t} \right) \tag{2.29}$$

Using the ansatz (2.23) in Eq. (2.29), we obtain

$$\rho^{\text{neutral}} = -\frac{\gamma \hbar}{m^* c^2} n_0 \dot{\theta} \tag{2.30}$$

From Eqs. (2.28) and (2.30), we obtain for the neutral time-dependent Ginzburg–Landau superfluid

$$\rho^{\text{neutral}} = -\frac{1}{\hbar}\frac{\partial \mathcal{L}^{\text{neutral}}}{\partial \dot{\theta}} \tag{2.31}$$

showing again that the variables ρ^{neutral} and $-\hbar\theta$ form a canonically conjugate pair.

The Euler–Lagrange equations for the classical field $\theta(\mathbf{r}, t)$ can be written in terms of the functional derivatives of the Lagrangian L

$$\frac{\partial}{\partial t}\left(\frac{\delta L}{\delta \dot{\theta}}\right) - \frac{\delta L}{\delta \theta} = 0 \tag{2.32}$$

We have in general

$$\frac{\delta L}{\delta \theta} = \frac{\partial \mathcal{L}}{\partial \theta} - \nabla \cdot \frac{\partial \mathcal{L}}{\partial (\nabla \theta)} \tag{2.33}$$

$$\frac{\delta L}{\delta \dot{\theta}} = \frac{\partial \mathcal{L}}{\partial \dot{\theta}} - \nabla \cdot \frac{\partial \mathcal{L}}{\partial (\nabla \dot{\theta})} \tag{2.34}$$

From Eq. (2.24) it follows that $\partial \mathcal{L}/\partial(\nabla \dot{\theta}) = 0$, so that Eqs. (2.26) and (2.34) yield

$$\frac{\delta L}{\delta \dot{\theta}} = \frac{\partial \mathcal{L}_m}{\partial \dot{\theta}} = -\frac{\hbar\rho}{e^*} \tag{2.35}$$

The form of the RHS of Eq. (2.24) also implies that

$$\frac{\delta L}{\delta \theta} = -\nabla \cdot \frac{\partial \mathcal{L}}{\partial (\nabla \theta)} = \left(\frac{\hbar c}{e^*}\right)\nabla \cdot \frac{\partial \mathcal{L}}{\partial \mathbf{A}} = \left(\frac{\hbar}{e^*}\right)\nabla \cdot \mathbf{j} \tag{2.36}$$

where we have used Eq. (2.22). Introducing the expressions (2.35) and (2.36) into the Euler–Lagrange equation (2.32), we recover the equation of charge conservation

$$\frac{\partial \rho}{\partial t} + \nabla \cdot \mathbf{j} = 0 \tag{2.37}$$

2.3.1.   Hamilton's equations

The Hamilton equations of motion for the canonical "momentum" ρ_v and the conjugate coordinate φ defined as

$$\varphi = -\hbar\theta \tag{2.38}$$

are

$$\dot{\varphi} = \frac{\delta H}{\delta \rho_v} \tag{2.39}$$

and

$$\dot{\rho}_v = -\frac{\delta H}{\delta \varphi} \tag{2.40}$$

where the Hamiltonian H is given by the spatial integral of the Hamiltonian density

$$\mathcal{H} = \rho_v \dot{\varphi} - \mathcal{L} \tag{2.41}$$

The explicit form of the equations (2.39) and (2.40) can be found by first inserting the ansatz (2.23) into Eq. (2.19), which yields for the canonical "momentum"

$$\rho_v = \frac{\gamma n_0}{m^* c^2} (\dot{\varphi} - e^* A_0) \tag{2.42}$$

Next, we write the Lagrangian density (2.24) in terms of φ as

$$\mathcal{L}_m = \frac{\gamma n_0}{2 m^* c^2} (\dot{\varphi} - e^* A_0)^2 + \widetilde{\mathcal{L}}_m \tag{2.43}$$

where $\widetilde{\mathcal{L}}_m$ are the terms that do *not* involve $\dot{\varphi}$ or A_0. Introducing Eqs. (2.42) and (2.43) into (2.41), we obtain for the Hamiltonian density of the whole system, composed of the superconductor and the electromagnetic field

$$\begin{aligned} \mathcal{H} &= \frac{\gamma n_0}{2 m^* c^2} (\dot{\varphi} - e^* A_0)^2 + \frac{\gamma n_0 e^* A_0}{m^* c^2} (\dot{\varphi} - e^* A_0) - \widetilde{\mathcal{L}}_m - \mathcal{L}_{EM} \\ &= \frac{m^* c^2}{2 \gamma n_0} \rho_v^2 + \rho_v e^* A_0 - \widetilde{\mathcal{L}}_m - \mathcal{L}_{EM} \end{aligned} \tag{2.44}$$

The RHS of Eq. (2.44) implies

$$\frac{\delta H}{\delta \rho_v} = \frac{\partial \mathcal{H}}{\partial \rho_v} = \frac{m^* c^2}{\gamma n_0} \rho_v + e^* A_0 = \dot{\varphi} \tag{2.45}$$

showing that the Hamilton's equation (2.39) is satisfied and is essentially equivalent to the relation (2.42). Equation (2.40) is equivalent to the continuity equation (2.37). In fact, from Eq. (2.41) we have with the use of Eqs. (2.33) and (2.36)

$$-\frac{\delta H}{\delta \varphi} = \frac{\delta L}{\delta \varphi} = -\boldsymbol{\nabla} \cdot \frac{\partial \mathcal{L}}{\partial (\boldsymbol{\nabla} \varphi)} = \frac{1}{\hbar} \boldsymbol{\nabla} \cdot \frac{\partial \mathcal{L}}{\partial (\boldsymbol{\nabla} \theta)} = -\frac{1}{e^*} \boldsymbol{\nabla} \cdot \mathbf{j} \tag{2.46}$$

Inserting this result into Eq. (2.40), we obtain Eq. (2.37).

2.4. Phase–number commutator

For a system described by the Lagrangian density (2.24), the field $\theta(\mathbf{r}, t)$ satisfies the classical equation of motion (2.32). The canonical momentum, conjugate to the variable $\varphi = -\hbar\theta$, is equal to ρ_v [see Eq. (2.26)]. According to the canonical quantization procedure, the system is quantized by imposing the following commutation relations (at equal time)

$$\begin{aligned} [\widehat{\rho}_v(\mathbf{r}, t), \varphi(\mathbf{r}', t)] &= -\hbar[\widehat{\rho}_v(\mathbf{r}, t), \theta(\mathbf{r}', t)] = -i\hbar\, \delta(\mathbf{r} - \mathbf{r}') \\ [\widehat{\rho}_v(\mathbf{r}, t), \widehat{\rho}_v(\mathbf{r}', t)] &= [\theta(\mathbf{r}, t), \theta(\mathbf{r}', t)] = 0 \end{aligned} \tag{2.47}$$

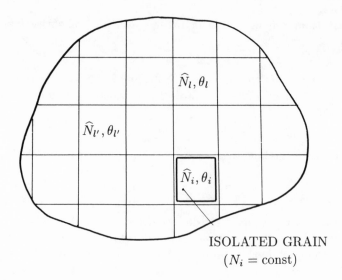

ISOLATED GRAIN
($N_i = \text{const}$)

FIG. 2.2.   A bulk superconductor partitioned into cells. Each cell is specified by an operator of the number of Cooper pairs and a phase variable averaged over the volume of the cell. The extra box around the ith cell represents a perfect insulation.

For granular system and arrays involving Josephson-coupled supercon-ductors of small size, it is important to know the commutator of the phase and number on a *particular grain*. This can be obtained from Eqs. (2.47) by the following *coarse-graining* procedure. We divide the volume of the superconductor into cells of a small volume $\Delta\Omega$, each containing a large number of Cooper pairs (see Fig. 2.2) The operator of the number of pairs in the lth cell of volume $\Delta\Omega_l = \Delta\Omega$ is

$$\widehat{N}_l = \int_{\Delta\Omega_l} d^3\mathbf{r}\, \widehat{\rho}_v(\mathbf{r}, t) \tag{2.48}$$

For the same volume, we define the average phase

$$\theta_l = \frac{1}{\Delta\Omega_l} \int_{\Delta\Omega_l} d^3\mathbf{r}\, \theta(\mathbf{r}, t) \tag{2.49}$$

The desired commutator of $\widehat{N}_l$ and θ_l is obtained by double volume inte-gration of Eq. (2.47)

$$\int_{\Delta\Omega_l} d^3\mathbf{r} \int_{\Delta\Omega_{l'}} d^3\mathbf{r}'\, [\rho_v(\mathbf{r}, t), \theta(\mathbf{r}', t)] = i \int_{\Delta\Omega_l} d^3\mathbf{r} \int_{\Delta\Omega_{l'}} d^3\mathbf{r}'\, \delta(\mathbf{r} - \mathbf{r}')$$
$$= i(\Delta\Omega_l)\delta_{l,l'} \tag{2.50}$$

Recalling Eqs. (2.48) and (2.49), we obtain from Eq. (2.50)

$$\left[\widehat{N}_l, \theta_{l'}\right] = i\delta_{l,l'} \tag{2.51}$$

So far this commutation relation has been derived from a cellular model of a homogeneous superconductor described by the Lagrangian density (2.7). For an inhomogeneous system with partitions between the cells (in Fig. 2.2) that are only "weakly transparent" to Cooper pairs, the Lagrangian density must be suitably modified. However, the *gauge invariance* of $\mathcal{L}$ imposes that the variables θ and A_0 appear again in the combination $(\hbar\dot{\theta}+e^*A_0)$, so the step (2.25) leading to the commutator (2.47) remains *unchanged*. Hence, the phase–number commutation relation (2.51) is expected to hold also for a cellular model of weakly coupled grains. The next question is what happens to the phase–number relation as the temperature is raised substantially above zero. In the limit $T \rightarrow T_c$, the dynamics is diffusionlike (see Tinkham, 1975), but the equation for the order parameter ψ involves the time-dependent term in a gauge-invariant form $[\partial/\partial t + (ie^*/\hbar)A_0]\psi$, so that we can again invoke the same argument, as previously, to show that Eqs. (2.25) and (2.47) are *valid* in this case as well. It should be noticed, however, that the *quantum dynamics* of the phase is now quite *different* from that based on the Lagrangian Eq. (2.7). In the latter dynamics, the quantum fluctuations of the phase variable stem from:

1. The first (time-dependent) term in Eq. (2.24) that is second order in $\dot{\theta}$;

2. The electric-field energy term $\mathbf{E}^2/8\pi$. Note that the electric field couples to the variable $\dot{\theta}$ through the Poisson equation [see Eq. (2.42)]

$$\mathbf{\nabla}\cdot\mathbf{E} = 4\pi\rho = -\frac{4\pi\gamma n_0 e^*}{m^*c}\left(\hbar\dot{\theta} + e^*A_0\right) \tag{2.52}$$

While, in bulk homogeneous superconductors, the source (2) is usually insignificant, it can play a major role in granular superconductors or arrays where relatively strong electric fields exist between the oppositely charged grains. Quantum effects in such systems are conveniently analyzed using lattice models, which are the subject of Chapter 3.

For $l = l'$, the commutation relation (2.51) can be recast into an equivalent form

$$\left[e^{i\theta_l}, \widehat{N}_l\right] = e^{i\theta_l} \tag{2.53}$$

For a proof of (2.53), we write Eq. (2.51) for $l = l'$ in the form (discarding the cell index l)

$$\left[\theta, \widehat{N}\right] = -i \tag{2.54}$$

By a repeated application of Eq. (2.54), we obtain

$$\left[\theta^k, \widehat{N}\right] = -ik\theta^{k-1} \tag{2.55}$$

Expanding the exponential function on the LHS of Eq. (2.53) and using Eq. (2.55), we have

$$\left[e^{i\theta}, \widehat{N}\right] = \sum_{k=0}^{\infty} \frac{1}{k!}\left[(i\theta)^k, \widehat{N}\right] = \sum_{k=1}^{\infty} \frac{(i\theta)^{k-1}}{(k-1)!} = \sum_{k'=0}^{\infty} \frac{(i\theta)^{k'}}{k'!} \tag{2.56}$$

Since the RHS of Eq. (2.56) is equal to $e^{i\theta}$, Eq. (2.53) is verified to be true, whenever Eq. (2.54) holds. The significance of the form (2.53) is that it is consistent with the *quantization* of the *number* N_l in integer values. To show this, consider an isolated subsystem (shown as the ith cell) of our cellular model. As no pairs can enter and leave this cell, the number N_i is a constant of motion. The corresponding number eigenstate for the ith cell is $|N_i\rangle$. Applying the operator equation (2.53) on $|N_i\rangle$, we have

$$e^{i\theta_i}\widehat{N}_i|N_i\rangle - \widehat{N}_ie^{i\theta_i}|N_i\rangle = e^{i\theta_i}|N_i\rangle \tag{2.57}$$

Since $|N_i\rangle$ is a charge eigenstate, it follows that

$$\widehat{N}_i|N_i\rangle = N_i|N_i\rangle \tag{2.58}$$

Using Eq. (2.58) in Eq. (2.57), we obtain

$$\widehat{N}_ie^{i\theta_i}|N_i\rangle = (N_i - 1)e^{i\theta_i}|N_i\rangle \tag{2.59}$$

Equation (2.59) shows that the state $e^{i\theta_i}|N_i\rangle$ is again a *number eigenstate* with the *eigenvalue* $N_i - 1$. The same steps can be repeated by first applying Eq. (2.53) on $e^{i\theta_i}|N_i\rangle$ and then showing that $e^{2i\theta_i}|N_i\rangle$ is a number eigenstate with the eigenvalue $N_i - 2$, etc. In this way, we see that $\widehat{N}_i$ is an integer-valued operator. It has been shown by Susskind and Glogower (1964) that the commutation relation (2.54) runs into difficulties when the operator $\widehat{N}$ has only positive eigenvalues (such as the photon number operator). This is irrelevant here, since the operator $\widehat{N}_l$ defined in Eq. (2.48) can take both the positive and negative eigenvalues. This follows from Eq. (2.19), which yields a charge density of both signs dependent upon the choice of the time-dependent factor $e^{\pm i\omega t}$ in the solution $\psi(\mathbf{r}, t)$. Physically, $\widehat{N}_l$ represents *a deviation from the average pair number* in an electrically *neutral* cell. Adding or withdrawing pairs from the neutral cell corresponds to positive or negative eigenvalues of $\widehat{N}_l$. Clearly, the average number of pairs in a cell must be large to allow for a sufficiently broad range of $|N_l|$.

Associated with the commutation relation (2.51), there is a corresponding uncertainty relation for $\Delta\theta_l$ and ΔN_l. If two operators $\widehat{A}$ and $\widehat{B}$ satisfy a commutation relation

$$\left[\widehat{A}, \widehat{B}\right] = iC \tag{2.60}$$

where C is a c number, then

$$\Delta A \, \Delta B \geq \frac{C}{2} \qquad (2.61)$$

Applying Eqs. (2.60) and (2.61) to the commutator (2.51), we obtain the uncertainty relation in the form

$$\Delta N_l \, \Delta \theta_{l'} \geq \frac{1}{2} \delta_{l,l'} \qquad (2.62)$$

This relation is very useful for qualitative assessments of the charging effects on the destruction of phase coherence in granular superconductors.

2.5. Quantum phase fluctuations in an isolated particle

We now consider space- and time-dependent fluctuations of the condensate wave function about the point A in Fig. 2.1. We therefore write

$$\psi(\mathbf{r}, t) = \left[\sqrt{\bar{n}_0} + \psi_1(\mathbf{r}, t) \right] e^{i\theta_1(\mathbf{r}, t)} \qquad (2.63)$$

where $\bar{n}_0$ =constant, and θ_1 and ψ_1 are small fluctuations. Introducing Eq. (2.63) into the Lagrangian density (2.27) and keeping only terms up to second order in θ_1 and ψ_1, we have

$$\mathcal{L}_m^{\text{neutral}} = \frac{\hbar^2 \bar{n}_0}{2m^*} \left[\frac{\gamma}{c^2} \left(\frac{\partial \theta_1}{\partial t} \right)^2 - (\boldsymbol{\nabla} \theta_1)^2 \right]$$
$$+ \frac{\hbar^2}{2m^*} \left[\frac{\gamma}{c^2} \left(\frac{\partial \psi_1}{\partial t} \right)^2 - (\boldsymbol{\nabla} \psi_1)^2 \right] \qquad (2.64)$$

We see that the amplitude and phase fluctuations become decoupled in this approximation. Let us return to the case of a subsystem, consisting of an isolated small particle, of volume $\Delta\Omega$, for which the spatially averaged phase $\bar{\theta}_1(t)$ is defined by Eq. (2.49). According to the uncertainty relation (2.62), we have $\Delta\theta_1 = \infty$, since perfect isolation implies $\Delta N = 0$. This result can be also understood by analyzing the evolution of the phase variable starting from Eq. (2.64). For simplicity, we disregard the amplitude fluctuations, use Eq. (2.64) for a spatially averaged phase ($\boldsymbol{\nabla}\bar{\theta}_1 = 0$) and obtain for the Lagrangian

$$L = \mathcal{L}_m^{\text{neutral}} \Delta\Omega = \frac{\hbar^2 \bar{n}_0 \gamma \Delta\Omega}{2m^* c^2} \left(\frac{\partial \bar{\theta}_1}{\partial t} \right)^2 \qquad (2.65)$$

The classical Hamiltonian corresponding to Eq. (2.65) is

$$H = \int_{\Delta\Omega} d^3 r \, \mathcal{H} = \frac{\hbar^2 \bar{n}_0 \gamma \Delta\Omega}{2m^* c^2} \left(\frac{\partial \bar{\theta}_1}{\partial t} \right)^2 \qquad (2.66)$$

which describes the dynamics of a *free planar rotator*. Since the number N and $-\hbar\bar{\theta}_1$ are canonically conjugate [see Eq. (2.26)], we have with the use of Eq. (2.65)

$$N = -\frac{\partial L}{\partial(\hbar\dot{\bar{\theta}}_1)} = -\frac{\hbar\bar{n}_0\gamma\Delta\Omega}{m^*c^2}\frac{\partial\bar{\theta}_1}{\partial t} \qquad (2.67)$$

Equation (2.67) enables us to express the Hamiltonian (2.66) in terms of N

$$H = \frac{m^*c^2}{2\bar{n}_0\gamma\Delta\Omega}N^2 \qquad (2.68)$$

The quantization of this system is done by imposing the commutation relation (2.54)

$$\left[\bar{\theta}_1,\widehat{N}\right] = -i \qquad (2.69)$$

which implies that $\widehat{N}$ is an operator of the form

$$\widehat{N} = i\frac{\partial}{\partial\bar{\theta}_1} \qquad (2.70)$$

From Eqs. (2.68) and (2.70), the Hamiltonian operator becomes

$$\widehat{H} = -\frac{m^*c^2}{2\bar{n}_0\gamma\Delta\Omega}\left(\frac{\partial}{\partial\bar{\theta}_1}\right)^2 \qquad (2.71)$$

The wave function $\psi(\theta,t)$ of this planar rotator satisfies the time-dependent Schrödinger equation

$$i\hbar\frac{\partial\psi(\theta,t)}{\partial t} = -\frac{m^*c^2}{2\bar{n}_0\gamma\Delta\Omega}\frac{\partial^2\psi(\theta,t)}{\partial\theta^2} \qquad (2.72)$$

where we write $\bar{\theta}_1 = \theta$, for simplicity. We solve this equation with the initial condition

$$\psi(\theta,t=0) = \delta(\theta) \qquad (2.73)$$

Taking into account of the fact that $\psi(\theta,t)$ must be a periodic function of θ, we make the ansatz

$$\psi(\theta,t) = \sum_{n=-\infty}^{\infty} c_n(t)e^{in\theta} \qquad (2.74)$$

Since the physical interval for the phase variable is $\theta \in [0,2\pi]$, the points $\theta = 0$ and 2π are *indistinguishable*. Such a *constraint distinguishes* the planar rotator from a *free* particle for which the sum (2.74) can be replaced by a Fourier integral over the *continuous* momentum variable. Introducing the expansion (2.74) into Eq. (2.72), we find a condition for c_n

$$\dot{c}_n = -iDn^2c_n \qquad (2.75)$$

where

$$D = \frac{m^* c^2}{2\bar{n}_0 \gamma \hbar \Delta \Omega} \tag{2.76}$$

Using the initial condition (2.73), we have $c_n(t = 0) = 1/2\pi$, and the solution of Eq. (2.75) becomes

$$c_n(t) = \frac{1}{2\pi} \exp\left(-iDn^2 t\right) \tag{2.77}$$

Inserting this result into Eq. (2.74), the wave function can be written as

$$\psi(\theta, t) = \frac{1}{2\pi} \sum_{n=-\infty}^{\infty} \exp\left(-iDn^2 t + in\theta\right) \tag{2.78}$$

The expression on the RHS of Eq. (2.78) defines the Jacobi theta function (Bellman, 1961). The summation over n can be executed by means of the Poisson formula (A4.15). Substituting $\alpha\beta = 2iDt$ for the product in the expression (A4.6), and noting the equivalence of Eqs. (A4.6) and (A4.14), we obtain from the latter formula

$$\psi(\theta, t) = \frac{1}{(4\pi iDt)^{\frac{1}{2}}} \sum_{l=-\infty}^{\infty} \exp\left(\frac{i}{4Dt}(\theta - 2\pi l)^2\right)$$

$$= \frac{\exp\left(i\theta^2/4Dt\right)}{(4\pi iDt)^{\frac{1}{2}}} \sum_{l=-\infty}^{\infty} \exp\left(\frac{i\pi^2}{Dt}l^2 - \frac{i\pi}{Dt}l\theta\right) \tag{2.79}$$

Comparing the l sum on the RHS of Eq. (2.79) with Eq. (2.78), we have

$$\psi(\theta, t) = \frac{\exp\left(i\theta^2/4Dt\right)}{(iDt/\pi)^{\frac{1}{2}}} \psi^*\left(\frac{\theta\pi}{Dt}, \frac{\pi^2}{D^2 t}\right) \tag{2.80}$$

which is the fundamental identity, satisfied by the Jacobi theta function (Bellman, 1961). Taking the absolute value squared of Eq. (2.80), we see that the width of the probability density, associated with $\psi(\theta, t)$, increases with time as

$$\Delta\theta = \frac{1}{\pi}Dt \tag{2.81}$$

From Eq. (2.81), it follows that the wave packet spreads over an interval $[0, 2\pi]$ within the "dephasing time" t_d given by

$$t_d \simeq \frac{2\pi^2}{D} = \frac{12\pi^2 \hbar \bar{n}_0 \Delta\Omega}{m^* v_F^2} = \frac{6\pi^2 \hbar \bar{n}_0 \Delta\Omega}{\epsilon_F} \tag{2.82}$$

where ϵ_F is the Fermi energy. This result suggests that rapid dephasing is expected in a particle with a small number of Cooper pairs and a large Fermi energy. For an orientational estimate of t_d, let us consider superconducting aluminum, for which $\bar{n}_0 \simeq 10^{23}$ and $\epsilon_F \simeq 12$ eV. Using these values in Eq. (2.82), we obtain $t_d = 3 \times 10^8$ s, for a bulk sample of volume $\Delta\Omega = 1$ cm^3. In a small Al-particle, 100 Å in diameter, the dephasing time is only 1.6×10^{-10} s. Nevertheless, for an isolated superconductor, such rigid phase motion is not expected to lead to any measurable effects. For instance, the diamagnetic moment, caused by intraparticle supercurrents, will not be affected by this rigid phase rotation. However, in a junction or array of coupled superconductors, the quantum phase diffusion acts to disrupt the global phase coherence of the system, very much like the charging energy in the self-charging model of Josephson-coupled superconducting arrays (see Chapter 3). Abrahams and Tsuneto (1966) point out that the $|\partial\psi/\partial t|^2$ term, in the Lagrangian (2.7), is due to the fact that superconductivity involves only electrons near the Fermi surface. This observation is related to the fact that the Fermi superfluid exhibits a finite compressibility determined by the Fermi velocity. This *mechanism* is expected to be operational in *all superfluid Fermi* systems. In fact, Efetof and Salomaa (1981) found a Lagrangian of the form (2.65) for the phase fluctuations in small volumes of liquid ^{3}He at $T = 0$.

2.6. Josephson-coupled superconductors

Let us consider two superconductors that can exchange Cooper pairs. The operators of the number of pairs, $\widehat{N}_1$ and $\widehat{N}_2$, and the phases, θ_1 and θ_2, satisfy the commutation relation (2.51). The operator of the total number of pairs, $\widehat{N} = \widehat{N}_1 + \widehat{N}_2$, is, clearly, a constant motion. It turns out that the phase difference, $\theta_1 - \theta_2$, commutes with the operator $\widehat{N}$. Using Eq. (2.51), we have

$$[\theta_1 - \theta_2, \widehat{N}] = [\theta_1, \widehat{N}_1] - [\theta_2, \widehat{N}_2] = 0 \qquad (2.83)$$

Equation (2.83) implies the possibility of a simultaneous sharp measurement of the phase difference $\theta_1 - \theta_2$ and the total number N. This fact is the basis of the phase-coherence phenomena in superconducting junctions, predicted by Josephson (1962) and first observed by Anderson and Rowell (1963).

In what follows, we derive the Josephson coupling energy, using the phenomenological approach of Ferrell and Prange (1963). A microscopic derivation of the Josephson coupling constant J is given in Section 5.4.1. The Hamiltonian, describing the exchange of the Cooper pairs, is

$$\widehat{H}_J = -\frac{J}{2}(\widehat{a}_1^\dagger \widehat{a}_2 + \widehat{a}_2^\dagger \widehat{a}_1) \qquad (2.84)$$

where the operator $\widehat{a}_1^\dagger \widehat{a}_2$ acts to transfer an electron pair from superconductor 2 to 1. We assume that the superconductors are of a large size, so that the Coulomb energy associated with the charge transfer can be neglected. Let $|N_1, N_2\rangle$ denote a state with well-defined numbers of pairs, N_1 and N_2, in superconductors 1 and 2, respectively. This state has the same energy as a state produced from the *original state* by a transfer of any number, n, of pairs. For simplicity, we assume that, before the transfer takes place, $N_1 = N_2 = \bar{N}$. The state, created by a transfer of n pairs, is then

$$|\bar{N} + n, \bar{N} - n\rangle = |n\rangle \tag{2.85}$$

where $n = 0, \pm 1, \pm 2, \ldots$. The degeneracy of the states (2.85) is removed by the operator (2.84), for which the matrix elements are assumed to be

$$\langle n|\widehat{H}_J|n \pm 1\rangle = -\frac{J}{2} \tag{2.86}$$

Following the standard degenerate perturbation theory, we make an ansatz

$$|\psi\rangle = \sum_{n=-\infty}^{\infty} c_n|n\rangle \tag{2.87}$$

and find the coefficients c_n by substituting Eq. (2.87) into the Schrödinger equation

$$(\widehat{H}_0 + \widehat{H}_J)|\psi\rangle = E|\psi\rangle \tag{2.88}$$

For the moment, we disregard all gauge fields, so that $\widehat{H}_0$ is the Hamiltonian of the two *uncoupled neutral* superconductors. The only contributions to $\widehat{H}_0$ that depend on N_1, and N_2 are those given in Eq. (2.68), yielding

$$\widehat{H}_0 = \widehat{h}_0 + \frac{m^*c^2}{\bar{n}_0\gamma\Delta\Omega}n^2 \tag{2.89}$$

where h_0 is the part of the Hamiltonian that is independent of the number of Cooper pairs. We note that N, in Eq. (2.68), has been replaced by n, giving rise to the second term in Eq. (2.89). This term acts to remove the degeneracy of the states $|n\rangle$, even in the absence of the coupling Hamiltonian (2.84). We shall assume that the volume $\Delta\Omega$ of the superconductors is so large that the n^2 term in Eq. (2.89) can be neglected. Then

$$\widehat{H}_0 = \widehat{h}_0 \tag{2.90}$$

and

$$\widehat{h}_0|n\rangle = E_0|n\rangle \tag{2.91}$$

Introducing Eq. (2.87) into Eq. (2.88), we obtain with use of Eqs. (2.86) and (2.91)

$$(E_0 - E)c_n - \frac{J}{2}(c_{n+1} + c_{n-1}) = 0 \tag{2.92}$$

This is a homogeneous difference equation of second order, which is solved by assuming

$$c_n = be^{-in\theta} \tag{2.93}$$

Inserting this ansatz into Eq. (2.92), we obtain the eigenvalues of the system energy in the form

$$E = E_0 - J\cos\theta \qquad (2.94)$$

The second term in Eq. (2.94) is the well-known Josephson coupling energy (Josephson, 1962). From Eqs. (2.87) and (2.93), the wave function is

$$|\psi\rangle = \text{const} \times \sum_{n=-\infty}^{\infty} |n\rangle e^{-in\theta} \qquad (2.95)$$

Eq. (2.95) describes a state with a well-defined phase θ. This can be verified by comparing the wave function $|\psi\rangle$ with the coherent state $|\psi_c\rangle$ which is an eigenstate of the destruction operator for the harmonic oscillator (Schiff, 1968). We have

$$|\psi_c(\theta, t)\rangle = \sum_n b_n e^{-in\theta} |\psi_n(t)\rangle \qquad (2.96)$$

where $|\psi_n(t)\rangle$ is the nth, time-dependent, energy eigenstate of the harmonic oscillator. The coefficient b_n is given by

$$b_n = (n!)^{-\frac{1}{2}} f_0^n e^{-\frac{f_0^2}{2}} \qquad (2.97)$$

where $f_0 = d(m\omega_0/2\hbar)^{\frac{1}{2}}$, d being the amplitude of the coherent oscillations of the wave packet $|\psi_c(\theta, t)\rangle$, and m and ω_0 are the mass and the resonant frequency of the oscillator, respectively. The proportion of the state $|n\rangle$ in the wave packet (2.96) (given by $|b_n|^2$) is, according to Eq. (2.97), given by the Poisson distribution P_n

$$|b_n|^2 = P_n = \frac{e^{-\bar{n}}(\bar{n})^n}{n!} \qquad (2.98)$$

where $\bar{n} = f_0^2$. The width of the function (2.98) (plotted versus n) is equal to $\bar{n}$. Thus a flat (n independent) distribution is obtained in the limit $f_0^2 \to \infty$. This corresponds physically to a wave packet, oscillating with an amplitude $d \gg (\delta x^2)^{\frac{1}{2}}$ where δx^2 is the mean-square amplitude of the zero-point motion. Eq. (2.95) can be matched against Eq. (2.96) in this limit of perfectly coherent motion with well-defined phase.

2.7.   Charging energy in a Josephson junction

When the two superconductors comprising a Josephson junction are small, the charging energy, associated with the transfer of a Cooper pair, may not be any more negligible in comparison with the Josephson coupling energy E_1. At the same time, there appears a new contribution (that has the same form as the charging energy) coming from the first term of Eq. (2.7). To derive a Hamiltonian for the combined system of the junction plus the electric field, we start from Eq. (2.44). We neglect the fluctuations of the superfluid density and put $n_0(\mathbf{r},t) = \bar{n}_0 = $ const. In the absence of the magnetic field, the contribution of the *kinetic* energy of supercurrents to $\widetilde{\mathcal{L}}_m$ *vanishes*, and we obtain from Eq. (2.24)

$$\mathcal{L}_{EM} + \widetilde{\mathcal{L}}_m = \mathcal{L}_{EM} - \alpha \bar{n}_0 - \frac{\beta}{2} \bar{n}_0^2 = \frac{\mathbf{E}^2}{8\pi} + \text{const} \qquad (2.99)$$

Using this result in Eq. (2.44), we obtain (in the absence of Josephson coupling)

$$\mathcal{H}_0 \simeq \frac{m^* c^2}{2\gamma \bar{n}_0} \rho_v^2(\mathbf{r},t) + \rho_v(\mathbf{r},t) e^* A_0(\mathbf{r},t) - \frac{\mathbf{E}^2(\mathbf{r},t)}{8\pi} \qquad (2.100)$$

Integrating this equation over the volume yields a Hamiltonian

$$H_0 = \int d^3r\, \mathcal{H}_0$$

$$= \int_{\Omega_1 + \Omega_2} d^3r \left[\frac{m^* c^2}{2\gamma \bar{n}_0} \rho_v^2 + \rho_v e^* A_0 \right] - \int_\Omega d^3r\, \frac{\mathbf{E}^2}{8\pi} \qquad (2.101)$$

where Ω_1 and Ω_2 denote the volumes of the first and second superconductors, respectively. The last term of Eq. (2.101) is integrated over the entire space. Assuming that n Cooper pairs are transferred, say, from the first to the second superconductor, we have

$$n = \int_{\Omega_1} d^3r\, \rho_v = -\int_{\Omega_2} d^3r\, \rho_v \qquad (2.102)$$

As a result of the screening of the electric fields inside the superconductors, the excess charge ne^* resides on the surface within a layer of thickness t, which is of the order of the Debye shielding length. Then Eqs. (2.101) and (2.102) imply, assuming two equal superconductors of the same size

$$H_0 = \frac{m^* c^2}{\gamma \bar{n}_0 St} n^2 + ne^* (A_0^{(1)} - A_0^{(2)}) - E_c \qquad (2.103)$$

where S is the surface area of one superconductor and $A_0^{(i)}$ is the electrostatic potential on the surface of the ith superconductor $(i = 1, 2)$. E_c is the electrostatic energy, needed to charge the capacitance C of the junction

$$E_c = \frac{(e^* n)^2}{2C} = \frac{2e^2}{C} n^2 \qquad (2.104)$$

We note that the second term in Eq. (2.103) is actually equal to $2E_c$ so that

$$H_0 = \left(\frac{2\epsilon_F}{3\bar{n}_0 St} + \frac{2e^2}{C} \right) n^2 = E_p + E_c \qquad (2.105)$$

When the pair transfer Hamiltonian is taken into account, the Josephson coupling term (2.94) must be included into the net Hamiltonian H and we obtain with the use of (2.105)

$$H = H_0 - E_1 \cos\theta = \frac{U}{2}n^2 - J\cos\theta \qquad (2.106)$$

where U is given by

$$\frac{1}{2}U = \frac{2\epsilon_F}{3\bar{n}_0 St} + \frac{2e^2}{C} \qquad (2.107)$$

In previous treatments, the Hamiltonian of the tunnel junction is also of the form (2.106), but the constant U is given just by the charging energy (Anderson, 1964). The correction term in Eq. (2.107) may be significant, if the superconductors comprising the junction are small enough. For instance, taking again the example of an Al particle, 100 Å in diameter, we obtain for the first term on the RHS of Eq. (2.107), using $t \sim k_F^{-1} \sim 0.57 \times 10^{-8}$ cm, a value of 4.5×10^{-3} eV. This is comparable with the charging energy of a junction with the capacitance of order 10^{-16} F. We note that capacitance of this magnitude is expected in microscopic junctions of such small particles, indicating the possibility of E_p and E_c being of a comparable magnitude. However, it should be kept in mind that the E_p term has been derived from the "relativistic" Lagrangian (2.7), which is essentially a *zero-temperature result*. The temperature dependence of this term is not known, but as temperature is raised towards the BCS transition temperature T_c, the wavelike dynamics is expected to *fade away*. This is because the calculations of Abrahams and Tsuneto (1966) show that, near T_c, the dynamics is *relaxational*.

The quantization of the Hamiltonian (2.106) proceeds by applying the commutation relations (2.51) to each superconductor. Since $\widehat{N}_1 = \bar{N} + \widehat{n}$ and $\widehat{N}_2 = \bar{N} - \widehat{n}$, we have

$$[\widehat{N}_1, \theta_1] = [\widehat{n}, \theta_1] = i \qquad (2.108)$$

$$[\widehat{N}_2, \theta_2] = -[\widehat{n}, \theta_2] = i \qquad (2.109)$$

Adding Eqs. (2.108) and (2.109) and letting $\theta_1 - \theta_2 = \theta$, we obtain

$$[\widehat{n}, \theta_1 - \theta_2] = [\widehat{n}, \theta] = 2i \qquad (2.110)$$

It follows from Eq. (2.110), that $\widehat{n}$ and the phase difference θ are conjugate variables, such that the operator for $\widehat{n}$ has the form

$$\widehat{n} = 2i\frac{\partial}{\partial\theta} \qquad (2.111)$$

Introducing this operator into Eq. (2.106), the Hamiltonian operator for the Josephson junction becomes

$$\widehat{H} = -2U\frac{\partial^2}{\partial\theta^2} - J\cos\theta \qquad (2.112)$$

This Hamiltonian corresponds to a planar rotator in an external periodic potential. The corresponding Schrödinger equation for the stationary state $\psi_n(\theta)$ is

$$\left(-2U \frac{\partial^2}{\partial \theta^2} - J \cos \theta\right) \psi_n(\theta) = E_n \psi_n(\theta) \qquad (2.113)$$

An important question is: "Is there a broken symmetry in the ground state?" Equation (2.113) is a Mathieu's equation, for which solutions can be found as expansions in terms of a small parameter $q = -J/U$. In particular, we have (for the ground state) to first order in q (Abramowitz and Stegun, 1964)

$$\psi_0(\theta) \simeq \frac{1}{\sqrt{2\pi}} \left(1 + \frac{J}{2U} \cos \theta\right) \qquad (2.114)$$

Using this result, we calculate the expectation value of the "order parameter," $\cos \theta$, to be

$$\langle \psi_0 | \cos \theta | \psi_0 \rangle = \frac{1}{\sqrt{2\pi}} \int_{-\pi}^{\pi} d\theta \left(1 + \frac{J}{2U}\right) \cos \theta = \frac{J}{2U} \qquad (2.115)$$

Equation (2.115) indicates the presence of the spontaneous symmetry breaking as long as the ratio J/U is finite. This is clearly a result of the periodic modulation of the wave function (2.114). Increasing the ratio J/U makes this modulation stronger, and a larger value of $\langle \psi_0 | \cos \theta | \psi_0 \rangle$ is expected. As the particle tunnels between the minima of the potential, there is obviously a *restoration* of broken symmetry (if such a restoration is *defined via the expectation value of* θ). For phase transitions in junction arrays, it is, however, the quantity $\langle \psi_0 | \cos \theta | \psi_0 \rangle$ that should be used to characterize the physical state of the broken symmetry.

2.8. Absence of symmetry breaking in 1D superconductors

The destruction of the ODLRO in $(1+1)$-dimensional systems, first demonstrated by Ferrell (1964), represents one of the first examples of the role played the quantum phase fluctuations in the restoration of broken gauge symmetry in superfluids. To present the essential arguments, we confine ourselves to a neutral superfluid. We start from the Lagrangian density (2.27) in which the phase and amplitude fluctuations contribute independently. Since we are interested only in a qualitative answer, we neglect the amplitude fluctuations. Then the Lagrangian density for the phase fluctuation $\theta_1(x,t)$ is

$$\mathcal{L} = \frac{\hbar^2 \bar{n}_0}{2m^*} \left[\frac{\gamma}{c^2} \left(\frac{\partial \theta_1}{\partial t}\right)^2 - \left(\frac{\partial \theta_1}{\partial x}\right)^2\right] \qquad (2.116)$$

where $\bar{n}_0$ is the average density of the superfluid, $(\bar{n}_0)^{\frac{1}{2}}$ specifying the distance of the point A, in Fig. 2.1, from the origin. The broken symmetry is characterized, at $T = 0$, by nonzero ground-state expectation value, $\langle 0| \cos \theta_1(x,t)|0\rangle$. The phase θ_1 is considered as an extended variable $\theta_1 \in [-\infty, \infty]$, to allow for fluctuations, in which the vector $(\bar{n}_0)^{\frac{1}{2}} e^{i\theta_1(x,t)}$ encircles the origin any number of times. It should be noted that the points $\theta_1 = 0$ and 2π are considered as *distinguishable*, in this case. This is in contrast to the case of a particle on a circle where these points are indistinguishable, leading to a topological constraint discussed in Appendix A3. The Lagrangian density (2.116) describes a free, scalar, and massless field for which the canonical quantization procedure is well known (Henley and Thirring, 1962). First, we find the classical conjugate-momentum density

$$\pi(x,t) = \frac{\partial \mathcal{L}}{\partial \dot{\theta}_1} = \frac{\hbar^2 \bar{n}_0 \gamma}{m^* c^2} \dot{\theta}_1 = -\hbar \rho^{\text{neutral}}(x,t) \tag{2.117}$$

where we have used Eqs. (2.31) and (2.116). The corresponding Hamiltonian density is

$$\mathcal{H}(x,t) = \frac{1}{2\mu} \pi^2(x,t) + \frac{\kappa}{2} \left(\frac{\partial \theta_1(x,t)}{\partial x} \right)^2 \tag{2.118}$$

where

$$\mu = \frac{\hbar^2 \bar{n}_0 \gamma}{m^* c^2}, \qquad \kappa = \frac{\hbar^2 \bar{n}_0}{m^*} \tag{2.119}$$

Introducing the Fourier expansions,

$$\theta_1(x,t) = L^{-\frac{1}{2}} \sum_k \theta_k(t) e^{ikx} \tag{2.120}$$

$$\pi(x,t) = L^{-\frac{1}{2}} \sum_k \pi_k(t) e^{ikx} \tag{2.121}$$

into Eq. (2.118), the Hamiltonian for the one-dimensional superfluid of length L becomes

$$H = \int_0^L dx \, \mathcal{H}(x,t) = \sum_k \left(\frac{1}{2\mu} \pi_k^* \pi_k + \frac{\kappa k^2}{2} \theta_k^* \theta_k \right) \tag{2.122}$$

where we used the reality conditions

$$\theta_k^* = \theta_{-k}, \qquad \pi_k^* = \pi_{-k} \tag{2.123}$$

The Hamiltonian (2.122) corresponds to a collection of harmonic oscillators, each of mass μ, κk^2 being the spring constant of the kth mode. The quantization of this scalar field proceeds by imposing the equal-time commutation relations for the canonically conjugate operators $\hat{\theta}_1(x)$ and $\hat{\pi}(x)$

$$[\hat{\theta}_1(x), \hat{\pi}(x')] = i\hbar \, \delta(x - x')$$
$$\left[\hat{\theta}_1(x), \hat{\theta}_1(x') \right] = [\hat{\pi}(x), \hat{\pi}(x')] = 0 \tag{2.124}$$

Inverting the Fourier transforms (2.120) and (2.121) and using the commutation relations (2.124), we obtain

$$[\widehat{\theta}_k, \widehat{\pi}_{k'}] = i\hbar\delta_{k,k'}$$
$$\left[\widehat{\theta}_k, \widehat{\theta}_{k'}\right] = [\widehat{\pi}_k, \widehat{\pi}_{k'}] = 0 \qquad (2.125)$$

Introducing the creation and annihilation operators, $\widehat{a}_k^\dagger$ and $\widehat{a}_k$, through the relations

$$\widehat{\theta}_k = \left(\frac{\hbar}{2\omega_k\mu}\right)^{\frac{1}{2}}\left(\widehat{a}_k + \widehat{a}_{-k}^\dagger\right)$$
$$\widehat{\pi}_k = i\left(\frac{\hbar\omega_k\mu}{2}\right)^{\frac{1}{2}}\left(\widehat{a}_k^\dagger - \widehat{a}_{-k}\right) \qquad (2.126)$$

the operator for the Hamiltonian (2.122) is diagonalized as follows

$$\widehat{H} = \sum_k \left(\frac{1}{2\mu}\widehat{\pi}_k^\dagger\widehat{\pi}_k + \frac{\kappa k^2}{2}\widehat{\theta}_k^\dagger\widehat{\theta}_k\right) = \sum_k \hbar\omega_k\left(\widehat{a}_k^\dagger\widehat{a}_k + \frac{1}{2}\right) \qquad (2.127)$$

where ω_k is the resonant frequency of the ith mode, given by Eq. (2.119)

$$\omega_k = \left(\frac{\kappa k^2}{\mu}\right)^{\frac{1}{2}} = k\left(\frac{c^2}{\gamma}\right)^{\frac{1}{2}} = k\left(\frac{v_F}{\sqrt{3}}\right) \qquad (2.128)$$

The mean-square fluctuation of θ_1 (at $T = 0$), defined as $\langle 0|\widehat{\theta}_1^2(x)|0\rangle$, is calculated using Eq. (2.120), as follows

$$\langle 0|\widehat{\theta}_1^2(x)|0\rangle = \frac{1}{L}\sum_{k,k'} e^{i(k+k')x}\langle 0|\widehat{\theta}_k\widehat{\theta}_{k'}|0\rangle$$
$$= \frac{1}{L}\sum_{k,k'} \langle 0|\widehat{\theta}_k\widehat{\theta}_{-k}|0\rangle\delta_{k,-k'}e^{i(k+k')x}$$
$$= \frac{1}{L}\sum_k \langle 0|\widehat{\theta}_k\widehat{\theta}_{-k}|0\rangle \qquad (2.129)$$

where we have imposed a translational invariance through the $\delta_{k,-k'}$ term. Introducing the decomposition (2.126), we calculate the quantity on the RHS of Eq. (2.129)

$$\langle 0|\widehat{\theta}_k\widehat{\theta}_{-k}|0\rangle = \left(\frac{\hbar}{2\omega_k\mu}\right)\langle 0|(\widehat{a}_k + \widehat{a}_{-k}^\dagger)(\widehat{a}_{-k} + \widehat{a}_k^\dagger)|0\rangle$$
$$= \left(\frac{\hbar}{2\omega_k\mu}\right)\langle 0|\widehat{a}_k\widehat{a}_k^\dagger|0\rangle$$
$$= \frac{\hbar}{2\omega_k\mu} \qquad (2.130)$$

Substituting this result into Eq. (2.129), we obtain with the use of Eq. (2.128)

$$\langle 0|\widehat{\theta}_1^2(x)|0\rangle = \frac{\hbar}{2L\mu}\sum_k \frac{1}{\omega_k} = \left(\frac{m^* v_F}{2\sqrt{3}\hbar\bar{n}_0}\right)\frac{1}{L}\sum_k \frac{1}{k} \tag{2.131}$$

Passing to the limit $L \to 0$, the k sum becomes an integral and Eq. (2.131) goes over to

$$\lim_{L\to\infty} \langle 0|\widehat{\theta}_1^2|0\rangle = \left(\frac{m^* v_F}{2\sqrt{3}\hbar\bar{n}_0}\right)\frac{1}{2\pi}\int_0^{k_{\max}} \frac{dk}{k} = \infty \tag{2.132}$$

The integral on the RHS of Eq. (2.132) exhibits a logarithmic divergence in the long-wavelength limit, which makes the mean-square phase fluctuation $\langle 0|\widehat{\theta}_1^2|0\rangle$ to diverge. That this implies the absence of broken gauge symmetry can be shown using the identity (see Ma and Rajaraman, 1975)

$$\langle 0| \cos\widehat{\theta}_1(x)|0\rangle = \exp\left[-\frac{1}{2}\langle 0|\widehat{\theta}_1^2(x)|0\rangle\right] \tag{2.133}$$

For a proof of this identity, we use Eqs. (2.120) and (2.126) to write $\widehat{\theta}_1(x)$ as

$$\widehat{\theta}_1(x) = \left(\frac{\hbar}{2L\omega_k\mu}\right)^{\frac{1}{2}}\sum_k \left(\widehat{a}_k e^{ikx} + \widehat{a}_k^\dagger e^{-ikx}\right) = \widehat{\theta}_1^-(x) + \widehat{\theta}_1^+(x) \tag{2.134}$$

Using Eq. (2.134), we have for the LHS of Eq. (2.133)

$$\langle 0| \cos\widehat{\theta}_1(x)|0\rangle = \mathrm{Re}\,\langle 0|e^{i\widehat{\theta}_1(x)}|0\rangle = \mathrm{Re}\,\langle 0|e^{i[\widehat{\theta}_1^+(x)+\widehat{\theta}_1^-(x)]}|0\rangle \tag{2.135}$$

The RHS of Eq. (2.134) can be written using the operator identity (Messiah, 1961)

$$e^{\widehat{A}+\widehat{B}} = e^{\widehat{A}}e^{-\frac{1}{2}[\widehat{A},\widehat{B}]}e^{\widehat{B}} \tag{2.136}$$

This identity holds for two noncommuting operators $\widehat{A}$ and $\widehat{B}$ that satisfy the conditions

$$[\widehat{A},[\widehat{A},\widehat{B}]] = [\widehat{B},[\widehat{A},\widehat{B}]] = 0 \tag{2.137}$$

From Eq. (2.136) we have, setting $\widehat{A} = i\widehat{\theta}_1^+(x)$, $\widehat{B} = i\widehat{\theta}_1^-(x)$,

$$e^{i[\widehat{\theta}_1^+(x)+\widehat{\theta}_1^-(x)]} = e^{i\widehat{\theta}_1^+(x)}e^{-\frac{1}{2}C}e^{i\widehat{\theta}_1^-(x)} \tag{2.138}$$

where

$$C = -[\widehat{\theta}_1^+(x),\widehat{\theta}_1^-(x)] = -\frac{\hbar}{2L\mu}\sum_{k,k'} \frac{1}{\omega_k}[\widehat{a}_k^\dagger,\widehat{a}_{k'}]e^{i(k'-k)x}$$

$$= \frac{\hbar}{2L\mu}\sum_k \frac{1}{\omega_k} = \langle 0|\widehat{\theta}_1^2(x)|0\rangle \tag{2.139}$$

where we have invoked Eq. (2.131) to obtain the last equality on the RHS of Eq. (2.139). Using this result in Eq. (2.138) and taking the ground-state expectation value of the latter, we have

$$\langle 0|e^{i[\widehat{\theta}_1^+(x)+\widehat{\theta}_1^-(x)]}|0\rangle = \langle 0|e^{i\widehat{\theta}_1^+(x)}e^{-\frac{1}{2}C}e^{i\widehat{\theta}_1^-(x)}|0\rangle = e^{-\frac{1}{2}C} \qquad (2.140)$$

where we used the property of the annihilation operator $\widehat{\theta}_1^-$, yielding

$$e^{-i\widehat{\theta}_1^-(x)}|0\rangle = \left[1 - i\widehat{\theta}_1^-(x) + \cdots\right]|0\rangle = |0\rangle \qquad (2.141)$$

Using Eqs. (2.135), (2.139), and (2.140), the identity (2.133) is verified. The divergence of $\langle 0|\widehat{\theta}_1^2|0\rangle$, shown in Eq. (2.132), yields, in conjunction with Eq. (2.133), the final result

$$\lim_{L\to\infty} \langle 0|\cos\widehat{\theta}_1(x)|0\rangle = 0 \qquad (2.142)$$

According to Eq. (2.142), the infinite fluctuation in the phase precludes the possibility of long-range order of the phase variable in one-dimensional systems. However, it turns out that these "Goldstone mode" fluctuations are not capable of destroying the superconductivity of one-dimensional samples, such as narrow strips and whiskers. The finite resistance in such systems, at finite temperatures below T_c, is a result of topology-changing fluctuations (phase slips). The probability for the formation of a phase slip center, due to a thermal tunneling process, is proportional to $\exp(-\delta F/k_B T)$, where δF is the difference in free energy between neighboring states. The theory of resistive transition in narrow superconducting channels, based on this mechanism, was first proposed by Langer and Ambegaokar (1967) and further developed by McCumber and Halperin (1970) and by Tucker and Halperin (1971). These theories seem to provide a satisfactory explanation of the experimental temperature dependence of the resistance in tin whiskers (Newbower et al., 1972). The thermal tunneling fluctuations become exponentially weak as $T \to 0$, and the remaining quantum phase fluctuations (of the Goldstone kind) do not produce any observable resistance, implying that the $T = 0$ superconductivity is not prevented by the absence of ODLRO.

2.9. Quantum critical phenomena

It is interesting to compare the phase fluctuations, given by Eq. (2.131), with their classical counterpart, which appears in the exponent of Eq. (A9.21). If the latter expression is specialized to two-dimensional systems, the k sum becomes, in the $\Omega \to \infty$ limit, a one-dimensional k integral

$$\frac{1}{\Omega}\sum_k \frac{1}{k^2} \to \frac{1}{(2\pi)^2}\int \frac{d^2k}{k^2} = \int \frac{dk}{k} \qquad (2.143)$$

which shows that, as far as the order parameter goes, the one-dimensional system at $T = 0$ is equivalent to a two-dimensional *classical* system. This property illustrates a general feature of time-dependent Ginzburg–Landau models, in which space and time enter the Lagrangian in the same way. Specifically, a d-dimensional system exhibits, at $T = 0$, a quantum critical behavior of the same nature as that of the $(1 + 1)$-dimensional classical systems (Hertz, 1976). This can be verified by considering the case of a d-dimensional neutral superconductor described by the Lagrangian density [see Eq. (2.7) with $e^* = 0$]

$$\mathcal{L} = \frac{3\hbar^2}{2m^* v_F^2} \left| \frac{\partial \psi}{\partial t} \right|^2 - f(\psi, \boldsymbol{\nabla}\psi) \tag{2.144}$$

where $f(\psi, \boldsymbol{\nabla}\psi)$ is the free-energy density of the time-independent Ginzburg–Landau model [see Eq. (2.1) with $e^* = 0$]

$$f(\psi, \boldsymbol{\nabla}\psi) = f_{no} + \alpha|\psi|^2 + \frac{\beta}{2}|\psi|^4 + \frac{\hbar^2}{2m^*}|\boldsymbol{\nabla}\psi|^2 \tag{2.145}$$

The partition function for this model is, according to Eq. (A2.24), given by the path integral over *space- and time*-dependent field $\psi(\mathbf{r}, \tau)$

$$Z = \int \mathcal{D}\psi(\mathbf{r}, \tau) \exp\left(-\frac{1}{\hbar} S[\psi, \dot{\psi}]\right) \tag{2.146}$$

where $S[\psi, \dot{\psi}]$ is the Euclidean action functional [see Eq. (A2.21)]

$$S[\psi, \dot{\psi}] = \int_{-\frac{\hbar\beta}{2}}^{\frac{\hbar\beta}{2}} d\tau \int d^d x \, \mathcal{L}_e[\psi, \dot{\psi}] \tag{2.147}$$

In Eq. (2.147) the τ-integration limits are written in a symmetrized form. $\mathcal{L}_e[\psi, \dot{\psi}]$ is the *Euclidean* Lagrangian density, given by

$$\mathcal{L}_e[\psi, \dot{\psi}] = \frac{3\hbar^2}{2m^* v_F^2} \left| \frac{\partial \psi}{\partial \tau} \right|^2 + f^{(d)}(\psi, \boldsymbol{\nabla}\psi) \tag{2.148}$$

where $f^{(d)}$ is the free-energy density of Eq. (2.145) involving a d-*dimensional* gradient of $\psi(\mathbf{r})$. An important feature of Eq. (2.148) is that the space and the time derivative enter $\mathcal{L}_e$ with the *same sign*. In the quantum $(T = 0)$ limit, the τ integration limits in Eq. (2.147) extend to $[-\infty, \infty]$, so that the action becomes a multiple integral of $\mathcal{L}_e$ over the $(d + 1)$-dimensional Euclidean space–time $\{x_0 = \tau, x_1, \ldots, x_d\}$

$$S^{(T=0)}[\psi, \dot{\psi}] = \int_{-\infty}^{\infty} dx_0 \int_{-\infty}^{\infty} dx_1 \cdots \int_{-\infty}^{\infty} dx_d \, \mathcal{L}_e[\psi, \dot{\psi}] \tag{2.149}$$

In the classical (high-temperature) limit, the partition function (2.146) becomes [see Eq. (A9.3)]

$$Z^{\text{class}} = \int \mathcal{D}\psi(\mathbf{r}) \, e^{-\beta F[\psi, \boldsymbol{\nabla}\psi]} \tag{2.150}$$

For a $(d+1)$-dimensional classical system, with the free-energy density f, we have

$$F[\psi, \boldsymbol{\nabla}\psi] = \int_{-\infty}^{\infty} dx_1 \cdots \int_{-\infty}^{\infty} dx_{d+1} \, f^{(d+1)}(\psi, \boldsymbol{\nabla}\psi) \tag{2.151}$$

We note that the RHS of Eq. (2.149) is equal to $f^{(d+1)}$, which implies the equivalence of expressions (2.149) and (2.151). Consequently, the partition function of the d-dimensional time-dependent model, in the $T = 0$ limit, given by

$$Z^{(T=0)} = \int \mathcal{D}\psi(\mathbf{r}, \tau) \, \exp\left(-\frac{1}{\hbar} S^{(T=0)}[\psi, \dot{\psi}]\right) \tag{2.152}$$

is equivalent to the partition function of a classical model in $(d+1)$ spatial dimensions. This equivalence provides us with a useful tool for the study of the critical phenomena caused by quantum fluctuations at $T = 0$. Such phase transitions are assumed to take place when some system parameter, such as a coupling constant, reaches a certain threshold value. The phase fluctuation model, discussed previously, does not exhibit any phase transition, because of the Gaussian nature of the Lagrangian (2.116). In the following chapters we shall study lattice models of superconducting networks, which can be described by Hamiltonians, involving the Josephson coupling energy with a nonlinear (periodic) functional dependence on the phase. These systems exhibit, at $T = 0$, a phase transition driven by the quantum fluctuations of the phase. The dependence of these phase transitions on the dimensionality of the networks can be studied by mapping the problem on a classical model in $(d+1)$ dimensions, for which the nature of the phase transition may be known. This mapping can be applied even to zero-dimensional systems ($d = 0$), such as a small superconductor or a microscopic tunnel junction. For two-dimensional granular superconductors with charging energy, this mapping has been derived by Doniach (1981) (see Section 6.8).

3

REGULAR ARRAYS WITHOUT DISSIPATION

3.1. Experimental background

Superconducting phase transitions in granular systems have been a target of intensive research for more than two decades. One class of such systems consists of bulk grains (diameter < 50 μm) weakly coupled through Josephson point contacts. Samples of this kind are made by pressing together, in an epoxy resin, slightly oxidized superconducting grains. Rosenblatt et al. (1970) have studied the resistivities of arrays, consisting of about 10^6 point contacts between the Nb grains. The results indicate a phase transition to a low-temperature coherent state at a temperature T_0, well below the Bardeen–Cooper–Schrieffer (BCS) transition temperature of the bulk Nb. The large size of the grains insures that the fluctuations of the magnitude of the superconducting order parameter in the grains are negligible. For the same reason, the charging energy U [see Eq. (2.107)] and the resulting quantum phase fluctuations do not play any measurable role. Hence the phase transition, taking place at temperature T_0, can be described by a classical xy model with fixed lengths of the spins and a spin–spin interaction, given by the Josephson-coupling energy (Rosenblatt et al., 1975). The phase transition in a regular three-dimensional array, described by this classical xy model, is discussed in Section 3.3. We realize that this is a highly idealized description of real samples which involve all kinds of disorder, such as fluctuations in Josephson coupling, grain size, or even the number of nearest neighbors (Rapaport, 1972). Nevertheless, we believe that the periodic array captures the essential physics describing the phase transition as a competition between the Josephson-coupling energy and the thermal energy.

Another class of granular systems involves thin granular films obtained by codeposition of a superconductor and an insulator. These materials consist of superconducting metallic grains. These grains are separated by oxide barriers that are thin enough to allow Josephson tunneling to take place. The size of the grains usually ranges from several hundreds to tenths of Å. There are many experimental examples of such films. Abeles and Hanak (1971) have investigated granular $Al - SiO_2$ films by measuring their resistivity, and the energy gap as a function of the concentration x of the SiO_2 component. The films, prepared by cosputtering Al and SiO_2, were

1000 Å thick and the average grain size varied from 100 to 30 Å (on increasing x from 0.1 to 0.4). The resistive transition curves were broadened, and the transition temperature obtained from these curves dropped to zero for $x > 0.4$. Moreover, in this concentration range, the energy gap, determined by tunneling, vanishes, and the resistivity shows a semiconducting behavior. Abeles and Hanak (1971) have proposed that the vanishing of the energy gap is a result of the destruction of the superconductivity in the grains at the *surface* of the sample. On the other hand, the superconductor-semiconductor transition in the bulk, probed by the resistive transition, has been attributed to a phase transition in a Josephson-coupled array of superconducting grains. The quenching of the Josephson coupling by thermal fluctuations, similar to the above-mentioned classical xy model, has been invoked to explain the destruction of the bulk superconductivity. Using the relation between the junction normal resistance R_N and the Josephson-coupling energy, Abeles and Hanak (1971) have established a theoretical upper limit of R_N, at which superconductivity vanishes. For a regular array of grains with the coordination number z, this *critical* normal resistance is [see Eq. (3.22) for derivation of Eq. (3.1)]

$$R_N^c = \frac{z\pi\hbar}{8e^2} \frac{\Delta(T_c)}{k_B T_c} \tag{3.1}$$

where $\Delta(T_c)$ is the energy gap at the temperature of the *superconductor–semiconductor* phase transition. Approximating the ratio $\Delta(T_c)/k_B T_c$ by a constant value 1.7, the expression (3.1) yields $R_N^c = 0.7(z\hbar/e^2)$. It is interesting that the theories of phase transitions in dissipative arrays, discussed in Chapter 6, yield a critical junction resistance with a similar dependence on $z\hbar/e^2$, though the proportionality constant is somewhat smaller. Chakravarty et al. (1986) considered a self-charging model with ohmic dissipation, which is assumed to persist down to $T = 0$, and derived a universal resistance for the onset of global coherence given by [see Eq. (6.89)]

$$R_N^c = \frac{\pi}{4}\left(\frac{z\hbar}{e^2}\right) \tag{3.2}$$

Although this theory has been developed mainly for the interpretation of data on ultrathin superconducting films (Jaeger et al., 1986), it seems to be also in reasonable agreement with experiments on thick $Al - SiO_2$ films. (Abeles and Hanak, 1971). Near the critical concentration, $x = 0.4$, the average grain size obtained from the electron micrographs is $d = 30$ Å. If the film is modeled by a regular array of closely packed cubes of size d, the critical resistivity ρ_n^c is [using the expression (3.2) with $z = 6$]

$$\rho_n^c = dR_N^c = 0.78 dz\left(\frac{\hbar}{e^2}\right) = 5.7 \times 10^{-3} \ \Omega \ \mathrm{cm} \tag{3.3}$$

which is close to the value of $\rho_n(x = 0.4)$, measured for the critical concentration of SiO_2 (Abeles and Hanak, 1971).

The original idea of the disruption of the phase coherence in granular superconductors, by the charging energy, is due to Abeles (1977). By considering the quenching of the Josephson coupling between just two adjacent grains, he derived a criterion for the existence of bulk granular superconductivity. His arguments were based on a simplified "quantum pendulum" model of the Josephson junction (Anderson, 1964). Subsequently, the effects of the charging energy were incorporated into models of regular arrays of superconducting grains. Giovannini and Weiss (1978) and independently de Gennes (1978) have proposed pseudospin models in which the charging energy enters as a uniaxial anisotropy term. The author (Šimánek, 1979) has studied the phase diagram of a three-dimensional array of junctions with the diagonal form of the charging energy (self-charging model). An extension of this model to include the off-diagonal Coulomb interactions between the grains has been studied, in the mean-field approximation, by Efetof (1980). A similar off-diagonal model has been investigated, in the Gaussian approximation, by McLean and Stephen (1979). These authors have shown that the criterion of Abeles (1977), based on a single-junction model, is too stringent. Perhaps the most interesting result of the mean-field calculations is the prediction of superconductor–semiconductor phase transitions, taking place as the ratio of the Josephson coupling to the charging energy reaches a critical value of $1/z$ (z is the coordination number of the array). This transition can be viewed as an analog of the well-known Mott–Hubbard metal–insulator transition (Mott, 1974). It is known that such a transition may be induced by changing the interatomic distance in a regular array of, say, hydrogen atoms. The disappearance of the superconductivity, observed experimentally in granular films as the thickness of oxide increases (Abeles and Hanak, 1971), is a superconducting version of such an effect.

Another aspect of the early mean-field theories is the prediction of a reentrant phase transition. Specifically, the "phase-ordered" superconducting state undergoes a second transition into a "phase-incoherent" state on further lowering the temperature. The self-charging model has been shown to exhibit a well-pronounced reentrant phase transition, originating from the inclusion of the 4π-periodic states in the thermal excitations of the intergranular Josephson junctions (Šimánek, 1979). A similar reentrant phase diagram has been obtained for a similar model, in which the phase variable is treated as an extended coordinate, $\theta \in [-\infty, \infty]$ (Šimánek, 1985). This interval is appropriate, when the states θ and $\theta + 2\pi$ are distinguishable, as a result of the coupling of these states to the environment (Likharev and Zorin, 1985). Granular arrays involving intergranular ohmic shunts belong to this category. The 4π-periodic states corresponding to single-electron excitations can be used for junctions involving the quasiparticle tunneling processes (Schön and Zaikin, 1990). For arrays involving only Josephson tunneling of pairs without the possibility of single-electron transfers, the correct phase interval should be $\theta \in [0, 2\pi)$. Efetof (1980] has suggested another mechanism of reentrance, based on the Debye-like screening of the intergrain Coulomb energy by thermally activated quasiparticles. On lowering the temperature, the number of thermal carriers decreases, leading to an increase of charging energy, producing the reentrant behavior.

The reentrant phase transition appears to manifest itself most convincingly in the measurements of the resistive transitions in thin films. Hebard and Vandenberg (1980) have observed a rapid broadening of the resistive transition in granular lead films when the sheet resistance approached the maximum metallic resistance of 30 kΩ. Subsequently, Hebard (1979) predicted that the resistance would, on lowering the temperature, flatten out and possibly start to increase, instead of dropping to zero. The actual observation of this effect has been reported by Kobayashi et al. (1980) in granular tin films. Minima in the resistance versus temperature have been observed by these authors on a group of films of larger normal sheet resistance. Maekawa et al. (1981) invoked the self-charging model (Šimánek, 1979) to interpret the resistance minima as a vortex unbinding caused by the reentrance into the phase-disordered state (see Section 7.6.4). A more recent observation of a reentrant phase transition has been made by Lin et al. (1984) in granular $BaPb_{0.75}Bi_{0.25}O_3$ superconductor. The occurrence of the resistance minima is usually associated with a negative temperature coefficient of the normal-state resistance, measured above T_c (for amorphous films, see Jaeger et al., 1986). This correlation indicates that the mechanism of reentrance is due to the effects of the charging energy on the phase ordering. At the present time, however, the precise origin of the reentrant behavior is not known.

3.2. Superconductivity in isolated grains

The basic building block of a granular superconductor is a small superconducting particle (grain) of a size that can range from about ten to a thousand Å. It is perhaps useful first to examine the role played by the finite size in superconducting ordering of an isolated grain. These are several effects to be considered:

1. The quantization of the electron levels, which produces a decrease of the transition temperature and eventually destroys the superconducting pairing.

2. The modification of the attractive interaction between the electrons, which can either increase or decrease the transition temperature. For instance, Hurault (1968) considered the enhancement of T_c due to the increase of electron–phonon interaction and the excitonic effects at the *surface* of a grain. However, Deutscher et al. (1973) presented a conclusive experimental evidence that excitonic effects are absent in $Al - Al_2O_3$ granular films.

3. Amplitude fluctuations of the order parameter, which are made more probable in small grains, owing to the low cost of the free energy associated with a fluctuation. These fluctuations tend to wash out various singularities, characteristic of the phase transition in a bulk superconductor.

In what follows, we concentrate on mechanisms (1) and (3) which are somewhat better understood than effects (2). An excellent recent review of the electronic properties of small metallic particles is by Kobayashi (1990).

A standard way of estimating the electron-level separation δ near the Fermi level, caused by random irregularities of the grain shape, is based on the expression due to Kubo (1962)

$$\delta = [N(0)\Omega]^{-1} \qquad (3.4)$$

where $N(0)$ is the density of states, per spin, at the Fermi energy, and Ω is the volume of the grain. For example, in an Al particle of diameter 100 Å, the level splitting is $\delta/k_B = 2$ K. The expression (3.4) predicts a rapid dependence of δ on grain size. Decreasing the diameter to about 50 Å yields $\delta/k_B = 16$ K, which is already one order of magnitude larger than the superconducting transition temperature of bulk aluminum. Energy splittings of this magnitude play a destructive role in the superconducting ordering (Anderson, 1959). This can be shown, starting from a simple mean-field argument due to Strongin et al. (1970). The self-consistent gap equation of the BCS theory yields a condition for superconducting transition temperature T_c of the form (Tinkham, 1975)

$$1 = V \sum_n \frac{1}{2\varepsilon_n} \tanh\left(\frac{\varepsilon_n}{2k_B T_c}\right) \qquad (3.5)$$

where V is the attractive interaction constant. The summation goes over the discrete one-electron energy levels ε_n measured with respect to the Fermi level. The values of ε_n are taken within a fixed interval $\varepsilon_n \in [-\hbar\omega_D, \hbar\omega_D]$, where ω_D is the Debye frequency. Thus, on increasing δ, the number of terms contributing to the sum in Eq. (3.5) decreases. Consequently, this equation can be satisfied only if the transition temperature T_c decreases on increasing δ. Strongin et al. (1970) have shown that for

$$\delta \geq 2\pi k_B T_c^{(0)} \qquad (3.6)$$

the sum is too small to satisfy the condition (3.5) with a finite real value of T_c, implying the breakdown of superconducting order. We note that $T_c^{(0)}$ in Eq. (3.6) is the BCS transition temperature of a bulk superconductor. A similar criterion, for a critical value of the level splitting, was predicted a long time ago by Anderson (1959). For an isolated lead particle it implies, on using Eq. (3.4), a critical radius of 22 Å, or about 750 electrons are needed to give rise to a spontaneous formation of superconducting order.

The criterion (3.6) should be taken with some caution, since it is based on a mean-field gap equation, which implies the existence of a conventional, second-order, phase transition. It is well known, however, that a system with a finite number of degrees of freedom cannot exhibit a true phase transition. In fact, thermal and quantum fluctuations of the order parameter take place that modify the mean-field picture. The influence of these fluctuations on the thermodynamic properties of small superconducting particles was studied, using functional-integration methods, by Mühlschlegel et al. (1972). For our purpose, it is adequate to discuss this problem from a simple-minded phenomenological approach (Schmidt, 1967). Above T_c, the fluctuations correspond to temporary formations of condensed pairs with an average size, given by the coherence length ξ. Fluctuations below T_c, on the other hand, are visualized as temporary normal regions in the condensate background. If the diameter of the grain is well below ξ, these fluctuations are spatially homogeneous throughout the volume of the grain. In this limit, the grain becomes a zero-dimensional superconductor. Its thermodynamic properties can be calculated starting from the Ginzburg–Landau free-energy functional (of the complex order parameter ψ)

$$F[\psi] = \Omega \left(\alpha |\psi|^2 + \frac{\beta}{2} |\psi|^4 \right) \tag{3.7}$$

where $\alpha = N(0)(T - T_c)T_c$ and $\beta \simeq 0.1 N(0)/(k_B T_c)^2$ (see Tinkham, 1975). We note that Eq. (3.7) follows from Eq. (2.1) (for a spatially constant ψ and in the absence of a magnetic field). The probability of a thermal fluctuation $P(\psi)$ is given by (Landau and Lifshitz, 1980)

$$P(\psi) = \frac{\exp\left(-F[\psi]/k_B T\right)}{\int \exp\left(-F[\psi]/k_B T\right) d^2\psi} \tag{3.8}$$

The mean-square order parameter is then

$$\langle |\psi|^2 \rangle = \int |\psi|^2 P(\psi) \, d^2\psi \tag{3.9}$$

Equations (3.8) and (3.9) could also be obtained from the path-integral approach in the classical limit [see Eq. (A9.3)]. The resulting $\langle |\psi|^2 \rangle$ is sketched in Fig. 3.1. The mean-field order parameter, obtained by minimizing the free energy (3.7), is given by

$$|\psi_0|^2 = \begin{cases} -\alpha/\beta, & \text{for } T \leq T_c \\ 0, & \text{for } T > T_c \end{cases} \tag{3.10}$$

The T dependence of $|\psi_0|^2$ is shown in Fig. 3.1 as the broken line. In contrast, the quantity $\langle |\psi|^2 \rangle$ exhibits a smooth temperature variation near

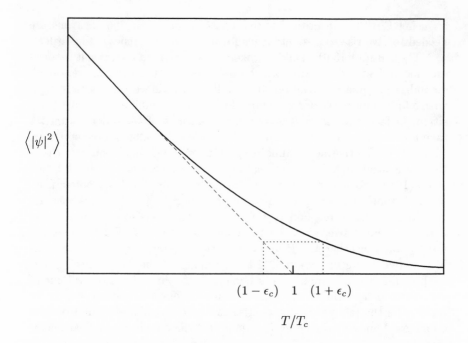

$$\langle|\psi|^2\rangle$$

$$(1 - \epsilon_c) \quad 1 \quad (1 + \epsilon_c)$$

$$T/T_c$$

FIG. 3.1.   The mean-square order parameter $\langle|\psi|^2\rangle$, plotted as a function of T/T_c, where T_c is the mean-field transition temperature of a small superconductor. The broken line represents the square of the mean-field order parameter, and ϵ_c is the critical fluctuation width.

T_c. The degree of departure of $\langle|\psi|^2\rangle$ from the mean-field result (3.10) is characterized by the critical fluctuation width ϵ_c (Ginzburg, 1960). As indicated in Fig. 3.1, ϵ_c is defined by setting $\langle|\psi|^2\rangle$ at $T/T_c = 1 + \epsilon_c$ equal to $|\psi_0|^2$ at $T/T_c = 1 - \epsilon_c$. For a rough estimate of ϵ_c, we can use $\langle|\psi|^2\rangle$ calculated from Eq. (3.9) within a Gaussian approximation. Taking $F[\psi] \simeq \Omega\alpha|\psi|^2$, we have

$$\langle|\psi|^2\rangle_{\text{Gauss}} = \frac{k_B T}{\Omega\alpha} \tag{3.11}$$

Comparing this result with Eq. (3.10), we obtain

$$\epsilon_c = \left[\frac{0.1}{\Omega N(0)k_B T_c}\right]^{\frac{1}{2}} \sim \left(\frac{\delta}{k_B T_c}\right)^{\frac{1}{2}} \tag{3.12}$$

where we have introduced the level splitting δ, given by Eq. (3.4).

   The expression on the RHS of Eq. (3.12) provides us with a useful criterion for the significance of the thermal fluctuations of the order parameter in an isolated grain. For instance, the granular films studied by

Hebard and Vandenberg (1980) contain lead grains with nominal sizes of 100–200 Å, implying a level splitting, ranging from 0.6 to $0.07k_BT_c$. Hence, for the 100 Å grains, the critical fluctuation width $\epsilon_c \simeq 0.8$. This implies that the magnitude of the order parameter $|\psi|$ in such grains cannot be regarded as a constant. On the other hand, the grains appear to be strongly coupled into larger clusters in which the fluctuations of $|\psi|$ are suppressed (Ward et al., 1978). In view of this, a better description of the granular lead films, in the experiments of Hebard and Vandenberg (1980), appears to be an array of weakly Josephson-coupled *clusters* of grains. The fluctuations of $|\psi|^2$, in these clusters, can then be neglected and the superconductor–semiconductor transition is due to the quantum phase fluctuations driven by the charging energy (Šimánek, 1980b). To summarize, the effects of level discreteness and thermodynamic fluctuations are controlled by the same ratio δ/k_BT_c and can be neglected when this ratio is well below one.

Superconducting films containing grains so small that the level splitting δ exceeds the criterion (3.6) by a wide margin represent a more difficult case (from a theoretical standpoint). As a starting point, one can model such materials as a collection of metal atoms randomly dispersed in the oxide matrix. A possible description of such a system may be based on a Hamiltonian previously used for superconducting transition metal alloys (Weinkauf and Zittartz, 1974)

$$H = \sum_{i,\sigma} \varepsilon_i c_{i\sigma}^\dagger c_{i\sigma} + \sum_{i,j,\sigma} t_{ij} c_{i\sigma}^\dagger c_{j\sigma} - \sum_{i,j} \lambda_{ij} c_{i\uparrow}^\dagger c_{i\downarrow}^\dagger c_{j\downarrow} c_{j\uparrow} \qquad (3.13)$$

where $c_{i\sigma}$ annihilates an electron of spin σ and energy ε_i at the ith site. The transfer matrix element between the sites i and j is denoted by t_{ij}, and λ_{ij} is the attractive coupling constant between these sites. For localized pairing ($\lambda_{ij} = \lambda_i \delta_{ij}$), the Hamiltonian (3.13) represents the negative U Hubbard model (Pincus et al., 1973). Such a model is perhaps applicable to the Al films studied by the Rutgers group (Worthington et al., 1978; Deutscher et al., 1980). Another class of materials, where the granular model does not apply, are the amorphous films (e.g., amorphous bismuth, amorphous indium oxide). The degradation of superconductivity with increasing disorder, in such materials, is generally believed to be due to Anderson localization (Anderson et al., 1983). The disorder-induced *enhancement* of the *Coulomb interaction* is the main mechanism of the destruction of superconductivity. Experimentally, this effect has been studied most closely in two dimensions, where disorder-induced changes of T_c are large and in reasonable agreement with the theory. Maekawa and Fukuyama (1982) and later Ebisawa et al. (1985) have calculated the leading correction to the transition temperature in the weak localization limit, in two dimensions. For three-dimensional systems, the experimental situation is less clear, due to disorder-induced changes of T_c that are of nonlocalization origin, for example, the broadening of the density of states peaks and changes in the phonon structure. There are also disagreements in the theoretical predictions for the depression of T_c, which may be possibly due to the subtleties in the handling of short-wavelength cutoffs (Fukuyama et al., 1984; Belitz, 1987). We will not review the details of these theories, as the subject lies outside our main theme. In closing this section, let us point out that the degradation of superconductivity due to Anderson localization may also be the explanation for the absence of the superconducting gap at the surface of the $Al - SiO_2$ films, observed by Abeles and Hanak (1971).

3.3.   3D arrays without Coulomb interaction

In this section, we consider three-dimensional periodic arrays of equivalent grains large enough that the finite-size effects considered in Section (3.2) can be neglected. According to Eqs. (3.6) and (3.12), this implies that the energy-level splitting δ is much less than $k_B T_c$. The grains are coupled by the Josephson energy, given in Eq. (2.94). Since the charging energy is neglected, all configurations generated by transfer of n pairs across a junction are degenerate. As shown in Section (2.6), this implies that the correct wave function is that of a *definite relative phase*. Thus the phase variables in such array are *c numbers* and the Hamiltonian is that of a *classical xy model*. Applying Eq. (2.94) to all intergranular junctions of the array, we arrive at the Hamiltonian

$$H = -J \sum_{\langle ij \rangle} \cos(\theta_i - \theta_j) = -J \sum_{\langle ij \rangle} \mathbf{S}_i \cdot \mathbf{S}_j \qquad (3.14)$$

where the summation is over the *nearest-neighbor grains*. The classical spin is constrained to rotate in the xy plane, with one degree of freedom θ_i

$$\mathbf{S}_i = \mathbf{e}_x \cos \theta_i + \mathbf{e}_y \sin \theta_i \qquad (3.15)$$

where $\mathbf{e}_x$ and $\mathbf{e}_y$ are unit vectors along the x and y axes, respectively. The choice of fixed unit spin length is justified when:

1. The thermodynamic fluctuations of the magnitude of the order parameter can be neglected.

2. The increase of $|\psi|$, induced by the Josephson coupling, is insignificant. This condition can be met if the temperature is well below the BCS transition temperature of the bulk material and if the Josephson coupling is weak.

The xy model described by the Hamiltonian (3.14) exhibits, in three dimensions, a conventional long-range order associated with a nonzero average $\langle \mathbf{S} \rangle$ (Wegner, 1967). The phase transition is characterized by the exponents $\alpha = 0$, $\beta = 1/3$, and $\gamma = 4/3$. Qualitative insight into this phase transition can be obtained by applying the mean-field approximation. Assuming that the spin ordering is taking place along the x axis, we have

$$\mathbf{S}_i \cdot \mathbf{S}_j = (\langle S_{xi}\rangle + \delta S_{xi})(\langle S_{xj}\rangle + \delta S_{xj}) + \delta S_{yi}\delta S_{yj}$$
$$\simeq \langle S_{xi}\rangle S_{xj} + \langle S_{xj}\rangle S_{xi} - \langle S_{xi}\rangle\langle S_{xj}\rangle \tag{3.16}$$

where we have neglected, in the spirit of the MFA, terms of second order in the fluctuations. Introducing this result into Eq. (3.14), we obtain the MFA Hamiltonian of the array

$$H_{\mathrm{MFA}} = -zJ\langle S_x\rangle \sum_i S_{xi} \tag{3.17}$$

where z is the coordination number. The thermal average $\langle S_x\rangle = \langle\cos\theta\rangle$ describes the long-range order of the phase (broken θ symmetry) and the phase transition is called the "phase-locking" transition. The order parameter $\langle\cos\theta\rangle$ can be evaluated using the standard formula of classical statistical mechanics [see also Eqs. (A5.8) and (A9.3)]

$$\langle\cos\theta\rangle = \frac{\int_0^{2\pi} d\theta\,\cos\theta\exp\left(-H_{\mathrm{MFA}}/k_B T\right)}{\int_0^{2\pi} d\theta\,\exp\left(-H_{\mathrm{MFA}}/k_B T\right)} \tag{3.18}$$

From this equation we can calculate the "phase-locking" transition temperature T_c^c (superscript c means classical). Near T_c^c, $\langle\cos\theta\rangle$ is small, which allows us to expand the exponentials in Eq. (3.18), and we obtain with the use of Eq. (3.17)

$$\langle\cos\theta\rangle = \frac{zJ\langle\cos\theta\rangle}{k_B T_c^c}\left(\frac{1}{2\pi}\int_0^{2\pi} d\theta\,\cos^2\theta\right) = \frac{zJ\langle\cos\theta\rangle}{2k_B T_c^c} \tag{3.19}$$

Equation (3.19) implies that the transition temperature T_c^c, in the mean-field approximation, is given by

$$T_c^c = \frac{zJ}{2k_B} \tag{3.20}$$

The Josephson-coupling constant J is given by the expression [Tinkham, 1975, and Eq. (5.146)]

$$J = \frac{\pi\hbar\Delta(T)}{4e^2 R_N}\tanh\frac{\beta\Delta(T)}{2} \tag{3.21}$$

where $\Delta(T)$ is the energy gap of the grains and R_N is the normal resistance of the junction. Introducing Eq. (3.21) into Eq. (3.20), we obtain a relation between T_c^c and R_N, which for $T_c^c \ll \Delta(T_c^c)/k_B$ takes the form

$$T_c^c = \frac{z\pi\hbar\Delta(T_c^c)}{8e^2 k_B R_N} \tag{3.22}$$

This relation has been used by Abeles and Hanak (1971) to determine the critical normal resistivity of the $Al - SiO_2$ granular films [see Eq. (3.1)]. For highly resistive granular metals, the Josephson-coupling constant is small enough to allow T_c^c to fall well below the bulk BCS transition temperature T_{c0} of the individual grains. As pointed out by Rosenblatt (1974), such arrays become superconducting in two stages. At the bulk transition temperature T_{c0}, the magnitude of the order parameter in each grain becomes nonzero. The second transition, at $T_c^c \ll T_{c0}$, corresponds to the formation of the long-range phase order. The often observed double transitions in the resistivity versus temperature curves are attributed to these two stages of superconducting ordering (Patton et al., 1979). The resistivity drop at T_{c0} is often too large to be caused by the individual grains becoming superconducting. Actually, a formation of large clusters of strongly coupled grains is believed to be responsible for this drop. This picture has emerged particularly clearly in the works on $Al - Ge$ systems (Deutscher and Rapaport, 1978) and granular lead films (Hebard and Vandenberg, 1980).

3.4.   Coulomb energy and capacitance matrix

Consider a system of electrical conductors formed by the metallic grains of an array. The charge is confined to the surface of each conductor, and its surface density is given by $\sigma_i(\mathbf{r})$ for the ith conductor. Then the Coulomb energy $E_{c,i}$ of the ith conductor is given by the surface integral

$$E_{c,i} = \frac{1}{2} \int_S da\, \sigma_i(\mathbf{r})V_i(\mathbf{r}) = \frac{1}{2}V_iQ_i \tag{3.23}$$

where $V_i(\mathbf{r})$ is the electrostatic potential and V_i denotes its value at the surface of the conductor. Q_i is the net charge on the ith conductor. In Eq. (3.23), we have taken V_i out of the integral, since it is a constant over the surface S of a metallic grain. For a system of N charged grains, the total Coulomb energy is then given by

$$E_c = \sum_{i=1}^{N} E_{c,i} = \frac{1}{2}\sum_{i=1}^{N} V_iQ_i \tag{3.24}$$

In applications to granular superconductivity, it is convenient to work with a form of Coulomb energy expressed as a quadratic function of the charges. This is achieved by calculating the potentials V_i in terms of the charges using a set of equations

$$V_i = \sum_{j=1}^{N} p_{ij}Q_j, \qquad i = 1, 2, \ldots, N \tag{3.25}$$

where p_{ij} are the *coefficients* of the *potential*. Introducing this equation into Eq. (3.24), we have

$$E_c = \frac{1}{2} \sum_{i=1}^{N} \sum_{j=1}^{N} p_{ij} Q_i Q_j \tag{3.26}$$

The set of the linear equations (3.25) can be solved for the charges with the result

$$Q_i = \sum_{j=1}^{N} \widehat{C}_{ij} V_j \tag{3.27}$$

where the coefficients $\widehat{C}_{ij}$ define the capacitance matrix. They are given by the inversion of the matrix p, (formed by the coefficients of the potential).

$$\widehat{C}_{ij} = \left(p^{-1}\right)_{ij} \tag{3.28}$$

Introducing Eq. (3.27) into the formula (3.24), we obtain another useful form of the Coulomb energy

$$E_c = \frac{1}{2} \sum_{i=1}^{N} \sum_{j=1}^{N} \widehat{C}_{ij} V_i V_j \tag{3.29}$$

By inverting the matrix equation (3.28), we obtain $p_{ij} = (\widehat{C}^{-1})_{ij}$, so that Eq. (3.26) can be written

$$E_c = \frac{1}{2} \sum_{i=1}^{N} \sum_{j=1}^{N} \left(\widehat{C}^{-1}\right)_{ij} Q_i Q_j \tag{3.30}$$

3.4.1. *Self-charging model*

Two limiting forms of the capacitance matrix have been used to model the charging effects in the granular superconductors (Doniach, 1984). One of them corresponds to the diagonal choice of the p matrix

$$p_{ij} = p_{ii} \delta_{ij} \tag{3.31}$$

According to Eq. (3.25), this choice implies that the potential V_i on the ith conductor receives only a contribution from its own charge Q_i. For this reason, the model based on Eq. (3.31) is called the *self-charging model*. Inverting the matrix equation (3.31), we calculate, using the relation (3.28), the capacitance matrix

$$\widehat{C}_{ij} = \widehat{C}_{ii} \delta_{ij} = C_0 \delta_{ij} = \frac{1}{p_{ii}} \delta_{ij} \tag{3.32}$$

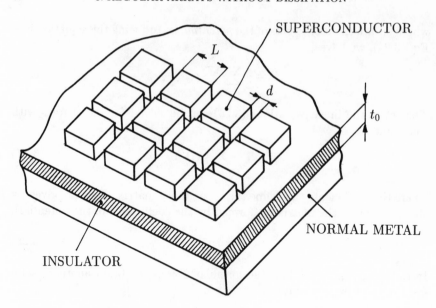

FIG. 3.2. A realization of the self-charging model, proposed by Bradley and Doniach (1984).

where $\widehat{C}_{ii} = C_0$, the *self-capacitance* of the ith site, is a constant for the regular array. Introducing this result into Eqs. (3.30) and (3.31), we obtain the Coulomb energy of the self-charging model in a diagonal form

$$E_c = \frac{1}{2C_0} \sum_{i=1}^{N} Q_i^2 \qquad (3.33)$$

This expression is analogous to the short-range Coulomb energy used in the Hubbard model. Kawabata (1977) has applied this form of the charging energy to study the metal–insulator transition in an array of normal junctions, and the author has used it in the mean-field theory of the phase ordering in a granular superconductor (Šimánek, 1979). A possible experimental realization of the self-charging model, proposed by Bradley and Doniach (1984), is shown in Fig. 3.2 The array is formed by thin square superconducting plates, of size L, on an insulating substrate, the latter being evaporated on the planar surface of a normal metal. The electrostatic energy of a charged array is confined to narrow gaps of width t_0, formed between the plates and the normal metal. The separation between the plates d must be small enough to allow for the Josephson coupling. On the other hand, the thickness t_0 must be adjusted so that the tunneling between the normal metal and the superconductor is negligible. It is also required that $L \gg t_0$, in order that the capacitance edge effects [which would split

μ-th bond n-th site C_0 C

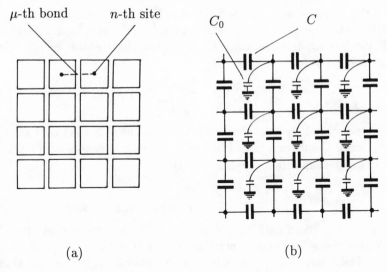

(a)   (b)

FIG. 3.3. Suggested realization of the nearest-neighbor charging model. (a) A planar square array, which minimizes electrostatic leakage beyond nearest-neighbor sites. (b) Equivalent network with lumped capacitances C_0 and C.

the condition (3.31)] can be neglected. The size L itself must not, however, exceed the bulk coherence length ξ of the superconductor, to satisfy the criterion of zero dimensionality. Such a realization, can be, of course, applied only to the one- or two-dimensional arrays. Nevertheless, the self-charging model is frequently used in the theories of three-dimensional arrays, where it is approximately valid for well-separated, small grains.

3.4.2. *Nearest-neighbor charging model*

The other limiting form of the capacitance matrix describes the so called nearest-neighbor charging model. This model is suitable for cases where metal grains occupy a large fraction of sample volume, ensuring suppression of the leakage fields beyond nearest neighbors. The electrostatic energy, for this model, is that of the capacitor network shown for the case of two-dimensional array in Fig. 3.3. The lattice sites of this network (grain centers) are labeled by n. The sites are connected by lattice vectors (links) denoted as μ. We emphasize that the capacitors C_0 are essential in producing the short range of the charge interaction of the nearest-neighbor charging model. In fact, as we show shortly, the range of the Coulomb matrix $\widehat{C}^{-1}$ becomes infinite if C_0 is set equal to zero. Even if the array of Fig. 3.3(a) is fabricated in such a way that the grains have a negligible capacitance with respect to ground, there is still an important contribution to the charging energy associated with the *capacitance of isolated grains.*

Since the grains occupy a large fraction of the volume, the capacitance C_0 is clearly not negligible, in the nearest-neighbor charging model, compared with the intergrain capacitance C. The total electrostatic energy of the network is given by

$$E_c = \frac{1}{2}C_0 \sum_n V_n^2 + \frac{1}{4}C \sum_{n,\mu} (V_{n+\mu} - V_n)^2 \qquad (3.34)$$

where the second term involves an extra $\frac{1}{2}$ factor to correct overcounting of the links. Comparing Eq. (3.34) with the general expression (3.29), we obtain for the capacitance matrix $\widehat{C}$

$$\widehat{C}_{ij} = \begin{cases} C_0 + zC, & \text{for } i = j \\ -C, & \text{for } \langle i, j \rangle = \text{nearest neighbors} \end{cases} \qquad (3.35)$$

where z is the coordination number of the network. We remark that this expression holds for periodic networks of any dimension ($d = 1, 2, 3$). Although the matrix (3.35) is of a nearest-neighbor kind, it need not always represent the nearest-neighbor charging model. Only if the *inverse capacitance matrix is of short range*, does this model ensue.

To invert the matrix (3.35), we use the method of Fourier transform. Because of the translational symmetry of the network, the matrices $\widehat{C}$ and $\widehat{C}^{-1}$ must be only functions of the relative position $\mathbf{R}_i - \mathbf{R}_j$ of the grains. This allows us to expand the matrix element $\widehat{C}_{ij}$ as follows:

$$\widehat{C}_{ij} = \widehat{C}(\mathbf{R}_i - \mathbf{R}_j) = \frac{1}{N} \sum_k \widehat{C}_\mathbf{k} e^{i\mathbf{k}\cdot(\mathbf{R}_i - \mathbf{R}_j)} \qquad (3.36)$$

where N is the number of grains in the network. Inverting the Fourier series (3.36), we have

$$\widehat{C}_\mathbf{k} = \sum_\mathbf{R} \widehat{C}(\mathbf{R}) e^{-i\mathbf{k}\cdot\mathbf{R}} \qquad (3.37)$$

The Fourier transforms of the capacitance matrix and of its inverse transform satisfy a useful relation

$$\left(\widehat{C}^{-1}\right)_\mathbf{k} = \left(\widehat{C}_\mathbf{k}\right)^{-1} \qquad (3.38)$$

To verify Eq. (3.38), we consider the matrix element of the product $\widehat{C}\widehat{C}^{-1} = \widehat{I}$, where $\widehat{I}$ is the unit matrix.

$$\begin{aligned} \delta_{il} = \left(\widehat{C}\widehat{C}^{-1}\right)_{il} &= \sum_j \widehat{C}_{ij}\widehat{C}_{jl}^{-1} \\ &= \frac{1}{N^2} \sum_{\mathbf{R}_j} \sum_\mathbf{k} \sum_{\mathbf{k}'} \widehat{C}_\mathbf{k} \left(\widehat{C}^{-1}\right)_{\mathbf{k}'} e^{i\mathbf{k}\cdot(\mathbf{R}_i - \mathbf{R}_j)} e^{i\mathbf{k}'\cdot(\mathbf{R}_j - \mathbf{R}_l)} \\ &= \frac{1}{N} \sum_\mathbf{k} \widehat{C}_\mathbf{k} \left(\widehat{C}^{-1}\right)_\mathbf{k} e^{i\mathbf{k}\cdot(\mathbf{R}_i - \mathbf{R}_l)} \end{aligned} \qquad (3.39)$$

Expressing δ_{il} on the LHS of this equation by the Fourier series

$$\delta_{il} = \frac{1}{N} \sum_{\mathbf{k}} e^{i\mathbf{k}\cdot(\mathbf{R}_i - \mathbf{R}_l)} \tag{3.40}$$

and comparing with the RHS of (3.39), we obtain Eq. (3.38).

Let us now calculate from Eq. (3.37) the Fourier transform of the capacitance matrix $\widehat{C}$ of the network. Using Eq. (3.35), we have

$$\widehat{C}_{\mathbf{k}} = (C_0 + zC) - C \sum_{\mathbf{h}} e^{i\mathbf{k}\cdot\mathbf{h}} \tag{3.41}$$

where the vectors $\mathbf{h}$ form a set of z lattice translation vectors (star), connecting a given site to its nearest neighbors.

3.4.3. Two-dimensional network

Consider a space lattice, such as the one in Fig. 3.3(a). In this case, we have $z = 4$, and Eq. (3.41) yields

$$\widehat{C}_{\mathbf{k}} = C_0 + 4C\left\{1 - \frac{1}{2}[\cos(k_x a) + \cos(k_y a)]\right\} \tag{3.42}$$

The approximate dependence of the inverse matrix on the intergrain distance is especially easy to determine by considering the long-wavelength limit $\mathbf{k}\cdot\mathbf{h} \ll 1$. Expanding the cosines in Eq. (3.42), we have in this limit

$$\widehat{C}_{\mathbf{k}} \simeq C_0 + Ca^2 k^2 \tag{3.43}$$

Using this expression, we find with the help of the relation (3.38) an approximate expression for the Coulomb matrix

$$\widehat{C}^{-1}(\mathbf{R}) = \frac{1}{N} \sum_{\mathbf{k}} \left(\widehat{C}_{\mathbf{k}}\right)^{-1} e^{i\mathbf{k}\cdot\mathbf{R}}$$

$$\simeq \frac{a^2}{(2\pi)^2} \int_0^\infty dk\, k \int_0^{2\pi} d\theta\, \frac{e^{ikR\cos\theta}}{C_0 + Ca^2 k^2} \tag{3.44}$$

where we introduced polar coordinates for the k integration. The upper cutoff of the k integral has been replaced by infinity. Hence, the expression (3.44) holds only in the limit $R \gg a$. Using the definitions of the Bessel functions J_0 and K_0 (Abramowitz and Stegun, 1964)

$$J_0(kR) = \frac{1}{2\pi} \int_0^{2\pi} d\theta\, e^{ikR\cos\theta} \tag{3.45}$$

$$K_0(\bar{\alpha}R) = \int_0^\infty dk\, \frac{J_0(kR)k}{k^2 + \bar{\alpha}^2} \tag{3.46}$$

we obtain from Eq. (3.44)

$$\widehat{C}^{-1}(\mathbf{R}) \simeq \frac{1}{2\pi C} K_0(\bar{\alpha}R) \qquad (3.47)$$

where

$$\bar{\alpha} = \frac{1}{a}\left(\frac{C_0}{C}\right)^{\frac{1}{2}} \qquad (3.48)$$

The modified Bessel function $K_0(\bar{\alpha}R)$ falls off exponentially for $\bar{\alpha}R \gg 1$. This implies that the range of the inverse capacitance matrix is given by $1/\bar{\alpha}$. As $C_0 \to 0$, the range goes to infinity. The nearest-neighbor charging model requires, according to Eq. (3.48), that $C_0 \gtrsim C$. A similar result for the range of the Coulomb matrix is obtained for one- and three-dimensional arrays. For $C_0 \ll C$, the range of the Coulomb matrix is much larger than the intergrain distance and there is an interval of distances satisfying the condition: $a \ll R \ll 1/\bar{\alpha}$. In this interval, the expression (3.47) can be approximated as follows (Abramowitz and Stegun, 1964)

$$\widehat{C}^{-1}(\mathbf{R}) \simeq -\frac{1}{2\pi C} \ln(\bar{\alpha}R) \qquad (3.49)$$

Recalling Eq. (3.30), this suggests that grain charges form a two-dimensional Coulomb gas with logarithmic interaction. Based on this observation, Yoshikawa et al. (1987) and Widom and Badjou (1988) have proposed the possibility of a charge-unbinding transition of the Kosterlitz–Thouless–Berezinskii variety in superconducting films. A similar effect has been considered for normal metal arrays by Mooij et al. (1990).

In a number of works, the Coulomb matrix of the nearest-neighbor charging model is obtained by writing the charging energy as follows (see Šimánek, 1983; Doniach, 1984; Bradley and Doniach, 1984)

$$E_c = \alpha_c \sum_{\langle ij \rangle} (Q_i - Q_j)^2 \qquad (3.50)$$

where α_c is inversely proportional to the intergrain capacitance C. Comparing this expression with the formula (3.30), one obtains for the inverse capacitance matrix a diagonal element that is positive and proportional to $1/C$ and a nearest-neighbor off-diagonal element with the same dependence on $1/C$ but with a negative sign. This result is inconsistent with the fact that putting a positive charge on a given grain always raises the potential on neighboring islands. Thus, Eq. (3.25) implies that $p_{ij} = \left(\widehat{C}^{-1}\right)_{ij}$ is always positive. The error is due to the fact that expression (3.50) does not take into account the electrostatic energy produced by configurations for which $Q_i = Q_j \neq 0$. However, such configurations do not affect the charge transfer between the nearest-neighbor grains. Hence, calculations based on the charging energy (3.50) should still yield correct results for the phase fluctuations.

Another method of matrix inversion uses the expansion in the inverse coordination number (Fishman, 1989). The starting point of this approach is to write the capacitance matrix (3.35) as follows

$$\widehat{C} = (C_0 + zC)\left(\widehat{I} - \frac{1}{z(1+\lambda)}\widehat{K}\right) \tag{3.51}$$

where $\lambda = C_0/zC$, $\widehat{I}$ is the unit matrix, and $\widehat{K}$ is a "nearest-neighbor Kronecker delta" matrix with matrix elements that are equal to one for nearest-neighbor sites $\langle ij \rangle$ and zero otherwise. The Coulomb matrix is obtained by expanding the inverse of (3.51)

$$\widehat{C}^{-1} = (C_0 + zC)^{-1}\left(\widehat{I} + \frac{1}{z(1+\lambda)}\widehat{K} + \frac{1}{z^2(1+\lambda)^2}\widehat{K}^2 + \cdots\right) \tag{3.52}$$

To lowest order in $1/z$, we obtain from this equation

$$(\widehat{C}^{-1})_{ij} \simeq \begin{cases} (C_0 + zC)^{-1}, & \text{for } i = j \\ C(C_0 + zC)^{-2}, & \text{for } \langle i, j \rangle = \text{nearest neighbors} \end{cases} \tag{3.53}$$

There will, of course, also be matrix elements of $\widehat{C}^{-1}$ for $(\mathbf{R}_i - \mathbf{R}_j)$ beyond the nearest-neighbor distances. However, for $C_0/C \gtrsim 1$, the range of the Coulomb matrix is of order a, and the expression (3.53) represents a reasonable description of the nearest-neighbor charging model. On the other hand, for $C_0/C \ll 1$, the parameter λ is small and the series (3.52) converges poorly. In this case, we resort to the method of matrix inversion by Fourier series. Equation (3.44) can still be used to calculate the diagonal element $\widehat{C}^{-1}(\mathbf{R} = 0)$ if a finite Debye cutoff k_D is used for the upper limit of the k integral. This cutoff is obtained by equating the number of k states in a circular zone of radius k_D to the number of sites in the planar network. Thus, we obtain for a square array a value of $k_D = 2\sqrt{\pi}/a$, where a is the lattice spacing.

3.5.   3D Josephson arrays with self-charging energy

The Coulomb energy of the array, described by the capacitance matrix $\widehat{C}_{ij}$, is given by Eq. (3.30). Expressing the charge on the ith grain Q_i via the excess pair number n_i, we have $Q_i = 2en_i$ so that the charging energy (3.30) becomes

$$E_c = 2e^2 \sum_{ij} \left(\widehat{C}^{-1}\right)_{ij} n_i n_j \tag{3.54}$$

The Josephson-coupling energy of a regular array is, according to Eq. (2.94), given by

$$E_J = -J \sum_{\langle ij \rangle} \cos(\theta_i - \theta_j) \tag{3.55}$$

where the summation is over the nearest-neighbor pairs of grains only ($i < j$ is implied to avoid overcounting).

In the special case of the self-charging model, the charging energy takes the diagonal form (3.33), and the total Hamiltonian of the array is, from Eqs. (3.54) and (3.55)

$$H = E_c + E_J = \frac{2e^2}{C_0} \sum_i n_i^2 - J \sum_{\langle ij \rangle} \cos(\theta_i - \theta_j) \qquad (3.56)$$

For very small grains, and near $T = 0$, the charging term should be augmented by the correction term, shown in Eq. (2.105). This amounts to replacing the constant $2e^2/C_0$ in (3.56), by the combinations given in Eq. (2.107)

$$\frac{2e^2}{C_0} \rightarrow \frac{2e^2}{C_0} + \frac{2\epsilon_F}{3\bar{n}_0 St} = 2U_a \qquad (3.57)$$

where we introduced the effective charging energy, U_a, of the array.

The Hamiltonian (3.56) is quantized by replacing the excess pair number n_i by an operator $\hat{n}_i$, satisfying the commutation relation [see Eq. (2.51)]

$$[\hat{n}_i, \theta_j] = i\delta_{i,j} \qquad (3.58)$$

which implies that

$$\hat{n}_i = i \frac{\partial}{\partial \theta_i} \qquad (3.59)$$

Using Eqs. (3.56)–(3.59), we obtain the Hamiltonian operator, for the array, in the form

$$\hat{H} = 2U_a \sum_i \hat{n}_i^2 - J \sum_{\langle ij \rangle} \cos(\theta_i - \theta_j) \qquad (3.60)$$

One possible way of treating the phase transition of this model is the self-consistent MFA (Šimánek, 1979). Assuming that the phase ordering is such that $\langle \cos \theta_i \rangle \neq 0$ and $\langle \sin \theta_i \rangle = 0$, we have in the MFA [see Eq. (3.16)]

$$\cos(\theta_i - \theta_j) = [\langle \cos \theta_i \rangle + \delta(\cos \theta_i)][\langle \cos \theta_j \rangle + \delta(\cos \theta_j)]$$
$$\simeq \langle \cos \theta_i \rangle \cos \theta_j + \langle \cos \theta_j \rangle \cos \theta_i - \langle \cos \theta_i \rangle \langle \cos \theta_j \rangle \quad (3.61)$$

Introducing the RHS of Eq. (3.61) into Eq. (3.60), we obtain the Hamiltonian of the array in the MFA

$$\hat{H}_{\text{MFA}} = - \sum_i \left(2U_a \frac{\partial^2}{\partial \theta_i^2} + zJ\langle \cos \theta \rangle \cos \theta_i \right) \qquad (3.62)$$

where z is the coordination number. We see that the sites are decoupled in the MFA, and the Hamiltonian, *at each site*, describes a quantum-mechanical particle in a periodic, $\cos \theta_i$ potential. The corresponding Schrödinger equation is

$$\left[2U_a \left(\frac{d}{d\theta} \right)^2 + zJ\langle \cos \theta \rangle \cos \theta \right] \psi_m(\theta) = -E_m \psi_m(\theta) \qquad (3.63)$$

The order parameter $\langle \cos \theta \rangle$ can be calculated from the equation

$$\langle \cos \theta \rangle = \frac{\sum_m e^{-E_m/k_B T} \langle \psi_m | \cos \theta | \psi_m \rangle}{\sum_m e^{-E_m/k_B T}} \tag{3.64}$$

Equation (3.63) is recognized to be the Mathieu equation. Near the transition temperature T_c, the order parameter $\langle \cos \theta \rangle$ is small and so is the coefficient of the $\cos \theta$ term in Eq. (3.63). Then the energies E_m and the wave function ψ_m can be *expanded* in terms of a *small* dimensionless parameter, q, defined as

$$q = \frac{-zJ\langle \cos \theta \rangle}{U_a} \tag{3.65}$$

To *order* q, the periodic wave functions ψ_m, of lowest m, are (Abramowitz and Stegun, 1964)

$$\psi_0(\theta) = \frac{1}{\sqrt{2\pi}}\left(1 - \frac{q}{2}\cos\theta\right) \tag{3.66}$$

$$\left.\begin{array}{l} \psi_1^e(\theta) = \frac{1}{\sqrt{2\pi}}\left(\cos\frac{\theta}{2} - \frac{q}{8}\cos\frac{3\theta}{2}\right) \\[2mm] \psi_1^o(\theta) = \frac{1}{\sqrt{2\pi}}\left(\sin\frac{\theta}{2} - \frac{q}{8}\sin\frac{3\theta}{2}\right) \end{array}\right\} \tag{3.67}$$

$$\left.\begin{array}{l} \psi_2^e(\theta) = \frac{1}{\sqrt{2\pi}}\left[\cos\theta - q\left(\frac{\cos 2\theta}{12} - \frac{1}{4}\right)\right] \\[2mm] \psi_2^o(\theta) = \frac{1}{\sqrt{2\pi}}\left(\sin\theta - q\frac{\sin 2\theta}{12}\right) \end{array}\right\} \tag{3.68}$$

The corresponding energy eigenvalues are, to order q

$$E_0 = 0; \quad E_1^e = \frac{U_a}{2}(1+q); \quad E_1^o = \frac{U_a}{2}(1-q); \quad E_2^e = E_2^e = 2U_a \tag{3.69}$$

The wave functions ψ_m^e and ψ_m^o are the even and odd combinations resulting from the splitting of the degenerate doublets $\psi_m^{(0)}$ that are eigenfunctions of (3.63) in the absence of the periodic potential. We have

$$\psi_m^{(0)} = \frac{1}{\sqrt{2\pi}} e^{\pm im\theta/2} \tag{3.70}$$

These wave functions are also eigenstates of the number operator (3.59) with the eigenvalues

$$n = \langle \psi_m^{(0)} | \hat{n} | \psi_m^{(0)} \rangle = \pm \frac{m}{2} \tag{3.71}$$

The $m = 2$ states are 2π periodic and correspond, according to Eq. (3.71), to $n = \pm 1$. On the other hand, the $m = 1$ states are 2π antiperiodic (or 4π periodic) and correspond to $n = \pm 1/2$, which can be thought of as single-electron excitations on a grain. As will be explained, the inclusion of the antiperiodic states in the expression (3.64) results in a phase diagram with pronounced reentrance. Strictly speaking, the 4π-periodic states can only be included if the states θ and $\theta + 2\pi$ are distinguishable, as a result of the coupling of these states to the environment. Since no coupling of this sort is present in the Hamiltonian (3.56), the states θ and $\theta + 2\pi$ are indistinguishable, and the Schrödinger equation (3.63) corresponds to a planar rotator with 2π-periodic eigenstates. Nevertheless, it is of interest to understand the mechanism of the reentrance, generated by the $m = 1$ states in this model, as it is, in fact, similar to that found in the systems with a weak environmental coupling. The matrix elements of $\cos\theta$ can be evaluated with the use of Eqs. (3.66)–(3.68), yielding to order q

$$\langle\psi_0| \cos\theta |\psi_0\rangle = -\frac{q}{2} \tag{3.72}$$

$$\left.\begin{array}{l} \langle\psi_1^e| \cos\theta |\psi_1^e\rangle = \frac{1}{2} - \frac{q}{8} \\[2mm] \langle\psi_1^o| \cos\theta |\psi_1^o\rangle = -\frac{1}{2} - \frac{q}{8} \end{array}\right\} \tag{3.73}$$

$$\left.\begin{array}{l} \langle\psi_2^e| \cos\theta |\psi_2^e\rangle = \frac{5q}{8} \\[2mm] \langle\psi_2^o| \cos\theta |\psi_2^o\rangle = \frac{-q}{12} \end{array}\right\} \tag{3.74}$$

Using Eqs. (3.69) and (3.72)–(3.74) in Eq. (3.64) and keeping only the terms proportional to $q \sim \langle\cos\theta\rangle$, we obtain the following self-consistency equation for the phase-ordering temperature T_c

$$1 = \alpha \frac{1 + \frac{1}{2}e^{-x} + 2xe^{-x} - \frac{2}{3}e^{-4x}}{1 + 2e^{-x} + 2e^{-4x}} \tag{3.75}$$

where

$$x = \frac{U_a}{2k_B T_c} \quad\text{and}\quad \alpha = \frac{zJ}{2U_a} \tag{3.76}$$

In Eq. (3.75), the terms, proportional to e^{-x}, come from the inclusion of the $m = 1$ doublet. Consequently, the self-consistency equation, obtained by including *only the $m = 2$ doublet* in Eq. (3.64), has the form

$$1 = \alpha \frac{1 - \frac{2}{3}e^{-4x}}{1 + 2e^{-4x}} \tag{3.77}$$

Solving numerically Eqs. (3.75) and (3.77) for x, we find the phase diagrams shown in Fig. 3.4, respectively. The solid curve (a), resulting from Eq. (3.75), shows a well-pronounced reentrant bulge extending to $\alpha_{\min} \simeq 0.78$, whereas the broken curve based on Eq. (3.77) shows no trace of the reentrance. On recalling the form of Eqs. (3.75) and (3.77), this implies that the reentrant behavior is generated by the presence of the $m = 1$ *doublet* in the excitation spectrum. The physics behind this effect can be understood as follows (Šimánek, 1982). To order q, the contribution of the $m = 1$ doublet to the numerator of Eq. (3.64) is

$$\langle\cos\theta\rangle_1 = \langle\psi_1^e| \cos\theta |\psi_1^e\rangle e^{-E_1^e/k_B T} + \langle\psi_1^o| \cos\theta |\psi_1^o\rangle e^{-E_1^o/k_B T}$$

$$\approx -q\left(\frac{U_a}{2k_B T} + \frac{1}{4}\right) \exp\left(-\frac{U_a}{2k_B T}\right) \tag{3.78}$$

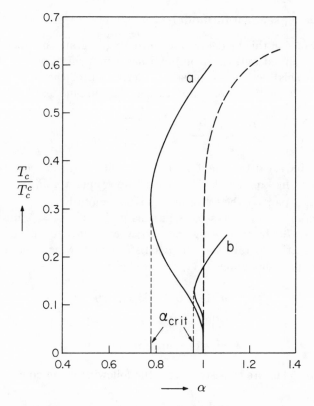

FIG. 3.4. Phase-ordering temperature ratio T_c/T_c^c of the self-charging model, plotted as a function of parameter $\alpha = zJ/2U_a$. Curve (a) and the broken curve are obtained by solving Eqs. (3.75) and (3.77), respectively. T_c^c is the classical phase-ordering temperature, given by Eq. (3.20). Curve (b) corresponds to the off-diagonal model, discussed at the end of Section (3.7) (Šimánek, 1982).

where we have used Eqs. (3.67) and (3.69). This contribution is *positive*, like the ground-state expectation value (3.72), and thus it *enhances the phase order*. As $T \to 0$, the expression (3.78) becomes exponentially small, so that the curve (a) and the broken curve of Fig. 3.4 coincide at the point $(T_c/T_c^c = 0, \alpha = 1)$. For small nonzero temperatures, the contribution (3.78) increases and the concomitant increase of $\langle \cos \theta \rangle$ causes the phase-ordered state to take place at *smaller values* of α. The contribution of the $m = 2$ doublet to the numerator of Eq. (3.64) is negative [see Eq. (3.77)] and it makes the curves bend over in the region of large T_c/T_c^c. At low values of T_c/T_c^c, the low-lying $m = 1$ excited doublet dominates the $m = 2$ contribution, which enables the reentrance to develop.

3.6. $S = 1$ pseudospin model

The relevance of this model to phase ordering in granular superconductors has been pointed out, in an unpublished communication, by de Gennes (1978). In this section, we present a derivation of the $S = 1$ model by starting with the Hamiltonian of the self-charging model (3.60)

$$H = 2U_a \sum_i \widehat{n}_i^2 - \frac{J}{2} \sum_{\langle ij \rangle} (a_i^\dagger a_j + a_j^\dagger a_i) \tag{3.79}$$

where we have written the Josephson-coupling term in second-quantized form shown in Eq. (2.84). The pair-number operator $\widehat{n}_i = a_i^\dagger a_i$, where $a_i^\dagger$ creates a pair on the ith grain. Let us denote by $|n_i\rangle$ the eigenstate, corresponding to n_i excess pairs on this grain. The $S = 1$ model results from Eq. (3.79) by truncating the basis vectors $|n_i\rangle$ to the states $|0\rangle$ and $|\pm 1\rangle$. Then the operators $a^\dagger$ and a satisfy the following algebra (dropping the subscript i, for simplicity)

$$
\begin{aligned}
a^\dagger|1\rangle &= 0 & a|1\rangle &= |0\rangle \\
a^\dagger|0\rangle &= |1\rangle & a|0\rangle &= |-1\rangle \\
a^\dagger|-1\rangle &= |0\rangle & a|-1\rangle &= 0
\end{aligned} \tag{3.80}
$$

Equations (3.80) are consistent with the following set of equations

$$
\begin{aligned}
(a^\dagger a - a a^\dagger)|1\rangle &= |1\rangle \\
(a^\dagger a - a a^\dagger)|0\rangle &= 0 \\
(a^\dagger a - a a^\dagger)|-1\rangle &= -|-1\rangle
\end{aligned} \tag{3.81}
$$

This set can be compactly written as a simple commutation relation

$$a^\dagger a - a a^\dagger = \widehat{n} \tag{3.82}$$

where the operator $\widehat{n}$ has the eigenvalues 0 and ± 1. If we let

$$a^\dagger = \frac{S^+}{\sqrt{2\hbar}}; \qquad a = \frac{S^-}{\sqrt{2\hbar}}; \qquad \widehat{n} = \frac{S_z}{\hbar} \tag{3.83}$$

then Eq. (3.82) is identical to the spin commutation relation

$$S^+ S^- - S^- S^+ = 2\hbar S_z \tag{3.84}$$

where $S^\pm = S_x \pm i S_y$. Introducing the spin operators of Eq. (3.83) into Eq. (3.79), we obtain the $S = 1$ pseudospin Hamiltonian

$$H = \frac{2U_a}{\hbar^2} \sum_i S_{zi}^2 - \frac{J}{2\hbar^2} \sum_{\langle ij \rangle} (S_{xi} S_{xj} + S_{yi} S_{yj}) \tag{3.85}$$

In the context of ferromagnetism, the first term in this equation corresponds to the crystal-anisotropy energy, and the second term describes the *ferromagnetic exchange* responsible for the formation of spin order in the xy plane. The model exhibits, at $T = 0$, a phase transition, driven by quantum fluctuations of the transverse spin components. The source of these fluctuations is the anisotropy energy, which is minimized by forcing the spin into the xy plane. By the uncertainty principle, the confinement of the z component to $S_z \simeq 0$ makes transverse fluctuations large, thus reducing the order parameter. Consequently, the internal field, responsible for the transverse spin ordering, is also reduced, leading to further increase of the transverse spin fluctuations. Eventually, when the parameter $\alpha = zJ/2U_a$ becomes smaller than one, the fluctuations grow without limit, leading to a complete destruction of spin order. Using the arguments of Section (2.9), this quantum critical phenomenon can be mapped onto a phase transition of a classical model. Specifically, the $T = 0$ behavior of the $S = 1$ model in d dimensions is equivalent to the phase transition of a $(d + 1)$-dimensional model. Since the MFA is known to be valid for $d \gtrsim 4$, the MFA treatment of the $T = 0$ transition in the three-dimensional $S = 1$ model seems justified.

In the MFA, the "exchange" term is approximated as follows [see Eq. (3.16)]:

$$S_{xi}S_{xj} + S_{yi}S_{yj} \simeq \langle S_{xi}\rangle S_{xj} + \langle S_{xj}\rangle S_{xi} - \langle S_{xi}\rangle\langle S_{xj}\rangle \qquad (3.86)$$

where we have assumed that the ordering takes place along the x axis ($\langle S_{yi}\rangle = 0$). Introducing Eq. (3.86) into Eq. (3.85) and discarding the constant terms, we obtain the pseudospin Hamiltonian in the MFA

$$H_{\mathrm{MFA}} = \frac{2U_a}{\hbar^2} \sum_i \left[S_{zi}^2 - \lambda S_{xi}\right] \qquad (3.87)$$

where

$$\lambda = \frac{zJ}{4U_a}\langle S_x\rangle \qquad (3.88)$$

The order parameter $\langle S_x\rangle$ is calculated from the expression

$$\langle S_x\rangle = \frac{\sum_n e^{-E_n/k_B T_c}\langle\psi_n|S_x|\psi_n\rangle}{\sum_n e^{-E_n/k_B T_c}} \qquad (3.89)$$

where ψ_n and E_n are the eigenstates and energy eigenvalues obtained by diagonalizing the Hamiltonian (3.87). Near T_c, the parameter $\lambda \sim \langle S_x\rangle$ is small and the matrix elements of S_x can be expanded in terms of λ, yielding the following self-consistency condition for T_c

$$1 = \alpha \frac{1 - e^{-4x}}{1 + 2e^{-4x}} \qquad (3.90)$$

where the parameters α and x are defined in Eq. (3.76). Although Eq. (3.90) differs from the "planar rotator" result (3.77), it leads to a qualitatively similar phase diagram. Specifically, the critical value of α is again equal to one, and although the transition temperatures T_c following from Eq. (3.90) lie below the broken curve of Fig. 3.4, the decrease is only significant for values of α well above 1. The reentrant behavior is again absent, in the $S = 1$ model, as expected by recalling that the states $|\pm 1\rangle$ correspond to an *integer number of excess pairs* on a grain.

3.7.   Models with off-diagonal Coulomb interactions

The Hamiltonian of an array involving the charging energy in the general form (3.54) can be written

$$\widehat{H} = 2 \sum_{\langle ij \rangle} U_{ij} \widehat{n}_i \widehat{n}_j - J \sum_{\langle ij \rangle} \cos(\theta_i - \theta_j) \tag{3.91}$$

where the pair-number operators $\widehat{n}_i$ satisfy the commutation relations (3.58). The Hamiltonian (3.91) is more realistic than the diagonal model (3.60) since the off-diagonal terms are usually not small. We note that, for a three-dimensional granular array, the nearest-neighbor interaction U_{12} is, typically, $U_{12} \simeq \frac{1}{2} U_{11}$ (Abeles et al., 1975).

In the MFA the Josephson-coupling term is replaced, in the same way as in Eq. (3.62), yielding

$$\begin{aligned}
\widehat{H}_{\mathrm{MFA}} &= 2 \sum_{\langle ij \rangle} U_{ij} \widehat{n}_i \widehat{n}_j - zJ\langle \cos\theta \rangle \sum_k \cos\theta_k \\
&= \widehat{H}_0 + \widehat{H}_1
\end{aligned} \tag{3.92}$$

Following Efetof (1980), we now derive the self-consistency condition for the phase-locking temperature T_c, by treating the Hamiltonian $\widehat{H}_1$, as a perturbation. Using Eq. (A6.13), the expectation value of $\cos\theta_l$ can be written

$$\langle \cos\theta_l \rangle = \frac{\mathrm{Tr}\left[e^{-\beta\widehat{H}_0}\left(1 - \frac{1}{\hbar}\int_0^{\hbar\beta} d\tau\, \widehat{H}_1[\tau]\right)\cos\theta_l[0]\right]}{\mathrm{Tr}\left(e^{-\beta\widehat{H}_0}\right)} \tag{3.93}$$

where $\widehat{H}_1[\tau]$ is, according to Eqs. (3.92) and (A6.5), given by

$$\widehat{H}_1[\tau] = -zJ\langle \cos\theta \rangle \sum_k \cos\theta_k[\tau] \tag{3.94}$$

where

$$\theta_k[\tau] = e^{\widehat{H}_0\tau/\hbar} \theta_k e^{-\widehat{H}_0\tau/\hbar} \tag{3.95}$$

Introducing Eq. (3.94) into Eq. (3.93), we have

$$1 = \frac{zJ \sum_i \frac{1}{\hbar}\int_0^{\hbar\beta} d\tau\, \mathrm{Tr}\left(e^{-\beta\widehat{H}_0}\cos\theta_i[\tau]\cos\theta_l[0]\right)}{\mathrm{Tr}\left(e^{-\beta\widehat{H}_0}\right)}$$

$$= \frac{1}{2} z J \sum_i \frac{1}{\hbar} \int_0^{\hbar \beta} d\tau \left\langle e^{i \theta_i [\tau]} e^{-i \theta_l [0]} \right\rangle_0 \tag{3.96}$$

where we have used the stationarity property of the correlation function to write

$$\left\langle e^{i \theta_i [\tau]} e^{-i \theta_l [0]} \right\rangle_0 = 2 \langle \cos \theta_i [\tau] \cos \theta_l [0] \rangle_0 \tag{3.97}$$

The correlator on the LHS of this equation is evaluated by first solving for the τ dependence of $\theta_k[\tau]$. Equation (3.95) implies the following equation of motion

$$\hbar \frac{\partial \theta_k [\tau]}{\partial \tau} = [\widehat{H}_0, \theta_k] = 2 \sum_{\langle ij \rangle} U_{ij} [\widehat{n}_i \widehat{n}_j, \theta_k] \tag{3.98}$$

Using the commutator (3.58), we have

$$[\widehat{n}_i \widehat{n}_j, \theta_k] = i (\widehat{n}_i \delta_{jk} + \widehat{n}_j \delta_{ik}) \tag{3.99}$$

Inserting this result into Eq. (3.98) and using the relation $U_{kj} = U_{jk}$, we obtain

$$\hbar \frac{\partial \theta_k [\tau]}{\partial \tau} = 4i \sum_j U_{kj} \widehat{n}_j [\tau] \tag{3.100}$$

This equation can be integrated to yield

$$\theta_k [\tau] = \theta_k [0] + \frac{4i\tau}{\hbar} \sum_j U_{kj} \widehat{n}_j [0] \tag{3.101}$$

Using this result and the operator identity (2.136), the commutator on the LHS of Eq. (3.97) is written as

$$\begin{aligned}
\left\langle e^{i \theta_i [\tau]} e^{-i \theta_l [0]} \right\rangle_0 &= \left\langle e^{i \theta_i - 4\tau / \hbar \sum_j U_{ij} \widehat{n}_j} e^{-i \theta_l} \right\rangle_0 \\
&= e^{2 \tau U_{ii} / \hbar} \left\langle e^{i \theta_i} e^{-4\tau / \hbar \sum_j U_{ij} \widehat{n}_j} e^{-i \theta_l} \right\rangle_0 \\
&= \frac{e^{2 \tau U_{ii} / \hbar}}{\mathrm{Tr} \left(e^{-\beta \widehat{H}_0} \right)} \sum_{\{n_k\}} e^{-2\beta \sum_{\langle pq \rangle} U_{pq} n_p n_q} \\
&\quad \times \langle \{n_k\} | e^{i \theta_i} e^{-4\tau / \hbar \sum_j U_{ij} \widehat{n}_j} e^{-i \theta_l} | \{n_k\} \rangle
\end{aligned} \tag{3.102}$$

where the symbol $\{n_k\}$ stands for a configuration characterized by excess pair numbers n_i. The matrix element on the RHS of Eq. (3.102) can be evaluated using the "number shifting" property of the number eigenstates $|n_i\rangle = e^{-in\theta_i} / \sqrt{2\pi}$, which allows us to write for a general configuration $\{n_k\}$

$$\begin{aligned}
e^{-i \theta_l} | \{n_k\} \rangle &= e^{-i \theta_l} |n_1, \dots, n_l, \dots, n_N\rangle \\
&= |n_1, \dots, n_l + 1, \dots, n_N\rangle
\end{aligned} \tag{3.103}$$

and

$$\langle\{n_k\}|e^{i\theta_i} = \langle n_1,\dots,n_i+1,\dots,n_N| \tag{3.104}$$

Using Eqs. (3.103) and (3.104), the matrix element on the RHS of Eq. (3.102) becomes

$$\langle\{n_k\}|e^{i\theta_i}e^{-4\tau/\hbar\sum_j U_{ij}\hat{n}_j}e^{-i\theta_l}|\{n_k\}\rangle$$

$$= \langle n_1,\dots,n_i+1,\dots,n_N|e^{-4\tau/\hbar\sum_j U_{ij}\hat{n}_j}|n_1,\dots,n_l+1,\dots,n_N\rangle$$

$$= \delta_{i,l}e^{-4\tau/\hbar\sum_j U_{ij}n_j}e^{-4\tau U_{ii}/\hbar} \tag{3.105}$$

where the last factor on the RHS of Eq. (3.105) results from the excess pair on the lth site. Introducing this result into Eq. (3.102), we have:

$$\left\langle e^{i\theta_i[\tau]}e^{-i\theta_l[0]}\right\rangle_0 = \frac{e^{-2\tau U_{ii}/\hbar}\delta_{i,l}}{\mathrm{Tr}\left(e^{-\beta\widehat{H}_0}\right)}$$

$$\times \sum_{\{n_k\}}\left(e^{-2\beta\sum_{\langle pq\rangle}U_{pq}n_pn_q}e^{-4\tau/\hbar\sum_j U_{ij}n_j}\right) \tag{3.106}$$

Using this result in Eq. (3.96), the self-consistency equation for T_c takes the final form

$$1 = \frac{zJ}{2\,\mathrm{Tr}\left(e^{-\beta_c\widehat{H}_0}\right)}$$

$$\times \sum_{\{n_k\}}\left(\frac{1-e^{-2\beta_c\left(U_{ii}+2\sum_j U_{ij}n_j\right)}}{2\left(U_{ii}+2\sum_j U_{ij}n_j\right)}e^{-2\beta_c\sum_{\langle pq\rangle}U_{pq}n_pn_q}\right) \tag{3.107}$$

As a simple application of this formula, let us consider the lowest integer-valued configurations, in which $n_i = 0,\pm1$. Assuming that $U_{ij} = \frac{1}{2}U_{ii} = \frac{1}{2}U_{11}$ (for nearest-neighbor i and j), we also calculate the Coulomb energy $E_c = 2\sum_{\langle pq\rangle}U_{pq}n_pn_q$ and the phase rotation frequency $\omega(n_i,n_j)$

$$\omega(n_i,n_j) = 2\left(U_{ii}+2\sum_j U_{ij}n_j\right) \tag{3.108}$$

which is equal to the coefficient of $-\tau$ in the exponent of Eq. (3.106). These numbers are collected in Table 3.1, along with the degeneracy D of each configuration. Introducing the values from Table 3.1 into Eq. (3.107), the equation for T_c is

$$1 = \alpha\frac{1+\left(2zx+\frac{z}{4}-\frac{5}{6}\right)e^{-4x}}{1+(2z+2)e^{-4x}} \tag{3.109}$$

where $\alpha = zJ/2U_{11}$ and $x = U_{11}/2k_BT_c$. The phase diagram resulting from Eq. (3.109) exhibits a very weak reentrance due to the presence of the term

n_i	n_j	E_c	$\omega(n_i, n_j)$	D
0	0	0	$2U_{11}$	1
1	0	$2U_{11}$	$6U_{11}$	1
-1	0	$2U_{11}$	$-2U_{11}$	1
1	-1	$2U_{11}$	$4U_{11}$	z
-1	1	$2U_{11}$	0	z

Table 3.1. Coulomb energies E_c, rotational frequencies $\omega(n_i, n_j)$, and degeneracies D associated with with the configuration (n_i, n_j) of pair numbers on nearest-neighbor grains.

$2zxe^{-4x}$, which comes from the dipolar configuration $(n_i = -1, n_j = 1)$, characterized by the *vanishing of the phase rotation* frequency ω. The reason that the reentrance is so weak is that the role of this term is masked by other terms generated by configurations of the same excitation energy but nonzero ω. A more detailed investigation of the MFA phase diagram of this model has been made by Fazekas (1982), who has included a set of higher integer-valued configurations (up to $n_i = \pm 4$). The phase diagram he obtains for $z = 3$ and $U_{ij} = \frac{1}{3}U_{ii}$ exhibits again a very weak reentrance ($\alpha_{min} \doteq 0.99917$). It is instructive to see that the condition (3.107) is consistent with Eq. (3.77) obtained for the self-charging model. In fact, the latter equation can be rederived from Eq. (3.107) by replacing the last two rows of Table 3.1 by the diagonal configurations $(n_i = 2, n_j = 0)$ and $(n_i = -2, n_j = 0)$. We note that these configurations are effectively present in the self-charging model, being admixed into the ψ_2^e and ψ_2^o wave functions as $\cos 2\theta$ and $\sin 2\theta$ terms, respectively [see Eqs. (3.68)]. We have also considered a model obtained by augmenting the configurations of Table 3.1 by those involving half-integer values of n_i (Šimánek, 1982). The resulting phase diagram shows a reentrance somewhat stronger than that obtained with Table 3.1, owing to a low-lying configuration $(n_i = -\frac{1}{2}, n_j = 0)$. But the reentrant bulge extends over a rather narrow interval [see curve (b) of Fig. 3.4]. This suppression of reentrance is presumably due to the presence of the off-diagonal configuration $(n_i = 1, n_j = -1)$. This makes us conclude that the reentrance caused by the diagonal single-electron excitations may be observable only when the condition $U_{ij} \ll U_{ii}$ is met.

3.8. Path-integral formulation of the MFA

The path-integral method of Feynman has become a standard tool used frequently in the studies of statistical mechanics of Josephson junctions and their arrays (Schön and Zaikin, 1990). More sophisticated applications involve quantum systems with dissipation, which are discussed in Chapters

4-6. In this section, we introduce this method by starting with a simple case: derivation of the phase diagram, in the MFA, for the self-charging model of a Josephson-coupled array. Besides serving as a warm-up exercise, this derivation sheds also some light on the relation between the reentrance and the choice of the interval for the phase variable (Šimánek, 1985).

Consider the MFA of the self-charging model described by the Hamiltonian (3.62). Treating the Josephson term as a perturbation, we can derive a self-consistency equation for the transition temperature of the form [see Eq. (A6.13)]

$$1 = \frac{zJ}{2\hbar} \int_0^{\hbar\beta} d\tau \left\langle e^{i\theta[\tau]} e^{-\theta[0]} \right\rangle_0 \tag{3.110}$$

Actually, this equation follows directly from the formula (3.96) by noting that the phase correlator is site diagonal [see Eq. (3.106)]. From Eq. (3.96), the explicit form of this correlator is

$$R(\tau) = \left\langle e^{i\theta[\tau]} e^{-i\theta[0]} \right\rangle$$
$$= \frac{\mathrm{Tr}\left(e^{-\beta\widehat{H}_0} e^{i\theta[\tau]} e^{-i\theta[0]}\right)}{\mathrm{Tr}\left(e^{-\beta\widehat{H}_0}\right)} \tag{3.111}$$

where $\widehat{H}_0$, in our case, is the diagonal charging energy

$$\widehat{H}_0 = 2U_a \sum_i \widehat{n}_i^2 = -2U_a \sum_i \frac{\partial^2}{\partial\theta_i^2} \tag{3.112}$$

Using the functional-integral formulation of the correlation function, given in Eq. (A7.10), we have

$$R(\tau) = Z^{-1} \int_{\mathrm{per}} \mathcal{D}\theta(0)\, e^{i[\theta(\tau)-\theta(0)]} e^{-\frac{1}{\hbar}S_0[\theta]} \tag{3.113}$$

where

$$Z = \int_{\mathrm{per}} \mathcal{D}\theta(\tau)\, e^{-\frac{1}{\hbar}S_0[\theta]} \tag{3.114}$$

The subscript on the path integrals indicates that only those paths are to be included, for which the periodic boundary condition, $\theta(0) = \theta(\hbar\beta)$, holds. The Euclidean action S_0 involves only the charging energy (3.112), so that

$$S_0 = \frac{\hbar^2}{8U_a} \int_0^{\hbar\beta} d\tau \left(\frac{\partial\theta}{\partial\tau}\right)^2 \tag{3.115}$$

Since the model of Eq. (3.62) does not include any dissipation, the phase θ is restricted to the interval $[0, 2\pi]$, which makes the action S_0 equivalent to that of a *free particle on a circle*. The partition function and the phase correlator for this case can, of course, be calculated much easier using straightforward quantum-statistical methods, as is done in Sections 3.5 and 3.7. In fact, the original Feynman formulation of the partition function (Feynman, 1972) suited for a particle moving in extended space $[-\infty, \infty]$ needs to be generalized, to cover the problem of a particle on circle. This extension is considered in Appendix A3, where we derive an expression for the partition function of a particle on a circle as a path integral, equivalent to Eq. (3.114), but involving paths $\theta(\tau)$, which are allowed to vary over the *extended interval*, $\theta(\tau) \in [-\infty, \infty]$. According to Eq. (A3.35), this expression is given by a sum over the *winding numbers l*

$$Z = \sum_{l=-\infty}^{\infty} \int_0^{2\pi} d\theta_0 \int_{\theta_0}^{\theta_0+2\pi l} \mathcal{D}\theta(\tau) \, e^{-\frac{1}{\hbar}S_0[\theta]} \tag{3.116}$$

We see that, for $l \neq 0$, the periodic boundary condition $\theta(0) = \theta(\hbar\beta)$ is replaced by $\theta(\hbar\beta) = \theta(0) + 2\pi l$. The phase correlator $R(\tau)$ is, in view of (3.116), given by [see Eqs. (A7.11) and (A7.17)]

$$R(\tau) = Z^{-1} \sum_{l=-\infty}^{\infty} \int_0^{2\pi} d\theta_0 \int_{\theta_0}^{\theta_0+2\pi l} \mathcal{D}\theta(\tau) \, e^{i[\theta(\tau)-\theta(0)]} e^{-\frac{1}{\hbar}S_0[\theta]} \tag{3.117}$$

Since S_0 is a Gaussian action, the path integrals in Eqs. (3.116) and (3.117) can be readily performed. In fact, the density matrix of a free particle on a circle is evaluated in Appendix A4, where we show the equivalence of Eq. (3.116) with the quantum-mechanical expression for the partition function of a free rotator. This implies that the phase diagram, calculated with use of Eq. (3.110), where $R(\tau)$ is given by the expression (3.117), is equivalent to that obtained from Eq. (3.64), in which the m sum involves all 2π-periodic (even m) states. In this case no new results are obtained by the path-integral method. When there is ohmic dissipation, the extended phase interval is appropriate, at the outset, and the action (3.115) corresponds to a particle moving in the one-dimensional *unrestricted space*. The original Feynman approach can then be directly applied, and the correlation function is given by the $l = 0$ component of Eq. (3.117)

$$R_0(\tau) = R_{l=0}(\tau) = Z^{-1} \int_{\text{per}} \mathcal{D}\theta(\tau) \, e^{i[\theta(\tau)-\theta(0)]} e^{-\frac{1}{\hbar}S_0[\theta]} \tag{3.118}$$

where per on the integral means to sum over all paths satisfying the condition $\theta(0) = \theta(\hbar\beta)$. The partition function Z is given by

$$Z = \int_0^{2\pi} d\theta_0 \int_{\theta(0)=\theta_0}^{\theta(\hbar\beta)=\theta_0} \mathcal{D}\theta(\tau) \, e^{-\frac{1}{\hbar}S_0[\theta]} = \int_{\text{per}} \mathcal{D}\theta(\tau) \, e^{-\frac{1}{\hbar}S_0[\theta]} \tag{3.119}$$

We assume that even a *small* dissipation is sufficient to change the phase interval to $[-\infty, \infty]$. Hence, $S_0[\theta]$ is still given by the undamped action of

Eq. (3.115). These path integrals (3.118) and (3.119) can be evaluated by expanding $\theta(\tau)$ into a Fourier series (Feynman and Hibbs, 1965)

$$\theta(\tau) = \sum_{n=-\infty}^{\infty} \theta_n e^{-i\omega_n \tau} \tag{3.120}$$

where $\omega_n = 2\pi n/\hbar\beta$ is a discrete frequency chosen so that $\theta(\tau)$ satisfies the periodic boundary condition $\theta(0) = \theta(\hbar\beta)$. Since $\theta(\tau)$ is real, the complex coefficients $\theta_n = \theta_n^r + i\theta_n^i$ satisfy the reality conditions

$$\theta_n^r = \theta_{-n}^r$$
$$\theta_n^i = -\theta_{-n}^i \tag{3.121}$$

The steps that follow are similar to those used in Appendix A9 to evaluate the **r**-dependent phase correlator. Introducing the expansion (3.120) into the action (3.115), we have

$$\frac{1}{\hbar}S_0 = \frac{\hbar^2\beta}{4U_a} \sum_{n=1}^{\infty} \omega_n^2 \left[(\theta_n^r)^2 + (\theta_n^i)^2\right] \tag{3.122}$$

Using Eqs. (3.120) and (3.121), we obtain, with a hint from Eq. (A9.15),

$$\theta(\tau) - \theta(0) = 2 \sum_{n=1}^{\infty} \left[\theta_n^r(\cos\omega_n\tau - 1) - \theta_n^i \sin\omega_n\tau\right] \tag{3.123}$$

The path integral (3.118) can be expressed, using Eqs. (3.122) and (3.123), and recalling Eq. (A9.17), as a multiple Gaussian integral

$$R_0(\tau) = Z^{-1} \prod_{n>0} \left[\int_{-\infty}^{\infty} d\theta_n^r \int_{-\infty}^{\infty} d\theta_n^i \, \exp\left(-\frac{\hbar^2\beta\omega_n^2}{4U_a}\left[(\theta_n^r)^2 + (\theta_n^i)^2\right]\right) \right.$$

$$\left. \times \exp\left[2i\theta_n^r(\cos\omega_n\tau - 1) - 2i\theta_n^i \sin\omega_n\tau\right]\right] \tag{3.124}$$

where

$$Z = \prod_{n>0} \left[\int_{-\infty}^{\infty} d\theta_n^r \int_{-\infty}^{\infty} d\theta_n^i \, \exp\left(-\frac{\hbar^2\beta\omega_n^2}{4U_a}\left[(\theta_n^r)^2 + (\theta_n^i)^2\right]\right)\right] \tag{3.125}$$

The integrations in Eqs. (3.124) and (3.125) can be done, using formula (A3.20), yielding

$$R_0(\tau) = \exp\left(-\frac{16U_a}{\hbar^2\beta} \sum_{n=1}^{\infty} \frac{\sin^2\left(\frac{\omega_n\tau}{2}\right)}{\omega_n^2}\right) = \exp\left(-\frac{2U_a}{\hbar^2\beta}\tau(\hbar\beta - \tau)\right) \tag{3.126}$$

For a simple derivation of this result, see Eqs. (A4.17)–(A4.23). Note that the periodicity property $R_0(\tau) = R_0(\hbar\beta - \tau)$ is satisfied by this correlator. Introducing Eq. (3.126) into Eq. (3.110), the equation for T_c is obtained in the form (Šimánek, 1985)

$$1 = \left(\frac{zJ}{2k_BT_c}\right)\frac{y(u)}{u} \qquad (3.127)$$

where $u = (U_a/2k_BT_c)^{\frac{1}{2}}$ and $y(u)$ is the Dawson integral (Abramowitz and Stegun, 1964)

$$y(u) = e^{-u^2}\int_0^u e^{t^2}\, dt \qquad (3.128)$$

Introducing the ratio $\alpha = zJ/2U_a$ defined in Eq. (3.76), Eq. (3.127) can be written

$$1 = 2\alpha u y(u) \qquad (3.129)$$

which is to be solved numerically for u. The phase-locking transition temperature T_c is then determined from the relation

$$\frac{T_c}{T_c^c} = \frac{1}{2\alpha u^2} \qquad (3.130)$$

where $T_c^c = zJ/2k_B$ is the phase-transition temperature of the classical $(U_a = 0)$ model. The phase diagram, obtained from Eqs. (3.129) and (3.130), is shown in Fig. 3.4. It practically coincides with the curve (a). A well-pronounced reentrant bulge is seen, which extends over the region $0.78 < \alpha < 1$. The numerical values $\alpha_{min} = 0.78$ and $(T_c/T_c^c)_{\alpha_{min}} = 0.29$ are in remarkable agreement with the phase diagram resulting from Eq. (3.75). It should be noted that the model leading to Eq. (3.75) involves the 4π-periodic states with discrete energy spectrum shown in Eq. (3.69). On the other hand, the energy spectrum of the Josephson oscillator, with phase variable extending over $\theta \in [-\infty, \infty]$, is continuous. The fact that both models exhibit very similar reentrant features indicates that the continuum of states (other that the 4π-periodic ones) presumably does not play any significant role in the mechanism of reentrance. Recently, Simkin (1991) has studied the self-charging model in the MFA, using a generalization of the expression (3.64) to a continuum of energy eigenstates. His results are similar to those obtained here with the path-integral method. In particular, the reentrant part of his phase diagram extends over $0.78 < \alpha < 1$ in close agreement with Fig. 3.4. The examples discussed in this chapter show how important it is to treat the indistinguishability of the states θ and $\theta + 2\pi$ with care. Likharev and Zorin (1985) suggest that these two states are distinguishable, due to the environmental coupling; however, they do not specify the form or strength of the dissipation required. Kampf and Schön (1987) argue that even a small ohmic dissipation is enough to break the 2π periodicity. This seems justified especially near the phase-locking transition temperature due to the smallness of the periodic potential $\sim \langle\cos\theta\rangle$. On increasing the dissipation, the quantum phase fluctuations and the concomitant reentrance become suppressed (Šimánek and Brown, 1986). Hence, the reentrant behavior shown in Fig. 3.4 should be observable on granular superconductors, with small, but nonzero, leakage currents between the grains or to the substrate. Experimentally, a quasireentrant behavior has been seen in the resistance versus temperature curves measured on thin granular films (for a review of these studies, see Kobayashi, 1992).

3.9. Self-consistent harmonic approximation

This approximation has been used in the early investigations of charging effects in regular undamped junction arrays (Šimánek, 1980a; Wood and Stroud, 1982). More recently, it has been applied to arrays with dissipation (Chakravarty et al., 1986). We introduce this method by considering the self-charging model of Eq. (3.60). The partition function of this system can be expressed as a path integral over the phases of the grains, θ_i, which we denote collectively as $\theta = \{\theta_i\}$ [see Eq. (A2.18)]

$$Z = \int_{\text{per}} \mathcal{D}\theta(\tau) \exp\left(-\frac{1}{\hbar} S[\theta]\right) \tag{3.131}$$

where $S[\theta]$ is the Euclidean action given by

$$S[\theta] = \int_0^{\hbar\beta} d\tau \left[\frac{\hbar^2}{8U_a} \sum_i \left(\frac{\partial\theta_i}{\partial\tau}\right)^2 + J\sum_{\langle ij\rangle}(1 - \cos\theta_{ij})\right] \tag{3.132}$$

where $\theta_{ij} = \theta_i - \theta_j$. We have added a constant to the Josephson-coupling term in (3.132) to follow a standard convention. The essence of the SCHA is to replace the action (3.132) by a trial harmonic action

$$S_{\text{tr}}[\theta] = \int_0^{\hbar\beta} d\tau \left[\frac{\hbar^2}{8U_a} \sum_i \left(\frac{\partial\theta_i}{\partial\tau}\right)^2 + \frac{1}{2}K\sum_{\langle ij\rangle} \theta_{ij}^2(\tau)\right] \tag{3.133}$$

where the "stiffness" constant K is to be determined by a variational approach.

   We start from the variational principle, which states that, for any trial action S_{tr} with associated free energy F_{tr}, the true free energy F satisfies the inequality (Feynman, 1972)

$$F \leq F_{\text{tr}} + \frac{1}{\hbar\beta Z_{\text{tr}}} \int_{\text{per}} \mathcal{D}\theta \, e^{-\frac{1}{\hbar}S_{\text{tr}}}(S - S_{\text{tr}})$$

$$= F_{\text{tr}} + \frac{1}{\hbar\beta}\langle S - S_{\text{tr}}\rangle_{\text{tr}} = F^* \tag{3.134}$$

where

$$Z_{\text{tr}} = \int_{\text{per}} \mathcal{D}\theta \, e^{-\frac{1}{\hbar}S_{\text{tr}}} \tag{3.135}$$

Using Eqs. (3.132) and (3.133), we obtain for the second term in Eq. (3.134)

$$\frac{1}{\hbar\beta}\langle S - S_{\text{tr}}\rangle_{\text{tr}} = \frac{1}{\hbar\beta}\int_0^{\hbar\beta} d\tau \sum_{\langle ij\rangle} \left\langle \left(J(1 - \cos\theta_{ij}) - \frac{1}{2}K\theta_{ij}^2 \right) \right\rangle_{\text{tr}}$$

$$= \frac{1}{\hbar\beta}\int_0^{\hbar\beta} d\tau \sum_{\langle ij\rangle}' \left[J\left(1 - e^{-\frac{1}{2}D_{ij}}\right) - \frac{1}{2}KD_{ij} \right] \tag{3.136}$$

where

$$D_{ij} = \langle\theta_{ij}^2\rangle_{\text{tr}} \tag{3.137}$$

Now we ask: "What is the best choice of K?" Changing $K \to K + \delta K$ and asserting that the upper bound F^* of the free energy in Eq. (3.134) is minimized, we obtain

$$\delta F^* = \delta\left(F_{\text{tr}} + \frac{1}{\hbar\beta}\langle S - S_{\text{tr}}\rangle_{\text{tr}}\right) = 0 \tag{3.138}$$

Since D_{ij} is also a function of K, the condition (3.138) can be written

$$\left(\frac{\partial F^*}{\partial K}\right)_{D_{ij}} + \left(\frac{\partial F^*}{\partial D_{ij}}\right)_K \left(\frac{\partial D_{ij}}{\partial K}\right) = 0 \tag{3.139}$$

First, let us calculate $\partial F_{\text{tr}}/\partial K$. Using Eq. (3.133), we see by definition

$$F_{\text{tr}} = -\frac{1}{\beta}\ln\int \mathcal{D}\theta \, \exp\left(-\frac{1}{\hbar}S_{\text{tr}}[\theta]\right)$$

$$= -\frac{1}{\beta}\ln\int \mathcal{D}\theta \, \exp\left(-\frac{S_0}{\hbar} - \int_0^{\hbar\beta} d\tau \frac{K}{2\hbar}\sum_{\langle ij\rangle}\theta_{ij}^2\right) \tag{3.140}$$

where we denoted the charging part of S_{tr} by S_0. Equation (3.140) implies

$$\frac{\partial F_{\text{tr}}}{\partial K} = \frac{1}{2\hbar\beta}\int_0^{\hbar\beta} d\tau \sum_{\langle ij\rangle}\langle\theta_{ij}^2\rangle_{\text{tr}} = \frac{1}{2\hbar\beta}\int_0^{\hbar\beta} d\tau \sum_{\langle ij\rangle}D_{ij} \tag{3.141}$$

where we have used the definition of D_{ij} given in Eq. (3.137). Next, we calculate, using Eq. (3.136), the partial derivative

$$\frac{1}{\hbar\beta}\left(\frac{\partial\langle S - S_{\text{tr}}\rangle_{\text{tr}}}{\partial K}\right)_{D_{ij}} = -\frac{1}{2\hbar\beta}\int_0^{\hbar\beta} d\tau \sum_{\langle ij\rangle}D_{ij} \tag{3.142}$$

Thus we have, using Eqs. (3.141) and (3.142),

$$\left(\frac{\partial F^*}{\partial K}\right)_{D_{ij}} = \frac{\partial F_{\mathrm{tr}}}{\partial K} + \frac{1}{\hbar\beta}\left(\frac{\partial\langle S - S_{\mathrm{tr}}\rangle_{\mathrm{tr}}}{\partial K}\right)_{D_{ij}} = 0 \qquad (3.143)$$

Consequently, only the second term on the LHS of Eq. (3.139) yields a nonzero contribution. From Eq. (3.133), we see that F_{tr} is not an explicit function of D_{ij}, so that

$$\left(\frac{\partial F^*}{\partial D_{ij}}\right)_K = \frac{1}{\hbar\beta}\left(\frac{\partial\langle S - S_{\mathrm{tr}}\rangle_{\mathrm{tr}}}{\partial D_{ij}}\right)_K$$

$$= \frac{1}{2\hbar\beta}\int_0^{\hbar\beta} d\tau \sum_{\langle ij\rangle} D_{ij}\left(Je^{-\frac{1}{2}D_{ij}} - K\right) \qquad (3.144)$$

where we have used Eq. (3.136). Inserting this result and Eq. (3.143) into Eq. (3.139), we obtain for the best value of the coupling constant, K, the condition

$$K = Je^{-\frac{1}{2}D_{ij}} \qquad (3.145)$$

Equations (3.145) and (3.137) form a pair of coupled self-consistent equations requiring a numerical solution for K. We now derive an explicit expression for D_{ij}. Expanding $\theta_i(\tau)$ into a Fourier series

$$\theta_i(\tau) = \frac{1}{N}\sum_{\mathbf{k}} \theta_{\mathbf{k}}(\tau)e^{i\mathbf{k}\cdot\mathbf{R}_i} \qquad (3.146)$$

we have

$$D_{ij} = \langle\theta_{ij}^2\rangle_{\mathrm{tr}}$$

$$= \frac{1}{N^2}\sum_{\mathbf{k},\mathbf{k}'} \left(e^{i\mathbf{k}\cdot\mathbf{R}_i} - e^{i\mathbf{k}\cdot\mathbf{R}_j}\right)$$

$$\times \left(e^{i\mathbf{k}'\cdot\mathbf{R}_i} - e^{i\mathbf{k}'\cdot\mathbf{R}_j}\right)\langle\theta_{\mathbf{k}}\theta_{\mathbf{k}'}\rangle_{\mathrm{tr}} \qquad (3.147)$$

Translational invariance of the array implies

$$\langle\theta_{\mathbf{k}}\theta_{\mathbf{k}'}\rangle_{\mathrm{tr}} = \langle\theta_{\mathbf{k}}\theta_{-\mathbf{k}}\rangle_{\mathrm{tr}}\delta_{\mathbf{k},-\mathbf{k}'} \qquad (3.148)$$

Using this result in Eq. (3.147), we obtain

$$D_{ij} = \frac{2}{N^2}\sum_{\mathbf{k}} [1 - \cos(\mathbf{k}\cdot\mathbf{R}_{ij})]\langle\theta_{\mathbf{k}}\theta_{-\mathbf{k}}\rangle_{\mathrm{tr}} \qquad (3.149)$$

where $\mathbf{R}_{ij} = \mathbf{R}_i - \mathbf{R}_j$.

To calculate the equal-time correlation function $\langle\theta_{\mathbf{k}}\theta_{-\mathbf{k}}\rangle_{\mathrm{tr}}$, we express the trial action (3.133) in terms of the Fourier components, $\theta_{\mathbf{k},n}$, defined by the expansion

$$\theta_{\mathbf{k}}(\tau) = \sum_{n=-\infty}^{\infty} \theta_{\mathbf{k},n}e^{-i\omega_n\tau} \tag{3.150}$$

where $\omega_n = (2\pi/\hbar\beta)n$. Introducing Eqs. (3.146) and (3.150) into the charging part of Eq. (3.133), we obtain

$$\begin{aligned}
S_0[\theta] &= \int_0^{\hbar\beta} d\tau \left(\frac{\hbar^2}{8U_aN^2} \sum_i \sum_{\mathbf{k},\mathbf{k}'} \dot\theta_{\mathbf{k}}\dot\theta_{\mathbf{k}'} e^{i(\mathbf{k}+\mathbf{k}')\cdot\mathbf{R}_i} \right) \\
&= \frac{\hbar^2}{8U_aN} \int_0^{\hbar\beta} d\tau \sum_{\mathbf{k}} \dot\theta_{\mathbf{k}}\dot\theta_{-\mathbf{k}} \\
&= -\frac{\hbar^2}{8U_aN} \int_0^{\hbar\beta} d\tau \sum_{\mathbf{k}} \sum_{n,n'} \theta_{\mathbf{k},n}\theta_{-\mathbf{k},n'}\omega_n\omega_{n'}e^{-i(\omega_n+\omega_{n'})\tau} \\
&= \frac{\hbar^3\beta}{8U_aN} \sum_{\mathbf{k}} \sum_n \theta_{\mathbf{k},n}\theta_{-\mathbf{k},-n}\omega_n^2 \tag{3.151}
\end{aligned}$$

Making similar substitutions in the second term of Eq. (3.133), we obtain with use of Eq. (3.147)

$$\begin{aligned}
S_h &= \frac{K}{2} \int_0^{\hbar\beta} d\tau \sum_{\langle ij\rangle} \theta_{ij}^2 \\
&= \frac{K}{2N^2} \int_0^{\hbar\beta} d\tau \sum_{\langle ij\rangle} \sum_{\mathbf{k},\mathbf{k}'} e^{i(\mathbf{k}+\mathbf{k}')\cdot\mathbf{R}_i} \left(1 - e^{i\mathbf{k}\cdot\mathbf{R}_{ji}}\right) \\
&\quad \times \left(1 - e^{i\mathbf{k}'\cdot\mathbf{R}_{ji}}\right) \theta_{\mathbf{k}}(\tau)\theta_{\mathbf{k}'}(\tau) \tag{3.152}
\end{aligned}$$

Performing the summation over $\mathbf{R}_i$ with use of the identity

$$\sum_i e^{i(\mathbf{k}+\mathbf{k}')\cdot\mathbf{R}_i} = N\delta_{\mathbf{k},-\mathbf{k}'} \tag{3.153}$$

and using the Fourier expansion (3.150), Eq. (3.152) yields

$$\begin{aligned}
S_h &= \frac{K}{2N} \int_0^{\hbar\beta} d\tau \sum_{\mathbf{k}} \sum_{j(i)} [1 - \cos(\mathbf{k}\cdot\mathbf{R}_{ji})]\theta_{\mathbf{k}}(\tau)\theta_{-\mathbf{k}}(\tau) \\
&= \frac{\hbar\beta K}{2N} \sum_{\mathbf{k}} \sum_n \sum_{j(i)} [1 - \cos(\mathbf{k}\cdot\mathbf{R}_{ji})]\theta_{\mathbf{k},n}\theta_{-\mathbf{k},-n} \tag{3.154}
\end{aligned}$$

where an extra $\frac{1}{2}$ factor is inserted to correct overcounting of the links. The summation over nearest neighbors, $j(i)$ of a grain site i, defines a structure factor of the lattice, given by

$$f(\mathbf{k}) = \sum_{j(i)} [1 - \cos(\mathbf{k} \cdot \mathbf{R}_{ji})] = z - 2 \sum_{j=1}^{d} \cos(k_j a) \qquad (3.155)$$

where z and d are the coordination number and the dimensionality of the lattice, respectively. From Eqs. (3.133), (3.151), (3.154), and (3.155), we obtain the trial action in a Fourier-transformed form

$$S_{\text{tr}} = S_0 + S_h = \frac{\hbar\beta}{N} \sum_{\mathbf{k}} \sum_{n} \left(\frac{\hbar^2}{8U_a} \omega_n^2 + \frac{1}{2} K f(\mathbf{k}) \right) |\theta_{\mathbf{k},n}|^2 \qquad (3.156)$$

where we have used the conditions of reality of phase variable $\theta(\tau)$

$$\theta_{\mathbf{k},n}^* = \theta_{-\mathbf{k},-n} \qquad (3.157)$$

Using Eq. (3.150) and invoking the stationary property of the correlation function, we have for the equal-time correlator on the RHS of Eq. (3.149)

$$\langle \theta_{\mathbf{k}} \theta_{-\mathbf{k}} \rangle_{\text{tr}} = \sum_{n} \langle \theta_{\mathbf{k},n} \theta_{-\mathbf{k},-n} \rangle_{\text{tr}} \qquad (3.158)$$

The averages on the RHS of this equation can be expressed as multiple Gaussian integrals over the real and imaginary parts of the Fourier components [see Eqs. (A9.14)–(A9.20) for a similar calculation]

$$\theta_{\mathbf{k},n} = \theta_{\mathbf{k},n}^r + i\theta_{\mathbf{k},n}^i \qquad (3.159)$$

In view of the constraint (3.157), the real and imaginary parts of (3.159), for a given $(\mathbf{k}, n)$, are not independent of those for $(-\mathbf{k}, -n)$. Hence, to avoid overcounting, the action (3.156) is written as a sum over the space of positive $(\mathbf{k}, n)$ values, and also the path integral for $\langle \theta_{\mathbf{k},n} \theta_{-\mathbf{k},-n} \rangle$ involves only variables in this space. Thus, we obtain for a given $\mathbf{k}_0 > 0$, $n_0 > 0$

$$\langle \theta_{\mathbf{k}_0,n_0} \theta_{-\mathbf{k}_0,-n_0} \rangle_{\text{tr}}$$

$$= \frac{1}{Z_{\text{tr}}} \prod_{\substack{k>0 \\ n>0}} \int_{-\infty}^{\infty} d\theta_{\mathbf{k},n}^r \int_{-\infty}^{\infty} d\theta_{\mathbf{k},n}^i \left[(\theta_{\mathbf{k}_0,n_0}^r)^2 + (\theta_{\mathbf{k}_0,n_0}^i)^2 \right]$$

$$\times \exp \left[-\frac{2\beta}{N} \left(\frac{\hbar^2}{8U_a} \omega_n^2 + \frac{1}{2} K f(\mathbf{k}) \right) \right.$$

$$\times \left. \left[(\theta_{\mathbf{k},n}^r)^2 + (\theta_{\mathbf{k},n}^i)^2 \right] \right] \qquad (3.160)$$

We note that the integrals over Fourier components with $\mathbf{k} \neq \mathbf{k}_0$ and $n \neq n_0$ cancel with corresponding integrals in Z_{tr}, so that we are left with a single integral

$$
\begin{aligned}
\langle \theta_{\mathbf{k}_0,n_0} &\theta_{-\mathbf{k}_0,-n_0} \rangle_{\mathrm{tr}} \\
&= 2\langle (\theta^r_{\mathbf{k}_0,n_0})^2 \rangle_{\mathrm{tr}} \\
&= 2\frac{\int_{-\infty}^{\infty} d\theta^r_{\mathbf{k}_0,n_0} (\theta^r_{\mathbf{k}_0,n_0})^2 \exp\left[-\frac{2\beta}{N}\left(\frac{\hbar^2}{8U_a}\omega^2_{n_0} + \frac{1}{2}Kf(\mathbf{k}_0)\right)(\theta^r_{\mathbf{k}_0,n_0})^2\right]}{\int_{-\infty}^{\infty} d\theta^r_{\mathbf{k}_0,n_0} \exp\left[-\frac{2\beta}{N}\left(\frac{\hbar^2}{8U_a}\omega^2_{n_0} + \frac{1}{2}Kf(\mathbf{k}_0)\right)(\theta^r_{\mathbf{k}_0,n_0})^2\right]} \\
&= \frac{N}{2\beta}\left(\frac{\hbar^2}{8U_a}\omega^2_{n_0} + \frac{1}{2}Kf(\mathbf{k}_0)\right)^{-1}
\end{aligned}
\tag{3.161}
$$

Introducing this result into Eq. (3.158), we obtain from Eq. (3.149)

$$
D_{ij} = \frac{1}{N\beta}\sum_{\mathbf{k}}\sum_{n}\frac{1 - \cos(\mathbf{k}\cdot\mathbf{R}_{ij})}{\frac{\hbar^2}{8U_a}\omega^2_n + \frac{1}{2}Kf(\mathbf{k})}
\tag{3.162}
$$

For further calculations, it is convenient to introduce a "bond-averaged" D_{ij} defined as

$$
\bar{D}_{ij} = \frac{1}{z}\sum_{j(i)} D_{ij} = \frac{1}{N\beta z}\sum_{\mathbf{k}}\sum_{n}\frac{f(\mathbf{k})}{\frac{\hbar^2}{8U_a}\left(\omega^2_n + \frac{4U_aKf(\mathbf{k})}{\hbar^2}\right)}
\tag{3.163}
$$

where $f(\mathbf{k})$ is the structure factor (3.155).

In what follows, we confine ourselves to the case of very low temperatures and consider the SCHA for a two-dimensional square lattice ($d = 2$, $z = 4$). For this case, Eq. (3.155) yields in the long-wavelength limit

$$
f(\mathbf{k}) = 4 - 2[\cos(k_x a) + \cos(k_y a)] \approx k^2 a^2
\tag{3.164}
$$

As $T \to 0$, the Matsubara sum in Eq. (3.163) is evaluated using [see Eq. (4.32)]

$$
\sum_{n=-\infty}^{\infty}\frac{1}{\omega^2_n + \frac{4U_aKf(\mathbf{k})}{\hbar^2}} \xrightarrow{T\to 0} \frac{\hbar^2\beta}{4[U_aKf(\mathbf{k})]^{\frac{1}{2}}}
\tag{3.165}
$$

From Eqs. (3.163)–(3.165), we can evaluate $\bar{D}_{ij}$ in the Debye approximation

$$
\begin{aligned}
\bar{D}_{ij} &\approx \frac{a^3}{2}\left(\frac{U_a}{K}\right)^{\frac{1}{2}}\int\frac{d^2k}{(2\pi)^2}k = \frac{a^3}{4\pi}\left(\frac{U_a}{K}\right)^{\frac{1}{2}}\int_0^{k_D} dk\,(k^2) \\
&= \frac{(k_D a)^3}{12\pi}\left(\frac{U_a}{K}\right)^{\frac{1}{2}}
\end{aligned}
\tag{3.166}
$$

Taking for the Debye wave vector k_D the value $2\sqrt{\pi}/a$ (see Section 3.4.3), the expression (3.166) yields $\bar{D}_{ij} \approx 1.2(U_a/K)^{\frac{1}{2}}$. Introducing this result

into Eq. (3.145), we obtain a self-consistent equation for the relative stiffness constant $k_1 = K/J$

$$k_1 = \exp\left[-0.6\left(\frac{U_a}{Jk_1}\right)^{\frac{1}{2}}\right] = \exp\left[-\frac{0.85}{(\alpha k_1)^{\frac{1}{2}}}\right] \qquad (3.167)$$

where we introduced the parameter $\alpha = zJ/2U_a$ [see Eq. (3.76)].

This equation can be solved numerically for k_1, regarding α as a parameter. The results show a discontinuous drop of k_1 as α approaches a value
$\alpha_c = 1.33$ from above. This is a manifestation of the instability of the array
with respect to the phase fluctuations which grow out of control for $\alpha = \alpha_c$.
The origin of the instability is the same as that found in the MFA approximation in Section 3.5. As pointed out in Section 2.9, the $T = 0$ properties
of the *two-dimensional* time-dependent Ginzburg–Landau model are identical with the finite-temperature properties of a classical time-independent
model in *three dimensions*. Since the phase transition of a model involving
a two-component order parameter in 3D exists, the instability found in the
SCHA at $\alpha = \alpha_c$ is confirmed, but the discontinuity may be an artifact of
this approximation. It should also be pointed out that the numerical value
of α_c is not very accurate, as it is based on the Debye-like approximation for
the **k** sum in Eq. (3.163), which fails when $|\mathbf{k}| \simeq 1/a$. Another weakness of
the SCHA is related to the replacement of Eq. (3.132) by the harmonic action (3.133). The phase variables in the former are confined to the interval
$(-\pi, \pi)$, while the latter involves phases that are allowed to vary over the
extended interval $\theta_i(\tau) \in [-\infty, \infty]$. According to Eq. (A3.35), this change
of interval implies that the partition function in the SCHA approximation
is actually a sum of path integrals with *various winding numbers l*. This
theory confines itself to taking only the $l = 0$ path integral. It should be
interesting to investigate the $l > 0$ corrections to the SCHA theory. An
alternative approach to this problem would be to consider a self-consistent
periodic harmonic approximation, which replaces the $\cos\theta$ potential by a
periodic function that is quadratic about each of its equivalent minima.

4

QUANTUM EFFECTS AND OHMIC DISSIPATION

4.1. General remarks

Up to this point, our attention was focused on problems of phase ordering in Josephson junction arrays, taking into account the effects of electrostatic charging energy. In general, quantum phase fluctuations associated with the charging energy were found to inhibit the phase order; this suggests the possibility of a fluctuation-driven phase transition. This and the following chapters are concerned, in the first place, with the role played by dissipation in such a phase transition. The discussion will, however, also include macroscopic quantum tunneling in the presence of dissipation. The first hints of the relevance of dissipation in phase-locking transition came from the experiments of Abeles and Hanak (1971) on the $Al - SiO_2$ granular superconductor. As discussed in Section 3.1, these authors have shown the existence of a *threshold* value of the normal junction resistance above which the system goes over to a semiconducting state. Two decades later, the experiments of the Minnesota group (Jaeger et al., 1986) have established the existence of a *universal threshold* resistance for films with *different microstructures* and made of *different metals*. These experiments imply that the *charging energy U_a* is entirely *absent* from the criterion for a phase-locking transition. This calls for a radical departure from the theoretical approach of Chapter 3, in which the ratio of Josephson coupling to *charging energy* plays a central role in the determination of the phase diagrams.

Since dissipation is a *macroscopic* concept, there has been little interest in a formalism of dissipative systems, during the history of quantum mechanics, dealing predominantly with microscopic phenomena. The situation has changed with the discovery of the Josephson effect, leading to the possibility of a physical realization of a *true macroscopic quantum system*. When a nonzero supercurrent I flows through a Josephson junction, the Hamiltonian (2.112) must be augmented by a term proportional to $I\theta$, and a corresponding mechanical analog is that of a particle moving in a "washboard" potential [see Eq. (4.101)]. The states of nonzero supercurrent are *metastable* owing to transitions to lower-lying minima of the potential. At sufficiently low temperatures, such transitions (phase slips) can be caused by macroscopic quantum tunneling (MQT) through the potential barrier.

This possibility was first pointed out for Josephson junctions by Ivanchenko and Zil'berman (1968). In the case of a superconducting quantum interference device (SQUID), the possibility of quantum tunneling between fluxoid states was first proposed by Scalapino (1969). The first convincing experimental observations of MQT were made in 1981 on Josephson junctions of low capacitance (Voss and Webb, 1981; Jackel et al., 1981). Since macroscopic systems are inherently dissipative, there arises a fundamental question of what is the effect of dissipation on the MQT. This question was first addressed by Caldeira and Leggett (1981) in a seminal paper containing a theoretical prediction of the reduction of MQT rate. This effect has been verified in experiments on low-capacitance junctions.

Another manifestation of the effects of dissipation occurs in the context of equilibrium statistical mechanics of fluctuations $\langle x^2 \rangle$ of the coordinate of a dissipative harmonic oscillator about the equilibrium position (Caldeira and Leggett, 1983) (see also Section 4.5). Whereas at the high-temperature (classical) limit, $\langle x^2 \rangle$ is unaffected by dissipation, at low temperatures, the dissipation tends to reduce this quantity compared with that of an undamped oscillator. Applying this concept to equilibrium fluctuations in a dissipative Josephson junction, we see that quantum phase fluctuations are expected to be suppressed by the dissipation. Consequently, we also expect some changes of the phase diagrams, discussed in Chapter 3, to take place due to the dissipation. These effects have been studied by Chakravarty et al. (1986) and independently by Šimánek and Brown (1986) using the effective action of a dissipative Josephson junction first derived by Ambegaokar et al. (1982). The theoretical significance of the latter work is that it represents a true first-principles derivation of the effective Euclidean action for quasiparticle tunneling. This action happens to be periodic in the phase variable, a feature leading to the quantization of tunneling charge. This is to be distinguished from the action for ohmic dissipation, which is nonperiodic (Gaussian) and represents a continuum of transferred charge. We saw in Chapter 3 that the periodicity of the effective action influences profoundly the choice of the interval for the phase variable θ and the phase diagram. Hence, one must carefully distinguish between the case of an array involving ohmically shunted junctions and that where dissipation is due to quasiparticle tunneling (Kampf and Schön, 1988).

4.2. Resistively shunted junction model

The conditions for observability of macroscopic quantum phenomena are usually met in ordered many-body systems, in which the classical motion of a coordinate variable, such as phase, is accompanied by a coherent motion of the system as a whole. Among systems known to exhibit such behavior are Josephson junctions, SQUIDS, superfluid vortices, and charge-density waves. The model of a resistively shunted Josephson junction (RSJ) is of primary interest for granular superconductors. In this model, the dissipa-

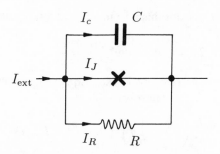

FIG. 4.1. Equivalent circuit for a resistively shunted junction. External current I_{ext} branches into displacement current I_c, Josephson current I_J, and the ohmic current I_R.

tion is due to an effective ohmic resistance R in parallel with the junction so that the equivalent circuit looks like that shown in Fig. 4.1.

The equation of motion for the phase difference θ is obtained by expressing the external current I_{ext} via Kirchhoff's law

$$I_C + I_J + I_R = I_{\text{ext}} \tag{4.1}$$

where I_C, I_J, and I_R are the currents through the capacitance, Josephson junction, and resistance, respectively. We introduce V for the voltage difference across the junction. Using the notation introduced in Eq. (2.103), we thus have

$$V = A_0^{(2)} - A_0^{(1)} \tag{4.2}$$

where $A_0^{(i)}$ is the potential on the surface of the ith superconductor. To relate the phase difference $\theta = \theta_1 - \theta_2$ to V, we use the Josephson relation [see Eqs. (2.38) and (2.45)]

$$\dot{\theta}_i = -\frac{e^*}{\hbar} A_0^{(i)} \tag{4.3}$$

and obtain from Eq. (4.2)

$$\dot{\theta} = \dot{\theta}_1 - \dot{\theta}_2 = \frac{e^*}{\hbar} V \tag{4.4}$$

Using Eq. (4.4), the displacement current is given by

$$I_C = C\dot{V} = \frac{\hbar C}{e^*} \ddot{\theta} \tag{4.5}$$

and the ohmic current is

$$I_R = \frac{V}{R} = \frac{\hbar}{e^*} \frac{1}{R} \dot{\theta} \tag{4.6}$$

The supercurrent I_J is obtained from Eqs. (2.38), (2.40), and (2.106) as follows

$$I_J = e^* \dot{n}_1 = -e^* \frac{\partial H}{\partial \varphi_1} = \frac{e^*}{\hbar} J \sin \theta \qquad (4.7)$$

Introducing expressions (4.5)–(4.7) into Eq. (4.1), we obtain the classical equation of motion for the phase variable θ

$$\frac{\hbar}{e^*}\left(C\ddot{\theta} + \frac{1}{R}\dot{\theta}\right) + I_c \sin \theta = I_{\text{ext}}(t) \qquad (4.8)$$

where

$$I_c = \frac{e^* J}{\hbar} \qquad (4.9)$$

Equation (4.9) defines the critical current of a Josephson junction. Next, we rewrite Eq. (4.8) as follows

$$\left(\frac{\hbar}{e^*}\right)^2 \left(C\ddot{\theta} + \frac{1}{R}\dot{\theta}\right) + \frac{\partial U_p(\theta)}{\partial \theta} = 0 \qquad (4.10)$$

where $U_p(\theta)$ is the potential energy given by

$$U_p(\theta) = -J\left(\cos\theta + \frac{I_{\text{ext}}}{I_c}\theta\right) \qquad (4.11)$$

Equations (4.10) and (4.11) describe a particle with mass $M = C(\hbar/e^*)^2$ and an ohmic damping constant $1/RC$, moving in the "tilted washboard" potential $U_p(\theta)$ shown in Fig. 4.2. The particle is initially confined to the well at $\theta \simeq \theta_0$. After the particle escapes from this metastable well due to thermal activation or tunneling, it slides down the washboard, picking up an additional phase difference, and a voltage appears across the junction [see Eq. (4.3)]. The potential, shown in Fig. 4.2, provides two frequency scales, important for the dynamics and MQT. First of all, there is the well frequency

$$\omega_0 = \left(\frac{U_p''(\theta_0)}{M}\right)^{\frac{1}{2}} \qquad (4.12)$$

which corresponds to the oscillations of the particle about quasiequilibrium state, $\theta \simeq \theta_0$. Then there is the barrier frequency

$$\omega_b = \left(-\frac{U_p''(\theta_b)}{M}\right)^{\frac{1}{2}} \qquad (4.13)$$

which characterizes the width of the parabolic top of the barrier separating the initial and final states. For a thermally activated decay of the metastable state, the *classical* formula for the decay rate in the limit of weak damping is (Kramers, 1940)

$$\tau_{\text{cl}}^{-1} = \frac{\omega_0}{2\pi} \exp\left(-\frac{\Delta U_p}{k_B T}\right) \qquad (4.14)$$

$U_p(\theta)$

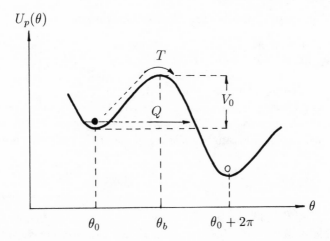

FIG. 4.2. Tilted washboard potential $U_p(\theta)$ versus phase difference θ. Shown are minima of $U_p(\theta)$ taking place at $\theta = \theta_0$ and $\theta \simeq \theta_0 + 2\pi$ and a maximum at $\theta = \theta_b$. The arrows T and Q indicate the escape routes via thermal activation and quantum tunneling, respectively.

where $\Delta U_p = V_0$ is the height of the barrier, shown in Fig. 4.2. We note that the attempt frequency ω_0 is, according to Eq. (4.12), equal to the Josephson plasma frequency $(2eI_c/\hbar C)^{\frac{1}{2}}$. The significance of the barrier frequency ω_b is that it defines the crossover temperature

$$T_0 = \frac{\hbar\omega_b}{k_B} \qquad (4.15)$$

below which *quantum tunneling* is the *dominant* decay mechanism. As shown in the following sections, dissipation can substantially modify these findings. In the classical regime, a moderately or strongly damped particle moves diffusively across the barrier, resulting in the reduction of the pre-exponential factor in Eq. (4.14). In fact, the attempt frequency is then changed from ω_0 to ω_a given by

$$\omega_a = \frac{1}{2\pi}\omega_0^2 RC \qquad (4.16)$$

For an overdamped junction, the crossover temperature T_0 of Eq. (4.15) is similarly reduced to $T_0 \simeq \hbar\omega_a/k_B$ (Larkin and Ovchinnikov, 1983). Before going into detailed discussion of the role of dissipation on the MQT rate, we first consider the microscopic description of the interaction between "macroscopic particle" and environment, which reproduces the ohmic dissipative term in the classical equation of motion (4.10).

4.3.   Microscopic model for ohmic dissipation

A quantitative determination of the effects of dissipation on MQT can be done only in terms of a fully quantum-mechanical description of the tunneling particle and all the environmental modes, which give rise to the relevant dissipation. In case of ohmic dissipation, when the power loss is proportional to the particle velocity squared, the environmental modes can be chosen to be harmonic oscillators with a continuously distributed resonant frequency. The Lagrangian of these modes is (in *real-time* formulation)

$$L_e^{(r)} = \frac{1}{2} \sum_\alpha m_\alpha \left[(\dot{x}^{(\alpha)})^2 - \omega_\alpha^2 (x^{(\alpha)})^2 \right] \qquad (4.17)$$

where m_α, $x^{(\alpha)}$, and ω_α are the mass, coordinate, and frequency of the αth oscillator. Denoting the coordinate of the tunneling particle by q, the Lagrangian of the particle is

$$L_p^{(r)} = \frac{1}{2} M \dot{q}^2 - V(q) \qquad (4.18)$$

where $V(q)$ is the potential energy exhibiting metastable minima (see Fig. 4.2). In the context of the RSJ model, Eq. (4.18) corresponds to a parallel combination of the junction and capacitance.

The dissipation is generated by interaction between the particle and the environmental oscillators. The Lagrangian of this interaction is of the form

$$L_i^{(r)} = q \sum_\alpha C_\alpha x^{(\alpha)} \qquad (4.19)$$

where C_α is the coupling constant for the αth mode. Using Eqs. (4.17)–(4.19), the Lagrangian of the whole interacting system is (Caldeira and Leggett, 1981)

$$L^{(r)} = L_e^{(r)} + L_p^{(r)} + L_i^{(r)} - q^2 \sum_\alpha \frac{C_\alpha^2}{2 m_\alpha \omega_\alpha^2} \qquad (4.20)$$

The last term in (4.20) is a counterterm introduced to make sure that $V(q)$ in Eq. (4.18) is the same as the potential in the phenomenological equation of motion [see Eq. (4.10)]. In fact, this counterterm cancels the renormalization of the potential V_q due to the interaction (4.19).

The algebraic derivation leading from Eq. (4.20) to a concrete expression for the MQT rate is rather lengthy and has not been presented in the literature in full detail. Thus it seems instructive to build up the theory from smaller conceptual units, where necessary derivations are worked out. We see from Eqs. (4.17)–(4.19) that the Lagrangian $L^{(r)}$ involves oscillator degrees of freedom $x^{(\alpha)}$ up to second order only. This means that the coordinates $x^{(\alpha)}$ can be *integrated out* and one can formulate the partition function of the whole system in terms of a *path integral* over the *coordinate* q. This leads to the concept of an *effective Euclidean action* $S_{\rm eff}[q]$ from which most of the physical quantities, such as $\langle q^2 \rangle$ or the MQT rate, can be calculated.

4.4. Particle coupled to an oscillator—Effective action

Before discussing the effective action of the whole system, we shall first consider the more elementary case of a *single* environmental oscillator, with coordinate $x^{(\alpha)} = x$, coupled to the particle of coordinate q. This subsystem is described by the Lagrangian (in real-time formulation)

$$L^{(\alpha)(r)} = \frac{1}{2} M \dot{q}^2 - V(q) + \frac{m_\alpha}{2} \left(\dot{x}^2 - \omega_\alpha^2 x^2 \right) + C_\alpha q x - \frac{C_\alpha^2 q^2}{2 m_\alpha \omega_\alpha^2} \qquad (4.21)$$

The partition function of this *two-body* system is obtained from Eq. (A2.18) in the form

$$Z^{(\alpha)} = \int_{-\infty}^{\infty} dq \int_{-\infty}^{\infty} dx \int_{q(0)=q}^{q(\hbar\beta)=q} \mathcal{D}q(\tau) \int_{x(0)=x}^{x(\hbar\beta)=x} \mathcal{D}x(\tau)$$

$$\times \exp\left(-\frac{1}{\hbar} \int_0^{\hbar\beta} d\tau \, L^{(\alpha)}[x, q; \dot{x}, \dot{q}] \right) \qquad (4.22)$$

where $L^{(\alpha)}[x, q; \dot{x}, \dot{q}]$ is the Euclidean Lagrangian, which follows from Eq. (4.21) by reverting the sign of the potential terms and replacing the real-time derivatives by imaginary-time ones. Thus,

$$L^{(\alpha)}[x, q; \dot{x}, \dot{q}] = \frac{1}{2} M \left(\frac{dq}{d\tau} \right)^2 + V(q) + \frac{C_\alpha^2 q^2}{2 m_\alpha \omega_\alpha^2}$$

$$+ \frac{m_\alpha}{2} \left[\left(\frac{dx}{d\tau} \right)^2 + \omega_\alpha^2 x^2 \right] - C_\alpha q x \qquad (4.23)$$

The effective action is defined by writing the partition function as

$$Z^{(\alpha)} = \int_{-\infty}^{\infty} dq \int_{q(0)=q}^{q(\hbar\beta)=q} \mathcal{D}q(\tau) \, \exp\left(-\frac{1}{\hbar} S_{\rm eff}^{(\alpha)}[q] \right) \qquad (4.24)$$

From Eqs. (4.22) and (4.24), we obtain for the effective action an equation

$$\exp\left(-\frac{1}{\hbar} S_{\rm eff}^{(\alpha)}[q] \right)$$

$$= \exp\left[-\frac{1}{\hbar} \int_0^{\hbar\beta} d\tau \left(\frac{1}{2} M \dot{q}^2 + V(q) + \frac{C_\alpha^2 q^2}{2 m_\alpha \omega_\alpha^2} \right) \right]$$

$$\times \int_{-\infty}^{\infty} dx \int_{x(0)=x}^{x(\hbar\beta)=x} \mathcal{D}x(\tau)$$

$$\times \exp\left[-\frac{1}{\hbar}\int_0^{\hbar\beta} d\tau \left(\frac{1}{2}m_\alpha \dot{x}^2 + \frac{1}{2}m_\alpha \omega_\alpha^2 x^2 - C_\alpha qx\right)\right] \quad (4.25)$$

The path integral over the oscillator coordinate $x(\tau)$ is evaluated by expanding $x(\tau)$ into a Fourier series [see Eqs. (3.119)–(3.125) for a similar calculation]

$$x(\tau) = \sum_{n=-\infty}^{\infty} x_n e^{-i\omega_n \tau} \quad (4.26)$$

where $\omega_n = 2\pi n/\hbar\beta$. The reality conditions for the complex Fourier coefficients $x_n = x_n^r + ix_n^i$ are

$$x_n^r = x_{-n}^r$$
$$x_n^i = -x_{-n}^i \quad (4.27)$$

Using Eqs. (4.26) and (4.27), the x-dependent part of the action in Eq. (4.25) is

$$\begin{aligned}
S_x^{(\alpha)} &= \int_0^{\hbar\beta} d\tau \left(\frac{1}{2}m_\alpha \dot{x}^2 + \frac{1}{2}m_\alpha \omega_\alpha^2 x^2 - C_\alpha qx\right) \\
&= \frac{1}{2}m_\alpha \hbar\beta \sum_{n=-\infty}^{\infty} [(x_n^r)^2 + (x_n^i)^2](\omega_n^2 + \omega_\alpha^2) \\
&\quad - C_\alpha \hbar\beta \sum_{n=-\infty}^{\infty} [(q_n^r)(x_n^r) + (q_n^i)(x_n^i)] \quad (4.28)
\end{aligned}$$

where $q_n = q_n^r + iq_n^i$ is the coefficient of the Fourier series for $q(\tau)$. Using Eq. (4.28), the path integral on the RHS of Eq. (4.25) becomes

$$\begin{aligned}
\rho^{(\alpha)}(x, x) &= \int_{x(0)=x}^{x(\hbar\beta)=x} \mathcal{D}x(\tau) \exp\left(-\frac{1}{\hbar}S_x^{(\alpha)}\right) \\
&= \int_{-\infty}^{\infty} dx_0^r \exp\left[-\left(\frac{1}{2}m_\alpha \beta \omega_\alpha^2 (x_0^r)^2 - C_\alpha \beta q_0^r x_0^r\right)\right] \\
&\quad \times \prod_{n=1}^{\infty} \int_{-\infty}^{\infty} dx_n^r \int_{-\infty}^{\infty} dx_n^i \exp\{-m_\alpha \beta(\omega_n^2 + \omega_\alpha^2)[(x_n^r)^2 + (x_n^i)^2] \\
&\quad + 2C_\alpha \beta(q_n^r x_n^r + q_n^i x_n^i)\} \\
&= \text{const} \times \exp\left(\frac{C_\alpha^2 \beta(q_0^r)^2}{2m_\alpha \omega_\alpha^2}\right) \prod_{n=1}^{\infty} \exp\left(\frac{C_\alpha^2 \beta}{m_\alpha} \frac{[(q_n^r)^2 + (q_n^i)^2]}{\omega_n^2 + \omega_\alpha^2}\right) \\
&= \text{const} \times \exp\left(\frac{C_\alpha^2 \beta}{2m_\alpha} \sum_{n=-\infty}^{\infty} \frac{q_n q_{-n}}{\omega_n^2 + \omega_\alpha^2}\right) \quad (4.29)
\end{aligned}$$

where the q_n-independent terms are lumped into a constant factor. Also note that the boundary value, $x(0) = x$, does not enter the result. Hence, the x integration in Eq. (4.25) leads to an irrelevant constant that does not affect the effective action. Taking the logarithm of Eq. (4.25) and using Eq. (4.29), the effective Euclidean action for the q particle (coupled to the αth environmental mode) is, up to a q-independent constant, given by

$$
S_{\text{eff}}^{(\alpha)}[q] = \int_0^{\hbar\beta} d\tau \left(\frac{1}{2} M\dot{q}^2 + V(q) + \frac{C_\alpha^2 q^2}{2m_\alpha \omega_\alpha^2} \right)
$$
$$
- \frac{C_\alpha^2 \hbar\beta}{2m_\alpha} \sum_{n=-\infty}^{\infty} \frac{q_n q_{-n}}{\omega_n^2 + \omega_\alpha^2} \quad (4.30)
$$

The last term on the RHS of Eq. (4.30) can be written as a double τ integral, using the Parseval theorem

$$
- \frac{C_\alpha^2 \hbar\beta}{2m_\alpha} \sum_{n=-\infty}^{\infty} \frac{q_n q_{-n}}{\omega_n^2 + \omega_\alpha^2} = - \frac{C_\alpha^2}{2m_\alpha \hbar\beta} \int_0^{\hbar\beta} d\tau \int_0^{\hbar\beta} d\tau'
$$
$$
\times q(\tau) F^{(\alpha)}(\tau - \tau') q(\tau') \quad (4.31)
$$

where

$$
F^{(\alpha)}(\tau) = \sum_{n=-\infty}^{\infty} f_n^{(\alpha)} e^{-i\omega_n \tau} = \sum_{n=-\infty}^{\infty} \frac{e^{-i\omega_n \tau}}{\omega_n^2 + \omega_\alpha^2} \simeq \frac{\hbar\beta}{2\omega_\alpha} e^{-\omega_\alpha |\tau|} \quad (4.32)
$$

where we have used the large β-approximation. The third term on the RHS of Eq. (4.30) can also be transformed to a double τ integral by noting

$$
\int_0^{\hbar\beta} d\tau \int_0^{\hbar\beta} d\tau' \, q^2(\tau) F^{(\alpha)}(\tau - \tau') = (\hbar\beta)^2 \sum_{n=-\infty}^{\infty} f_0^{(\alpha)} q_n q_{-n}
$$
$$
= \frac{(\hbar\beta)^2}{\omega_\alpha^2} \sum_{n=-\infty}^{\infty} q_n q_{-n}
$$
$$
= \frac{\hbar\beta}{\omega_\alpha^2} \int_0^{\hbar\beta} d\tau \, q^2(\tau) \quad (4.33)
$$

so that

$$
\frac{C_\alpha^2}{2m_\alpha \omega_\alpha^2} \int_0^{\hbar\beta} d\tau \, q^2(\tau) = \frac{C_\alpha^2}{4m_\alpha \hbar\beta} \int_0^{\hbar\beta} d\tau \int_0^{\hbar\beta} d\tau'
$$
$$
\times [q^2(\tau) + q^2(\tau')] F^{(\alpha)}(\tau - \tau') \quad (4.34)
$$

Combining the RHS of this equation with Eq. (4.31), the effective action (4.30) becomes

$$
S_{\text{eff}}^{(\alpha)}[q] = \int_0^{\hbar\beta} d\tau \left[\frac{1}{2} M\dot{q}^2 + V(q) \right]
$$
$$
+ \frac{C_\alpha^2}{4m_\alpha \hbar\beta} \int_0^{\hbar\beta} d\tau \int_0^{\hbar\beta} d\tau' \, [q(\tau) - q(\tau')]^2 F^{(\alpha)}(\tau - \tau') \quad (4.35)
$$

We note that this result can be also derived by the method of the forced harmonic oscillator, introduced by Feynman (1972) and used by Caldeira and Leggett (1981).

4.5.   Effective action for ohmic dissipation

Expression (4.35) can be easily generalized to the case of a particle coupled to the entire ensemble of environmental oscillators. From the form of the Lagrangian (4.20) it follows that the partition function (A2.18) involves, in this case, a product of terms, such as that in Eq. (4.29). Consequently, the effective action becomes

$$
S_{\text{eff}}[q] = \int_0^{\hbar\beta} d\tau \left[\frac{1}{2} M\dot{q}^2 + V(q) \right]
$$
$$
+ \sum_\alpha \frac{C_\alpha^2}{8 m_\alpha \omega_\alpha} \int_0^{\hbar\beta} d\tau \int_0^{\hbar\beta} d\tau' \, [q(\tau) - q(\tau')]^2 e^{-\omega_\alpha |\tau - \tau'|} \quad (4.36)
$$

where the RHS of Eq. (4.32) has been used for $F^\alpha(\tau - \tau')$. In case of ohmic dissipation, specific values of the constants C_α, m_α, and ω_α need not be determined, instead one proceeds by requiring that the last term on the RHS of Eq. (4.36) reproduces the ohmic dissipation in the classical equation of motion for $q(\tau)$. To see what sort of constraint this imposes on the constants, we write the dissipative part of Eq. (4.36) in terms of the Fourier components q_n [see Eqs. (4.30) and (4.33)]

$$
S_D[q] = \hbar\beta \sum_n \omega_n^2 q_n q_{-n} \sum_\alpha \frac{C_\alpha^2}{2 m_\alpha \omega_\alpha^2 (\omega_n^2 + \omega_\alpha^2)} \quad (4.37)
$$

Following Caldeira and Leggett (1981), we introduce the effective spectral density of environmental modes

$$
J(\omega) = \frac{\pi}{4} \sum_\alpha \frac{C_\alpha^2}{m_\alpha \omega_\alpha} \delta(\omega - \omega_\alpha) \quad (4.38)
$$

which allows us to express the α sum in Eq. (4.37) by an ω integral

$$
\sum_\alpha \frac{C_\alpha^2}{2 m_\alpha \omega_\alpha^2 (\omega_n^2 + \omega_\alpha^2)} = \frac{2}{\pi} \int_0^\infty d\omega \, \frac{J(\omega)}{\omega(\omega_n^2 + \omega^2)} \quad (4.39)
$$

The general form of the dissipative action for ohmic friction is (see Section 4.6)

$$
S_D^{\text{ohmic}}[q] = \eta \hbar\beta \sum_n |\omega_n| q_n q_{-n} \quad (4.40)
$$

Comparing this equation with Eq. (4.37), we see from Eq. (4.39) that the spectral density for ohmic dissipation is required to have the form

$$
J^{\text{ohmic}}(\omega) = \eta \omega \quad (4.41)
$$

We now use the formulas (4.38) and (4.41) to evaluate the α sum in the dissipative part of the action (4.36). We have

$$
\begin{aligned}
\sum_\alpha \frac{C_\alpha^2}{8 m_\alpha \omega_\alpha} e^{-\omega_\alpha |\tau - \tau'|} &= \frac{1}{2\pi} \int_0^\infty d\omega \, J^{\text{ohmic}}(\omega) e^{-\omega |\tau - \tau'|} \\
&= \frac{\eta}{2\pi} \int_0^\infty d\omega \, \omega e^{-\omega |\tau - \tau'|} \\
&= \frac{\eta}{2\pi (\tau - \tau')^2}
\end{aligned}
\tag{4.42}
$$

Using this result in Eq. (4.36), the effective Euclidean action of a particle with ohmic friction is given by

$$
\begin{aligned}
S_{\text{eff}}[q] = &\int_0^{\hbar\beta} d\tau \left[\frac{1}{2} M \dot{q}^2 + V(q) \right] \\
&+ \frac{\eta}{2\pi} \int_0^{\hbar\beta} d\tau \int_0^{\hbar\beta} d\tau' \left(\frac{q(\tau) - q(\tau')}{\tau - \tau'} \right)^2
\end{aligned}
\tag{4.43}
$$

Starting from this expression, one can determine the effect of dissipation on the MQT rate, using the so-called "bounce" technique (see Section 4.8). This effective action can be also used to discuss the "phase-locking" phase transition in arrays of Josephson junctions with ohmic dissipation (see Chapter 6).

4.6. Fluctuations in damped harmonic oscillators

As mentioned in Section 4.1, quantum fluctuations tend to be suppressed by dissipation. This can be demonstrated, for instance, by considering the mean-square fluctuation of the coordinate of a damped harmonic oscillator. We start from the effective action (4.36), where the potential $V(q)$ is replaced by $\frac{1}{2} M \omega_0^2 q^2$, ω_0 being the well frequency (4.12). Using Eq. (4.40) for the dissipative action, we have

$$
S[q] = \hbar\beta \sum_{n=-\infty}^{\infty} \left[\frac{1}{2} M (\omega_n^2 + \omega_0^2) + \eta |\omega_n| \right] q_n q_{-n}
\tag{4.44}
$$

To calculate the mean-square fluctuation $\langle q^2 \rangle$, we use the method of the generating functional (see Appendix A7). According to Eq. (A7.17), the equal-time correlation function for the displacement $q(\tau)$ is given by

$$
\langle q^2 \rangle = \left(\hbar^2 Z^{-1}[\zeta] \frac{\delta^2}{\delta \zeta^2(\tau)} Z[\zeta(\tau)] \right)_{\zeta = 0}
\tag{4.45}
$$

where

$$
Z[\zeta(\tau)] = \int_{-\infty}^{\infty} dq \int_{q(0)=q}^{q(\hbar\beta)=q} \mathcal{D}q(\tau) \, e^{-\frac{1}{\hbar} S_\zeta[q]}
\tag{4.46}
$$

$S_\zeta[q]$ is the action of the oscillator interacting with external field $\zeta(\tau)$

$$S_\zeta[q] = S[q] - \int_{-\hbar\beta/2}^{\hbar\beta/2} d\tau \, \zeta(\tau)q(\tau) = S[q] - \zeta q(0) \qquad (4.47)$$

where $S[q]$ is given by Eq. (4.44), and the last term on the RHS of Eq. (4.47) is generated by assuming $\zeta(\tau) = \zeta\,\delta(\tau)$ [see Eq. (A7.15)]. Introducing Eq. (4.44) into Eq. (4.47) and expanding $q(\tau)$ into a Fourier series, we have

$$\frac{1}{\hbar}S_\zeta[q] = \frac{1}{2}\beta M\omega_0^2(q_0^r)^2 - \frac{\zeta}{\hbar}q_0^r + \beta\sum_{n=1}^{\infty}\left\{\left[M(\omega_n^2 + \omega_0^2) + 2\eta|\omega_n|\right]\right.$$
$$\left. \times \left[(q_n^r)^2 + (q_n^i)^2\right] - \frac{2\zeta}{\hbar\beta}q_n^r\right\} \qquad (4.48)$$

The partition function $Z[\zeta(\tau)]$ can be evaluated, using Eq. (4.48), by multiple Gaussian integrations that are quite similar to those in Eq. (4.29). We obtain, by lumping the ζ-independent terms into a constant factor,

$$Z[\zeta] = \text{const} \times \exp\left(\frac{\zeta^2}{2\hbar^2\beta M\omega_0^2}\right)\prod_{n=1}^{\infty}\exp\left(\frac{\zeta^2}{\hbar^2\beta[M(\omega_n^2 + \omega_0^2) + 2\eta|\omega_n|]}\right)$$
$$= \text{const} \times \exp\left(\frac{\zeta^2}{2\hbar^2\beta}\sum_{n=-\infty}^{\infty}\frac{1}{M(\omega_n^2 + \omega_0^2) + 2\eta|\omega_n|}\right) \qquad (4.49)$$

Introducing this result into Eq. (4.45) and noting that, in view of (4.47), the second functional derivative reduces to ordinary second derivative with respect to the τ-independent parameter ζ, we have for the mean-square displacement

$$\langle q^2\rangle = \frac{1}{\beta}\sum_{n=-\infty}^{\infty}\frac{1}{M(\omega_n^2 + \omega_0^2) + 2\eta|\omega_n|} \qquad (4.50)$$

The expression (4.50) coincides with the formula obtained from the fluctuation-dissipation theorem (Landau and Lifshitz, 1980)

$$\langle q^2\rangle = \frac{\hbar}{\pi}\int_{-\infty}^{\infty}d\omega\,\frac{\text{Im}\,\chi(\omega)}{1 - e^{-\hbar\beta\omega}} \qquad (4.51)$$

where χ is the dynamical susceptibility of the *classical damped oscillator*, given by

$$\chi(\omega) = \frac{1}{M(\omega_0^2 - \omega^2) - 2i\eta\omega} \qquad (4.52)$$

The equivalence of Eqs. (4.50) and (4.51) can be proven by evaluating the ω integral by contour integration and reducing (4.51) to a sum of residues over the poles of the function $\left(1 - e^{-\hbar\beta\omega}\right)^{-1}$ (Grabert et al., 1984). It provides a substantiation for the choice of the dissipative action, given in Eq. (4.40). Explicit formulas for $\langle q^2 \rangle$ were given by Caldeira and Leggett (1983) in the limit of low and high temperatures. Detailed derivation of these formulas, starting from expression (4.50), has been presented by Grabert et al. (1984). Following their method, we first rewrite expression (4.50) as follows

$$\langle q^2 \rangle = \frac{1}{\beta M \omega_0^2} + \frac{2}{\beta} \sum_{n=1}^{\infty} \frac{1}{M(\omega_n^2 + \omega_0^2) + 2\eta \omega_n}$$

$$= \frac{1}{\beta M \omega_0^2} + \frac{2}{\beta M (\lambda_1 - \lambda_2)} \sum_{n=1}^{\infty} \left(\frac{1}{\omega_n + \lambda_2} - \frac{1}{\omega_n + \lambda_1} \right) \quad (4.53)$$

where λ_1 and λ_2 are given by

$$\lambda_{1,2} = \frac{\eta}{M} \pm \sqrt{\left(\frac{\eta}{M} \right)^2 - \omega_0^2} = \frac{\eta}{M} \pm \delta \quad (4.54)$$

The n summation on the RHS of Eq. (4.53) can be performed using the identity

$$\psi(z_2) - \psi(z_1) = \sum_{n=0}^{\infty} \left(\frac{1}{n + z_1} - \frac{1}{n + z_2} \right) \quad (4.55)$$

where $\psi(z)$ is the digamma function (Abramowitz and Stegun, 1964). Letting $\omega_n = 2\pi n / \hbar \beta$ in Eq. (4.53), we obtain with use of Eq. (4.55)

$$\langle q^2 \rangle = \frac{1}{\beta M \omega_0^2} + \frac{\hbar}{2\pi M \delta} \sum_{n=0}^{\infty} \left[\frac{1}{n + 1 + \frac{\lambda_2 \hbar \beta}{2\pi}} - \frac{1}{n + 1 + \frac{\lambda_1 \hbar \beta}{2\pi}} \right]$$

$$= \frac{1}{\beta M \omega_0^2} + \frac{\hbar}{2\pi M \delta} \left[\psi(1 + \frac{\lambda_1 \hbar \beta}{2\pi}) - \psi(1 + \frac{\lambda_2 \hbar \beta}{2\pi}) \right] \quad (4.56)$$

To see how dissipation affects the zero-point fluctuations of the oscillator, we consider the low-temperature limit of Eq. (4.56). Using the asymptotic expression of $\psi(z)$ for large z

$$\psi(z) = \ln z - \frac{1}{2z} + \cdots \quad (4.57)$$

the expression (4.56) becomes in the limit of $\beta = 1/k_B T \to \infty$

$$\langle q^2 \rangle_{T=0} = \frac{\hbar}{2\pi M \delta} \ln \frac{\lambda_1}{\lambda_2} = \frac{\hbar}{2\pi M \delta} \ln \left(\frac{\eta/M + \delta}{\eta/M - \delta} \right) \quad (4.58)$$

Introducing the damping coefficient κ

$$\kappa = \frac{\eta}{M \omega_0} \quad (4.59)$$

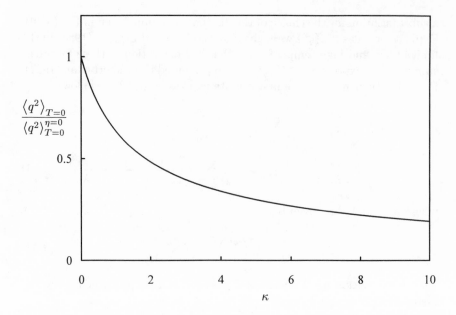

FIG. 4.3. The ratio $\langle q^2 \rangle_{T=0} / \langle q^2 \rangle_{T=0}^{\eta=0}$, given by Eqs. (4.60) and (4.61), plotted versus the damping coefficient $\kappa = \eta / M\omega_0$.

into the RHS of Eq. (4.58), we have

$$\langle q^2 \rangle_{T=0} = \frac{\hbar}{2\pi M\omega_0 (\kappa^2 - 1)^{\frac{1}{2}}} \ln \left(\frac{\kappa + (\kappa^2 - 1)^{\frac{1}{2}}}{\kappa - (\kappa^2 - 1)^{\frac{1}{2}}} \right) \qquad (4.60)$$

For an underdamped oscillator ($\kappa < 1$), expression (4.60) can be rewritten in terms of a real quantity $(1 - \kappa^2)^{\frac{1}{2}}$

$$\langle q^2 \rangle_{T=0} = \frac{\langle q^2 \rangle_{T=0}^{\eta=0}}{(1 - \kappa^2)^{\frac{1}{2}}} \left[1 - \frac{2}{\pi} \arctan \left(\frac{\kappa}{(1 - \kappa^2)^{\frac{1}{2}}} \right) \right] \qquad (4.61)$$

where

$$\langle q^2 \rangle_{T=0}^{\eta=0} = \frac{\hbar}{2M\omega_0} \qquad (4.62)$$

is the quantum-mechanical *ground-state expectation* value of q^2. In Fig. 4.3 we show the dependence of $\langle q^2 \rangle_{T=0}$ on the damping coefficient κ, as it follows from Eqs. (4.60) and (4.61).

4.7. Path-integral method for MQT rate

Quantum tunneling of a particle coupled to a reservoir is most conveniently treated by the functional integral method, first proposed by t'Hooft (1976),

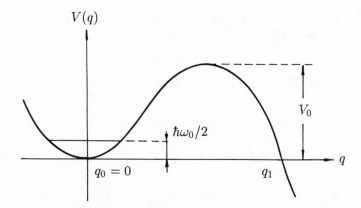

FIG. 4.4. Potential $V(q)$ with a single metastable minimum at $q_0 = 0$. The height of the barrier is V_0, the classical particle energy is $E = 0$, and ω_0 is the frequency of oscillations about $q = 0$.

Callan and Coleman (1977), and Stone (1977). By working with the imaginary time, these authors have calculated the decay rate of the metastable state of vacuum due to quantum tunneling. Their method is also suitable for the quantum tunneling of vortices, as discussed in Chapter 8. Mathematical techniques, used by the authors cited, were originally developed by Langer (1967) in his paper on thermal nucleation. Before we consider the dissipative tunneling, we introduce these techniques by calculating the decay rate for a *single* particle without coupling to reservoir. In real-time formulation, the Lagrangian of the particle is given by

$$L^{(r)} = \frac{1}{2} M \left(\frac{dq}{dt} \right)^2 - V(q) \tag{4.63}$$

where $V(q)$ is a one-dimensional potential that has a single metastable minimum, like that shown in Fig. 4.4. We assume that the potential barrier to be crossed has height V_0 which is much *larger* than the ground-state energy $\hbar\omega_0/2$ of the particle in the potential well. Then a semiclassical treatment of the barrier penetration can be used, yielding for the transmission coefficient the well-known WKB formula (Landau and Lifshitz, 1958)

$$D = \exp\left(-\frac{2}{\hbar} \int_{q_0}^{q_1} dq \, [2MV(q)]^{\frac{1}{2}} \right) \tag{4.64}$$

Let us now show how to obtain the tunneling rate from a path-integral formalism. First, we define the quantum-mechanical transition amplitude $K(q, t; q', t')$ for real times by the relation (Feynman and Hibbs, 1965)

$$\psi(q, t) = \int_{-\infty}^{\infty} dq' \, K(q, t; q', t') \psi(q', t') \tag{4.65}$$

Assuming that the particle is at q_1 at time $t = t_1$, we have $\psi(q', t_1) = \delta(q' - q_1)$ and Eq. (4.65) yields for $t > t_1$

$$\psi(q, t) = K(q, t; q_1, t_1) \tag{4.66}$$

which shows that $K(q, t; q_1, t_1)$ is the amplitude for finding the particle at q at time t, if it were at q_1 at time t_1. In Appendix A10 we show that

$$
\begin{aligned}
K(q, t; q', t') &= \langle q | e^{-\frac{i}{\hbar}(t - t')\widehat{H}} | q' \rangle \\
&= \int_{q(t')=q'}^{q(t)=q} \mathcal{D}q(t) \, \exp\left(\frac{i}{\hbar} \int_{t'}^{t} dt'' \, L^{(r)}[q, \dot{q}] \right)
\end{aligned}
\tag{4.67}
$$

where $L^{(r)}[q, \dot{q}]$ is the real-time Lagrangian given in Eq. (4.63). In the semi-classical limit, the path integral in Eq. (4.67) is dominated by the extremal path $\bar{q}$, for which

$$\delta S = \delta \int_{t'}^{t} dt'' \, L^{(r)}[\bar{q}, \dot{\bar{q}}] = 0 \tag{4.68}$$

Equation (4.68) yields an Euler–Lagrange equation for $\bar{q}$

$$M \frac{d^2 \bar{q}}{dt^2} + \frac{\partial V(\bar{q})}{\partial \bar{q}} = 0 \tag{4.69}$$

This equation describes the *classical* equation of motion for the potential V of Fig. 4.4. Since the energy of the particle is zero, there is *no classical path* leading from q_0 to q_1. This point provides a motivation for introducing the *imaginary-time* variable $\tau = it$, in terms of which Eq. (4.69) reads

$$- M \frac{d^2 \bar{q}}{d\tau^2} + \frac{\partial V}{\partial \bar{q}} = 0 \tag{4.70}$$

Equation (4.70) is the equation of motion for a particle of mass M moving in the *inverted potential*, $-V(\bar{q})$ (see Fig. 4.5) We see from Fig. 4.5 that, in *imaginary* time, there *will be a path* leading from q_0 to q_1, which is a *nontrivial* solution of Eq. (4.70). Assuming that the particle starts at $q_0 = 0$ when $\tau = 0$, it will for $\tau > 0$ roll down the hill, *bounce off* the potential wall at $q = q_1$, and return to q_0. This solution, called the bounce, is also the extremal path (saddle point) that dominates the path integral for the *imaginary-time transition amplitude* obtained from (4.67) by letting $(t - t') = -i\hbar\beta$

$$
\begin{aligned}
K(q, q'; \beta) &= \langle q | e^{-\beta \widehat{H}} | q' \rangle \\
&= \int_{q(0)=q'}^{q(\hbar\beta)=q} \mathcal{D}q(\tau) \, \exp\left(-\frac{1}{\hbar} \int_0^{\hbar\beta} d\tau \, L[q, \dot{q}] \right)
\end{aligned}
\tag{4.71}
$$

where $L[q, \dot{q}]$ is the imaginary-time Lagrangian [see Eq. (A1.21)] given by

$$L[q, \dot{q}] = \frac{1}{2} M \left(\frac{dq}{d\tau} \right)^2 + V(q) \tag{4.72}$$

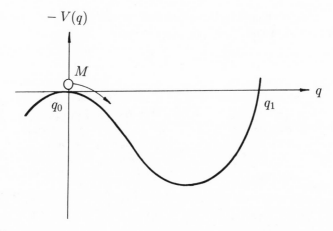

$-V(q)$

M

q_0

q_1

q

FIG. 4.5.   Motion of a particle of mass M in an inverted potential, taking place in imaginary time.

$K(q, q', \beta)$ is the quantum-mechanical transition amplitude for the particle to go from the coordinate q' at time zero to q at time $\hbar\beta$. We note that the RHS of Eq. (4.71) is nothing else than the density matrix derived in Eq. (A1.22).

The expression for the decay rate can be obtained by considering the expansion of $K(q, q', \beta)$ in terms of the energy eigenstates $|n\rangle$, satisfying

$$\widehat{H}|n\rangle = E_n|n\rangle \tag{4.73}$$

Inserting a complete set of these states, we obtain for $\beta \to \infty$ and $q = q'$

$$K(q, q; \beta) = \langle q|e^{-\beta\widehat{H}}|q\rangle = \sum_n \langle q|n\rangle e^{-\beta E_n} \langle n|q\rangle$$

$$\xrightarrow[\beta \to \infty]{} |\psi_0(q)|^2 e^{-\beta E_0} \tag{4.74}$$

For the limiting process on the RHS of Eq. (4.74) to be valid, it is essential that $|\psi_0(q)|^2$ be as large as possible to minimize the contributions of the $n > 0$ terms. This is especially important for small but *nonzero* temperatures. Since $\psi_0(q)$ describes the ground state in the metastable well, this condition is best met if we take $q = 0$. We then have from Eq. (4.74)

$$\lim_{\beta \to \infty} K(0, 0; \beta) = |\psi_0(0)|^2 e^{-\beta E_0} \tag{4.75}$$

The prefactor on the RHS of Eq. (4.75) gives the probability of the particle in the potential well, and the $e^{-\beta E_0}$ term yields information about the ground-state energy. Due to quantum tunneling, the ground-state energy develops a *small imaginary* part that can be extracted from Eq. (4.75) by taking $\ln K$. The decay rate is then given by

$$\Gamma = -2 \operatorname{Im} \frac{E_0}{\hbar} \simeq \frac{2}{\hbar\beta} \operatorname{Im} \left[\ln K(0,0;\beta)\right] \tag{4.76}$$

Equation (4.76) is, of course, valid only for very large values of β.

We now consider the evaluation of the RHS of Eq. (4.71). In the semi-classical limit, this path integral is dominated by *extremal* paths $\bar{q}$, which are solutions of the equation of motion (4.70). Therefore, we put

$$q(\tau) = \bar{q}(\tau) + \xi(\tau) \tag{4.77}$$

where $\xi(\tau)$ represents the fluctuation about the extremal path. We expand $\xi(\tau)$ into a complete set of *real* orthonormal functions $\xi_s(\tau)$

$$\xi(\tau) = \sum_s \eta_s \xi_s(\tau) \tag{4.78}$$

where

$$\int_0^{\hbar\beta} d\tau \, \xi_s(\tau)\xi_{s'}(\tau) = \hbar\beta\delta_{s,s'} \tag{4.79}$$

The boundary conditions $q(0) = q'$ and $q(\hbar\beta) = q$ translate into the following boundary conditions for $\bar{q}$ and $\xi_s(\tau)$

$$\bar{q}(\tau = 0) = q'; \qquad \bar{q}(\tau = \hbar\beta) = q \tag{4.80}$$

and

$$\xi_s(\tau = 0) = \xi_s(\tau = \hbar\beta) = 0, \qquad \text{for all } s \tag{4.81}$$

Introducing the ansatz (4.77) into the Euclidean action, we have with use of Eq. (4.72)

$$
\begin{aligned}
S &= \int_0^{\hbar\beta} d\tau \left[\frac{1}{2}M(\dot{\bar{q}} + \dot{\xi})^2 + V(\bar{q} + \xi)\right] \\
&= \int_0^{\hbar\beta} d\tau \left[\frac{1}{2}M\dot{\bar{q}}^2 + V(\bar{q})\right] + \int_0^{\hbar\beta} d\tau \left[\frac{1}{2}M\dot{\xi}^2 + \frac{1}{2}V''(\bar{q})\xi^2\right] \\
&\quad + \int_0^{\hbar\beta} d\tau \left[M\ddot{\bar{q}}\dot{\xi} + V'(\bar{q})\xi\right]
\end{aligned}
\tag{4.82}
$$

We note that the last integral on the RHS of this equation *vanishes*. This can be shown by integrating by parts the term

$$M\int_0^{\hbar\beta} d\tau \, \dot{\bar{q}}\dot{\xi} = -M\int_0^{\hbar\beta} d\tau \, \ddot{\bar{q}}\xi + M\dot{\bar{q}}\xi\Big|_0^{\hbar\beta} = -M\int_0^{\hbar\beta} d\tau \, \ddot{\bar{q}}\xi \tag{4.83}$$

where we have used the boundary condition (4.81). Using the equation of motion (4.70), we see that the RHS of Eq. (4.83) *cancels* the integral of the $V'(\bar{q})\xi$ term in Eq. (4.82).

There are two extremal paths $\bar{q}$ that have to be included in the calculation of $K(0,0,\beta)$. One is the trivial path, $\bar{q} = 0$, for which the action (4.82) reduces to (replacing ξ by ζ)

$$S_\zeta = \frac{1}{2}M \int_0^{\hbar\beta} d\tau \left(\dot{\zeta}^2 + \omega_0^2 \zeta^2 \right) \tag{4.84}$$

where $\omega_0 = [V''(0)/M]^{\frac{1}{2}}$ is the well frequency. Equation (4.84) is just the Euclidean action of a harmonic oscillator. The other extremal path is the bounce trajectory, $\bar{q}(\tau) = q_B(\tau)$, representing a nontrivial solution of the equation of motion (4.70) with the initial conditions (4.80). The action, corresponding to $q_B(\tau)$ and fluctuations about it, is, according to Eq. (4.82),

$$S_1 = S_B + S_\xi \tag{4.85}$$

where

$$S_B = \int_0^{\hbar\beta} d\tau \left[\frac{1}{2}M\dot{q}_B^2 + V(q_B) \right] \tag{4.86}$$

and

$$S_\xi = \frac{1}{2}M \int_0^{\hbar\beta} d\tau \left(-\ddot{\xi} + \frac{1}{M}V''(q_B)\xi \right)\xi \tag{4.87}$$

where we have applied integration by parts in the first term of Eq. (4.87).

Following Langer (1967), the fluctuation action S_ξ can be diagonalized by assuming that the eigenfunctions $\xi_s(\tau)$ satisfy the differential equation

$$\left(-\frac{d^2}{d\tau^2} + \frac{1}{M}V''[q_B(\tau)] \right)\xi_s(\tau) = \lambda_s \xi_s(\tau) \tag{4.88}$$

Introducing the expansion (4.78) into Eq. (4.87) and using Eq. (4.88), we have

$$S_\xi = \frac{1}{2}M \sum_{s,s'} \int_0^{\hbar\beta} d\tau \left[\lambda_s \xi_s(\tau)\eta_s\eta_{s'}\xi_{s'}(\tau) \right] = \frac{1}{2}M\hbar\beta \sum_s \eta_s^2 \lambda_s \tag{4.89}$$

where the orthogonality condition (4.79) has been used.

From Eqs. (4.84)–(4.87), the path integral (4.71) can be written as

$$K(q,q;\beta) = K_0 + K_B \tag{4.90}$$

where

$$K_0 = \int_{\zeta(0)=\zeta(\hbar\beta)=0} \mathcal{D}\zeta(\tau) \, \exp\left(-\frac{1}{\hbar}S_\zeta \right) \tag{4.91}$$

where ζ is a fluctuation about $\bar{q} = 0$, to be distinguished from $\xi(\tau)$, which we use for fluctuation about $\bar{q} = q_B(\tau)$. The bounce contribution K_B is given by

$$
K_B = \int_{q(0)=q(\hbar\beta)=0} \mathcal{D}q(\tau) \exp\left(-\frac{1}{\hbar}(S_B + S_\xi)\right)
$$

$$
= \exp\left(-\frac{1}{\hbar}S_B\right) \int_{\xi(0)=\xi(\hbar\beta)=0} \mathcal{D}\xi(\tau) \exp\left(-\frac{1}{\hbar}S_\xi\right) \qquad (4.92)
$$

The evaluation of K_0 is done by expanding $\zeta(\tau)$ into a complete set of real orthonormal functions $\xi_m^{(0)}(\tau)$

$$
\zeta(\tau) = \sum_m \zeta_m \xi_m^{(0)}(\tau) \qquad (4.93)
$$

where the coefficients ζ_m are real and the functions $\xi_m^{(0)}(\tau)$ satisfy [see Eq. (4.79)]

$$
\int_0^{\hbar\beta} d\tau\, \xi_m^{(0)}(\tau)\xi_{m'}^{(0)}(\tau) = \hbar\beta\delta_{m,m'} \qquad (4.94)
$$

In keeping with Eq. (4.84), $\xi_m^{(0)}(\tau)$ satisfy the differential equations [see Eq. (4.88)]

$$
\left(-\frac{d^2}{d\tau^2} + \omega_0^2\right)\xi_m^{(0)}(\tau) = \lambda_m^{(0)}\xi_m^{(0)}(\tau) \qquad (4.95)
$$

The solutions of this equation, satisfying the boundary condition $\xi_m^{(0)}(0) = \xi_m^{(0)}(\hbar\beta) = 0$ and the normalization condition (4.94), are

$$
\xi_m^{(0)}(\tau) = \sqrt{2}\sin\left(\frac{m\pi}{\hbar\beta}\tau\right) \qquad (4.96)
$$

Introducing Eq. (4.96) into Eq. (4.95), the eigenvalues $\lambda_m^{(0)}$ are given by

$$
\lambda_m^{(0)} = \left(\frac{m\pi}{\hbar\beta}\right)^2 + \omega_0^2 \qquad (4.97)
$$

The action (4.84) becomes, after inserting the expansion (4.93) and using Eqs. (4.94)–(4.97)

$$
S_\zeta = \frac{1}{2}M\hbar\beta \sum_{m=1}^{\infty} \lambda_m^{(0)}\zeta_m^2 \qquad (4.98)
$$

Using this result, the path integral (4.91) can be written as a multiple Gaussian integral

$$
K_0 = \prod_{m=1}^{\infty} \int_{-\infty}^{\infty} d\zeta_m \exp\left(-\frac{1}{2}M\beta\lambda_m^{(0)}\zeta_m^2\right) = \prod_{m=1}^{\infty} \left(\frac{2\pi}{M\beta\lambda_m^{(0)}}\right)^{\frac{1}{2}} \qquad (4.99)
$$

The bounce contribution K_B is evaluated in Appendix A12 for a special choice of the potential $V(q)$

$$V(q) = \frac{1}{2} M \omega_0^2 (q^2 - q^3/q_1) \tag{4.100}$$

where q_1 is the second zero of $V(q)$ (see Fig. 4.4). Besides being a reasonably realistic model potential for MQT, the function (4.100) allows us to find the solution for $q_B(\tau)$ in analytic form. From Eq. (A12.27) we obtain for K_B a purely *imaginary* value

$$K_B = i\pi \left(\frac{\hbar S_B}{\beta M^3 |\lambda_0| \lambda_2} \right)^{\frac{1}{2}} Z_c \exp \left(-\frac{S_B}{\hbar} \right) \tag{4.101}$$

where λ_0 and λ_2 are the eigenvalues of two bound states of Eq. (4.88), given by [see Eq. (A11.20)]

$$\lambda_0 = -\frac{5\omega_0^2}{4} \qquad \lambda_2 = \frac{3\omega_0^2}{4} \tag{4.102}$$

Z_c is the fluctuation contribution to K_B, produced by the continuum ($n \geq 3$) eigenstates of Eq. (4.88)

$$Z_c = \prod_{n=3}^{\infty} \left(\frac{2\pi}{\beta M \lambda_n} \right)^{\frac{1}{2}} \tag{4.103}$$

Using Eq. (4.90) and assuming that $|K_B| \ll K_0$, we obtain from Eq. (4.76) the rate of decay in the form

$$\Gamma = \frac{2}{\hbar \beta} \operatorname{Im} [\ln(K_0 + K_B)] \simeq \frac{2}{\hbar \beta} \operatorname{Im} \left(\frac{K_B}{K_0} \right) \tag{4.104}$$

Introducing into the RHS of this equation the expressions (4.99) and (4.101)–(4.103), we obtain

$$\Gamma = \frac{2\pi}{\omega_0^2 \beta} \left(\frac{S_B}{\hbar} \right)^{\frac{1}{2}} \left(\frac{16}{15\beta M^3} \right)^{\frac{1}{2}} \exp \left(-\frac{S_B}{\hbar} \right) \frac{Z_c}{K_0} \tag{4.105}$$

The ratio Z_c/K_0 on the RHS of this equation is evaluated by first factorizing the expression (4.99) as follows

$$K_0 \simeq \left(\frac{2\pi}{\beta M \omega_0^2} \right)^{\frac{3}{2}} \prod_{m=4}^{\infty} \left(\frac{2\pi}{M \beta \lambda_m^{(0)}} \right)^{\frac{1}{2}} \tag{4.106}$$

where the first three eigenvalues $\lambda_m^{(0)} < 4$ are approximated by ω_0^2. This introduces a negligible error for $\beta \to \infty$ [see Eq. (4.97)]. In this way, the product of the $m \geq 4$ terms in Eq. (4.106) can be matched with the continuum product (4.103). Setting $m = n + 1$, we have from Eqs. (4.103) and (4.106)

$$\frac{Z_c}{K_0} = \left(\frac{\beta M \omega_0^2}{2\pi}\right)^{\frac{3}{2}} R \qquad (4.107)$$

where R is the ratio produced by the continuum of states and evaluated by Langer (1967) with the result

$$R = \prod_{n=3}^{\infty} \left(\frac{\lambda_n^{(0)}}{\lambda_n}\right)^{\frac{1}{2}} = \frac{15}{2} \qquad (4.108)$$

Introducing the expressions (4.107) and (4.108) into (4.105), we obtain

$$\Gamma = \omega_0 \left(\frac{30 S_B}{\pi \hbar}\right)^{\frac{1}{2}} \exp\left(-\frac{S_B}{\hbar}\right) \qquad (4.109)$$

Using the expression (A13.1) for the bounce action, we obtain from Eq. (4.109)

$$\Gamma = 12\omega_0 \left(\frac{3 V_0}{2\pi \hbar \omega_0}\right)^{\frac{1}{2}} \exp\left(-\frac{36 V_0}{5 \hbar \omega_0}\right) \qquad (4.110)$$

where V_0 is the barrier height.

The expression (4.110) is the decay rate for tunneling of a particle, without dissipation, out of a metastable minimum (Larkin and Ovchinnikov, 1983). It could be, of course, obtained much easier by solving the Schrödinger equation in the WKB approximation. This method is expected to yield a decay rate that is a product of the pre-exponential factor times the transmission coefficient D, given in Eq. (4.64). On comparing Eqs. (4.64) and (4.109), we thus expect that

$$S_B = 2 \int_0^{q_1} dq_B \, [2 M V(q_B)]^{\frac{1}{2}} \qquad (4.111)$$

To prove this equality, we start from expression (4.86). The integrand of this expression can be simplified by eliminating $V(q_B)$ using the relation

$$\frac{1}{2} M \dot{q}_B^2 = V(q_B) \qquad (4.112)$$

Equation (4.112) can be derived by multiplying Eq. (4.70) by $\dot{q}_B$ and integrating over τ

$$M \int_0^{\hbar\beta} d\tau \, \ddot{q}_B \dot{q}_B = \frac{1}{2} M \int_0^{\hbar\beta} d(\dot{q}_B^2) = \int_0^{\hbar\beta} d\tau \left(\frac{\partial V}{\partial q_B}\right) \dot{q}_B \qquad (4.113)$$

Equation (4.113) implies

$$\frac{1}{2}M\dot{q}_B^2 = V(q_B) + C \tag{4.114}$$

where C is a constant of integration. Applying the boundary condition

$$q_B(\tau = 0) = \dot{q}_B(\tau = 0) = 0 \tag{4.115}$$

to Eq. (4.114), we obtain $C = 0$, and Eq. (4.112) then follows from Eq. (4.114). Using Eq. (4.112), the action (4.86) can be written

$$S_B = M \int_0^{\hbar\beta} d\tau\, \dot{q}_B^2(\tau) \tag{4.116}$$

This expression can be transformed to an integral over $q_B \in [0, q_1]$. Since $q_B(\tau)$ is symmetric about $\tau = \hbar\beta/2$, we have

$$\begin{aligned}
S_B &= M \int_{-\hbar\beta/2}^{\hbar\beta/2} d\tau\, \dot{q}_B^2 \\
&= M \left(\int_0^{q_1} dq_B\, \dot{q}_B + \int_{q_1}^{0} (-dq_B)\, \dot{q}_B \right) \\
&= 2M \int_0^{q_1} dq_B\, \dot{q}_B
\end{aligned} \tag{4.117}$$

Using Eq. (4.112), we have

$$\dot{q}_B = \left(\frac{2V(q_B)}{M} \right)^{\frac{1}{2}} \tag{4.118}$$

Substituting this result on the RHS of Eq. (4.117), we obtain the desired Eq. (4.111).

4.8. Dissipative quantum tunneling

Let us now consider quantum tunneling in a dynamical system consisting of the q particle coupled to a set of environmental oscillators $\{x^{(\alpha)}\}$. The Lagrangian (in real-time formulation) of this system is given in Eq. (4.20). The methods of Section 4.7 can be generalized to calculate the effect of the environmental modes on the tunneling rate of the q particle.

Quantum tunneling in such a system is, strictly speaking, a many-body transition for which the transition amplitude is given by the following generalization of Eq. (4.71)

$$\begin{aligned}
&K(q, \{x^{(\alpha)}\}; q', \{x^{(\alpha)'}\}; \beta) \\
&= \langle q, \{x^{(\alpha)}\} | e^{-\beta\widehat{H}} | q', \{x^{(\alpha)'}\} \rangle
\end{aligned}$$

$$= \int_{q(0)=q'}^{q(\hbar\beta)=q} \mathcal{D}q(\tau) \int_{\{x^{(\alpha)}(0)\}=\{x^{(\alpha)'}\}}^{\{x^{(\alpha)}(\hbar\beta)\}=\{x^{(\alpha)}\}} \mathcal{D}x^{(\alpha)}(\tau)$$

$$\times \exp\left(-\frac{1}{\hbar}\int_0^{\hbar\beta} d\tau \, L[q, \{x^{(\alpha)}\}; \dot{q}, \{\dot{x}^{(\alpha)}\}]\right) \qquad (4.119)$$

where $\widehat{H}$ is the Hamiltonian of the system and $L[q, \{x^{(\alpha)}\}; \dot{q}, \{\dot{x}^{(\alpha)}\}]$ is the Euclidean version of the Lagrangian (4.20) [see Eq. (4.23)]

$$L[q, \{x^{(\alpha)}\}; \dot{q}, \{\dot{x}^{(\alpha)}\}] = \frac{1}{2}M\left(\frac{dq}{d\tau}\right)^2 + V(q)$$

$$+ \frac{1}{2}\sum_\alpha m_\alpha \left[\left(\frac{dx^{(\alpha)}}{d\tau}\right)^2 + \omega_\alpha^2 (x^{(\alpha)})^2\right]$$

$$- \sum_\alpha \left(qC_\alpha x^{(\alpha)} - \frac{C_\alpha^2 q^2}{2m_\alpha \omega_\alpha^2}\right) \qquad (4.120)$$

Equation (4.119) describes the quantum-mechanical transition amplitude for the system (q particle plus environment) to go from coordinates q', $\{x^{(\alpha)'}\}$, at time zero, to q, $\{x^{(\alpha)}\}$, at time $\hbar\beta$. Comparing Eq. (4.119) with Eq. (A2.14), we see that

$$K(q, \{x^{(\alpha)}\}; q', \{x^{(\alpha)'}\}; \beta) = \rho(q, \{x^{(\alpha)}\}; q', \{x^{(\alpha)'}\}) \qquad (4.121)$$

where the RHS of Eq. (4.121) is the density matrix for a many-body system. For large β, the spectral expansion of the *diagonal* transition amplitude $K(q, \{x^{(\alpha)}\}; q, \{x^{(\alpha)}\}; \beta)$ is dominated by the lowest-energy eigenvalue that develops a small imaginary part due to the quantum decay. In dissipative quantum tunneling we are only interested in measurements on the q particle, regardless of the initial positions of the environmental coordinates $\{x^{(\alpha)}\}$. Then the tunneling rate for the q particle is to be calculated from the *reduced* diagonal transition amplitude $\widetilde{K}(q, q, \beta)$, obtained from Eq. (4.121) by integrating over the coordinate set $\{x^{(\alpha)}\}$

$$\widetilde{K}(q, q; \beta) = \prod_\alpha \int_{-\infty}^{\infty} dx^{(\alpha)} \, K(q, \{x^{(\alpha)}\}; q, \{x^{(\alpha)}\}; \beta)$$

$$= \prod_\alpha \int_{-\infty}^{\infty} dx^{(\alpha)} \, \rho(q, \{x^{(\alpha)}\}; q, \{x^{(\alpha)}\})$$

$$= \left(\prod_\alpha \int_{-\infty}^{\infty} dx^{(\alpha)}\right) \int_{q(0)=q}^{q(\hbar\beta)=q} \mathcal{D}q(\tau)$$

$$\times \int_{\{x^{(\alpha)}(0)\}=\{x^{(\alpha)}\}}^{\{x^{(\alpha)}(\hbar\beta)\}=\{x^{(\alpha)}\}} \prod_\alpha \mathcal{D}x^{(\alpha)}(\tau)$$

$$\times \exp\left(-\frac{1}{\hbar}\int_0^{\hbar\beta} d\tau \, L[q, \{x^{(\alpha)}\}; \dot{q}, \{\dot{x}^{(\alpha)}\}]\right) \qquad (4.122)$$

where we used for K the path-integral formulation given in Eq. (4.119). Recalling Eqs. (4.22)–(4.24), the RHS of Eq. (4.122) can be expressed as an *effective one-body* transition amplitude

$$\widetilde{K}(q,q;\beta) = \int_{q(0)=q}^{q(\hbar\beta)=q} \mathcal{D}q(\tau)\, \exp\left(-\frac{1}{\hbar}S_{\text{eff}}[q]\right) \qquad (4.123)$$

where $S_{\text{eff}}[q]$ is the effective action defined by

$$\exp\left(-\frac{1}{\hbar}S_{\text{eff}}[q]\right) = \left(\prod_\alpha \int_{-\infty}^{\infty} dx^{(\alpha)}\right) \int_{\{x^{(\alpha)}(0)\}=\{x^{(\alpha)}\}}^{\{x^{(\alpha)}(\hbar\beta)\}=\{x^{(\alpha)}\}} \prod_\alpha \mathcal{D}x^{(\alpha)}(\tau)$$

$$\times \exp\left(-\frac{1}{\hbar}\int_0^{\hbar\beta} d\tau\, L[q,\{x^{(\alpha)}\};\dot{q},\{\dot{x}^{(\alpha)}\}]\right) \quad (4.124)$$

We see that the RHS of this equation is a straightforward generalization of Eq. (4.25) to a *set* of environmental modes $\{x^{(\alpha)}\}$, which has been given in Eq. (4.36). For ohmic dissipation, the effective action takes the explicit form (4.43).

Using Eq. (4.123), we obtain from Eq. (4.76) the tunneling rate

$$\Gamma \simeq \frac{2}{\hbar\beta}\,\text{Im}\left[\ln\widetilde{K}(0,0;\beta)\right] \qquad (4.125)$$

The RHS of expression (4.125) can be evaluated by applying the bounce technique introduced (for the undamped case) in Section 4.7. The details of the calculations are, however, more complicated owing to the presence of the dissipative term in the effective action. Closed analytic expressions for Γ can be obtained in two limiting cases: weak and strong dissipation (Caldeira and Leggett, 1983; Larkin and Ovchinnikov, 1983).

4.8.1. *Weak ohmic dissipation*

This case is of conceptual interest, as it clearly demonstrates that dissipation tends to *decrease* the tunneling rate. Using Eqs. (4.40) and (4.43), we first express the effective action, via Parseval's theorem, as follows

$$S_{\text{eff}}[q] = \int_0^{\hbar\beta} d\tau \left[\frac{1}{2}M\dot{q}^2 + V(q)\right] + \eta\hbar\beta \sum_n |\omega_n| q_n q_{-n}$$

$$= \int_0^{\hbar\beta} d\tau \left[\frac{1}{2}M\dot{q}^2 + V(q)\right]$$

$$+ \frac{1}{2}\int_0^{\hbar\beta} d\tau \int_0^{\hbar\beta} d\tau'\, q(\tau)K(\tau-\tau')q(\tau') \qquad (4.126)$$

where

$$K(\tau) = \sum_n K_n e^{-i\omega_n\tau} \qquad (4.127)$$

and

$$K_n = \frac{2\eta|\omega_n|}{\hbar\beta} \tag{4.128}$$

The bounce trajectory $q_B(\tau)$ satisfies the Euler–Lagrange equation obtained from the variational principle $\delta S_{\text{eff}}[q] = 0$. Taking the variation of Eq. (4.126), we thus obtain

$$M\frac{d^2 q_B}{d\tau^2} - \frac{\partial V}{\partial q_B} - \int_0^{\hbar\beta} d\tau'\, K(\tau - \tau')q_B(\tau') = 0 \tag{4.129}$$

We shall again consider the potential $V(q)$ given in Eq. (4.100). Then Eq. (4.129) takes the form

$$M\left[\frac{d^2 q_B}{d\tau^2} - \omega_0^2\left(q_B - \frac{3q_B^2}{2q_1}\right)\right] - \int_0^{\hbar\beta} d\tau'\, K(\tau - \tau')q_B(\tau') = 0 \tag{4.130}$$

Introducing into Eq. (4.130) the Fourier expansion

$$q_B(\tau) = \sum_n \widetilde{q}_n e^{-i\omega_n \tau} \tag{4.131}$$

and using Eqs. (4.127) and (4.128), we obtain the following equation for the coefficients $\widetilde{q}_n$

$$-M\left((\omega_n^2 + \omega_0^2)\widetilde{q}_n - \frac{3\omega_0^2}{2q_1}\sum_m \widetilde{q}_m \widetilde{q}_{n-m}\right) - 2\eta|\omega_n|\widetilde{q}_n = 0 \tag{4.132}$$

The criterion for weak dissipation can be obtained from this equation by comparing the dissipative term with the first term. For $|\omega_n| = \omega_0$, we see that these terms are of the same magnitude when the parameter $\kappa = \eta/M\omega_0 \simeq 1$. This implies that the weak-dissipation limit corresponds to the case of $\kappa \ll 1$. In this case, the last term in Eq. (4.130) can be treated as a small perturbation. To calculate the bounce action, we limit ourselves to the lowest-order iteration by first solving Eq. (4.130) without the dissipative term and then substituting the result in Eq. (4.43). In this way, we obtain a perturbation formula for the bounce action

$$S_B(\eta) = S_B(0) + \Delta S_B \tag{4.133}$$

where $S_B(0)$ is the bounce action in the absence of dissipation [see Eqs. (A13.1) and (A13.3)]

$$S_B(0) = \int_0^{\hbar\beta} d\tau \left[\frac{1}{2}M\dot{q}_B^2 + V(q_B)\right] = \frac{36V_0}{5\omega_0} = \frac{8M\omega_0 q_1^2}{15} \tag{4.134}$$

and ΔS_B is the correction due to dissipation

$$\Delta S_B = \frac{\eta}{2\pi}\int_0^{\hbar\beta} d\tau \int_0^{\hbar\beta} d\tau' \left(\frac{q_B(\tau) - q_B(\tau')}{\tau - \tau'}\right)^2 \tag{4.135}$$

In Appendix A14 we evaluate the expression (4.135) with the result [see Eq. (A14.7)]

$$\Delta S_B = \frac{24}{\pi^3} \zeta(3) \eta q_1^2 \simeq \eta q_1^2 \qquad (4.136)$$

From Eqs. (4.136) and (4.134) we calculate the ratio

$$\frac{\Delta S_B}{S_B(0)} = \frac{15\eta}{8M\omega_0} = \frac{15}{8}\kappa \ll 1 \qquad (4.137)$$

where we have applied the weak-damping condition, $\kappa \ll 1$. Introducing the expression (4.133) into Eq. (4.109), we obtain, expanding in a small parameter $\Delta S_B / S_B(0)$, the decay rate

$$\Gamma(\eta) \simeq \omega_0 \left(\frac{30 S_B(0)}{\pi\hbar}\right)^{\frac{1}{2}} \left(1 + \frac{1}{2}\frac{\Delta S_B}{S_B(0)}\right) \exp\left(-\frac{1}{\hbar}[S_B(0) + \Delta S_B]\right)$$
$$\simeq \Gamma(\eta = 0) \exp\left(-\frac{1}{\hbar}\Delta S_B\right) \qquad (4.138)$$

where $\Gamma(\eta = 0)$ is the decay rate in the *absence* of dissipation given by Eq. (4.110). On the RHS of Eq. (4.138), the correction to the prefactor is neglected, as it is small compared with the correction to the exponential term. From Eqs. (4.136) and (4.138), we have for the ratio

$$\frac{\Gamma(\eta)}{\Gamma(\eta = 0)} \simeq \exp\left(-\frac{\eta q_1^2}{\hbar}\right) < 1 \qquad (4.139)$$

which shows that dissipation always tends to reduce the tunneling rate.

4.8.2. *Strong ohmic dissipation*

We now turn to the case of *strongly overdamped* systems characterized by $\kappa \gg 1$. In this case, both the exponent and the prefactor in the expression for $\Gamma(\eta)$ become significantly modified in comparison with the undamped case. We confine ourselves to deriving only the exponential term, $\exp\left[-(1/\hbar)S_B(\eta)\right]$, following the method of Larkin and Ovchinnikov (1983).

Substituting the expansion (4.131) in Eq. (4.126), we obtain for the bounce action the expression

$$S_B(\eta) = S_{\text{eff}}[q_B]$$
$$= \hbar\beta \sum_n \left[\frac{1}{2}M(\omega_n^2 + \omega_0^2) + \eta|\omega_n|\right]\tilde{q}_n\tilde{q}_{-n}$$
$$- \frac{\hbar\beta M\omega_0^2}{2q_1} \sum_{n,l} \tilde{q}_n\tilde{q}_l\tilde{q}_{-n-l} \qquad (4.140)$$

where the coefficients $\tilde{q}_n$ satisfy the nonlinear equation of motion (4.132). For $|\omega_n| \lesssim \omega_0$, the condition for strong overdamping $\kappa \gg 1$, allows us to

neglect the first inertial term in (4.132). We note that the width of the
bounce (in the absence of dissipation) is of order ω_0^{-1} [see Eq. (A14.2)],
and the damping is expected to increase the width [see Eq. (A15.15)]. This
justifies the assumption $|\omega_n| \lesssim \omega_0$. Thus, in the limit $\eta/M\omega_0 \gg 1$, we can
replace Eq. (4.132) by

$$- M\left(\omega_0^2 \widetilde{q}_n - \frac{3\omega_0^2}{2q_1}\sum_n \widetilde{q}_m \widetilde{q}_{n-m}\right) - 2\eta|\omega_n|\widetilde{q}_n = 0 \qquad (4.141)$$

Larkin and Ovchinnikov (1983) have shown that this equation can be solved
for $\widetilde{q}_n$ exactly. The details of the solution are deferred to Appendix A15.
From Eqs. (A15.2), (A15.11), and (A15.12), we have

$$\widetilde{q}_n = \alpha e^{-b|n|} = \frac{8\pi\eta q_1}{3\hbar\beta M\omega_0^2}e^{-b|n|} \qquad (4.142)$$

where

$$b = \operatorname{arctanh}\left(\frac{4\pi\eta}{\hbar\beta M\omega_0^2}\right) \qquad (4.143)$$

We now present a derivation of $S_B(\eta)$ under the assumption that the temperature T is close to (but below) T_0, where T_0 is the critical temperature
given by (A15.13). Then the parameter b is very large, and we may confine
ourselves to terms $|n| \leq 1$. At the same time, we can replace the constant
α in Eq. (4.142) by $\alpha_0 = \frac{2}{3}q_1$ [see Eq. (A15.15)]. Thus, we make in Eq.
(4.140) the substitution

$$\widetilde{q}_n = \begin{cases} \alpha_0 e^{-b|n|}, & \text{for } |n| = 0, 1 \\ 0, & \text{for } |n| > 1 \end{cases} \qquad (4.144)$$

and obtain, upon neglecting the inertial term

$$S_B(\eta) \simeq \frac{1}{2}\alpha_0^2\hbar\beta M\omega_0^2\left[1 + 2\frac{1-x}{1+x} - \frac{\alpha_0}{q_1}\left(1 + 6\frac{1-x}{1+x}\right)\right]$$
$$+ 4\pi\eta\alpha_0^2\frac{1-x}{1+x} \qquad (4.145)$$

where we used the result

$$\widetilde{q}_1^2 = \alpha_0^2 e^{-2b} = \alpha_0^2 \exp\left(-\ln\frac{1+x}{1-x}\right) = \alpha_0^2\left(\frac{1-x}{1+x}\right) \qquad (4.146)$$

where $x = 3\alpha_0/2q_1$ is the argument of the arctanh function in Eq. (4.143).
Using the expression (A15.12) for α_0, we have

$$\frac{1}{2}\hbar\beta M\omega_0^2 = \frac{4\pi\eta q_1}{3\alpha_0} \qquad (4.147)$$

Introducing this relation into Eq. (4.145), we obtain for the bounce action (for $b \gg 1$)

$$S_B(\eta) \simeq 2\pi\eta\alpha_0^2\left(1 - \frac{\alpha_0}{q_1}\right) = \frac{8\pi}{27}\eta q_1^2 \approx \eta q_1^2 \qquad (4.148)$$

where we have used $\alpha_0 = \frac{2}{3}q_1$ to obtain the RHS of (4.148). Introducing this result into the expression for tunneling rate, we obtain for strong dissipation

$$\Gamma(\eta) \simeq A_0(\eta)\exp\left(-\frac{\eta q_1^2}{\hbar}\right) \qquad (4.149)$$

We note that the exponential factor in Eq. (4.149) and that on the RHS of Eq. (4.139) are the same. But there is an essential difference between these results, since the action $S_B(0)$ is absent in $S_B(\eta)$. This implies that the exponent of (4.149) is entirely independent of the parameters M and ω_0. Riseborough et al. (1985) have shown that $S_B(\eta)$ increases by a factor of 1.42 as T varies from T_0 to zero, indicating that the decay rate will be the lowest at $T = 0$. The prefactor $A_0(\eta)$ is of order $\omega_0(\eta/M\omega_0)^{\frac{7}{2}}$. We refer the reader to the original paper for its derivation (Larkin and Ovchinnikov, 1983). In our applications of the foregoing theory of dissipative tunneling to vortices, we focus our attention on the exponential term that often dominates the behavior of the rate (see Section 8.8).

5

MICROSCOPIC THEORY OF
SUPERCONDUCTING TUNNELING

5.1. Introduction

In Chapter 4 we discussed the effect of dissipation on the MQT by introducing the concept of effective Euclidean action. For ohmic friction, this action was determined by a phenomenological model in which the tunneling particle is coupled to an ensemble of environmental harmonic oscillators. Such a model provides a suitable description of a resistively shunted Josephson junction for which the MQT takes place in the space of the *phase variable*. This phase variable couples to "dissipative" currents in the shunting resistor via the Josephson relation (4.4); according to this relation, the voltage drop across the junction is proportional to the rate of change of the phase. Even in the absence of a shunting resistor, this voltage drop will produce a dissipation, as it will cause the *quasiparticle* currents to flow between the superconductors (for gapless materials or at finite temperatures). The derivation of the effective action of a Josephson junction with quasiparticle dissipation has been given by Ambegaokar et al. (1982) (AES), based on a microscopic model of superconducting tunneling. The dissipative part of their effective action differs in an important manner from the ohmic case. Specifically, the quadratic function of $[q(\tau) - q(\tau')]$, given in Eq. (4.43), is now replaced by a square of the sine function of the phase variable. This result has a number of consequences for the physics of Josephson junctions and their arrays (see Chapter 6).

This chapter contains a description of some of the techniques involved in the derivation of the effective action from a microscopic model. The method of AES is a sophisticated version of the Stratonovich transformation (Stratonovich, 1958), which replaces the two-body interaction between the particles by a one-body interaction with a random potential. The application of this method to superconductors has been developed by Hubbard (1959) and Mühlschlegel (1962). Rice (1967) has used this method to derive the Ginzburg–Landau free-energy functional for a superconductor. In Section 5.2 we introduce the Stratonovich method by considering first the model of a neutral fermion s-wave superfluid. In spite of its simplicity, this model is not devoid of some interesting physics related to phase fluctuations. In fact, in Section 5.3 we derive an effective Euclidean action for a small particle, formed by such neutral fermions, and find a phase-inertial term, proportional to the square of the time derivative of the phase variable. This result provides a microscopic justification for the phenomenological Lagrangian, given in Eq. (2.65).

Section 5.4 considers a model of a pair of such particles coupled by tunneling. It contains a derivation of the dissipative action, Josephson coupling constant, and effective phase inertia due to quasiparticle tunneling.

Finally, Section 5.5 is devoted to a more realistic model of coupled superconductors, including the charging energy. Except for some small simplifications, this model is essentially identical to that studied by AES. We derive the partition function as a path integral over the phase variables and a random voltage across the junction. By means of Gaussian path integration, the voltage variable is eliminated, and we obtain an effective Euclidean action which is a functional of the phase variables only. The part of this action involving phase-inertial terms represents a generalization of the action derived by AES. Consistent with the phenomenological result (2.105), there are two competing sources of quantum phase fluctuations. When the intergrain capacitance C is small, the charging energy $2e^2/C$ dominates the expression (2.105). In this case the action agrees with that derived by AES, assuming that the phase variables are pinned to the potentials. On the other hand, when the capacitance C is large, the first term on the RHS of Eq. (2.105) may dominate the charging energy, provided the grains are sufficiently small. This term is of the same physical origin as the phase-inertial term derived for an isolated particle in Section 5.3. It provides a new source of phase fluctuations, which is operational in the absence of charging energies. It may play a role in the phase ordering of arrays of superconducting grains imbedded in a metal or in a matrix of high dielectric constant. For two-dimensional arrays, the diagonal charging terms lead to a phason mode with an acoustic dispersion relation, which is essential for obtaining a nonzero dissipation of a vortex near $T = 0$ (see Section 8.7.4). The consequences of this dissipation mechanism for quantum tunneling of a vortex have been discussed by Larkin et al. (1988).

5.2. Partition function for a neutral superfluid

5.2.1. *Model Hamiltonian*

Consider a system of N fermions, in a sample of volume Ω, interacting via a short-range attraction of strength g. The grand canonical Hamiltonian of this system is given by (Abrikosov et al., 1965)

$$K = H - \mu N = \sum_k \xi_k n_k - \frac{g}{2\Omega} \sum_{k,k',Q,\sigma} a^\dagger_{k+Q,\sigma} a^\dagger_{-k,-\sigma} a_{-k',-\sigma} a_{k'+Q,\sigma}$$

$$= H_0 - \mu N + H_1 \tag{5.1}$$

where

$$n_k = \sum_\sigma a^\dagger_{k,\sigma} a_{k,\sigma} \tag{5.2}$$

At this point, we are dropping the carets over all operators to simplify the notation. Later on, we reintroduce them to indicate matrices and spinors in Nambu notation. $a^\dagger_{k,\sigma}$ and $a_{k,\sigma}$ are the creation and annihilation operators of one-fermion states $|k,\sigma\rangle$, where $\mathbf{k}$ and σ are the momentum and spin quantum numbers, respectively. We note that Eq. (5.1) is appropriate for s-wave pairing with total spin of the Cooper pair equal to zero. Let us introduce the "pair" creation operators

$$b^\dagger_Q = \sum_k a^\dagger_{k+Q,\uparrow} a^\dagger_{-k,\downarrow} \tag{5.3}$$

and their Hermitian conjugate b_Q. Then the interaction part of the Hamiltonian (5.1) can be written as

$$H_1 = -\frac{g}{\Omega} \sum_Q b^\dagger_Q b_Q \tag{5.4}$$

so that

$$K = K_0 + H_1 = K_0 - \frac{g}{\Omega} \sum_Q b^\dagger_Q b_Q \tag{5.5}$$

This form of the grand canonical Hamiltonian allows us to apply the Stratonovich method directly.

5.2.2.   Stratonovich transformation

The grand partition function Z_G, for the fermion system described, is

$$Z_G = \mathrm{Tr}\left(e^{-\beta K}\right) = \mathrm{Tr}\left(e^{-\beta(K_0+H_1)}\right) \tag{5.6}$$

Using Eq. (A6.3), we disentangle the operators in the exponent on the RHS of Eq. (5.6) as follows

$$e^{-\beta(K_0+H_1)} = e^{-\beta K_0} \sigma(\beta) \tag{5.7}$$

where $\sigma(\beta)$ is the imaginary-time evolution operator [see Eq. (A6.8)]

$$\sigma(\beta) = \mathrm{T}_\tau \exp\left(-\frac{1}{\hbar} \int_0^{\hbar\beta} d\tau\, H_1[\tau]\right) \tag{5.8}$$

The operator $H_1[\tau]$ is defined as [see Eq. (A6.5)]

$$H_1[\tau] = e^{K_0\tau/\hbar} H_1 e^{-K_0\tau/\hbar} \tag{5.9}$$

Using the Stratonovich method, the evolution operator (5.8) can be written as a path integral over a Gaussian random field, coupled to the pair operators b_Q and $b_Q^\dagger$. This method is based on the identity (valid for complex numbers: $b = b_1 + ib_2$ and $x = x_1 + ix_2$)

$$\exp |b|^2 = \frac{1}{\pi} \int_{-\infty}^{\infty} dx_1 \int_{-\infty}^{\infty} dx_2 \exp\left(-|x|^2 - bx^* - b^*x\right) \qquad (5.10)$$

We note that Eq. (5.10) follows simply by multiplying the equation [see Eq. (A3.20)]

$$e^{b_i^2} = \frac{1}{\sqrt{\pi}} \int_{-\infty}^{\infty} dx_i \exp\left(-x_i^2 - 2b_i x_i\right) \qquad (5.11)$$

with $i = 1$, by the same equation with $i = 2$. To apply Eq. (5.10) to the operator (5.8), we first discretize the time variable by writing

$$\frac{1}{\hbar} \int_0^{\hbar\beta} d\tau \, H_1[\tau] = \lim_{\Delta\tau \to 0} \frac{1}{\hbar} \sum_{i=1}^{\bar{N}} \Delta\tau H_1[\tau_i] \qquad (5.12)$$

In Eq. (5.12), we have divided the interval $[0, \hbar\beta]$ into $\bar{N}$ intervals, each of length $\Delta\tau$. The τ-ordering operator enables us to treat the operators $b_Q^\dagger$ and b_Q as c numbers, so that the identity (5.10) can be applied directly. By letting

$$|b|^2 = \left(\frac{g\Delta\tau}{\hbar\Omega}\right) b_Q^\dagger b_Q \qquad (5.13)$$

$$|x|^2 = \frac{\Delta\tau}{\hbar} |x(Q)|^2 \qquad (5.14)$$

Eqs. (5.8) and (5.12) lead us to an expression for $\sigma(\beta)$ in the form of a product of Eqs. (5.10), for each (Q, τ_i)

$$\sigma(\beta) = T_\tau \exp\left(-\frac{1}{\hbar} \int_0^{\hbar\beta} d\tau \, H_1[\tau]\right)$$

$$= \lim_{\Delta\tau \to 0} T_\tau \exp\left(\frac{g\Delta\tau}{\hbar\Omega} \sum_{Q,\tau_i} b_Q^\dagger[\tau_i] b_Q[\tau_i]\right)$$

$$= \lim_{\Delta\tau \to 0} T_\tau \prod_{Q,\tau_i} \left(\frac{\Delta\tau}{\pi\hbar}\right) \int_{-\infty}^{\infty} dx_1(Q, \tau_i) \int_{-\infty}^{\infty} dx_2(Q, \tau_i)$$

$$\times \exp\left\{-\frac{\Delta\tau}{\hbar}\left[|x(Q, \tau_i)|^2\right.\right.$$

$$\left.\left. + \left(\frac{g}{\Omega}\right)^{\frac{1}{2}} \left(b_Q^\dagger[\tau_i] x(Q, \tau_i) + \text{H.c.}\right)\right]\right\} \qquad (5.15)$$

It is convenient to express $\sigma(\beta)$ as a path integral over the random fields defined in *real* space $\mathbf{r}$. Thus we introduce the Fourier transforms

$$x(\mathbf{Q}, \tau) = \frac{1}{\sqrt{\Omega}} \int d^3r \, e^{-i\mathbf{Q}\cdot\mathbf{r}} x(\mathbf{r}, \tau) \tag{5.16}$$

$$a_{\mathbf{k},\sigma} = \frac{1}{\sqrt{\Omega}} \int d^3r \, e^{-i\mathbf{k}\cdot\mathbf{r}} \psi_\sigma(\mathbf{r}) \tag{5.17}$$

Furthermore, we define a rescaled random field $\Delta(\mathbf{r}, \tau)$ with a dimension of the superconducting gap

$$\Delta(\mathbf{r}, \tau) = \sqrt{g}\, x(\mathbf{r}, \tau) \tag{5.18}$$

Making the substitutions (5.16)–(5.18), Eq. (5.15) can be written as a path integral over the *complex* Gaussian field $\Delta(\mathbf{r}, \tau)$

$$\sigma(\beta) = \int \mathcal{D}^2 \Delta(\mathbf{r}, \tau) \exp\left(-\frac{1}{\hbar} \int_0^{\hbar\beta} d\tau \int d^3r \, \frac{|\Delta(\mathbf{r}, \tau)|^2}{g}\right)$$

$$\times \, \mathrm{T}_\tau \exp\left(-\frac{1}{\hbar} \int_0^{\hbar\beta} d\tau\right.$$

$$\left.\times \int d^3r \left\{\psi_\uparrow^\dagger[\mathbf{r}, \tau]\psi_\downarrow^\dagger[\mathbf{r}, \tau]\Delta(\mathbf{r}, \tau) + \mathrm{H.c.}\right\}\right) \tag{5.19}$$

The grand partition function (5.6) can be written, using Eqs. (5.7) and (5.19)

$$Z_G = \int \mathcal{D}^2 \Delta(\mathbf{r}, \tau) \, \mathrm{Tr}\left[\mathrm{T}_\tau \exp\left(-\frac{1}{\hbar} \int_0^{\hbar\beta} d\tau \, H_{\mathrm{eff}}(\tau)\right)\right] \tag{5.20}$$

where

$$H_{\mathrm{eff}}(\tau) = K_0 + \int d^3r \, \psi_\uparrow^\dagger[\mathbf{r}, \tau]\psi_\downarrow^\dagger[\mathbf{r}, \tau]\Delta(\mathbf{r}, \tau) + \mathrm{H.c.}$$

$$+ \frac{1}{g} \int d^3r \, |\Delta(\mathbf{r}, \tau)|^2 \tag{5.21}$$

5.2.3. *Nambu matrix notation*

The trace over the fermion variables in Eq. (5.20) can be expressed in terms of the anomalous propagator for a single fermion moving in the random field $\Delta(\mathbf{r}, \tau)$. At this point, it is convenient to introduce the Nambu notation. This enables us to treat, on the same footing, the complex random fields $\Delta(\mathbf{r}, \tau)$ as well as real random fields that result from applying the Stratonovich transformation to Coulomb interactions. This will become especially useful when we consider the effective action of a microscopic tunnel junction including the charging energy. The grand partition function is then given by a "double" path integral over the complex field $\Delta(\mathbf{r}, \tau)$ and the real field $V(\tau)$, the latter interacting with the charge difference between the electrodes (see Section 5.5).

Following Nambu (1960), we introduce a spinor-field operator

$$\widehat{\psi}(\mathbf{r}) = \begin{pmatrix} \psi_\uparrow(\mathbf{r}) \\ \psi_\downarrow^\dagger(\mathbf{r}) \end{pmatrix} \tag{5.22}$$

and its Hermitian adjoint

$$\widehat{\psi}^\dagger(\mathbf{r}) = \begin{pmatrix} \psi_\uparrow^\dagger(\mathbf{r}), & \psi_\downarrow(\mathbf{r}) \end{pmatrix} \tag{5.23}$$

Using these operators, the operator $H_{\text{eff}}(\tau)$ of Eq. (5.21) can be written as

$$H_{\text{eff}}(\tau) = K_0 + D(\tau) + \frac{1}{g} \int d^3r \, |\Delta(\mathbf{r}, \tau)|^2 \tag{5.24}$$

where

$$K_0 = \int d^3r \, \widehat{\psi}^\dagger \widehat{k}_0 \widehat{\psi} = -\int d^3r \, \widehat{\psi}^\dagger \widehat{\tau}_3 \left(\frac{\hbar^2}{2m} \nabla^2 + \mu \right) \widehat{\psi} \tag{5.25}$$

and

$$D(\tau) = \int d^3r \, \widehat{\psi}^\dagger[\mathbf{r}, \tau] \widehat{D}(\mathbf{r}, \tau) \widehat{\psi}[\mathbf{r}, \tau] \tag{5.26}$$

where

$$\widehat{D}(\mathbf{r}, \tau) = \Delta^r \widehat{\tau}_1 - \Delta^i \widehat{\tau}_2 \tag{5.27}$$

In Eq. (5.27), Δ^r and Δ^i are the real imaginary parts of $\Delta(\mathbf{r}, \tau)$, respectively, and we have introduced Pauli matrices

$$\widehat{\tau}_1 = \begin{pmatrix} 0 & 1 \\ 1 & 0 \end{pmatrix}, \qquad \widehat{\tau}_2 = \begin{pmatrix} 0 & -i \\ i & 0 \end{pmatrix}, \qquad \widehat{\tau}_3 = \begin{pmatrix} 1 & 0 \\ 0 & -1 \end{pmatrix} \tag{5.28}$$

The spinor-field operators (5.22) and (5.23) satisfy the "matrix" anticommutation relation

$$[\widehat{\psi}(\mathbf{r}), \widehat{\psi}^\dagger(\mathbf{r}')]_+ = \begin{pmatrix} [\psi_\uparrow(\mathbf{r}), \psi_\uparrow^\dagger(\mathbf{r}')]_+ & [\psi_\uparrow(\mathbf{r}), \psi_\downarrow(\mathbf{r}')]_+ \\ [\psi_\downarrow^\dagger(\mathbf{r}), \psi_\uparrow^\dagger(\mathbf{r}')]_+ & [\psi_\downarrow(\mathbf{r}), \psi_\downarrow^\dagger(\mathbf{r}')]_+ \end{pmatrix}$$

$$= \delta(\mathbf{r} - \mathbf{r}') \begin{pmatrix} 1 & 0 \\ 0 & 1 \end{pmatrix} \tag{5.29}$$

where we used the fermion anticommutation relations for ψ_σ and $\psi_\sigma^\dagger$. In a similar way, we find the remaining anticommutators to be

$$[\widehat{\psi}(\mathbf{r}), \widehat{\psi}(\mathbf{r}')]_+ = [\widehat{\psi}^\dagger(\mathbf{r}), \widehat{\psi}^\dagger(\mathbf{r}')]_+ = 0 \tag{5.30}$$

The Matsubara Green's function, in Nambu matrix notation, is defined as the average of a time-ordered product of spinor-field operators

$$\widehat{G}(\mathbf{r}, \tau; \mathbf{r}', \tau') = -\left\langle T_\tau \, \widehat{\psi}(\mathbf{r}, \tau) \widehat{\psi}^\dagger(\mathbf{r}', \tau') \right\rangle \tag{5.31}$$

where

$$\langle \cdots \rangle = \frac{\text{tr}\left(e^{-\beta K} \cdots\right)}{\text{tr}\left(e^{-\beta K}\right)} \tag{5.32}$$

The symbol tr indicates a trace with the *Nambu space excluded*. The field operators in Eq. (5.31) are given by

$$\widehat{\psi}(\mathbf{r}, \tau) = e^{K\tau/\hbar} \widehat{\psi}(\mathbf{r}) e^{-K\tau/\hbar}$$
$$\widehat{\psi}^\dagger(\mathbf{r}, \tau) = e^{K\tau/\hbar} \widehat{\psi}^\dagger(\mathbf{r}) e^{-K\tau/\hbar} \tag{5.33}$$

According to Eqs. (5.20) and (5.24), the grand partition function Z_G can be written as a path integral

$$Z_G = \int \mathcal{D}^2 \Delta(\mathbf{r}, \tau) \exp\left(-\frac{1}{g\hbar} \int_0^{\hbar\beta} d\tau \int d^3 r \, |\Delta(\mathbf{r}, \tau)|^2\right) Z_G[\Delta(\mathbf{r}, \tau)] \tag{5.34}$$

where

$$Z_G[\Delta(\mathbf{r}, \tau)] = \text{Tr}\left[e^{-\beta K_0} \, T_\tau \exp\left(-\frac{1}{\hbar} \int_0^{\hbar\beta} d\tau \, D(\tau)\right)\right] \tag{5.35}$$

This grand partition "functional" of $\Delta(\mathbf{r}, \tau)$ can be reduced to an expression that involves Nambu Green's function (5.31) in the presence of a space- and time-dependent field $\Delta(\mathbf{r}, \tau)$. This is done in Appendix A16. There, we derive a cumulant expansion theorem which allows us to express $Z_G[\Delta(\mathbf{r}, \tau)]$ as follows [see Eq. (A16.15)]

$$Z_G[\Delta] = Z_{G_0} \exp\left(-\int_0^1 \frac{d\lambda}{\lambda} \left\langle \frac{1}{\hbar} \int_0^{\hbar\beta} d\tau \, D_\lambda(\tau) \right\rangle_\lambda\right) \tag{5.36}$$

where

$$Z_{G_0} = \text{Tr}\left(e^{-\beta K_0}\right) \tag{5.37}$$

and

$$\langle \cdots \rangle_\lambda = \frac{\text{tr}\left(e^{-\beta K_\lambda} \cdots\right)}{\text{tr}\left(e^{-\beta K_\lambda}\right)} \tag{5.38}$$

In Eq. (5.38) we define K_λ by introducing the variable coupling constant λ into the interaction term of Eq. (5.24)

$$K_\lambda = K_0 + D_\lambda(\tau) = K_0 + \lambda D(\tau) \tag{5.39}$$

where $\lambda \in [0, 1]$.

5.2.4. *Expressing* $\left\langle \frac{1}{\hbar} \int_0^{\hbar\beta} d\tau \, D_\lambda(\tau) \right\rangle_\lambda$ *in terms of* $\widehat{G}_\lambda$

According to Eqs. (5.26) and (5.36)–(5.38), we have for the integrand in the exponent of Eq. (5.36)

$$
\begin{aligned}
I_\lambda &= \frac{1}{\hbar} \int_0^{\hbar\beta} d\tau \, \langle D_\lambda(\tau) \rangle_\lambda \\
&= \frac{\lambda}{\hbar} \int_0^{\hbar\beta} d\tau \int d^3r \left\langle \widehat{\psi}^\dagger[\mathbf{r}, \tau] \widehat{D}(\mathbf{r}, \tau) \widehat{\psi}[\mathbf{r}, \tau] \right\rangle_\lambda \\
&= \frac{\lambda}{\hbar} \int_0^{\hbar\beta} d\tau \int d^3r \, \frac{\mathrm{Tr}\left(e^{-\beta K_\lambda} \widehat{\psi}^\dagger[\mathbf{r}, \tau] \widehat{D}(\mathbf{r}, \tau) \widehat{\psi}[\mathbf{r}, \tau]\right)}{\mathrm{Tr}\left(e^{-\beta K_\lambda}\right)}
\end{aligned} \tag{5.40}
$$

where

$$
e^{-\beta K_\lambda} = e^{-\beta K_0} \, T_\tau \exp\left(\frac{1}{\hbar} \int_0^{\hbar\beta} d\tau \, D_\lambda(\tau)\right) = e^{-\beta K_0} \sigma_\lambda(\beta) \tag{5.41}
$$

We note that the trace in the numerator on the RHS of Eq. (5.40) *includes* also the *Nambu space*. Using the cyclic invariance of trace, we have

$$
\begin{aligned}
\mathrm{Tr}&\left(e^{-\beta K_\lambda} \widehat{\psi}^\dagger[\mathbf{r}, \tau] \widehat{D}(\mathbf{r}, \tau) \widehat{\psi}[\mathbf{r}, \tau]\right) \\
&= \mathrm{Tr}\left(\widehat{D}(\mathbf{r}, \tau) \widehat{\psi}[\mathbf{r}, \tau] e^{-\beta K_\lambda} \widehat{\psi}^\dagger[\mathbf{r}, \tau]\right) \\
&= \mathrm{Tr}^{(N)}\left[\widehat{D}(\mathbf{r}, \tau) \, \mathrm{tr}\left(e^{-\beta K_\lambda} \widehat{\psi}^\dagger[\mathbf{r}, \tau] \widehat{\psi}[\mathbf{r}, \tau]\right)\right]
\end{aligned} \tag{5.42}
$$

where $\mathrm{Tr}^{(N)}$ indicates a partial trace over the *Nambu* spinor components, and tr is the trace (over the fermion variables) used in Eq. (5.32) to define the Green's function matrix $\widehat{G}$. From Eqs. (5.40)–(5.42) we have

$$
\begin{aligned}
\left\langle \widehat{\psi}^\dagger[\mathbf{r}, \tau] \widehat{D}(\mathbf{r}, \tau) \widehat{\psi}[\mathbf{r}, \tau] \right\rangle_\lambda &= \mathrm{Tr}^{(N)}\left(\widehat{D}(\mathbf{r}, \tau) \frac{\mathrm{tr}\left(e^{-\beta K_\lambda} \widehat{\psi}^\dagger[\mathbf{r}, \tau] \widehat{\psi}[\mathbf{r}, \tau]\right)}{\mathrm{tr}\left(e^{-\beta K_\lambda}\right)}\right) \\
&= - \mathrm{Tr}^{(N)}\left[\widehat{D}(\mathbf{r}, \tau) \widehat{G}_\lambda(\mathbf{r}, \tau; \mathbf{r}, \tau)\right]
\end{aligned} \tag{5.43}
$$

where we have used the definition of the Green's function $\widehat{G}_\lambda$, given in Eqs. (A17.3)–(A17.5), in the following manner

$$
\begin{aligned}
\widehat{G}_\lambda(\mathbf{r}, \tau; \mathbf{r}', \tau') &= -\frac{\mathrm{tr}\left[T_\tau e^{-\beta K_\lambda} \widehat{\psi}(\mathbf{r}, \tau) \widehat{\psi}^\dagger(\mathbf{r}', \tau')\right]}{\mathrm{tr}\left(e^{-\beta K_\lambda}\right)} \\
&= -\frac{\mathrm{tr}\left\{T_\tau e^{-\beta K_\lambda} \sigma_\lambda^{-1}(\tau) \widehat{\psi}[\mathbf{r}, \tau] \sigma_\lambda(\tau) \sigma_\lambda^{-1}(\tau') \widehat{\psi}^\dagger[\mathbf{r}', \tau'] \sigma_\lambda(\tau')\right\}}{\mathrm{tr}\left(e^{-\beta K_\lambda}\right)} \\
&\xrightarrow[\substack{\tau \to \tau' \\ \mathbf{r} \to \mathbf{r}'}]{} -\frac{\mathrm{tr}\left(T_\tau e^{-\beta K_\lambda} \widehat{\psi}[\mathbf{r}, \tau] \widehat{\psi}^\dagger[\mathbf{r}, \tau]\right)}{\mathrm{tr}\left(e^{-\beta K_\lambda}\right)}
\end{aligned} \tag{5.44}
$$

In Eq. (5.44), we have used the property of the time-ordering operator, which allows us to cancel the time-evolution operators in the numerator. Introducing the RHS of Eq. (5.44) into Eq. (5.40), we have

$$I_\lambda = \text{Tr}^{(N)} \left(\frac{1}{\hbar} \int_0^{\hbar\beta} d\tau \int d^3r \, \widehat{D}_\lambda(\mathbf{r}, \tau) \widehat{G}_\lambda(\mathbf{r}, \tau; \mathbf{r}, \tau) \right) \tag{5.45}$$

Using the Fourier transforms (A17.10) and (A17.12), we can rewrite Eq. (5.45) as follows

$$I_\lambda = \text{Tr}^{(N)} \left(\sum_{m,m'} \sum_{\mathbf{k},\mathbf{k}'} \widehat{D}_\lambda(\mathbf{k}' - \mathbf{k}, m' - m) \widehat{G}_\lambda(\mathbf{k}, m; \mathbf{k}', m') \right) \tag{5.46}$$

This expression can be evaluated, using Eq. (A17.13). By setting $\mathbf{k}' = \mathbf{k}$ and $n' = n$, Eq. (A17.13) yields

$$\frac{\widehat{G}_\lambda(\mathbf{k}, n; \mathbf{k}, n)}{\widehat{G}_0(\mathbf{k}, n)} - \widehat{1} = \sum_{\mathbf{k}',n} \widehat{D}_\lambda(\mathbf{k} - \mathbf{k}', n - n') \widehat{G}_\lambda(\mathbf{k}', n'; \mathbf{k}, n) \tag{5.47}$$

Summing over $\mathbf{k}, n$ and taking $\text{Tr}^{(N)}$, the RHS of Eq. (5.47) now coincides with the RHS of Eq. (5.46), yielding

$$\begin{aligned}
I_\lambda &= \text{Tr}^{(N)} \sum_{\mathbf{k},n} \left(\frac{\widehat{G}_\lambda(\mathbf{k}, n; \mathbf{k}, n)}{\widehat{G}_0(\mathbf{k}, n)} - \widehat{1} \right) \\
&= \sum_{\mathbf{n}} \left(\frac{\widehat{G}_\lambda(\mathbf{n}, \mathbf{n})}{\widehat{G}_0(\mathbf{n})} - \widehat{1} \right) \\
&= \frac{1}{\Delta(\lambda)} \text{Tr} \left[\widehat{\Delta}(\lambda) \right]
\end{aligned} \tag{5.48}$$

where we have reintroduced the condensed notation $\mathbf{n} \equiv (\mathbf{k}, n, \sigma)$, which incorporates the Nambu space σ. The RHS of Eq. (5.48) is obtained by using the Fredholm solution (A17.17). The expression for I_λ can be further simplified with the use of the relation (A17.18), yielding

$$I_\lambda = -\frac{\lambda}{\Delta(\lambda)} \frac{d\Delta(\lambda)}{d\lambda} \tag{5.49}$$

Using this result in Eq. (5.36), we obtain

$$Z_G[\Delta] = Z_{G_0} \exp \left(\int_{\Delta(0)}^{\Delta(1)} \frac{d\Delta(\lambda)}{\Delta(\lambda)} \right) = Z_{G_0} \exp \left[\ln \Delta(1) \right] \tag{5.50}$$

where we have used the property of the Fredholm determinant: $\Delta(\lambda = 0) = 1$. From Eq. (A17.19), we have

$$\ln \Delta(1) = \text{Tr} \ln(\widehat{1} - \widehat{T}) \tag{5.51}$$

where the trace is over the variables $\mathbf{n} \equiv (\mathbf{k}, n, \sigma)$. The RHS of this equation can be written in terms of $\widehat{G} = \widehat{G}_{\lambda=1}$ and $\widehat{G}_0$. By letting $\lambda = 1$ in Eq. (A17.14), we have

$$(\widehat{1} - \widehat{T})\widehat{G} = \widehat{G}_0 \tag{5.52}$$

which implies

$$\ln(\widehat{1} - \widehat{T}) = \ln(\widehat{G}_0\widehat{G}^{-1}) = \ln \widehat{G}_0 + \ln \widehat{G}^{-1} \tag{5.53}$$

Introducing this result into Eq. (5.51) and using the latter in Eq. (5.50), we obtain finally

$$Z_G[\Delta] = Z_{G_0} \exp\left[\text{Tr}(\ln \widehat{G}_0 + \ln \widehat{G}^{-1})\right] \tag{5.54}$$

5.3. Effective action for a small particle

As a simple application of the foregoing theory, we consider the effective action of a small superfluid particle. We assume that the diameter of the particle is *considerably smaller* than the zero-temperature coherence length, so that we may neglect the spatial gradients of ψ. On the other hand, the particle is assumed to be large enough, so that the fluctuations of $|\psi|$ are suppressed. Then only time-dependent fluctuations of the phase θ remain in the effective action. Under these circumstances, the effective Lagrangian is expected to take the form given in Eq. (2.65). The main purpose of this section is to present a derivation of this Lagrangian from a microscopic theory. We confine ourselves to temperatures near absolute zero.

According to Eqs. (5.34) and (5.54), the grand canonical partition function of our superfluid particle is given by the path integral

$$Z_G = Z_{G_0} \int \mathcal{D}^2 \Delta(\mathbf{r}, \tau) \exp\left(-\frac{1}{\hbar} S_{\text{eff}}[\Delta]\right) \tag{5.55}$$

where the effective action is given by

$$S_{\text{eff}}[\Delta] = \frac{1}{g} \int_0^{\hbar\beta} d\tau \int d^3r \, |\Delta(\mathbf{r}, \tau)|^2 - \hbar \, \text{Tr}(\ln \widehat{G}_0 + \ln \widehat{G}^{-1}) \tag{5.56}$$

In this expression, $\widehat{G}_0$ is the Green's function for free electrons and $\widehat{G}$ satisfies the equation of motion [see Eq. (A17.6)]

$$\left(-\hbar \frac{\partial}{\partial \tau} - \widehat{k}_0 - \widehat{D}\right) \widehat{G}(\mathbf{r}, \tau; \mathbf{r}', \tau') = \hbar \, \delta(\mathbf{r} - \mathbf{r}') \, \delta(\tau - \tau') \tag{5.57}$$

where $\widehat{k}_0$ and $\widehat{D}$ are Nambu matrices defined in Eqs. (5.25)–(5.27).

The standard approach used to derive the effective action near the superconducting transition temperature is to expand $\ln \widehat{G}^{-1}$ in terms of

$\widehat{D}$. The resulting action is then of a Ginzburg–Landau form with a dynamics of diffusion type (Abrahams and Tsuneto, 1966). The problem of phase fluctuations at $T = 0$ must be treated differently, since the system is characterized by a large magnitude of the order parameter. In the complex ψ plane, the fluctuating order parameter of constant magnitude is a two-dimensional vector, moving along the circle of radius Δ_0. For the derivation of the effective action, it is convenient to perform a gauge transformation that eliminates the phase factor of $\Delta(\tau)$

$$\Delta(\tau) = \Delta_0 e^{i\theta(\tau)} \to \Delta_0 \tag{5.58}$$

where Δ_0 is real. This is reminiscent of a transformation to coordinates "rotating with the vector" $\Delta(\tau)$. Equation (5.58) results from the following transformation of the spinor-field operator (5.22)

$$\widehat{\psi}(\mathbf{r}, \tau) \to \widehat{U}\widehat{\psi}(\mathbf{r}, \tau) = \widehat{\psi}'(\mathbf{r}, \tau) \tag{5.59}$$

where $\widehat{U}$ is a unitary matrix

$$\widehat{U} = \begin{pmatrix} e^{-i\theta(\tau)/2} & 0 \\ 0 & e^{i\theta(\tau)/2} \end{pmatrix} \tag{5.60}$$

Since the trace is invariant to gauge transformations, the last term on the RHS of Eq. (5.56) can be evaluated with $\widehat{G}$ replaced by the gauge-transformed Green's function $\widehat{G}'$. Our problem now is to obtain an equation of motion for $\widehat{G}'$ defined as

$$\widehat{G}'(\mathbf{r}, \tau; \mathbf{r}', \tau') = -\left\langle T_\tau\, \widehat{\psi}'(\mathbf{r}, \tau)\widehat{\psi}'^\dagger(\mathbf{r}', \tau') \right\rangle \tag{5.61}$$

We start with the equation of motion for $\widehat{\psi}'$. From Eqs. (5.59) and (5.60), we obtain

$$\frac{\partial \widehat{\psi}'}{\partial \tau} = \frac{\partial \widehat{U}}{\partial \tau}\widehat{\psi} + \widehat{U}\frac{\partial \widehat{\psi}}{\partial \tau} = -\frac{i}{2}\frac{\partial \theta}{\partial \tau}\widehat{\tau}_3\widehat{\psi}' + \frac{\widehat{U}}{\hbar}[\widehat{K}, \widehat{\psi}] \tag{5.62}$$

The second term on the RHS can be evaluated using Eqs. (5.25)–(5.30) to yield

$$\frac{\widehat{U}}{\hbar}[\widehat{K}, \widehat{\psi}] = -\frac{1}{\hbar}(\widehat{k}_0 + \widehat{D}_0)\widehat{\psi}' \tag{5.63}$$

where

$$\widehat{D}_0 = \Delta_0\widehat{\tau}_1 \tag{5.64}$$

We note that the simple result (5.63) is true only when the phase θ does not depend on $\mathbf{r}$. Equations (5.63) and (5.64) imply that $\widehat{G}'$ satisfies the following equation of motion

$$\left(-\hbar\frac{\partial}{\partial \tau} - \widehat{k}_0 - \widehat{D}_0 - \widehat{D}_1(\tau)\right)\widehat{G}'(\mathbf{r}, \tau; \mathbf{r}', \tau') = \hbar\, \delta(\mathbf{r} - \mathbf{r}')\, \delta(\tau - \tau') \tag{5.65}$$

where $\widehat{D}_1 = i(\hbar/2)\widehat{\tau}_3\dot{\theta}$. We proceed by expanding $(\widehat{G}')^{-1}$ about $\widehat{G}_s^{-1}$, the equation of motion for $\widehat{G}_s$ being

$$\left(-\hbar\frac{\partial}{\partial\tau} - \widehat{k}_0 - \widehat{D}_0\right)\widehat{G}_s(\mathbf{r} - \mathbf{r}', \tau - \tau') = \hbar\,\delta(\mathbf{r} - \mathbf{r}')\,\delta(\tau - \tau') \qquad (5.66)$$

In this equation, we are neglecting all inhomogeneities produced by the finite size of the particle; henceforth, $\widehat{G}_s$ describes a BCS superconductor with a static, homogeneous, and real order parameter Δ_0. Using Eq. (5.66), the equation of motion (5.65) can be transformed to an integral equation [see Eq. (A17.13)]

$$\begin{aligned}\widehat{G}'(\mathbf{k}, n; \mathbf{k}', n') = {}&\delta_{\mathbf{k},\mathbf{k}'}\delta_{n,n'}\widehat{G}_s(\mathbf{k}, n)\\ &+ \sum_{\mathbf{k}'',n''}\widehat{G}_s(\mathbf{k}, n)\widehat{D}_1(n - n'')\widehat{G}'(\mathbf{k}'', n''; \mathbf{k}', n')\end{aligned} \qquad (5.67)$$

which can be written as a matrix equation

$$(\widehat{1} - \widehat{T})\widehat{G}' = \widehat{G}_s \qquad (5.68)$$

where

$$\widehat{T} = \widehat{G}_s\widehat{D}_1 \qquad (5.69)$$

The matrices in Eqs. (5.68) and (5.69) are now defined in the $(\mathbf{k}, n, \sigma)$ space, where σ labels the Nambu space. Solving Eq. (5.68) for $(\widehat{G}')^{-1}$, we have with the use of (5.69)

$$(\widehat{G}')^{-1} = \widehat{G}_s^{-1} - \widehat{D}_1 \qquad (5.70)$$

Expanding the logarithm of Eq. (5.70) to order $\widehat{D}_1^2$, we obtain

$$\mathrm{Tr}\left[\ln(\widehat{G}')^{-1}\right] = \mathrm{Tr}\left(\ln\widehat{G}_s^{-1}\right) - \mathrm{Tr}\left(\widehat{G}_s\widehat{D}_1\right) - \frac{1}{2}\mathrm{Tr}\left(\widehat{G}_s\widehat{D}_1\right)^2 \qquad (5.71)$$

Using this result in Eq. (5.56), we obtain with use of Eq. (5.58)

$$\begin{aligned}S_{\mathrm{eff}}[\Delta] = {}&-\hbar\,\mathrm{Tr}\left(\ln\widehat{G}_0\right) + \frac{\Omega\hbar\beta}{g}\Delta_0^2 + \hbar\,\mathrm{Tr}\left(\ln\widehat{G}_s^{-1}\right)\\ &+ \hbar\,\mathrm{Tr}\left(\widehat{G}_s\widehat{D}_1\right) + \frac{\hbar}{2}\mathrm{Tr}\left(\widehat{G}_s\widehat{D}_1\right)^2\end{aligned} \qquad (5.72)$$

where Ω is the volume of the superfluid particle. The second and third terms on the RHS of Eq. (5.72) give the mean-field (BCS) action of a superconductor with a constant gap Δ_0.

The term that is first order in $\widehat{D}_1$ is evaluated by performing the σ trace first and the $\mathbf{k}, n$ trace next, yielding

$$\begin{aligned}A = \hbar\,\mathrm{Tr}\left(\widehat{G}_s\widehat{D}_1\right) &= \frac{i\hbar}{2}\mathrm{tr}\left[\mathrm{Tr}^{(N)}\left(\widehat{G}_s\widehat{\tau}_3\right)\dot{\theta}\right]\\ &= \frac{\hbar}{2}\sum_{\mathbf{k},m}[G_s^{11}(\mathbf{k}, m) - G_s^{22}(\mathbf{k}, m)]\omega_m\theta(m)\end{aligned} \qquad (5.73)$$

where G_s^{11} and G_s^{22} are the components of $\widehat{G}_s$, given by (see Fetter and Walecka, 1971)

$$G_s^{11}(\mathbf{k}, m) = -G_s^{22}(\mathbf{k}, -m) = \frac{-\hbar(i\hbar\omega_m + \xi_k)}{\hbar^2\omega_m^2 + \xi_k^2 + \Delta_0^2} \qquad (5.74)$$

Performing the $\mathbf{k}$ sum in Eq. (5.73), we obtain with use of Eq. (5.74)

$$\sum_{\mathbf{k}} \left[G_s^{11}(\mathbf{k}, m) - G_s^{22}(\mathbf{k}, m) \right]$$

$$= -N(0) \int_{-\hbar\omega_D}^{\hbar\omega_D} d\xi_k \frac{2\hbar\xi_k}{\hbar^2\omega_m^2 + \xi_k^2 + \Delta_0^2} = 0 \qquad (5.75)$$

where $N(0) = N(\xi = 0)$ is the density of states (per spin) at the Fermi level. In view of Eq. (5.75), the RHS of Eq. (5.73) vanishes, so that

$$A = 0 \qquad (5.76)$$

as expected from the time-reversal invariance of the effective Lagrangian.

Let us now consider the last term on the RHS of Eq. (5.72)

$$B = \frac{\hbar}{2} \operatorname{Tr} \left(\widehat{G}_s \widehat{D}_1 \right)^2 = \frac{\hbar}{2} \operatorname{tr} \left[\operatorname{Tr}^{(N)} \left(\widehat{G}_s \widehat{D}_1 \right)^2 \right]$$

$$= -\frac{\hbar^3}{8} \operatorname{tr} \left(G_s^{11}\dot{\theta}G_s^{11}\dot{\theta} + G_s^{22}\dot{\theta}G_s^{22}\dot{\theta} - G_s^{12}\dot{\theta}G_s^{21}\dot{\theta} - G_s^{21}\dot{\theta}G_s^{12}\dot{\theta} \right)$$

$$= -\frac{\hbar}{4} \int_0^{\hbar\beta} d\tau \int d^3r \int_0^{\hbar\beta} d\tau' \int d^3r' \, \dot{\theta}(\tau)$$

$$\times \left[G_s^{11}(\mathbf{r} - \mathbf{r}', \tau - \tau')G_s^{11}(\mathbf{r}' - \mathbf{r}, \tau' - \tau) \right.$$

$$\left. - G_s^{12}(\mathbf{r} - \mathbf{r}', \tau - \tau')G_s^{12}(\mathbf{r}' - \mathbf{r}, \tau' - \tau) \right] \dot{\theta}(\tau') \qquad (5.77)$$

where we have used the following properties of $\widehat{G}_s$ (see Fetter and Walecka, 1971)

$$G_s^{11}(\mathbf{r}, \tau) = -G_s^{22}(-\mathbf{r}, -\tau) \qquad (5.78)$$

and

$$G_s^{12}(\mathbf{r}, \tau) = G_s^{21}(\mathbf{r}, \tau) \qquad (5.79)$$

We assume that $\theta(\tau)$ varies slowly on the time scale of $\hbar/\Delta_0$, which allows us to expand $\dot{\theta}(\tau')$ into a Taylor series. Keeping only the zero-order term, we have $\dot{\theta}(\tau') = \dot{\theta}(\tau)$, and Eq. (5.77) yields

$$B = -\frac{\Omega\Lambda_0\hbar}{4} \int_0^{\hbar\beta} d\tau \, [\dot{\theta}(\tau)]^2 \qquad (5.80)$$

where

$$\Lambda_0 = \int_0^{\hbar\beta} d\tau' \int d^3r' \, [G_s^{11}(\mathbf{r} - \mathbf{r}', \tau - \tau')G_s^{11}(\mathbf{r}' - \mathbf{r}, \tau' - \tau)$$

$$- G_s^{12}(\mathbf{r} - \mathbf{r}', \tau - \tau')G_s^{12}(\mathbf{r}' - \mathbf{r}, \tau' - \tau)] \qquad (5.81)$$

The constant Λ_0 can be evaluated by transforming the RHS of Eq. (5.81) back to the $(\mathbf{k}, n)$ space

$$
\begin{aligned}
\Lambda_0 &= \frac{1}{\hbar\beta\Omega} \sum_m \left\{ [G_s^{11}(\mathbf{k}, m)]^2 - [G_s^{12}(\mathbf{k}, m)]^2 \right\} \\
&= \frac{\hbar}{\beta} \sum_m \int \frac{d^3k}{(2\pi)^3} \frac{(i\hbar\omega_m + \xi)^2 - \Delta_0^2}{(\hbar^2\omega_m^2 + \xi_k^2 + \Delta_0^2)^2}
\end{aligned}
\tag{5.82}
$$

where we have used Eq. (5.74) and the expression

$$
G_s^{12}(\mathbf{k}, m) = \frac{\hbar\Delta_0}{\hbar^2\omega_m^2 + \xi_k^2 + \Delta_0^2}
\tag{5.83}
$$

Evaluating the sum over the Matsubara frequencies ω_m, Eq. (5.82) yields

$$
\Lambda_0 = -\hbar N(0) \int_{-\hbar\omega_D}^{\hbar\omega_D} d\xi \left[\frac{\Delta_0^2}{2E^3} \tanh \frac{\beta E}{2} - \frac{\beta\xi^2}{4E^2} \left(\cosh \frac{\beta E}{2} \right)^{-2} \right]
\tag{5.84}
$$

where ω_D is the Debye cutoff frequency. For $T = 0$, the expression (5.84) reduces to

$$
\Lambda_0 = -\hbar\Delta_0^2 N(0) \int_0^\infty \frac{d\xi}{(\xi^2 + \Delta_0^2)^{\frac{3}{2}}} = -\hbar N(0) = -\frac{3n\hbar}{4\epsilon_F}
\tag{5.85}
$$

where ϵ_F and n are the Fermi energy and the electron density, respectively. Introducing this result into Eq. (5.80), the term B becomes

$$
B = \frac{3n\Omega\hbar^2}{16\epsilon_F} \int_0^{\hbar\beta} d\tau \, [\dot{\theta}(\tau)]^2
\tag{5.86}
$$

Using Eq. (5.86) in the effective action (5.72), the grand partition function (5.55) becomes

$$
Z_G = \text{const} \times \int \mathcal{D}\theta(\tau) \exp\left[-\frac{3n\Omega\hbar^2}{16\epsilon_F} \int_0^{\hbar\beta} d\tau \left(\frac{\partial\theta}{\partial\tau} \right)^2 \right]
\tag{5.87}
$$

where the constant includes all contributions independent of the phase variable $\theta(\tau)$. The effective Euclidean Lagrangian, associated with the effective action in Eq. (5.87), is

$$
L = \frac{3n\Omega\hbar^2}{16\epsilon_F} \left(\frac{\partial\theta}{\partial\tau} \right)^2
\tag{5.88}
$$

This expression is in agreement with the real-time Lagrangian of a neutral superfluid particle, given in Eq. (2.65). This can be seen by noting that the density of pairs and the effective mass in Eq. (2.65) satisfy $\bar{n}_0 = n/2$ and $m^* = 2m$, respectively. The expression (5.86) can be given a simple physical interpretation. Using the expressions (3.4) and (5.85), the coefficient of the $\dot\theta^2$ can be written as $\hbar^2/4\delta$. On comparing this with the charging action (3.115), we see that the effective action (5.86) is associated with an effective charging energy $U_a^{\text{eff}} = \frac{1}{2}\delta$. For this reason, we shall refer to this contribution as the "δ effect." It should, however, be emphasized that, in contrast to the one-electron splitting (3.4), this contribution is a true many-body effect. Its origin can be traced to the energy associated with "squeezing" the quantum of a superfluid (with finite compressibility) into the small volume of the grain.

We note that Eq. (5.87) can be also derived directly by expanding $\widehat{G}^{-1}$ in terms of small phase fluctuations without the use of gauge transformation. Such an approach is similar to the functional-integral method applied to a bulk superconductor (near the transition temperature) by Rice (1967). It uses expansion of a slowly varying order parameter in a Taylor series. The first-order derivative with respect to time then yields the diffusive term, which, near the transition temperature, dominates over the wavelike term stemming from the second-order derivative (see also Abrahams and Tsuneto, 1966). If this method is applied to the problem of Josephson-coupled superconductors, the second-order term is still present, leading to an action involving terms such as (5.86). However, difficulties arise with the derivation of the dissipative part of the effective action. It is not possible to obtain a phase-periodic form of the latter. Thus, the method of gauge transformation, introduced in this section, is essential to obtain the proper gauge-invariant action for the case of coupled superconductors.

5.4.   Neutral Fermi superfluids coupled by tunneling

As a next step towards a microscopic theory of superconducting tunneling, let us consider two identical neutral superfluid particles forming a Josephson tunnel junction. The grand canonical Hamiltonian of this system is given by

$$K = K_L + K_R + \sum_{k,k'}(T_{kk'}a_{k,\sigma}^\dagger c_{k',\sigma} + \text{H.c.}) \tag{5.89}$$

where K_L and K_R are the grand canonical Hamiltonians of the particle on the left and right, respectively. They both have the form given in Eq. (5.1). The operators creating electrons on the left and right are denoted by $a_{k,\sigma}^\dagger$ and $c_{k,\sigma}^\dagger$, respectively. The last term on the RHS of Eq. (5.89) is the tunneling Hamiltonian, $T_{kk'}$ being the tunneling matrix element.

The attractive fermion–fermion interaction terms in K_L and K_R are again treated by Stratonovich transformation. We introduce the pair creation operators

$$b_{L,Q}^\dagger = \sum_k a_{k+Q,\uparrow}^\dagger a_{-k,\downarrow}^\dagger \tag{5.90}$$

and

$$b_{R,Q}^\dagger = \sum_{k'} c_{k'+Q,\uparrow}^\dagger c_{-k',\downarrow}^\dagger \tag{5.91}$$

for particles on the left and right, respectively. Then, following Eq. (5.5), the grand canonical Hamiltonian K can be written as

$$K = K_0 + H_1 \tag{5.92}$$

where

$$K_0 = \sum_{k,\sigma} \xi_k a_{k,\sigma}^\dagger a_{k,\sigma} + \sum_{k',\sigma} \xi_{k'} c_{k',\sigma}^\dagger c_{k',\sigma} + \sum_{k,k'} (T_{kk'} a_{k,\sigma}^\dagger c_{k',\sigma} + \text{H.c.}) \tag{5.93}$$

and

$$H_1 = -\frac{g}{\Omega} \sum_Q (b_{L,Q}^\dagger b_{L,Q} + b_{R,Q}^\dagger b_{R,Q}) \tag{5.94}$$

The grand partition function for this system can be written, using the steps shown in Eqs. (5.6)–(5.21), as follows

$$Z_G = \int \mathcal{D}^2 \Delta_L(\mathbf{r},\tau)\, \mathcal{D}^2 \Delta_R(\mathbf{r},\tau)$$

$$\times \text{Tr}\left[T_\tau \exp\left(-\frac{1}{\hbar} \int_0^{\hbar\beta} d\tau\, H_{\text{eff}}(\tau) \right) \right] \tag{5.95}$$

where Δ_L and Δ_R are complex functions of $\mathbf{r}$ and τ and the effective Hamiltonian is given by

$$H_{\text{eff}} = K_0 + \int_{\Omega_L} d^3 r \left\{ \psi_{L\uparrow}^\dagger[\mathbf{r},\tau] \psi_{L\downarrow}^\dagger[\mathbf{r},\tau] \Delta_L(\mathbf{r},\tau) + \text{H.c.} \right\}$$

$$+ \frac{1}{g} \int_{\Omega_L} d^3 r\, |\Delta_L(\mathbf{r},\tau)|^2 + (L \leftrightarrow R) \tag{5.96}$$

where Ω_L and Ω_R denote the volumes of integration for the left- and right-hand particles, respectively. Since the left- and right-hand field operators ψ_L and ψ_R are mixed by tunneling, it is convenient to augment the spinor field (5.22) to include *both sides* of the junction. Thus, following AES, we introduce the four-component operator

$$\widehat{\psi}(\mathbf{r}) = \begin{pmatrix} \psi_{L\uparrow}(\mathbf{r}) \\ \psi_{L\downarrow}^\dagger(\mathbf{r}) \\ \psi_{R\uparrow}(\mathbf{r}) \\ \psi_{R\downarrow}^\dagger(\mathbf{r}) \end{pmatrix} \tag{5.97}$$

and its Hermitian adjoint $\widehat{\psi}^\dagger(\mathbf{r})$. With use of these operators, the effective Hamiltonian (5.96) can be expressed as follows

$$H_{\text{eff}}(\tau) = \int_{\Omega_L} d^3 r \int_{\Omega_R} d^3 r'\, \widehat{\psi}^\dagger(\mathbf{r}) \underline{\widehat{k}}_0 \widehat{\psi}(\mathbf{r}') + \int_{\Omega_L + \Omega_R} d^3 r\, \widehat{\psi}^\dagger[\mathbf{r},\tau] \underline{\widehat{D}}(\mathbf{r},\tau) \widehat{\psi}[\mathbf{r},\tau]$$

$$+ \frac{1}{g} \left(\int_{\Omega_L} d^3 r\, |\Delta_L(\mathbf{r},\tau)|^2 + (L \leftrightarrow R) \right) \tag{5.98}$$

where

$$\widehat{\underline{k}}_0 = \begin{pmatrix} k_0 & 0 & T(\mathbf{r}, \mathbf{r}') & 0 \\ 0 & -k_0 & 0 & -T^*(\mathbf{r}, \mathbf{r}') \\ T^*(\mathbf{r}, \mathbf{r}') & 0 & k_0 & 0 \\ 0 & -T(\mathbf{r}, \mathbf{r}') & 0 & -k_0 \end{pmatrix} \tag{5.99}$$

The diagonal matrix elements are the single-particle kinetic energy operators

$$k_0 = \delta(\mathbf{r} - \mathbf{r}') \left(-\frac{\hbar^2}{2m} \nabla^2 - \mu \right) \tag{5.100}$$

and the off-diagonal ones are defined by Fourier transforming the tunnel matrix elements $T_{kk'}$ to real space. The matrix $\widehat{\underline{D}}$ is given by

$$\widehat{\underline{D}} = \begin{pmatrix} 0 & \Delta_0 e^{i\theta_L(\tau)} & 0 & 0 \\ \Delta_0 e^{-i\theta_L(\tau)} & 0 & 0 & 0 \\ 0 & 0 & 0 & \Delta_0 e^{i\theta_R(\tau)} \\ 0 & 0 & \Delta_0 e^{-i\theta_R(\tau)} & 0 \end{pmatrix} \tag{5.101}$$

where Δ_0 is again assumed independent of $\mathbf{r}$ and τ. The second term on the RHS of Eq. (5.98) is now treated as a perturbation acting on the free-electron system described by the matrix $\widehat{\underline{k}}_0$. Using the cumulant expansion theorem of Appendix A16, we can express the grand partition function (5.95) as

$$Z_G = Z_{G_0} \int \mathcal{D}^2 \Delta_L(\mathbf{r}, \tau) \, \mathcal{D}^2 \Delta_R(\mathbf{r}, \tau) \exp\left(-\frac{1}{\hbar} S_{\text{eff}}[\Delta_L, \Delta_R] \right) \tag{5.102}$$

The effective action is given by

$$S_{\text{eff}}[\Delta_L, \Delta_R] = \frac{1}{g} \int_0^{\hbar\beta} d\tau \left[\int_{\Omega_L} d^3r \, |\Delta_L(\mathbf{r}, \tau)|^2 + \int_{\Omega_R} d^3r \, |\Delta_R(\mathbf{r}, \tau)|^2 \right]$$
$$- \hbar \, \text{Tr} \left(\ln \widehat{\underline{G}}_0 + \ln(\widehat{\underline{G}}^{-1}) \right) \tag{5.103}$$

where $\widehat{\underline{G}}_0$ and $\widehat{\underline{G}}$ are 4×4 matrix Green's functions. They satisfy the equations of motion

$$\left(-\hbar \frac{\partial}{\partial \tau} - \widehat{\underline{k}}_0 \right) \widehat{\underline{G}}_0(\mathbf{r}, \tau; \mathbf{r}', \tau') = \hbar \, \delta(\mathbf{r} - \mathbf{r}') \, \delta(\tau - \tau') \tag{5.104}$$

and

$$\left(-\hbar \frac{\partial}{\partial \tau} - \widehat{\underline{k}}_0 - \widehat{\underline{D}} \right) \widehat{\underline{G}}(\mathbf{r}, \tau; \mathbf{r}', \tau') = \hbar \, \delta(\mathbf{r} - \mathbf{r}') \, \delta(\tau - \tau') \tag{5.105}$$

Similar to the previous section, we perform a gauge transformation, to eliminate the phase factors on both sides of the junction

$$\widehat{\psi}(\mathbf{r}, \tau) \rightarrow \widehat{U}\widehat{\psi}(\mathbf{r}, \tau) \tag{5.106}$$

where

$$\widehat{U} = \begin{pmatrix} e^{-i\theta_L(\tau)/2} & 0 & 0 & 0 \\ 0 & e^{i\theta_L(\tau)/2} & 0 & 0 \\ 0 & 0 & e^{-i\theta_R(\tau)/2} & 0 \\ 0 & 0 & 0 & e^{i\theta_R(\tau)/2} \end{pmatrix} \tag{5.107}$$

The gauge-transformed Green's function $\widehat{G}'$ satisfies an equation of motion that is obtained by generalizing Eq. (5.65) to the combined left- and right-hand space

$$\left(-\hbar\frac{\partial}{\partial\tau} - \widetilde{\underline{k}}_0' - \widehat{\underline{D}}_0 - \widehat{\underline{D}}_1(\tau) \right)\widehat{\underline{G}}'(\mathbf{r}, \tau; \mathbf{r}', \tau') = \hbar\,\delta(\mathbf{r} - \mathbf{r}')\,\delta(\tau - \tau') \tag{5.108}$$

where

$$\widehat{\underline{D}}_0 = \widehat{U}\,\underline{\widehat{D}}\,\widehat{U}^{-1} = \begin{pmatrix} 0 & \Delta_0 & 0 & 0 \\ \Delta_0 & 0 & 0 & 0 \\ 0 & 0 & 0 & \Delta_0 \\ 0 & 0 & \Delta_0 & 0 \end{pmatrix} \tag{5.109}$$

and

$$\widetilde{\underline{k}}_0' = \widehat{U}\widehat{\underline{k}}_0\widehat{U}^{-1}$$

$$= \begin{pmatrix} k_0 & 0 & Te^{i\theta_{RL}/2} & 0 \\ 0 & -k_0 & 0 & -T^*e^{-i\theta_{RL}/2} \\ T^*e^{-i\theta_{RL}/2} & 0 & k_0 & 0 \\ 0 & -Te^{i\theta_{RL}/2} & 0 & -k_0 \end{pmatrix} \tag{5.110}$$

where

$$\theta_{RL} = \theta_R(\tau) - \theta_L(\tau) \quad \text{and} \quad T = T(\mathbf{r}, \mathbf{r}') \tag{5.111}$$

The matrix $\widehat{\underline{D}}_1$ is a 4×4 generalization of the $\widehat{D}_1$ matrix, defined in Eq. (5.65).

$$\widehat{\underline{D}}_1(\tau) = \frac{i}{2}\begin{pmatrix} \hbar\dot{\theta}_L & 0 & 0 & 0 \\ 0 & -\hbar\dot{\theta}_L & 0 & 0 \\ 0 & 0 & \hbar\dot{\theta}_R & 0 \\ 0 & 0 & 0 & -\hbar\dot{\theta}_R \end{pmatrix} \tag{5.112}$$

Considering $\widehat{\underline{D}}_1$ and the tunneling part of (5.110) as a perturbation, we now expand $(\widehat{\underline{G}}')^{-1}$ about $\widehat{\underline{G}}_s^{-1}$. $\widehat{\underline{G}}_s$ satisfies the equation of motion

$$\left(-\hbar\frac{\partial}{\partial\tau} - \widehat{\underline{k}}_0(T=0) - \widehat{D}_0\right)\widehat{G}_s(\mathbf{r},\tau;\mathbf{r}',\tau') = \hbar\,\delta(\mathbf{r}-\mathbf{r}')\,\delta(\tau-\tau') \quad (5.113)$$

Thus $\widehat{G}_s$ describes two uncoupled particles, each with an order parameter fixed to a real constant Δ_0. Using Eq. (5.113), the differential equation (5.108) can be transformed to an integral equation

$$\begin{aligned}
\widehat{\underline{G}}'(\mathbf{k},n;\mathbf{k}',n') = {}& \delta_{k,k'}\delta_{n,n'}\widehat{\underline{G}}_s(\mathbf{k},n) \\
& + \sum_{\mathbf{k}'',n''} \widehat{\underline{G}}_s(\mathbf{k},n)\widehat{\underline{V}}_1(\mathbf{k}-\mathbf{k}'',n-n'') \\
& \times \widehat{\underline{G}}'(\mathbf{k}'',n'';\mathbf{k}',n')
\end{aligned} \quad (5.114)$$

where

$$\widehat{\underline{V}}_1(\mathbf{k}-\mathbf{k}'',n-n'') = \widehat{\underline{D}}_1(n-n'') + \widehat{\underline{T}}_1(\mathbf{k}-\mathbf{k}'',n-n'') \quad (5.115)$$

$\widehat{\underline{T}}_1$ is a 4×4, gauge-transformed tunneling matrix, given by the off-diagonal elements of the matrix (5.110). Equation (5.114) can be written as a matrix equation

$$(\widehat{\underline{G}}')^{-1} = \widehat{\underline{G}}_s^{-1} - \widehat{\underline{V}}_1 \quad (5.116)$$

where the matrices are now defined in the $(\mathbf{k},n,\sigma)$ space, with the variable σ denoting the 4×4 space. In analogy to Eq. (5.71), we obtain, from Eq. (5.116), the following expansion to order $\widehat{\underline{V}}_1^2$

$$\mathrm{Tr}\left\{\ln(\widehat{\underline{G}}')^{-1}\right\} = \mathrm{Tr}\left(\ln\widehat{\underline{G}}_s^{-1}\right) - \mathrm{Tr}\left(\widehat{\underline{G}}_s\widehat{\underline{V}}_1\right) - \frac{1}{2}\mathrm{Tr}\left(\widehat{\underline{G}}_s\widehat{\underline{V}}_1\right)^2 \quad (5.117)$$

The matrix $\widehat{\underline{G}}_s$ has, according to Eq. (5.113), the form

$$\widehat{\underline{G}}_s = \begin{pmatrix} G_L^{11} & G_L^{12} & 0 & 0 \\ G_L^{21} & G_L^{22} & 0 & 0 \\ 0 & 0 & G_R^{11} & G_R^{12} \\ 0 & 0 & G_R^{21} & G_R^{22} \end{pmatrix} \quad (5.118)$$

where G_L^{ij} and G_R^{ij} are the Gorkov Green's functions written down in Eqs. (5.74) and (5.83). Using Eqs. (5.112) and (5.118), we obtain with help of Eq. (5.75)

$$\mathrm{Tr}\left(\widehat{\underline{G}}_s\widehat{\underline{D}}_1\right) = 0 \quad (5.119)$$

Moreover, the form of the matrix $\widehat{T}_1$ implies that $\mathrm{Tr}(\widehat{\underline{G}}_s\widehat{\underline{T}}_1) = 0$. Hence, the first-order term in $\widehat{V}_1$ on the RHS of Eq. (5.117) is

$$\mathrm{Tr}\left(\widehat{\underline{G}}_s\widehat{\underline{V}}_1\right) = 0 \tag{5.120}$$

The second-order term can be written

$$\mathrm{Tr}\left(\widehat{\underline{G}}_s\widehat{\underline{V}}_1\right)^2 = \mathrm{Tr}\left(\widehat{\underline{G}}_s\widehat{\underline{D}}_1\right)^2 + \mathrm{Tr}\left(\widehat{\underline{G}}_s\widehat{\underline{T}}_1\right)^2 \tag{5.121}$$

On the RHS of this equation, we have used the fact that the σ trace of cross-products, such as $\widehat{\underline{G}}_s\widehat{\underline{D}}_1\widehat{\underline{G}}_s\widehat{\underline{T}}_1$, vanishes. The evaluation of the first term on the RHS of Eq. (5.121) closely resembles that given in Section 5.3 for the case of single particle. Using Eqs. (5.112) and (5.118), we obtain with the help of Eqs. (5.77) and (5.86)

$$B_D = \frac{\hbar}{2}\,\mathrm{Tr}\left(\widehat{\underline{G}}_s\widehat{\underline{D}}_1\right)^2 = \frac{3n\Omega\hbar^2}{16\epsilon_F}\int_0^{\hbar\beta}d\tau\,(\dot{\theta}_L^2 + \dot{\theta}_R^2) \tag{5.122}$$

The term in (5.121) that is second order in $\widehat{T}_1$ becomes, after performing the σ trace and using the cyclic invariance of the trace

$$\begin{aligned}
B_T &= \frac{\hbar}{2}\,\mathrm{Tr}\left(\widehat{\underline{G}}_s\widehat{\underline{T}}_1\right)^2 \\
&= \hbar\,\mathrm{tr}\,[G_L^{11}T'G_R^{11}(T')^* - G_L^{12}(T')^*G_R^{21}(T')^* \\
&\quad + G_L^{22}(T')^*G_R^{22}T' - G_L^{21}T'G_R^{12}T']
\end{aligned} \tag{5.123}$$

where T' is the Fourier transform of matrix element $(\widehat{\underline{k}}_0')_{13}$, which, according to Eq. (5.110), is

$$T'(\mathbf{k}-\mathbf{k}',n-n') = \frac{T_{kk'}}{\hbar^2\beta}\int_0^{\hbar\beta}d\tau\,\exp\left(i(\omega_n - \omega_{n'})\tau + \frac{i}{2}\theta_{RL}(\tau)\right) \tag{5.124}$$

where the extra $\hbar$ factor can be traced to Eqs. (A17.9) and (A17.13).

To illustrate the method of evaluation of the expression (5.123), we consider, in some detail, the first term. Fourier transforming the frequency trace, we obtain, using Eq. (5.124)

$$\begin{aligned}
\mathrm{tr}\,[G_L^{11}T'G_R^{11}(T')^*] &= \frac{1}{\hbar^2}\int_0^{\hbar\beta}d\tau\int_0^{\hbar\beta}d\tau'\sum_{k,k',k'',k'''}G_L^{11}(\mathbf{k},\mathbf{k}';\tau-\tau')T_{k'k''} \\
&\quad\times e^{i\theta_{RL}(\tau')/2}G_R^{11}(\mathbf{k}'',\mathbf{k}''';\tau'-\tau)T_{k'''k}^*e^{-i\theta_{RL}(\tau)/2} \\
&= \frac{1}{\hbar^2}\int_0^{\hbar\beta}d\tau\int_0^{\hbar\beta}d\tau'\sum_{k,k'}G_L^{11}(\mathbf{k},\tau-\tau')e^{i\theta_{RL}(\tau')/2} \\
&\quad\times G_R^{11}(\mathbf{k}',\tau'-\tau)e^{-i\theta_{RL}(\tau)/2}|T_{kk'}|^2
\end{aligned} \tag{5.125}$$

where we have used the property of translational invariance for the Green's functions

$$G_i^{11}(\mathbf{k}, \mathbf{k}'; \tau) = \delta_{\mathbf{k}, \mathbf{k}'} G_i^{11}(\mathbf{k}, \tau), \qquad i = L, R \qquad (5.126)$$

which holds approximately, as long as the finite-size (single-electron) effects of the small particles do not play a significant role. On the RHS of Eq. (5.125), the summations over $\mathbf{k}$ and $\mathbf{k}'$ belong to the left- and right-hand particles, respectively. Working out in a similar way the remaining terms in Eq. (5.123), we obtain

$$B_T = \frac{1}{\hbar} |\mathcal{T}|^2 \int_0^{\hbar\beta} d\tau \int_0^{\hbar\beta} d\tau' \int \frac{d^3k}{(2\pi)^3} \int \frac{d^3k'}{(2\pi)^3}$$

$$\times \left[G_L^{11}(k, \tau - \tau') G_R^{11}(k', \tau' - \tau) \exp\left(-\frac{i}{2} [\theta_{RL}(\tau) - \theta_{RL}(\tau')] \right) \right.$$

$$- G_L^{12}(k, \tau - \tau') G_R^{12}(k', \tau' - \tau) \exp\left(-\frac{i}{2} [\theta_{RL}(\tau) + \theta_{RL}(\tau')] \right)$$

$$\left. + (L \leftrightarrow R) \right] \qquad (5.127)$$

where

$$|\mathcal{T}|^2 = \Omega^2 \langle |T_{kk'}|^2 \rangle \qquad (5.128)$$

The average on the RHS of Eq. (5.128) is over the Fermi surfaces of the particles. In Eq. (5.127) we have used the properties

$$G_i^{22}(\mathbf{k}, \tau) = -G_i^{11}(\mathbf{k}, -\tau), \qquad i = L, R \qquad (5.129)$$

$$G_i^{21}(\mathbf{k}, \tau) = G_i^{12}(\mathbf{k}, \tau), \qquad i = L, R \qquad (5.130)$$

which follow from Eqs. (5.78) and (5.79) by Fourier transforming the $\mathbf{r}$ variable.

From Eqs. (5.103), (5.120), (5.121), and (5.127), it follows that the phase-dependent part of the effective action is given by

$$S_{\text{eff}} \to S_{\text{eff}} = B_D + B_T$$

$$= \frac{3n\Omega\hbar^2}{16\epsilon_F} \int_0^{\hbar\beta} d\tau \left(\dot{\theta}_L^2 + \dot{\theta}_R^2 \right) - \int_0^{\hbar\beta} d\tau \int_0^{\hbar\beta} d\tau'$$

$$\times \left(\alpha(\tau - \tau') \cos \frac{\theta_{RL}(\tau) - \theta_{RL}(\tau')}{2} \right.$$

$$\left. - \beta(\tau - \tau') \cos \frac{\theta_{RL}(\tau) + \theta_{RL}(\tau')}{2} \right) \qquad (5.131)$$

where

$$\alpha(\tau) = -\frac{2}{\hbar} |\mathcal{T}|^2 \int \frac{d^3k}{(2\pi)^3} G_L^{11}(k, \tau) \int \frac{d^3k'}{(2\pi)^3} G_R^{11}(k', -\tau) \qquad (5.132)$$

and

$$\beta(\tau) = -\frac{2}{\hbar}|\mathcal{T}|^2 \int \frac{d^3k}{(2\pi)^3} G_L^{12}(k,\tau) \int \frac{d^3k'}{(2\pi)^3} G_R^{12}(k',-\tau) \qquad (5.133)$$

The terms in Eq. (5.131) involving $\alpha(\tau)$ and $\beta(\tau)$ correspond to the dissipative action and Josephson tunneling, respectively. The phase-inertial terms are diagonal in (left–right) position. However, as shown in Eq. (5.167), the quasiparticle tunneling generates an off-diagonal phase-inertial term proportional to $\dot{\theta}_{RL}^2$. In Section 5.5 we discuss the conditions under which the diagonal terms may dominate, leading to a realization of an effective "self-charging" model.

5.4.1. *Josephson coupling energy*

The foregoing theory allows us to give a microscopic derivation of the Josephson coupling constant J [defined phenomenologically in Eq. (2.84)]. The part of the effective action (5.131) responsible for Josephson coupling is

$$S_J[\Delta_L, \Delta_R] = \int_0^{\hbar\beta} d\tau \int_0^{\hbar\beta} d\tau' \, \beta(\tau - \tau') \cos \frac{\theta_{RL}(\tau) + \theta_{RL}(\tau')}{2} \qquad (5.134)$$

For a symmetric junction, we have $G_L^{12} = G_R^{12}$, so that the function $\beta(\tau)$, defined by Eq. (5.133), can be written

$$\beta(\tau) = -\frac{2}{\hbar}|\mathcal{T}|^2 f(\tau) f(-\tau) \qquad (5.135)$$

where

$$f(\tau) = N(0) \int_{-\hbar\omega_D}^{\hbar\omega_D} d\xi \, G_L^{12}(k,\tau) \qquad (5.136)$$

Fourier transforming Eq. (5.83), we obtain

$$\begin{aligned}
G_L^{12}(k,\tau) &= \frac{1}{\hbar\beta} \sum_m \frac{\hbar\Delta_0 e^{-i\omega_m \tau}}{\hbar^2\omega_m^2 + \xi_k^2 + \Delta_0^2} \\
&= -\frac{\Delta_0}{2E_k}\left\{ \Theta(\tau)\left[ne^{E_k\tau/\hbar} + (n-1)e^{-E_k\tau/\hbar}\right] \right. \\
&\quad \left. + \Theta(-\tau)\left[ne^{-E_k\tau/\hbar} + (n-1)e^{E_k\tau/\hbar}\right] \right\}
\end{aligned} \qquad (5.137)$$

where Θ is a unit step function and $n = n(E_k)$ is the Fermi function. In what follows, we confine ourselves to temperatures well below the BCS bulk transition temperature, in which case we have

$$n(E_k) \simeq \Theta(-E_k) \qquad (5.138)$$

and

$$1 - n(E_k) \simeq \Theta(E_k) \tag{5.139}$$

Using Eqs. (5.137)–(5.139), the expression (5.136) yields

$$f(\tau) = F(\tau) + F(-\tau) \tag{5.140}$$

where

$$F(\tau) \simeq \Theta(\tau)N(0)\Delta_0 \int_{\Delta_0}^{\infty} dE\,(E^2 - \Delta_0^2)^{-\frac{1}{2}} e^{-E\tau/\hbar}$$

$$= \Theta(\tau)N(0)\Delta_0 K_0(\Delta_0\tau/\hbar) \tag{5.141}$$

Since $\hbar\omega_D \gg \Delta_0$, the upper limit of the energy integration has been extended to infinity. The function $K_0(x)$ is the modified Bessel function of the second kind of order zero. Introducing this result into Eq. (5.140), we obtain from Eq. (5.135)

$$\beta(\tau) = -\frac{2|\mathcal{T}|^2}{\hbar} N^2(0)\Delta_0^2 \big[\theta(\tau)K_0^2(\Delta_0\tau/\hbar) + \theta(-\tau)K_0^2(-\Delta_0\tau/\hbar)\big] \tag{5.142}$$

Since $K_0(x)$ diverges logarithmically for small x, the function $\beta(\tau - \tau')$ has a sharp maximum for $|\tau - \tau'| = 0$. This allows us to approximate the integration in Eq. (5.134) as follows

$$S_J[\Delta_L, \Delta_R] \simeq \int_0^{\hbar\beta} d\tau\,\cos\theta_{RL}(\tau) \int_0^{\hbar\beta} d\tau'\,\beta(\tau - \tau')$$

$$\simeq \int_0^{\hbar\beta} d\tau\,\cos\theta_{RL}(\tau)\beta(\omega_m = 0)$$

$$= -J \int_0^{\hbar\beta} d\tau\,\cos\theta_{RL}(\tau) \tag{5.143}$$

where

$$\beta(\omega_m) = \int_0^{\hbar\beta} d\tau\,\beta(\tau)e^{i\omega_m\tau} \tag{5.144}$$

Performing the integral on the RHS of this equation with use of Eqs. (5.135)–(5.137), we have

$$J = -\beta(\omega_m = 0)$$

$$= \frac{2}{\beta}|\mathcal{T}|^2 N^2(0)\Delta_0^2 \sum_n \left(\int_{-\hbar\omega_D}^{\hbar\omega_D} \frac{d\xi}{\hbar^2\omega_n^2 + \Delta_0^2 + \xi^2} \right)^2$$

$$\simeq \frac{2\pi^2}{\beta}|\mathcal{T}|^2 N^2(0)\Delta_0^2 \sum_n \frac{1}{\hbar^2\omega_n^2 + \Delta_0^2} \tag{5.145}$$

where we have replaced the Debye cutoff energies by infinity. Performing the Matsubara summation on the RHS of this equation, we obtain the following expression for the Josephson coupling energy at finite temperatures (Ambegaokar and Baratoff, 1963)

$$J(T) \simeq \pi^2 |\mathcal{T}|^2 N^2(0) \Delta_0(T) \tanh \frac{\beta \Delta_0(T)}{2} \tag{5.146}$$

At low temperatures, such that $k_B T \ll \Delta_0(T)$, this equation simplifies to

$$J(T) \simeq \pi^2 |\mathcal{T}|^2 N^2(0) \Delta_0(T) = \frac{\pi \hbar \Delta_0(T)}{4e^2 R_N} \tag{5.147}$$

where R_N is the tunneling resistance of a symmetric junction in the *normal state*, given by

$$\frac{1}{R_N} = \frac{4\pi e^2}{\hbar} |\mathcal{T}|^2 N^2(0) \tag{5.148}$$

We note that expressions (5.147) and (5.148) have been used in Chapter 3 in connection with the problem of phase ordering in granular arrays of junctions.

5.4.2. *Calculation of kernel $\alpha(\tau)$*

For a symmetric junction the expression (5.132) can be written

$$\alpha(\tau) = -\frac{2}{\hbar} |\mathcal{T}|^2 \bar{f}(\tau) \bar{f}(-\tau) \tag{5.149}$$

where

$$\bar{f}(\tau) = N(0) \int_{-\hbar\omega_D}^{\hbar\omega_D} d\xi \, G_L^{11}(k, \tau) \tag{5.150}$$

Fourier transforming the expression (5.74), we obtain

$$
\begin{aligned}
G_L^{11}(k, \tau) &= G_s^{11}(k, \tau) \\
&= \frac{1}{2} e^{-E_k \tau/\hbar} \left(1 + \frac{\xi_k}{E_k}\right) [n(E_k) - \Theta(\tau)] \\
&\quad - \frac{1}{2} e^{E_k \tau/\hbar} \left(1 - \frac{\xi_k}{E_k}\right) [n(E_k) - \Theta(-\tau)]
\end{aligned}
\tag{5.151}
$$

We assume again that the temperature is well below the bulk transition temperature, so that the Fermi function can be approximated as shown in Eqs. (5.138) and (5.139). Using Eq. (5.151), Eq. (5.150) yields

$$\bar{f}(\tau) = \bar{F}(\tau) - \bar{F}(-\tau) \tag{5.152}$$

where

$$
\begin{aligned}
\bar{F}(\tau) &\simeq -\Theta(\tau) N(0) \int_{\Delta_0}^{\infty} dE \, (E^2 - \Delta_0^2)^{-\frac{1}{2}} E e^{-E\tau/\hbar} \\
&= -\Theta(\tau) N(0) \Delta_0 K_1(\Delta_0 \tau/\hbar)
\end{aligned}
\tag{5.153}
$$

where $K_1(x)$ is the modified Bessel function of the second kind of first order. Introducing this result into Eq. (5.149), the kernel $\alpha(\tau)$ (at $T \simeq 0$) is given by the expression

$$\alpha(\tau) = \frac{2}{\hbar}|\mathcal{J}|^2 N^2(0)\Delta_0^2 K_1^2(\Delta_0|\tau|/\hbar) = \frac{\Delta_0^2 K_1^2(\Delta_0|\tau|/\hbar)}{2\pi e^2 R_N} \qquad (5.154)$$

where we have used the expression (5.148) for the normal tunneling resistance. Using the asymptotic forms of the modified Bessel function K_1, we obtain from Eq. (5.154)

$$\alpha(\tau) = \frac{\hbar^2}{2\pi e^2 R_N}\begin{cases} 1/\tau^2, & \text{for } |\tau| \ll \hbar/\Delta_0 \\ \frac{\pi}{2}\left(\frac{\Delta_0}{\hbar}\right)^2 \frac{\hbar}{\Delta_0|\tau|}e^{-2\Delta_0|\tau|/\hbar}, & \text{for } |\tau| \gg \hbar/\Delta_0 \end{cases} \qquad (5.155)$$

From Eq. (5.155) we see that the ohmiclike kernel $\alpha(\tau) \sim 1/\tau^2$ is obtained only for times τ short on the scale of the inverse gap. According to Eqs. (4.40)–(4.43), a true ohmic dissipative action is obtained if the Fourier transform of $\alpha(\tau)$ defined in Eq. (4.41) as $J^{\mathrm{ohmic}}(\omega)$ is proportional to ω for small ω. Such behavior is guaranteed only if $\alpha(\tau)$ remains proportional to $1/\tau^2$ for $\tau \to \infty$. Equation (5.155) suggests that this is only possible if $\Delta_0 = 0$. This is a formal restatement of the physical fact, that a real (energy-conserving) transfer of quasiparticles (at zero temperature) is possible only between *normal* Fermi systems. For coupled superconductors with a nonzero superconducting gap, quasiparticles can be formed only by breaking Cooper pairs, which requires an energy of at least $2\Delta_0$. In the presence of a pair-breaking mechanism, caused either by magnetic impurities (de Gennes, 1966) or by fluctuations of the order parameter in very small superconductors, the normal quasiparticle conductance channel may still exist down to zero temperature. We note that, for zero temperature and nonzero gap, there are still *virtual* (energy-nonconserving) processes in which quasiparticles tunnel temporarily back and forth across the junction. In case of charged Fermi systems, these processes generate virtual electric dipoles, which tend to enhance the capacitance of the junction, as shown by Eckern et al. (1984). A useful way to picture this effect is to recall the uncertainty relation (2.62) between the phase variable and the number of particles transferred across the junction. Virtual transfers of quasiparticles lead to increased fluctuations in the number, causing a reduction of the phase fluctuation. In terms of the path-integral formalism, such *mitigation* of quantum phase fluctuation is provided by a *phase-inertial* term, which is proportional to the square of the time derivative of the phase difference across the junction. As shown in the following section, this term is formally equivalent to that produced by the electrostatic charging energy of the junction capacitance.

5.4.3. *Phase inertia induced by tunneling*

Let us now consider the part of the effective action (5.131) involving the kernel $\alpha(\tau)$

$$S_D^{(\alpha)}[\Delta_L, \Delta_R] = -\int_0^{\hbar\beta} d\tau \int_0^{\hbar\beta} d\tau'\, \alpha(\tau - \tau') \cos \frac{\theta_{RL}(\tau) - \theta_{RL}(\tau')}{2} \quad (5.156)$$

At zero temperature, the RHS of this equation yields a truly dissipative action only when the gap vanishes. For a nonzero gap, it is possible to reduce Eq. (5.156) to a τ integral of the square of the time derivative of the phase difference. Following Eckern et al. (1984), we now derive this effective inertial term.

Assuming that the phase varies slowly on the scale of the inverse gap, we expand the cosine in Eq. (5.156), yielding

$$S_D^{(\alpha)}[\Delta_L, \Delta_R] \simeq -\int_0^{\hbar\beta} d\tau \int_0^{\hbar\beta} d\tau'\, \alpha(\tau - \tau') \left\{ 1 - \frac{1}{8}[\theta_{RL}(\tau) - \theta_{RL}(\tau')]^2 \right\}$$

$$\simeq \frac{1}{8} \int_0^{\hbar\beta} d\tau\, \dot\theta_{RL}^2(\tau)$$

$$\times \int_0^{\hbar\beta} d\tau'\, \alpha(\tau - \tau')(\tau - \tau')^2 + \text{const} \quad (5.157)$$

The integration over the imaginary time τ' can be done as follows. We introduce a new variable $u = \tau - \tau'$, where $u \in (\tau, \tau - \hbar\beta)$, and obtain

$$I(\tau) = \int_0^{\hbar\beta} d\tau'\, \alpha(\tau - \tau')(\tau - \tau')^2 = \int_{\tau - \hbar\beta}^{\tau} du\, \alpha(u)u^2 \quad (5.158)$$

The integral $I(\tau)$ depends weakly on τ. Since the integrand is an even function of u, we have $I(0) = I(\hbar\beta)$. A reasonable estimate of $I(\tau)$ is obtained by taking $\tau = \hbar\beta/2$, in which case Eq. (5.158) yields, for $T \to 0$

$$I(\tau) \sim I(\hbar\beta/2) \xrightarrow[T\to 0]{} 2\int_0^{\infty} du\, \alpha(u)u^2 \quad (5.159)$$

Introducing the expression (5.154) into (5.159) and substituting the resulting $I(\tau)$ into Eq. (5.157), we obtain

$$S_D^{(\alpha)}[\Delta_L, \Delta_R] = \left(\frac{1}{2\hbar}|\mathcal{J}|^2 N^2(0)\Delta_0^2 \int_0^{\infty} du\, K_1^2(\Delta_0 u/\hbar)u^2 \right) \int_0^{\hbar\beta} d\tau\, \dot\theta_{RL}^2(\tau)$$

$$= \frac{1}{2} I_{\text{eff}} \int_0^{\hbar\beta} d\tau\, \dot\theta_{RL}^2(\tau) \quad (5.160)$$

where I_{eff} is the effective "moment of inertia" for the phase variable θ_{RL}, given by

$$I_{\text{eff}} = \frac{3\pi^2\hbar^2}{32\Delta_0}|\mathcal{J}|^2 N^2(0) \quad (5.161)$$

We can now summarize the results obtained for the effective action of coupled neutral superfluid particles near $T = 0$. From Eqs. (5.131), (5.143), and (5.160), we have

$$S_{\text{eff}}[\Delta_L, \Delta_R] = \frac{3n\Omega\hbar^2}{16\epsilon_F} \int_0^{\hbar\beta} d\tau \left[\dot{\theta}_L^2(\tau) + \dot{\theta}_R^2(\tau)\right] + \frac{1}{2} I_{\text{eff}} \int_0^{\hbar\beta} d\tau \, \dot{\theta}_{RL}^2(\tau)$$
$$- J \int_0^{\hbar\beta} d\tau \, \cos\theta_{RL}(\tau) \tag{5.162}$$

The last two terms on the RHS of this equation are also present in the case of Josephson-coupled charged superfluid, except that the effective moment of inertia is replaced by a quantity proportional to junction capacitance.

5.5.   Josephson junction with charging energy

Let us now consider a model of two identical Josephson-coupled superconductors with the charging energy, as investigated originally by Ambegaokar et al. (1982). The grand canonical Hamiltonian of this system is obtained by adding to Eq. (5.89) the charging Hamiltonian

$$H_c = \frac{1}{8C}(Q_L - Q_R)^2 \tag{5.163}$$

where C is the capacitance of the junction and $Q_{L(R)}$ is the operator for the charge on the left (right) superconductors, which can be written as

$$Q_L = e \int_{\Omega_L} d^3r \, \psi_{L\sigma}^\dagger(\mathbf{r})\psi_{L\sigma}(\mathbf{r}) \tag{5.164}$$

A similar expression holds for the operator Q_R. Thus the grand canonical Hamiltonian is, according to Eqs. (5.92)–(5.94) and (5.163), given by

$$K = K_0 + H_1 \tag{5.165}$$

where K_0 is given by Eq. (5.93), and H_1 is a perturbation due to the attractive pairing plus the charging energy

$$H_1 = -\frac{g}{\Omega} \sum_Q \left(b_{L,Q}^\dagger b_{L,Q} + b_{R,Q}^\dagger b_{R,Q}\right) + H_c \tag{5.166}$$

According to Eqs. (5.6)–(5.9), the grand partition function can be written as

$$Z_G = \text{Tr}\left[e^{-\beta K_0}\sigma(\beta)\right] \tag{5.167}$$

where $\sigma(\beta)$ can be written, with the help of Eq. (5.15), as

$$\sigma(\beta) = T_\tau \exp\left(-\frac{1}{\hbar}\int_0^{\hbar\beta} d\tau \, H_1[\tau]\right)$$

$$= \lim_{\Delta\tau \to 0} T_\tau \exp \left(\frac{g\Delta\tau}{\hbar\Omega} \sum_{Q,\tau_i} \left\{ b^\dagger_{L,Q}[\tau_i] b_{L,Q}[\tau_i] + (L \leftrightarrow R) \right\} \right.$$

$$\left. - \frac{\Delta\tau}{8\hbar C} \sum_{\tau_j} (Q_L[\tau_j] - Q_R[\tau_j])^2 \right) \tag{5.168}$$

The first sum in the exponent of this equation is treated by the same Stratonovich transformation as shown in Eqs. (5.15)–(5.19). The transformation of the second sum is performed using Eq. (5.11). Thus we obtain for the jth summand an identity

$$\exp \left(-\frac{\Delta\tau}{8C\hbar} (Q_L[\tau_j] - Q_R[\tau_j])^2 \right)$$

$$= \left(\frac{\Delta\tau C}{2\pi\hbar} \right)^{\frac{1}{2}} \int_{-\infty}^{\infty} dV(\tau_j) \exp \left[-\frac{\Delta\tau}{\hbar} \left(\frac{1}{2} CV^2(\tau_j) \right. \right.$$

$$\left. \left. - \frac{i}{2} V(\tau_j)(Q_L[\tau_j] - Q_R[\tau_j]) \right) \right] \tag{5.169}$$

where $V(\tau_j)$ is a real Stratonovich (random) potential across the junction. Introducing this identity into Eq. (5.168), we obtain with use of Eqs. (5.95)–(5.98) and (5.167)

$$Z_G = \int \mathcal{D}^2 \Delta_L(\mathbf{r},\tau) \, \mathcal{D}^2 \Delta_R(\mathbf{r},\tau) \, \mathcal{D} V(\tau)$$

$$\times \, \text{Tr} \left[T_\tau \exp \left(-\frac{1}{\hbar} \int_0^{\hbar\beta} d\tau \, H_{\text{eff}}(\tau) \right) \right] \tag{5.170}$$

where

$$H_{\text{eff}} = \int_{\Omega_L + \Omega_R} d^3r \, \widehat{\psi}^\dagger(\mathbf{r}) \underline{\widehat{k}}_0 \widehat{\psi}(\mathbf{r}) + \int d^3r \, \widehat{\psi}^\dagger[\mathbf{r},\tau] \left[\underline{\widehat{D}}(\mathbf{r},\tau) + \underline{\widehat{M}}(\tau) \right] \widehat{\psi}[\mathbf{r},\tau]$$

$$+ \frac{1}{g} \left(\int_{\Delta_L} d^3r \, |\Delta_L(\mathbf{r},\tau)|^2 + \int_{\Delta_R} d^3r \, |\Delta_R(\mathbf{r},\tau)|^2 \right)$$

$$+ \frac{1}{2} CV^2(\tau) \tag{5.171}$$

where $\widehat{\psi}(\mathbf{r})$ is the four-component spinor (5.95), and $\underline{\widehat{k}}_0$ and $\underline{\widehat{D}}$ are the matrices defined by Eqs. (5.99)–(5.101), respectively. The term on the RHS of Eq. (5.171), representing the coupling of the random potential to the charge transferred across the junction, involves the matrix $\widehat{M}$ defined as

$$
\widehat{M} = -\frac{ie}{2} V(\tau)
\begin{pmatrix}
1 & 0 & 0 & 0 \\
0 & -1 & 0 & 0 \\
0 & 0 & -1 & 0 \\
0 & 0 & 0 & 1
\end{pmatrix}
\tag{5.172}
$$

Terms involving matrices $\widehat{D}$ and $\widehat{M}$ in Eq. (5.171) are now treated as a perturbation acting on a system described by the matrix $\widehat{k}_0$. With use of the cumulant expansion theorem of Appendix A16, the grand partition function (5.170) can be written as follows [see also Eq. (5.103)]

$$
Z_G = Z_{G_0} \int \mathcal{D}^2 \Delta_L(\mathbf{r}, \tau)\, \mathcal{D}^2 \Delta_R(\mathbf{r}, \tau)\, \mathcal{D}V_1(\tau)
$$

$$
\times \exp\left(-\frac{1}{\hbar} S_{\text{eff}}[\Delta_L, \Delta_R] \right)
\tag{5.173}
$$

where

$$
S_{\text{eff}}[\Delta_L, \Delta_R] = \frac{1}{g} \int_0^{\hbar\beta} d\tau \left(\int_{\Delta_L} d^3r\, |\Delta_L(\mathbf{r}, \tau)|^2 + \int_{\Delta_R} d^3r\, |\Delta_R(\mathbf{r}, \tau)|^2 \right)
$$

$$
+ \frac{1}{2} C \int_0^{\hbar\beta} d\tau\, V^2(\tau) - \hbar \operatorname{Tr}\left(\ln \widehat{G}_0 + \ln \widehat{G}^{-1} \right)
\tag{5.174}
$$

where $\widehat{G}_0$ satisfies the differential equation (5.104). The Green's function $\widehat{G}$ satisfies the following equation

$$
\left(-\hbar \frac{\partial}{\partial \tau} - \widehat{k}_0 - \widehat{D} - \widehat{M} \right) \widehat{G}(\mathbf{r}, \tau; \mathbf{r}', \tau') = \hbar\, \delta(\mathbf{r} - \mathbf{r}')\, \delta(\tau - \tau')
\tag{5.175}
$$

where the matrix $\widehat{k}_0$ is given in Eq. (5.99). Performing the gauge transformation (5.106), we are led to a transformed Green's function $\widehat{G}'$ that satisfies an equation of motion [see Eq. (5.108)]

$$
\left(-\hbar \frac{\partial}{\partial \tau} - \widehat{k}_0' - \widehat{D}_0 - \widehat{D}_1(\tau) - \widehat{M} \right) \widehat{G}'(\mathbf{r}, \tau; \mathbf{r}', \tau')
$$

$$
= \hbar\, \delta(\mathbf{r} - \mathbf{r}')\, \delta(\tau - \tau')
\tag{5.176}
$$

where the matrices $\widehat{k}_0'$ and $\widehat{D}_1$ are defined in Eqs. (5.110) and (5.112), respectively. Regarding $\widehat{D}_1 + \widehat{M}$ and the tunneling part of $\widehat{k}_0'$ as a perturbation, we expand $(\widehat{G}')^{-1}$ about the inverse of the Green's function $\widehat{G}_s$ that satisfies Eq. (5.113). Thus we may use the expansion shown in Eqs.

(5.116)–(5.121), except that the matrix $\underline{\widehat{D}}_1$ of Eq. (5.112) is now replaced by

$$\underline{\widehat{D}}_1 + \underline{\widehat{M}}$$

$$= \frac{i}{2} \begin{pmatrix} \hbar\dot{\theta}_L - eV & 0 & 0 & 0 \\ 0 & -\hbar\dot{\theta}_L + eV & 0 & 0 \\ 0 & 0 & \hbar\dot{\theta}_R + eV & 0 \\ 0 & 0 & 0 & -\hbar\dot{\theta}_R - eV \end{pmatrix} \quad (5.177)$$

As a result of this replacement, the expansions (5.117) and (5.121) are changed to

$$\text{Tr}\left[\ln(\underline{\widehat{G}}')^{-1}\right] = \text{Tr}\left(\ln\underline{\widehat{G}}_s^{-1}\right) - \frac{1}{2}\text{Tr}\left[\underline{\widehat{G}}_s(\underline{\widehat{D}}_1 + \underline{\widehat{M}})\right]^2$$

$$- \frac{1}{2}\text{Tr}\left(\underline{\widehat{G}}_s\underline{\widehat{T}}_1\right)^2 \quad (5.178)$$

With the help of Eq. (5.112), we find that the second term on the RHS of this equation makes a following contribution to the effective action (5.174)

$$B_{D+M} = \frac{\hbar}{2}\text{Tr}\left[\underline{\widehat{G}}_s(\underline{\widehat{D}}_1 + \underline{\widehat{M}})\right]^2$$

$$= \frac{3n\Omega\hbar^2}{16\epsilon_F}\int_0^{\hbar\beta} d\tau\left[\left(\dot{\theta}_L - \frac{e}{\hbar}V\right)^2 + \left(\dot{\theta}_R + \frac{e}{\hbar}V\right)^2\right] \quad (5.179)$$

Since the contribution of the last term on the RHS of Eq. (5.178) is the same as in the case of coupled uncharged superfluids, we obtain, using Eqs. (5.131), (5.174), and (5.179), the following effective Euclidean action

$$S_{\text{eff}}[\Delta_L, \Delta_R] = B_{D+M} + B_T \quad (5.180)$$

We have assumed that the magnitude of the order parameter does not fluctuate, so the path integral for the partition function involves only the phase variables and the potential $V(\tau)$. Equations (5.173) and (5.131) then yield

$$Z_G \to \int \mathcal{D}\theta_L(\tau)\,\mathcal{D}\theta_R(\tau)\,\mathcal{D}V(\tau)\,\exp\left(-\frac{1}{\hbar}S_{\text{eff}}[\theta_L,\theta_R]\right) \quad (5.181)$$

where

$$S_{\text{eff}}[\theta_L,\theta_R] = \int_0^{\hbar\beta} d\tau\left\{\frac{1}{2}CV^2(\tau) + \frac{3n\Omega\hbar^2}{16\epsilon_F}\left[\left(\dot{\theta}_L - \frac{e}{\hbar}V\right)^2 + \left(\dot{\theta}_R + \frac{e}{\hbar}V\right)^2\right]\right\}$$

$$- \int_0^{\hbar\beta} d\tau \int_0^{\hbar\beta} d\tau'\left(\alpha(\tau-\tau')\cos\frac{\theta_{RL}(\tau)-\theta_{RL}(\tau')}{2}\right.$$

$$\left. - \beta(\tau-\tau')\cos\frac{\theta_{RL}(\tau)+\theta_{RL}(\tau')}{2}\right) \quad (5.182)$$

The phase-inertial terms in the braces should be compared with the phenomenological Lagrangian, derived in Eqs. (2.43) and (2.99) for the TDGL model.

5.5.1. *Elimination of $V(\tau)$*

Since the $V(\tau)$ appears in the effective action (5.182) as a quadratic functional, it can be eliminated from (5.181) by a Gaussian integration. We thus consider the path integral

$$
I[\dot{\theta}_L, \dot{\theta}_R] = \int \mathcal{D}V(\tau) \exp\left\{ -\frac{1}{\hbar} \int_0^{\hbar\beta} d\tau \left[\frac{1}{2} C V^2(\tau) \right.\right.
$$
$$
\left.\left. + a_\delta \left(\frac{\hbar}{e} \dot{\theta}_L - V \right)^2 + a_\delta \left(\frac{\hbar}{e} \dot{\theta}_R + V \right)^2 \right] \right\} \tag{5.183}
$$

where

$$
a_\delta = \frac{3n\Omega e^2}{16\epsilon_F} \tag{5.184}
$$

We expand $V(\tau)$ into a Fourier series

$$
V(\tau) = \sum_{n=-\infty}^{\infty} v_n e^{-i\omega_n \tau} \tag{5.185}
$$

where $\omega_n = 2\pi n/\hbar\beta$. Since $V(\tau)$ is a real function of τ, we have

$$
v_n^r = v_{-n}^r
$$
$$
v_n^i = -v_{-n}^i \tag{5.186}
$$

where v_n^r and v_n^i are the real and imaginary parts of v_n. We also expand the time derivatives of the phase variables

$$
\frac{\hbar}{e}\dot{\theta}_L(\tau) = F_L(\tau) = \sum_{n=-\infty}^{\infty} f_{Ln} e^{-i\omega_n \tau}
$$
$$
\frac{\hbar}{e}\dot{\theta}_R(\tau) = F_R(\tau) = \sum_{n=-\infty}^{\infty} f_{Rn} e^{-i\omega_n \tau} \tag{5.187}
$$

Using Eqs. (5.185)–(5.187), the path integral (5.183) is reduced to a multiple Gaussian integral [see Eq. (4.29) for a similar calculation]

$$
I = I_0 \int_{-\infty}^{\infty} dv_0 \exp\left\{ -\beta \left[\left(\frac{1}{2}C + 2a_\delta \right) v_0^2 - 2a_\delta(f_{L0} - f_{R0})v_0 \right] \right\}
$$
$$
\times \prod_{n=1}^{\infty} \int_{-\infty}^{\infty} dv_n^r \int_{-\infty}^{\infty} dv_n^i \exp\left\{ -\beta(C + 4a_\delta)\left[(v_n^r)^2 + (v_n^i)^2 \right] \right.
$$
$$
\left. -4a_\delta\beta\left[(f_{Ln}^r - f_{Rn}^r)v_n^r + (f_{Ln}^i - f_{Rn}^i)v_n^i \right] \right\} \tag{5.188}
$$

where

$$I_0 = \exp\left[-\frac{1}{\hbar}\int_0^{\hbar\beta} d\tau \left(\frac{a_\delta \hbar^2}{e^2}\left(\dot\theta_L^2 + \dot\theta_R^2\right)\right)\right] \tag{5.189}$$

The integrals on the RHS of Eq. (5.188) can be performed with use of formula (A3.20), yielding

$$I[\dot\theta_L, \dot\theta_R] \sim I_0 \exp\left(\frac{2\beta a_\delta^2}{C + 4a_\delta} \sum_{n=-\infty}^{\infty}\left[\left(f_{Ln}^r - f_{Rn}^r\right)^2 + \left(f_{Ln}^i - f_{Rn}^i\right)^2\right]\right)$$

$$= \exp\left[-\frac{1}{\hbar}\int_0^{\hbar\beta} d\tau \left(\frac{a_\delta \hbar^2}{e^2}\left(\dot\theta_L^2 + \dot\theta_R^2\right)\right.\right.$$

$$\left.\left.-\frac{2a_\delta^2 \hbar^2}{e^2(C + 4a_\delta)}\dot\theta_{RL}^2\right)\right] \tag{5.190}$$

where the pre-exponential factor (independent of $\dot\theta_L$ and $\dot\theta_R$) is disregarded. It is interesting to consider this expression in two extreme limits: $C \gg a_\delta$ and $C \ll a_\delta$.

5.5.2. Functional $I[\dot\theta_L, \dot\theta_R]$ for $C \gg a_\delta$

This limit corresponds to a strong suppression of charging effects. For $C \gg a_\delta$, the second term in the exponent on the RHS of Eq. (5.190) is negligible in comparison to the first one. Thus we obtain, with use of Eq. (5.184),

$$I[\dot\theta_L, \dot\theta_R] \sim \exp\left[-\frac{1}{\hbar}\int_0^{\hbar\beta} d\tau \left(\frac{3n\Omega\hbar^2}{16\epsilon_F}\left(\dot\theta_L^2 + \dot\theta_R^2\right)\right)\right] \tag{5.191}$$

Introducing this result into Eq. (5.182), the effective action of coupled neutral superfluid particles is recovered. This result is also expected by qualitative inspection of Eq. (5.183). For large C, the charging energy in the exponent of the integrand tends to pin the fluctuations of $V(\tau)$ to a nearly zero value. This makes the path integral dominated by a trivial path: $V(\tau) = 0$, reducing the expression (5.183) to Eq. (5.191).

5.5.3. $I[\dot\theta_L, \dot\theta_R]$ for $C \ll a_\delta$

In this limit, the charging energy dominates the phase inertia. This can be shown by expanding the coefficient of the $\dot\theta_{RL}^2$ term in Eq. (5.190) in the ratio $C/4a_\delta$

$$\frac{2a_\delta^2 \hbar^2}{e^2(C + 4a_\delta)} = \frac{a_\delta \hbar^2}{2e^2}\left(1 - \frac{C}{4a_\delta} + \cdots\right) \tag{5.192}$$

Introducing this expansion into Eq. (5.190), we obtain to order C/a_δ

$$I[\dot\theta_L, \dot\theta_R] \sim \exp\left[-\frac{C\hbar}{8e^2}\int_0^{\hbar\beta} d\tau \left(\frac{4a_\delta}{C}\left(\dot\theta_L + \dot\theta_R\right)^2 + \dot\theta_{RL}^2\right)\right] \tag{5.193}$$

Since $a_\delta/C \gg 1$, the first term in the exponent of this equation enforces a constraint: $\dot\theta_L(\tau) = -\dot\theta_R(\tau)$, so that the path integral (5.181) can be written

$$Z_G \to \int \mathcal{D}\theta_L(\tau)\, \mathcal{D}\theta_R(\tau)\, I[\dot\theta_L, \dot\theta_R] \exp\left(-\frac{B_T}{\hbar}\right)$$

$$\sim \int \mathcal{D}\theta_L(\tau)\, \mathcal{D}\theta_R(\tau)\, \delta[\dot\theta_L + \dot\theta_R] \exp\left[-\left(\frac{C\hbar}{8e^2}\int_0^{\hbar\beta} d\tau\, \dot\theta_{RL}^2 + \frac{B_T}{\hbar}\right)\right]$$

$$\sim \int \mathcal{D}\theta_{RL}(\tau) \exp\left[-\left(\frac{C\hbar}{8e^2}\int_0^{\hbar\beta} d\tau\, \dot\theta_{RL}^2 + \frac{B_T}{\hbar}\right)\right] \tag{5.194}$$

The RHS of Eq. (5.194) agrees with the partition function of Ambegaokar et al. (1982), who obtained it by applying the constraints (pinning)

$$\dot\theta_L = eV/\hbar$$
$$\dot\theta_R = -eV/\hbar \tag{5.195}$$

If we include in the expression (5.192) terms of order C^2/a_δ^2, then the expression on the RHS of (5.194) involves an effective capacitance that corresponds to a series connection of C and the capacitance $C_\delta = 4a_\delta$.

5.5.4. Renormalization of junction capacitance

In Eq. (5.160) we have derived the contribution to the effective action in the form of a phase-inertial term induced by quasiparticle tunneling. Exactly the same mechanism is working in the case of charged superconductors. Since the effective action (5.160) is of the same form as the charging energy term in Eq. (5.194), we can define a renormalized capacitance C_{ren} by combining these two contributions

$$C_{\text{ren}} = C + \frac{4e^2}{\hbar^2} I_{\text{eff}} \tag{5.196}$$

Introducing on the RHS of this equation the expression (5.161), we obtain using Eq. (5.148)

$$C_{\text{ren}} = C + \frac{3\pi\hbar}{32\Delta_0 R_N} = C + \Delta C \tag{5.197}$$

As mentioned in Section 3.1, the capacitance increase ΔC, given by this expression, plays an important role in phase transitions of granular super-conducting arrays (Chakravarty et al., 1987; Ferrell and Mirhashem, 1988).

In a microscopic, strongly coupled junction ($R_N \Delta_0$ small), the increment ΔC may exceed the "bare" capacitance C, so that

$$C_{\text{ren}} \simeq \frac{3\pi\hbar}{32\Delta_0 R_N} \tag{5.198}$$

It is interesting to investigate for what size of the superconducting particles the "δ effect," shown in Eq. (5.191), starts to play a significant role. From the expansion (5.192) we see that this takes place when the ratio $C/4a_\delta$ is larger than one. At this point, it is important to realize that it is only the *geometric junction capacitance C* which becomes affected (reduced) by the δ effect [see Eq. (5.182)]. The quasiparticle capacitance ΔC is *connected in parallel with this reduced geometric capacitance*. Hence, if $\Delta C \gg C$, the overall capacitance C_{ren} remains to be given by (5.198), irrespective of the δ effect. When $C/a_\delta \gg 1$, the geometric capacitance is renormalized to $C_\delta = 4a_\delta$. If ΔC is much greater than C_δ, the overall capacitance is equal to ΔC. Under such conditions, the geometric capacitance disappears entirely from the picture. This fact may have a bearing on the universality of the sheet resistance at the onset of superconductivity in granular films (see Chapter 6). Thus, even when the geometric capacitance $C \simeq \Delta C$, Eq. (5.198) holds as long as $\Delta C/C_\delta \gg 1$. Using Eqs. (5.184) and (5.198), this condition can be written

$$\frac{\Delta C}{4a_\delta} = \frac{3\pi\hbar\delta}{32\Delta_0 R_N e^2} \gg 1 \qquad (5.199)$$

where δ is the electron-level separation in a small grain of volume Ω [see Eq. (3.4)]. Since we are mostly interested in the problem of phase transitions in granular arrays, it is useful to consider the case when R_N *is equal to the universal critical resistance* (3.2). Inserting this value into Eq. (5.199), we obtain

$$\delta \gg 2.65z\Delta_0 \simeq 4.4zk_B T_c^{(0)} \qquad (5.200)$$

where z is the coordination number of the array and $T_c^{(0)}$ is the BCS transition temperature of the bulk superconductor. Comparing this result with Eq. (3.6), we see that the level separation must exceed the value for which the internal superconducting ordering of the grain begins to break down. It should be pointed out, however, that a considerably less stringent condition than (5.200) is obtained, if one takes into account *screening* of the electric fields inside the superconductors, as discussed in Section 2.7. This makes the excess Cooper pair *charges reside in a thin layer on* the grain surface. Thus we should replace the grain volume Ω in Eq. (5.182) by the product St, where S and t are the surface area and the thickness of the layer, respectively [see Eq. (2.105)]. If the same replacement is made in Eqs. (5.184) and (5.199), the condition (5.200) is changed to

$$\delta \gg \left(\frac{4.4zSt}{\Omega}\right)k_B T_c^{(0)} \qquad (5.201)$$

Introducing Eq. (3.4) for δ, Eq. (5.201) translates into a criterion for the surface S of the grains

$$S \ll \left[4.4zk_B T_c^{(0)} N(0)t\right]^{-1} \qquad (5.202)$$

Assuming that $t \simeq k_F^{-1}$, we have

$$N(0)t = \frac{mk_F t}{2\pi^2 \hbar^2} \simeq \frac{m}{2\pi^2 \hbar^2} \qquad (5.203)$$

Using this result on the RHS of Eq. (5.202), we obtain a criterion

$$S \ll \frac{2\pi^2 \hbar^2}{4.4 z m k_B T_c^{(0)}} = S_0 \qquad (5.204)$$

Let us evaluate the RHS of this equation for a square array of aluminum particles with $z = 4$ and $T_c^{(0)} \simeq 1.2$ K. Using these values, we obtain from Eq. (5.204) a characteristic surface area $S_0 \simeq 8 \times 10^{-12}$ cm^2, which corresponds to a spherical particle of diameter 160 Å. The energy-level splitting in particle of this size is $\delta/k_B = 0.53$ K, which is well *below* the *critical* value $2\pi T_c^{(0)} = 7.53$ K (above which the internal superconducting ordering begins to break down). Hence, for particles of diameter under 160 Å, we expect the δ effect to start obliterating the role of the geometric junction capacitance.

6

PHASE TRANSITIONS IN DISSIPATIVE ARRAYS

6.1. Introduction

We are now in a position to discuss the effect of dissipation on phase transitions in granular superconductors. The primary theoretical stimulus for such studies came from the work of Caldeira and Leggett (1981), showing how dissipation tends to suppress the macroscopic quantum tunneling. Subsequently, Chakravarty (1982) and, independently, Bray and Moore (1982) have considered the tunneling of a particle in a (double-well potential) in the presence of dissipation. The dissipative part of the action [see Eq. (4.43)] produces a *long-range interaction* between the bounces representing the hopping path between the two wells. In the theory of Caldeira and Leggett (1981), the MQT rate is calculated within the single-bounce approximation. A reduction of the tunneling rate is predicted, as shown in Sections 4.8.1 and 4.8.2. The *logarithmic interaction* between the bounces has the *additional* effect of allowing the system to undergo a quantum *phase transition* from a disordered state, in which the average particle position (phase) is zero, to a broken-symmetry state in which the particle stays *pinned* to one potential well. These ideas have been extended by Schmid (1983) and Fisher and Zwerger (1985) to a particle in a cosine potential (and coupled to an ohmic bath), thus making a contact with dissipative Josephson junction more explicit. Again, on increasing the strength of ohmic dissipation, the behavior changes from diffusion to localization of the particle. The presence of such quantum phase transitions in a single Josephson junction suggests that arrays of Josephson junctions, which are phase disordered due to the presence of charging energies, *will undergo a transition* into a phase-locked state at some critical value of the dissipation. This idea has been exploited by Orr et al. (1986) to explain their observation of a universal critical resistance for the onset of superconductivity in ultrathin films.

The first theories of dissipative phase transitions in Josephson-junction arrays were set up independently by Chakravarty et al. (1986) and Šimánek and Brown (1986). The former authors have considered the self-charging model with a dissipative term taken in the harmonic approximation. The anharmonicity of the Josephson coupling has been treated by the SCHA (see Section 3.9). The latter authors have used an SCHA approximation for the anharmonic damping action, whereas the Josephson coupling was treated within the MFA.

In this chapter we consider in detail the mathematical aspects of these two works. Due to different approximations, there are differences in the results. For example, the reentrant features (which are usually associated with the MFA) are not seen in the SCHA phase diagram obtained by Chakravarty et al. (1986). On the other hand, their work shows clearly the existence of a threshold sheet resistance above which the array loses its global coherence. This critical resistance can also be seen in the phase diagram of Šimánek and Brown (1986); however, it seems to be linked to the breakdown of the Coulomb gap in the normal tunnel junctions (see Brown and Šimánek, 1986). The common assumption of both theories is that a *dissipative conducting* channel *exists* down to *zero temperature*. In the presence of a nonzero superconducting gap, the energy-conserving quasiparticle tunneling is blocked at $T = 0$. Only virtual processes leading to the renormalization of junction capacitances remain. This effect has been considered as a mechanism of universal critical resistance for the onset of global coherence by Chakravarty et al. (1987) and Ferrell and Mirhashem (1988). By extending the effective action (5.182) to an array of junctions, we find a modification of the theory of the authors cited, which takes into account the extra phase fluctuations due to the δ effect. In light of these results, we discuss the experiments on ultrathin films in which a universal critical sheet resistance has been found. We discuss, in detail, the weak nonuniversality stemming from the logarithmic terms in the theory of Chakravarty et al. (1987). The dissipation induced by bosonic degrees of freedom, following the ideas of Fisher et al. (1990), is discussed briefly as a possible mechanism for universal sheet resistance. In the last section, the coarse-graining procedure of Doniach (1981) is discussed, by means of which the problem of arrays with charging energy is mapped onto a quantum Ginzburg–Landau model. In the framework of this model, we elucidate the relation of the MFA method used by Chakravarty et al. (1987) and that of Efetof (1980). We also discuss the corrections to the MFA predictions due to the quartic terms in the Ginzburg–Landau functional. The chapter is concluded by a discussion of the correlation length in Josephson-coupled arrays.

6.2.   Dissipative array with Coulomb interaction

In this section we generalize the single-junction results of Section 5.5 to the case of a regular array of Josephson junctions. The Coulomb energy of the array is taken in the general form (3.30). Assuming that the array is described by the capacitance matrix $\widehat{C}_{ij}$, we write the charging Hamiltonian of a system of N superconducting grains as

$$H_c = \frac{1}{2} \sum_{i=1}^{N} \sum_{j=1}^{N} (\widehat{C}^{-1})_{ij} Q_i Q_j \qquad (6.1)$$

where Q_i is the operator for the charge on the ith superconductor. The grand canonical Hamiltonian for the array is obtained by generalizing the single-junction result (5.165)

$$K = K_0 + H_1 \tag{6.2}$$

where K_0 is the free-electron Hamiltonian, including tunneling between grains, and H_1 is the pairing and Coulomb interaction. Recalling Eq. (5.93), we thus have

$$K_0 = \sum_{k_i,\sigma} \xi(k_i) a^\dagger(k_i,\sigma) a(k_i,\sigma)$$

$$+ \sum_{k_i,k_j,\sigma} \left[T(k_i,k_j) a^\dagger(k_i,\sigma) a(k_j,\sigma) + \text{H.c.} \right] \tag{6.3}$$

where k_i is the fermion wave vector on the ith grain. The Hamiltonian H_1 is given by [see Eq. (5.166)]

$$H_1 = -\frac{g}{\Omega} \sum_j \sum_Q b^\dagger_{j,Q} b_{j,Q} + H_c \tag{6.4}$$

where $b^\dagger_{j,Q}$ and $b_{j,Q}$ are the pair creation and destruction operators on the jth grain, respectively.

The grand partition function is, according to Eqs. (5.6)–(5.9), given by

$$Z_G = \text{Tr}\left[e^{-\beta K_0} \sigma(\beta) \right] \tag{6.5}$$

where the imaginary time-evolution operator is written, following Eq. (5.168), in a discretized form

$$\sigma(\beta) = \text{T}_\tau \exp\left(-\frac{1}{\hbar} \int_0^{\hbar\beta} d\tau \, H_1[\tau] \right)$$

$$= \lim_{\Delta\tau \to 0} \text{T}_\tau \exp\left(\frac{g\Delta\tau}{\hbar\Omega} \sum_{Q,\tau_k} \sum_j b^\dagger_{j,Q}[\tau_k] b_{j,Q}[\tau_k] \right.$$

$$\left. -\frac{\Delta\tau}{2\hbar} \sum_{\tau_k} \sum_{ij} (\hat{C}^{-1})_{ij} Q_i[\tau_k] Q_j[\tau_k] \right) \tag{6.6}$$

The charging term in the exponent of Eq. (6.6) can be rewritten with the use of the identity (Friedman, 1964)

$$\exp\left(-\frac{\Delta\tau}{2\hbar} \sum_{ij} (\hat{C}^{-1})_{ij} Q_i Q_j \right) = \text{const} \times \int \int \int \cdots dV_1 \, dV_2 \, dV_3$$

$$\times \exp\left(-\frac{\Delta\tau}{2\hbar} \sum_{ij} \hat{C}_{ij} V_i V_j - \frac{i\Delta\tau}{\hbar} \sum_i V_i Q_i \right) \tag{6.7}$$

where the explicit form of the proportionality constant is irrelevant for further calculations. We note that this expression holds for $V_i = V_i[\tau_k]$ and $Q_i = Q_i[\tau_k]$ at each discrete time τ_k. Introducing (6.7) into (6.6), we obtain for the partition function (6.5) the following generalization of (5.170)

$$Z_G = \int \prod_i \mathcal{D}^2 \Delta_i(\tau)\, \mathcal{D}V_i(\tau)\, \mathrm{Tr}\left[\mathrm{T}_\tau \exp\left(-\frac{1}{\hbar}\int_0^{\hbar\beta} d\tau\, H_{\mathrm{eff}}(\tau) \right) \right] \quad (6.8)$$

where

$$\begin{aligned}
H_{\mathrm{eff}} = &\int_{\Omega_1} d^3 r_1 \cdots \int_{\Omega_N} d^3 r_N\, \widehat{\psi}^\dagger(\mathbf{r})\widehat{k}_0\widehat{\psi}(\mathbf{r}) \\
&+ \int d^3 r\, \widehat{\psi}^\dagger[\mathbf{r},\tau]\left[\widehat{\underline{D}}_1(\mathbf{r},\tau) + \widehat{\underline{M}}(\mathbf{r},\tau) \right]\widehat{\psi}[\mathbf{r},\tau] \\
&+ \frac{1}{g}\sum_i \int_{\Omega_i} d^3 r\, |\Delta_i(\mathbf{r},\tau)|^2 + \frac{1}{2}\sum_{ij} \widehat{C}_{ij} V_i(\tau) V_j(\tau) \quad (6.9)
\end{aligned}$$

The spinor $\widehat{\psi}(\mathbf{r})$ in Eq. (6.9) includes fermion operators of all N grains:

$$\widehat{\psi}(\mathbf{r}) = \begin{pmatrix} \psi_{1\uparrow}(\mathbf{r}) \\ \psi_{1\downarrow}^\dagger(\mathbf{r}) \\ \vdots \\ \psi_{N\uparrow}(\mathbf{r}) \\ \psi_{N\downarrow}^\dagger(\mathbf{r}) \end{pmatrix} \quad (6.10)$$

Consequently, $\widehat{\underline{k}}_0$ is now an $N \times N$ matrix generalizing the single-junction matrix (5.99) to an array of N coupled superconductors. The matrix $\widehat{\underline{M}}$ represents the $\sum_i V_i Q_i$ term in Eq. (6.7), and $\widehat{\underline{D}}_1$ is an $N \times N$ version of Eq. (5.112). We thus obtain, in the representation (6.10),

$$\widehat{\underline{D}}_1 + \widehat{\underline{M}} =$$

$$\frac{i}{2}\begin{pmatrix} \hbar\dot{\theta}_1 - 2eV_1 & 0 & \cdots & 0 & 0 \\ 0 & -\hbar\dot{\theta}_1 + 2eV_1 & \cdots & 0 & 0 \\ \vdots & \vdots & \ddots & \vdots & \vdots \\ 0 & 0 & \cdots & \hbar\dot{\theta}_N - 2eV_N & 0 \\ 0 & 0 & \cdots & 0 & -\hbar\dot{\theta}_N + 2eV_N \end{pmatrix}$$

$$(6.11)$$

The same manipulations that led to Eqs. (5.181) and (5.182) can be applied to perform the trace in Eq. (6.8), and we obtain

$$Z_G \to \prod_i \int \mathcal{D}^2 \Delta_i(\tau) \, \mathcal{D}V_i(\tau) \, \exp\left(-\frac{1}{\hbar} S_{\text{eff}}[\{\theta_i\}, \{V_i\}]\right) \tag{6.12}$$

where the effective action is a functional of the sets $\{\theta_i\}$ and $\{V_i\}$ of the form:

$$
\begin{aligned}
S_{\text{eff}}&[\{\theta_i\}, \{V_i\}] \\
&= \int_0^{\hbar\beta} d\tau \left[\frac{1}{2} \sum_{ij} \widehat{C}_{ij} V_i V_j + \frac{3n\Omega\hbar^2}{16\epsilon_F} \sum_i \left(\dot{\theta}_i - \frac{2e}{\hbar} V_i \right)^2 \right] \\
&\quad - \int_0^{\hbar\beta} d\tau \int_0^{\hbar\beta} d\tau' \sum_{\langle ij \rangle} \left[\alpha_{ij}(\tau - \tau') \cos\left(\frac{\theta_{ij}(\tau) - \theta_{ij}(\tau')}{2} \right) \right. \\
&\qquad \left. - \beta_{ij}(\tau - \tau') \cos\left(\frac{\theta_{ij}(\tau) + \theta_{ij}(\tau')}{2} \right) \right]
\end{aligned}
\tag{6.13}
$$

In this equation, we define $\theta_{ij} = \theta_i - \theta_j$, and the functions α_{ij} and β_{ij} are [see (5.132) and (5.133)]

$$\alpha_{ij} = -\frac{2}{\hbar} |\mathfrak{J}|^2 \int \frac{d^3 k_i}{(2\pi)^3} G_i^{11}(k_i, \tau) \int \frac{d^3 k_j}{(2\pi)^3} G_j^{11}(k_j, -\tau) \tag{6.14}$$

and

$$\beta_{ij} = -\frac{2}{\hbar} |\mathfrak{J}|^2 \int \frac{d^3 k_i}{(2\pi)^3} G_i^{12}(k_i, \tau) \int \frac{d^3 k_j}{(2\pi)^3} G_j^{12}(k_j, -\tau) \tag{6.15}$$

where G_i^{11} and G_i^{12} are the Gorkov Green's functions for the ith grain.

6.2.1. Path integral over $V_i(\tau)$

Let us now consider the part of Eq. (6.12) involving path integration over the N variables $V_i(\tau)$. This is a functional of the phase variables θ_i of the form

$$
\begin{aligned}
I[\{\dot{\theta}_i\}] = \prod_i \int \mathcal{D}V_i(\tau) \exp \left\{ -\frac{1}{\hbar} \int_0^{\hbar\beta} d\tau \left[\frac{1}{2} \sum_{ij} \widehat{C}_{ij} V_i V_j \right. \right. \\
\left. \left. + a_\delta \sum_i \left(\frac{\hbar \dot{\theta}_i}{e} - 2V_i \right)^2 \right] \right\}
\end{aligned}
\tag{6.16}
$$

where a_δ is given by Eq. (5.184). We note that a single-junction version of this Gaussian path integral was performed in Section 5.5 by expanding $V(\tau)$ into a Fourier series. To evaluate Eq. (6.16), we use a more expedient method that replaces the τ integral in the exponent by a discrete sum and then applies an identity, similar to (6.7), to each term. Proceeding in this way, we obtain

$$
\begin{aligned}
I[\{\dot{\theta}_i\}] &= \lim_{\Delta\tau\to 0} \prod_i \int \mathcal{D}V_i(\tau) \exp\left\{-\frac{\Delta\tau}{\hbar}\sum_k\left[\frac{1}{2}\sum_{ij}\widehat{C}_{ij}V_i(\tau_k)V_j(\tau_k)\right.\right. \\
&\qquad\left.\left. + a_\delta\sum_i\left(\frac{\hbar}{e}\dot{\theta}_i(\tau_k)-2V_i(\tau_k)\right)^2\right]\right\} \\
&= \lim_{\Delta\tau\to 0}\prod_k\left\{\prod_i\int dV_i(\tau_k)\exp\left[-\frac{\Delta\tau}{2\hbar}\sum_{ij}\widehat{C}_{ij}V_i(\tau_k)V_j(\tau_k)\right.\right. \\
&\qquad\left.\left. - \frac{\Delta\tau a_\delta}{\hbar}\sum_i\left(\frac{\hbar}{e}\dot{\theta}_i(\tau)-2V_i(\tau_k)\right)^2\right]\right\} \\
&= \lim_{\Delta\tau\to 0}\prod_k F[\{\dot{\theta}_i(\tau_k)\}]
\end{aligned}
\tag{6.17}
$$

On the RHS of this equation we see a product of Gaussian path integrals, each taken at a *fixed* time τ_k.

6.2.2. *Evaluation of $F[\{\dot{\theta}_i(\tau_k)\}]$*

From Eq. (6.17) we have

$$
\begin{aligned}
F[\{\dot{\theta}_i(\tau_k)\}] &= \prod_i\int dV_i(\tau_k)\exp\left\{-\frac{\Delta\tau}{\hbar}\left[\sum_i\frac{a_\delta\hbar^2}{e^2}\dot{\theta}_i^2(\tau_k)\right.\right. \\
&\qquad + \frac{1}{2}\sum_{ij}\bar{C}_{ij}V_i(\tau_k)V_j(\tau_k) \\
&\qquad\left.\left. - \frac{4a_\delta\hbar}{e}\sum_i\dot{\theta}_i(\tau_k)V_i(\tau_k)\right]\right\}
\end{aligned}
\tag{6.18}
$$

where

$$
\bar{C}_{ij} = \widehat{C}_{ij} + 8a_\delta\delta_{ij}
\tag{6.19}
$$

Using the identity (Friedman, 1964)

$$
\exp\left(-\frac{1}{2}\sum_{ij}\left(A^{-1}\right)_{ij}k_ik_j\right)
$$

$$
= \text{const}\times\prod_i\int dV_i\exp\left(-\frac{1}{2}\sum_{ij}A_{ij}V_iV_j - i\sum_i k_iV_i\right)
\tag{6.20}
$$

where $A_{ij} = (\Delta\tau/\hbar)\bar{C}_{ij}$ and $k_i = (4ia_\delta\Delta\tau/e)\dot{\theta}_i$, we obtain from Eq. (6.18) the functional F in the form

$$
F[\{\dot{\theta}_i(\tau_k)\}] = \exp\left[-\frac{\Delta\tau}{\hbar}\left(\frac{a_\delta\hbar^2}{e^2}\sum_i\dot{\theta}_i^2(\tau_k)\right.\right.
$$
$$
\left.\left. -\frac{8a_\delta^2\hbar^2}{e^2}\sum_{ij}(\bar{C}^{-1})_{ij}\dot{\theta}_i(\tau_k)\dot{\theta}_j(\tau_k)\right)\right] \tag{6.21}
$$

Introducing Eq. (6.21) into (6.17) and assuming the magnitudes of Δ_i to be fixed, the partition function (6.12) becomes, with the help of (6.13), a path integral over the phase variables alone

$$
Z_G = \prod_i\int\mathcal{D}\theta_i\,\exp\left(-\frac{1}{\hbar}S_{\text{eff}}[\{\theta_i\}]\right) \tag{6.22}
$$

where

$$
S_{\text{eff}}[\{\theta_i\}] = \int_0^{\hbar\beta}d\tau\left(\frac{a_\delta\hbar^2}{e^2}\sum_i\dot{\theta}_i^2(\tau) - \frac{8a_\delta^2\hbar^2}{e^2}\sum_{ij}(\bar{C}^{-1})_{ij}\dot{\theta}_i(\tau)\dot{\theta}_j(\tau)\right)
$$
$$
-\int_0^{\hbar\beta}d\tau\int_0^{\hbar\beta}d\tau'\sum_{\langle ij\rangle}\left[\alpha_{ij}(\tau-\tau')\cos\left(\frac{\theta_{ij}(\tau)-\theta_{ij}(\tau')}{2}\right)\right.
$$
$$
\left. -\beta_{ij}(\tau-\tau')\cos\left(\frac{\theta_{ij}(\tau)+\theta_{ij}(\tau')}{2}\right)\right] \tag{6.23}
$$

6.3. Effective action for a dissipative self-charging model

Introducing into Eq. (6.19) the capacitance matrix of the self-charging model given by Eq. (3.32), we obtain

$$
\bar{C}_{ij} = (C_0 + 8a_\delta)\delta_{ij} \tag{6.24}
$$

where C_0 is the self-capacitance. Inverting the matrix (6.24) and introducing the result into Eq. (6.23), we obtain for the effective action

$$
S_{\text{eff}}[\{\theta_i\}] = \int_0^{\hbar\beta}d\tau\left[\frac{\hbar^2}{e^2}\left(\frac{C_0a_\delta}{C_0 + 8a_\delta}\right)\sum_i\dot{\theta}_i^2(\tau)\right] + S_D[\{\theta_i\}] \tag{6.25}
$$

where $S_D[\{\theta_i\}]$ is the dissipative action given by the last two terms in Eq. (6.23). We see that Eq. (6.25) involves an effective capacitance resulting from a *series connection* of the geometric capacitance C_0 and the effective capacitance $C_\delta = 8a_\delta$, representing the δ effect (see Section 5.5.3 for a similar effect on the junction capacitance). The standard dissipative self-charging model is obtained from expression (6.25) by taking the limit $a_\delta \gg$

C_0. In this case we may expand the coefficient of the $\dot{\theta}_i^2$ term in powers of $C_0/8a_\delta$:

$$\frac{C_0 a_\delta}{8a_\delta(1 + C_0/8a_\delta)} \simeq \frac{C_0}{8}\left(1 - \frac{C_0}{8a_\delta}\right) = \frac{C_{\text{eff}}}{8} \qquad (6.26)$$

Thus, to first order in $C_0/8a_\delta$, we recover the charging energy in the usual diagonal form (with an effective self-capacitance C_{eff} slightly reduced compared to C_0). On the other hand, if $a_\delta \ll C_0$, the effective capacitance is obtained by expanding the coefficient of $\dot{\theta}_i^2$ in (6.25) in powers of $8a_\delta/C_0$

$$\frac{C_0 a_\delta}{C_0(1 + 8a_\delta/C_0)} \simeq a_\delta\left(1 - \frac{8a_\delta}{C_0}\right) \qquad (6.27)$$

To lowest order in a_δ/C_0, the expression (6.25) becomes, with use of (6.27)

$$S_{\text{eff}}[\{\theta_i\}] = \int_0^{\hbar\beta} d\tau \frac{\hbar^2}{e^2} a_\delta \sum_i \dot{\theta}_i^2 + S_D[\{\theta_i\}] \qquad (6.28)$$

Thus, in the limit $a_\delta \ll C_0$, the phase inertial terms are entirely *independent* of the *geometric capacitance*. The experimental realizability of this limit can be examined in the context of the self-charging model, proposed by Bradley and Doniach (1984) (see Fig. 3.2). The thin square superconducting plates of size L forming the array have a capacitance, C_0, given by

$$C_0 = \frac{\varepsilon_0 L^2}{t_0} \qquad (6.29)$$

where t_0 is the width of the gaps between the plates and the normal metal substrate, and ε_0 is the dielectric constant of the insulator, filling these gaps. Using Eq. (5.184) for the plate, the constant a_δ can be expressed as follows:

$$a_\delta = \frac{3n\Omega e^2}{16\epsilon_F} \simeq \frac{3ne^2 L^2 t}{8\epsilon_F} \simeq \frac{3ne^2 L^2}{8\epsilon_F k_F} \qquad (6.30)$$

where we have used, for the effective volume Ω, the approximate formula $\Omega \simeq 2L^2 t$, where $t \sim k_F^{-1}$ is the Debye-screening length. From Eqs. (6.29) and (6.30) we obtain a ratio

$$\frac{C_0}{8a_\delta} = \frac{\pi^2 \varepsilon_0 \epsilon_F}{e^2 t_0 k_F^2} \qquad (6.31)$$

Although this expression does not explicitly involve the plate size L, it is only valid when $L \gg t_0$. Using Eq. (6.31), the condition $C_0/8a_\delta \gg 1$ translates into the requirement

$$t_0 \ll t_c \qquad (6.32)$$

where

$$t_c = \frac{\pi^2 \varepsilon_0 \epsilon_F}{e^2 k_F^2} \qquad (6.33)$$

For an orientational estimate of t_c, let us consider an array of aluminum plates, in which case we have $\epsilon_F \simeq 11.6$ eV and $k_F \simeq 1.7 \times 10^8$ cm^{-1}. Taking $\varepsilon_0 = 10$, we obtain with these values a characteristic gap width $t_c \simeq 25$ Å. Technologically, this is about the lower limit that the width can reach. Hence, we do not expect that the limit for which expression (6.28) is valid is experimentally realizable. However, as the expression on the LHS of Eq. (6.27) indicates, a significant reduction of the effective capacitance C_{eff} is expected when $t_0 \simeq t_c$. For $t_0 \lesssim 10$Å, one expects quasiparticle tunneling to take place between the superconducting plates and the normal substrate. This should lead to an effective capacitance ΔC (caused by virtual transfers of quasiparticles), which acts in parallel with the geometric capacitance C_0.

6.4. Mean-field phase diagram

In this section, we consider the modifications of the mean-field phase diagram of the self-charging model due to dissipation (Šimánek and Brown, 1986). For simplicity, we assume a temperature-independent normal resistance R_N for the Josephson junctions comprising the array. It is, in general, difficult to justify the assumption of a normal conducting channel that persists down to zero temperature and yet is simply determined by normal conductance for temperatures above the bulk BCS transition temperature $T_c^{(0)}$. In fact, this problem has motivated further theoretical investigations, which take into account the renormalization of junction capacitance by virtual quasiparticle processes, which do not require finite normal conductance below $T_c^{(0)}$ (Chakravarty et al., 1987; Ferrell and Mirhashem, 1988). However, if the dissipation is produced by quasiparticle tunneling, a nonzero conductance, existing down to $T = 0$, is still possible if the superconductors in the array are gapless. We note that, in very small grains, fluctuations of the magnitude of the order parameter may lead to effective gaplessness (see Section 3.2). Using Eqs. (6.23) and (6.25), the effective action for the dissipative regular array is given by

$$S_{\text{eff}}[\{\theta_i\}] = \int_0^{\hbar\beta} d\tau \, L[\{\theta_i\}] \tag{6.34}$$

where the Euclidean Lagrangian $L[\{\theta_i\}]$ is

$$L[\{\theta_i\}] = \frac{\hbar^2 C_{\text{eff}}}{8e^2} \sum_i \dot{\theta}_i^2(\tau) + J \sum_{\langle ij \rangle} [1 - \cos\theta_{ij}(\tau)]$$
$$- \sum_{\langle ij \rangle} \int_0^{\hbar\beta} d\tau' \, \alpha_{ij}(\tau - \tau') \cos\left(\frac{\theta_{ij}(\tau) - \theta_{ij}(\tau')}{2}\right) \tag{6.35}$$

The constant J is the Josephson coupling energy and

$$\alpha_{ij}(\tau) = \alpha(\tau) = \frac{R_0}{2\pi R_N \hbar} \left(\frac{\pi}{\beta}\right)^2 \frac{1}{\sin^2(\pi\tau/\hbar\beta)} \tag{6.36}$$

where $R_0 = \hbar/e^2 = 4.11 \text{ k}\Omega$ and R_N is approximated by a T-independent normal resistance. The 4π periodicity of the dissipative term in Eq. (6.35) signifies tunneling of discretely charged quasiparticles.

We proceed by introducing the mean-field approximation, which replaces the nearest-neighbor interactions in Eq. (6.35) by an effective Gorkov field and a damping acting on a *single* superconductor. This can be achieved, formally, by using a variational principle for free energy (Feynman, 1972):

$$F \leq F_0 + \frac{1}{\hbar\beta Z_0} \int \mathcal{D}\theta \, \exp\left(-\frac{1}{\hbar}S_0[\theta]\right)(S_{\text{eff}} - S_0) = F_t \tag{6.37}$$

where S_{eff} is the effective action (6.34) and

$$Z_0 = e^{-\beta F_0} = \int \mathcal{D}\theta \, \exp\left(-\frac{1}{\hbar}S_0[\theta]\right) \tag{6.38}$$

The path integrals in Eqs. (6.37) and (6.38) are written in a simplified notation: $\{\theta_i\}$ being replaced by θ. The trial action $S_0[\theta] = \int_0^{\hbar\beta} d\tau \, L_0$ is chosen as follows

$$L_0 = \sum_i \left[\frac{\hbar^2 C_{\text{eff}}}{8e^2}\dot{\theta}_i^2(\tau) - z\gamma J \cos\theta_i(\tau) \right.$$
$$\left. - z\int_0^{\hbar\beta} d\tau' \, \Gamma(\tau - \tau')\alpha(\tau - \tau')\cos\left(\frac{\theta_i(\tau) - \theta_i(\tau')}{2}\right) \right] \tag{6.39}$$

where z is the coordination number of the array. The variational parameters γ and $\Gamma(\tau)$ are determined from Eq. (6.37) by requiring that $\delta F_t = 0$ as $\gamma \to \gamma + \delta\gamma$ and $\Gamma \to \Gamma + \delta\Gamma$. In this way, we obtain with the use of Eq. (6.39)

$$\gamma = \frac{1}{Z_0}\int \mathcal{D}\theta \, \exp\left(-\frac{1}{\hbar}S_0[\theta]\right)\cos\theta = \langle\cos\theta\rangle_0 \tag{6.40}$$

and

$$\Gamma(\tau - \tau') = \frac{1}{Z_0}\int \mathcal{D}\theta \, \exp\left(-\frac{1}{\hbar}S_0[\theta]\right)\cos\left(\frac{\theta(\tau) - \theta(\tau')}{2}\right) \tag{6.41}$$

We note that, owing to the assumption of a perfect periodic array, we may drop the subscript of the phase variable θ.

Near the phase-ordering transition temperature T_c, the expectation value $\langle\cos\theta\rangle_0$ is small, and the path integral (6.40) can be calculated by expanding $S_0[\theta]$ in parameter γ. We note that a similar perturbation expansion has been used in obtaining Eq. (3.96). In the present case, this results in the following self-consistent equation for T_c

$$1 = \frac{zJ}{2\hbar} \int_0^{\hbar\beta} d\tau \, R(\tau) \tag{6.42}$$

where $R(\tau)$ is the phase correlation function defined by

$$R(\tau) = \frac{1}{Z_u} \int \mathcal{D}\theta \, \exp\left(-\frac{1}{\hbar} S_u[\theta]\right) \exp\{i[\theta(\tau) - \theta(0)]\} \tag{6.43}$$

where

$$S_u[\theta] = \int_0^{\hbar\beta} d\tau \, L_u[\theta] \tag{6.44}$$

The Lagrangian $L_u[\theta]$ is obtained from Eq. (6.39) by discarding the Josephson coupling term. Thus, dropping the site summation

$$L_u[\theta] = \frac{\hbar^2 C_{\text{eff}}}{8e^2} \dot{\theta}^2 - z \int_0^{\hbar\beta} d\tau' \, \Gamma_u(\tau-\tau')\alpha(\tau-\tau') \cos\left(\frac{\theta(\tau) - \theta(\tau')}{2}\right) \tag{6.45}$$

where

$$\Gamma_u(\tau - \tau') = \frac{1}{Z_u} \int \mathcal{D}\theta \, \exp\left(-\frac{1}{\hbar} S_u[\theta]\right) \cos\left(\frac{\theta(\tau) - \theta(\tau')}{2}\right) \tag{6.46}$$

and

$$Z_u = \int \mathcal{D}\theta \, \exp\left(-\frac{1}{\hbar} S_u[\theta]\right) \tag{6.47}$$

The non-Gaussian form of the dissipative term in Eq. (6.45) forces us to make further approximations, in order to evaluate the path integrals (6.43) and (6.46). Calculations of the phase correlator for a normal junction indicate that the self-consistent harmonic approximation yields reasonably good results (Brown and Šimánek, 1986, 1992). Hence, we replace $L_u[\theta]$ as follows

$$L_u \to L_u^{\text{SCHA}} = \frac{\hbar^2 C_{\text{eff}}}{8e^2} \dot{\theta}^2 + \frac{z}{8} \int_0^{\hbar\beta} d\tau' \, \Gamma^{\text{SCHA}}(\tau - \tau')$$
$$\times \alpha(\tau - \tau')G(\tau - \tau')[\theta(\tau) - \theta(\tau')]^2 \tag{6.48}$$

where $G(\tau - \tau')$ is a new, SCHA variational parameter (see Section 3.9 for a simple version of the SCHA)

$$G(\tau - \tau') = \exp\left\{-\frac{1}{2}\langle[\theta(\tau) - \theta(\tau')]^2\rangle_{\text{SCHA}}\right\} \tag{6.49}$$

Consistent with Eq. (6.48), we also replace Γ_u by Γ^{SCHA} given by

$$\Gamma^{\mathrm{SCHA}}(\tau - \tau') = \left\langle \cos\left(\frac{\theta(\tau) - \theta(\tau')}{2}\right)\right\rangle_{\mathrm{SCHA}}$$

$$= \exp\left\{-\frac{1}{8}\langle[\theta(\tau) - \theta(\tau')]^2\rangle_{\mathrm{SCHA}}\right\} \qquad (6.50)$$

which is obtained from Eq. (6.46) by replacing S_u by

$$S_u^{\mathrm{SCHA}}[\theta] = \int_0^{\hbar\beta} d\tau\, L_u^{\mathrm{SCHA}}[\theta] \qquad (6.51)$$

The second term in Eq. (6.48) involves the product of Eqs. (6.49) and (6.50), which allows us to define a "combined" SCHA parameter

$$\Pi(\tau - \tau') = \Gamma^{\mathrm{SCHA}}(\tau - \tau')G(\tau - \tau')$$

$$= \exp\left\{-\frac{5}{8}\langle[\theta(\tau) - \theta(\tau')]^2\rangle_{\mathrm{SCHA}}\right\} \qquad (6.52)$$

Equation (6.52) allows us to write the Euclidean Lagrangian (6.48) as follows

$$L_u^{\mathrm{SCHA}} = \frac{\hbar^2 C_{\mathrm{eff}}}{8e^2}\dot\theta^2 + \frac{z}{8}\int_0^{\hbar\beta} d\tau'\, \Pi(\tau - \tau')\alpha(\tau - \tau')[\theta(\tau) - \theta(\tau')]^2 \qquad (6.53)$$

The phase correlator (6.43), evaluated within the SCHA approximation, is, according to Eq. (6.53), a Gaussian path integral, for which we may write

$$R(\tau) = \langle\exp\{i[\theta(\tau) - \theta(0)]\}\rangle_{\mathrm{SCHA}}$$

$$= \exp\left\{-\frac{1}{2}\langle[\theta(\tau) - \theta(0)]^2\rangle_{\mathrm{SCHA}}\right\} = [\Pi(\tau)]^{\frac{4}{5}} \qquad (6.54)$$

where we have used Eq. (6.52). The path integral in the exponent of $\Pi(\tau)$ can be performed by expanding $\theta(\tau)$ into a Fourier series [see Eqs. (3.120)–(3.126)]

$$\theta(\tau) = \sum_{n=-\infty}^{\infty} \theta_n e^{-i\omega_n\tau} \qquad (6.55)$$

where $\omega_n = 2\pi n/\hbar\beta$. Introducing (6.55) into (6.53) and performing the multiple Gaussian integration using formula (A3.20), we obtain the following integral equation for $\Pi(\tau)$:

$$\Pi(\tau) = \exp\Bigg\{-\frac{5U_{\mathrm{eff}}\beta_c}{2\pi^2}\sum_{n=1}^{\infty}\Bigg[[1 - \cos(\omega_n\tau)]$$

$$\Bigg/\left(n^2 + \frac{zU_{\mathrm{eff}}\beta_c^2}{2\pi^2}\int_0^{\hbar\beta_c} d\tau\, \Pi(\tau)\alpha(\tau)[1 - \cos(\omega_n\tau)]\right)\Bigg]\Bigg\} \qquad (6.56)$$

where we introduced the effective charging energy

$$U_{\text{eff}} = \frac{e^2}{C_{\text{eff}}} \tag{6.57}$$

Inserting the expression (6.36) into (6.56) and changing the imaginary time τ to a dimensionless variable $x = 2\pi\tau/\hbar\beta$, we obtain

$$\Pi(x) = \exp\left(-a\sum_{n=1}^{\infty} \frac{1 - \cos(nx)}{n^2 + ag\int_0^\pi dx\, \Pi(x)\frac{1-\cos(nx)}{1-\cos x}}\right) \tag{6.58}$$

where

$$a = \frac{5U_{\text{eff}}\beta_c}{2\pi^2} \tag{6.59}$$

and g is the dimensionless conductance

$$g = \frac{zR_0}{5R_N} \tag{6.60}$$

Equation (6.58) has been solved numerically for $\Pi(x)$ by the method of successive approximations (Šimánek and Brown, 1986). Once $\Pi(x)$ is known, we can obtain an implicit equation for T_c by using Eqs. (6.42) and (6.54)

$$1 = \frac{zJ}{2\pi}\beta_c \int_0^\pi dx\, [\Pi(x)]^{\frac{4}{5}} \tag{6.61}$$

Following Eq. (3.76), we define the parameter

$$\alpha = \frac{zJ}{2U_{\text{eff}}} \tag{6.62}$$

Using Eqs. (6.59) and (6.62), we can rewrite Eq. (6.61) as follows:

$$\frac{1}{\alpha} = \frac{2\pi a}{5} \int_0^\pi dx\, [\Pi(x)]^{\frac{4}{5}} \tag{6.63}$$

We note that the integral on the RHS of Eq. (6.63) is an implicit function of the parameters a and g. Regarding these parameters as fixed, the equation (6.63) is solved for a. The transition temperature T_c for the phase ordering is obtained from Eqs. (6.59) and (6.62) [see Eq. (3.130)]

$$\frac{T_c}{T_c^c} = \frac{5}{2\pi^2 a\alpha} \tag{6.64}$$

where $T_c^c = zJ/2k_B$ is the transition temperature for phase ordering of the classical model. The results for T_c/T_c^c as a function of α, for various values of the coupling parameter g, are plotted in Fig. 6.1. The changes in the phase diagram due to increasing g are consistent with the general

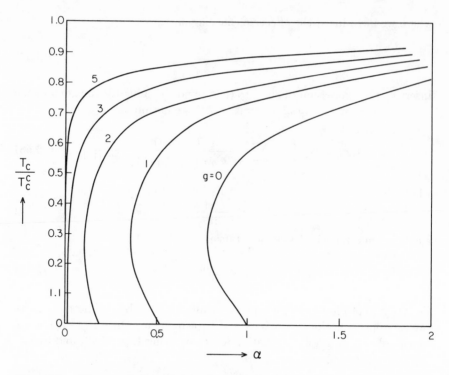

FIG. 6.1.   Phase-ordering temperature T_c/T_c^c, plotted as a function of the parameter $\alpha = zJ/2U_{\text{eff}}$. The dimensionless conductance g is defined by Eq. (6.60) (Šimánek and Brown, 1986).

expectation that dissipation tends to mitigate the effects associated with quantum fluctuations of the phase variables. First of all, we see that the phase boundaries move towards smaller values of α on increasing g. Considering the phase ordering at $T = 0$, we see that the critical value of $\alpha(T_c = 0) = \alpha_c$ decreases smoothly with g until $g \simeq 3$, where α_c drops precipitously to a nearly zero value. For $g \approx 1$, the expression (6.60) yields a critical junction resistance of order

$$R_N^c = \frac{zR_0}{5} \qquad (6.65)$$

which quantitatively disagrees with the universal critical junction resistance proposed by Abeles and Hanak (1971) [see Eq. (3.1)]. In a variety of ultrathin films the critical resistance is found to be $h/4e^2 \approx 6.45$ kΩ (Orr et al., 1985; Jaeger et al., 1986). The expression (6.65) yields (for $z = 4$) a value $R_N^c \approx 3.3$ kΩ, which also disagrees, though to a lesser degree, with the thin film data. It is interesting, in this context, to point out the recent studies of superconductor–insulator transition in granular tin films by Yamada et al. (1992). These authors found that, for a film with 280Å grains, the sheet resistance is close to $h/4e^2$, while, for that with 1200 Å, it is in the region of 20–30 kΩ. It is tempting to suggest that films with smaller grains fit into the category of the self-charging model, for which the result (6.65) has been obtained, though the assumption of gaplessness is hard to justify and the predicted R_N^c is too small.

We must now discuss further the origin of the precipitous transition of the critical value α_c taking place near $g \approx 3$. The condition for the onset of phase ordering, given in Eq. (6.61), involves the phase correlator $\Pi(x)$ of Eq. (6.58). The origin of the precipitous transition is the *anharmonicity* of the dissipative action, which is responsible for the g-proportional term on the RHS of (6.58). This term provides the feedback needed to produce *criticality* of the self-consistent solutions for $\Pi(x)$. We note that a phase correlator, of a form similar to Eq. (6.58), appears in the theory of effective conductance of normal tunnel junctions (Brown and Šimánek, 1986). The result of this theory is that a rapid crossover from activated to ohmic conductance has been obtained for the junction resistance $R_N \simeq (\pi/2)R_0/g_s$, where g_s is the dimensionless conductance that is close to 3. Though a direct quantitative comparison of this result with expression (6.65) is blurred by the *cooperative* phase correlations in an array (induced by Josephson coupling), there is little doubt that the precipitous drop of α_c, shown in Fig. 6.1, is related to the *metal–insulator transition in a single junction* that is in a *normal* state (due to the assumed gaplessness).

Kampf and Schön (1988) have also investigated the phase transition in dissipative Josephson junction in the mean-field approximation. In contrast to the work discussed in this section, they have replaced the anharmonic dissipative action by a quadratic functional of the form (4.43). The mean-field phase diagram, obtained in their work, is in qualitative agreement with that displayed in Fig. 6.1. In particular, the gradual disappearance of the reentrant features with increasing dissipation is apparent in their data. We remark, in this context, that the MFA is generally preferable to the SCHA because it preserves the reentrance due to the infinite phase interval in a dissipative array (see Section 3.5).

6.5. Dissipative self-charging model using the SCHA

Independent of the work described in Section 6.4, Chakravarty et al. (1986) investigated the model of Eq. (6.35) using a somewhat different approach. They replaced the anharmonic dissipative term by a harmonic one and studied the onset of global phase coherence using the SCHA. The idea of the SCHA was introduced in Section 3.9, and we only have to consider the modification produced by the presence of the dissipative term. According to Eq. (6.23), this term becomes (in the harmonic approximation)

$$S_D[\theta] \approx \frac{1}{8} \int_0^{\hbar\beta} d\tau \int_0^{\hbar\beta} d\tau' \sum_{\langle ij \rangle} \alpha(\tau - \tau')[\theta_{ij}(\tau) - \theta_{ij}(\tau')]^2 \qquad (6.66)$$

where the kernel $\alpha(\tau)$ is given by Eq. (5.154). Assuming an ohmiclike dissipation, we have from Eq. (5.155)

$$\alpha(\tau) = \frac{\hbar^2}{2\pi e^2 R_N \tau^2} \qquad (6.67)$$

where R_N is the shunting resistance across the nearest-neighbor grains. Introducing Eq. (6.67) into (6.66), we obtain, for the dissipative action, an expression of a form already encountered in Eq. (4.43). The same dissipative action, in a Fourier-transformed form, is given in Eq. (4.40). Using this equation, we can write Eq. (6.66) as

$$S_D^{\text{ohmic}}[\theta] = \frac{\hbar^3\beta}{8e^2 R_N} \sum_{\langle ij \rangle} \sum_n |\omega_n| \theta_{ij}(n)\theta_{ij}(-n) \qquad (6.68)$$

where $\theta_{ij}(n) = \theta_i(n) - \theta_j(n)$. Using Eq. (3.146), the expression (6.68) can be Fourier transformed to a $\mathbf{k}$ sum

$$\theta_{ij}(n) = \frac{1}{N} \sum_{\mathbf{k}} \theta_{\mathbf{k},n} \left(e^{i\mathbf{k}\cdot\mathbf{R}_i} - e^{i\mathbf{k}\cdot\mathbf{R}_j} \right) \qquad (6.69)$$

Introducing this result into Eq. (6.68), we have [see Eq. (3.152)]

$$S_D^{\text{ohmic}}[\theta] = \frac{\hbar^3\beta}{8e^2 R_N N^2} \sum_{\langle ij \rangle} \sum_{\mathbf{k},\mathbf{k}'} \sum_n e^{i(\mathbf{k}+\mathbf{k}')\cdot\mathbf{R}_i} \left(1 - e^{i\mathbf{k}\cdot\mathbf{R}_{ji}} \right)$$
$$\times \left(1 - e^{i\mathbf{k}'\cdot\mathbf{R}_{ji}} \right) \theta_{\mathbf{k},n}\theta_{\mathbf{k}',-n}|\omega_n| \qquad (6.70)$$

Performing the site summation, we obtain with the help of identity (3.153)

$$S_D^{\text{ohmic}} = \frac{\hbar^3\beta}{8e^2 R_N N} \sum_{\mathbf{k}} \sum_n f(\mathbf{k})\theta_{\mathbf{k},n}\theta_{-\mathbf{k},-n}|\omega_n| \qquad (6.71)$$

where $f(\mathbf{k})$ is the structure factor given by Eq. (3.155). Adding this result to expression (3.156), we obtain the trial action of the dissipative self-charging model in the form

$$S_{\text{tr}} = \frac{\hbar\beta}{N} \sum_{\mathbf{k}} \sum_{n=-\infty}^{\infty} \left[\frac{\hbar^2}{8U_a}\omega_n^2 + \left(\frac{1}{2}K + \frac{\hbar^2|\omega_n|}{8e^2 R_N} \right) f(\mathbf{k}) \right] |\theta_{\mathbf{k},n}|^2 \qquad (6.72)$$

We can now use this result to calculate the correlator $\langle \theta_{\mathbf{k}} \theta_{-\mathbf{k}} \rangle_{\text{tr}}$ that is needed for the evaluation of the quantity D_{ij} of Eq. (3.149). Using the formulas (3.158)–(3.163), we obtain from Eq. (6.72) the expression for "bond-averaged" D_{ij}

$$\bar{D}_{ij} = \frac{2}{N\beta d} \sum_{\mathbf{k}} \sum_{n=0}^{\infty} \frac{f(\mathbf{k})}{\left[\frac{\hbar^2}{4U_a}\omega_n^2 + \left(K + \frac{\hbar^2 \omega_n}{4e^2 R_N} \right) f(\mathbf{k}) \right]} \tag{6.73}$$

where we have *used the relation* $z = 2d$. The effective stiffness K is, according to Eq. (3.145), given by

$$K = Je^{-\frac{1}{2}\bar{D}_{ij}} \tag{6.74}$$

Following Chakravarty et al. (1986), we now derive from Eqs. (6.73) and (6.74) a self-consistent equation for the constant K. The n sum on the RHS of Eq. (6.73) is evaluated by making the "cutoff" approximation

$$\sum_{n=0}^{\infty} \frac{f(\mathbf{k})}{\left[\frac{\hbar^2}{4U_a}\omega_n^2 + \left(K + \frac{\hbar^2 \omega_n}{4e^2 R_N} \right) f(\mathbf{k}) \right]} \approx \sum_{n=0}^{n_0} \frac{1}{K + \frac{\hbar^2 \omega_n}{4e^2 R_N}} \tag{6.75}$$

where n_0 is the cutoff integer. It is assumed that the Josephson coupling is weak and the temperature is so low that $K \ll U_a$ and $U_a \gg 1/\beta$. Under these conditions, the cutoff n_0 is determined by setting the charging term in the denominator on the LHS of Eq. (6.75) equal to the dissipative term (where $\omega_n = \omega_{n_0}$)

$$\frac{\hbar^2}{4U_a}\omega_{n_0}^2 = \frac{\hbar^2 \omega_{n_0}}{4e^2 R_N} f(\mathbf{k}) \approx \frac{1}{4}\hbar\omega_{n_0} \tag{6.76}$$

On the RHS of this equation, we assumed $(\hbar/e^2 R_N)f(\mathbf{k})$ to be of order 1. We note that for $K \ll U_a$, the phase transition takes place for values of $\hbar/e^2 R_N \approx 1$. The form factor is also of order one, except near the origin of the Brillouin zone [see Eq. (3.164)]. Since the RHS of Eq. (6.75) is independent of $\mathbf{k}$, this region makes a relatively small contribution to the $\mathbf{k}$ sum in Eq. (6.73). Equation (6.76) implies for the cutoff integer a value

$$n_0 = \frac{\beta U_a}{2\pi} \tag{6.77}$$

Introducing Eq. (6.75) into (6.73) and using the fact that $(1/N) \sum_{\mathbf{k}} = 1$, we obtain

$$\bar{D}_{ij} = \frac{2}{\beta d} \sum_{n=0}^{n_0} \frac{1}{\left(\frac{\alpha_d \hbar \omega_n}{2\pi} + K \right)} = \frac{2}{d\alpha_d} \sum_{n=0}^{n_0} \frac{1}{\left(n + \frac{K\beta}{\alpha_d} \right)} \tag{6.78}$$

where we introduced the dimensionless dissipative parameter α_d

$$\alpha_d = \frac{\pi}{2}\left(\frac{\hbar}{e^2 R_N}\right) = \frac{\pi}{2}\left(\frac{R_0}{R_N}\right) \tag{6.79}$$

The summation on the RHS of Eq. (6.78) can be performed with use of the digamma function introduced in Section 4.6. Using the identity (4.55), we have

$$
\begin{aligned}
\sum_{n=0}^{n_0} \frac{1}{n+z_1} &= \sum_{n=0}^{\infty} \frac{1}{n+z_1} - \sum_{n=n_0+1}^{\infty} \frac{1}{n+z_1} \\
&= \sum_{n=0}^{\infty} \left(\frac{1}{n+z_1} - \frac{1}{n+n_0+1+z_1}\right) \\
&= \psi(n_0+1+z_1) - \psi(z_1)
\end{aligned}
\tag{6.80}
$$

where the dimensionless parameter

$$z_1 = \frac{K\beta}{\alpha_d} \tag{6.81}$$

Equations (6.77) and (6.81) imply, noting that $U_a \gg K$,

$$\frac{n_0}{z_1} = \frac{U_a \alpha_d}{2\pi K} \gg 1 \tag{6.82}$$

Moreover, since $U_a \gg 1/\beta$, we have from Eq. (6.77) that $n_0 \gg 1$. Hence, we can use the following approximation [see Eq. (4.57)]

$$\psi(n_0+1+z_1) \simeq \psi(n_0) \approx \ln n_0 \tag{6.83}$$

Introducing Eq. (6.83) into Eq. (6.80), we obtain from Eq. (6.78)

$$\bar{D}_{ij} = \frac{2}{d\alpha_d}\left[\ln n_0 - \psi\left(\frac{K\beta}{\alpha_d}\right)\right] \tag{6.84}$$

Inserting this result into Eq. (6.74), we obtain an implicit equation for the effective stiffness constant K

$$
\begin{aligned}
\frac{K}{J} &= (n_0)^{-\frac{1}{d\alpha_d}} \exp\left[\frac{1}{d\alpha_d}\psi\left(\frac{K\beta}{\alpha_d}\right)\right] \\
&= \left(\frac{2\pi}{\beta U_a}\right)^{\frac{1}{d\alpha_d}} \exp\left[\frac{1}{d\alpha_d}\psi\left(\frac{K\beta}{\alpha_d}\right)\right]
\end{aligned}
\tag{6.85}
$$

We now consider the solution of this self-consistent equation in the zero-temperature limit ($\beta \to \infty$). In this case, we can again use the asymptotic formula (4.57) to write

$$\psi\left(\frac{K\beta}{\alpha_d}\right) \xrightarrow[\beta \to \infty]{} \ln \frac{K\beta}{\alpha_d} \tag{6.86}$$

and Eq. (6.85) becomes

$$\frac{K}{J} = \left(\frac{2\pi}{\beta U_a}\right)^{\frac{1}{d\alpha_d}} \left(\frac{K\beta}{\alpha_d}\right)^{\frac{1}{d\alpha_d}} \tag{6.87}$$

Solving this equation for K, we have

$$K = \left[J\left(\frac{2\pi}{\alpha_d U_a}\right)^{\frac{1}{d\alpha_d}}\right]^{\frac{d\alpha_d}{d\alpha_d - 1}} = U_a \left(\frac{2\pi}{\alpha_d}\right)^{\frac{1}{d\alpha_d - 1}} \left(\frac{J}{U_a}\right)^{\frac{d\alpha_d}{d\alpha_d - 1}} \tag{6.88}$$

The expression on the RHS of Eq. (6.88) shows that the criticality of the stiffness constant K is determined by the product $\alpha_d d$. As $\alpha_d d$ approaches 1 from above, K increases, diverging at $\alpha_d d = 1$. On the other hand, when $\alpha_d d$ approaches 1 from below, K rapidly decreases to zero. Thus the transition from the globally coherent (K finite) to the incoherent ($K = 0$) state takes place at $\alpha_d = 1/d$ (in d dimensions). From Eq. (6.14) we also see that the size of the charging energy U_a does not affect the critical value of α_d, it only sets the energy scale via the cutoff condition (6.77). For $d = 2$, Eq. (6.79) implies that the sole critical parameter is the junction resistance, a value of $R_N^c = \pi R_0 \approx 12.9$ kΩ. On the other hand, the experimental data of Orr et al. (1985) and Jaeger et al. (1986) show a critical resistance close to $h/4e^2 = 6.45$ kΩ. Moreover, this value is found independent of the structure of the film. If these films are modeled by regular junction arrays, then a correct explanation of these experiments must yield R_N^c *independent of the coordination number* z. Since the condition $\alpha_d d = 1$ does not involve z, one would expect that the SCHA approach of this section is superior to that of Section 6.4, where R_N^c is found proportional to z [see Eq. (6.65)]. This is, however, illusory, since the dimensionality has entered the expression (6.73) by replacing z in the original formula (3.163) by $2d$. Consequently, the correct critical resistance, according to the SCHA theory of Chakravarty et al. (1986), should be given by the condition $\alpha_d = 2/z$, yielding

$$R_N^c \simeq \frac{\pi z}{4} R_0 \tag{6.89}$$

This expression shows a similar dependence on the coordination number as the expression (6.65). Hence, neither the MFA theory of Section 6.4 nor the present SCHA approach can explain the z independence of the experimental data on thin films. It should be noted, however, that the measured sheet resistance also depends on the coordination number (only for $z = 4$ is it equal to the link resistance). This may mitigate the z dependence produced by Eq. (6.89). Moreover, the reentrant features exhibited by the MFA are swept away in the SCHA approach.

6.6.   Dissipative nearest-neighbor charging model

In Section 3.4 we have shown that the nearest-neighbor charging model, for a regular array, can be described by the capacitance matrix (3.35), when the network capacitances C_0 and C satisfy the condition: $C_0 \gtrsim C$. From Eqs. (3.35) and (6.19) we obtain

$$\bar{C}_{ij} = \begin{cases} (C_0 + zC + 8a_\delta), & \text{for } i = j \\ -C, & \text{for } \langle i, j \rangle = \text{nearest neighbors} \end{cases} \tag{6.90}$$

Similar to Eq. (3.51), we write Eq. (6.90) in the matrix form

$$\widehat{\bar{C}} = (C_0 + zC + 8a_\delta)\left(\widehat{I} - \frac{1}{z(1 + \bar{\lambda})}\widehat{K}\right) \tag{6.91}$$

where

$$\bar{\lambda} = \frac{C_0 + 8a_\delta}{zC} \tag{6.92}$$

If $\bar{\lambda} \gg 1$, then the matrix (6.91) can be inverted by expanding in the parameter $1/[z(1 + \bar{\lambda})]$, yielding

$$(\bar{C})^{-1} = (C_0 + zC + 8a_\delta)^{-1}$$
$$\times \left(\widehat{I} + \frac{1}{z(1 + \bar{\lambda})}\widehat{K} + \frac{1}{z^2(1 + \bar{\lambda})^2}\widehat{K}^2 + \cdots\right) \tag{6.93}$$

Keeping only the first two terms, we obtain from this expansion the elements of Coulomb matrix in the form

$$(\bar{C})_{ij}^{-1} = \begin{cases} (C_0 + zC + 8a_\delta)^{-1}, & \text{for } i = j \\ C(C_0 + zC + 8a_\delta)^{-2}, & \text{for } \langle i, j \rangle = \text{nearest neighbors} \end{cases} \tag{6.94}$$

Inserting this matrix into Eq. (6.23), we obtain for the effective action:

$$S_{\text{eff}}[\{\theta_i\}] = \int_0^{\hbar\beta} d\tau \left[\left(\frac{a_\delta\hbar^2}{e^2} - \frac{8a_\delta^2\hbar^2}{e^2}(C_0 + zC + 8a_\delta)^{-1}\right)\sum_i \dot{\theta}_i^2(\tau)\right.$$
$$\left. - \frac{16a_\delta^2\hbar^2C}{e^2}(C_0 + zC + 8a_\delta)^{-2}\sum_{\langle ij \rangle}\dot{\theta}_i(\tau)\dot{\theta}_j(\tau)\right]$$
$$+ S_D[\{\theta_i\}] \tag{6.95}$$

where $S_D[\{\theta_i\}]$ is the same dissipative action as that in Eq. (6.25).

6.6.1. Standard capacitance network

Let us first consider the regime of $a_\delta \gg (C_0 + zC)/8$. Expanding the terms, multiplying the sums in Eq. (6.95) in the small parameter $(C_0 + zC)/8a_\delta$, we have to lowest order

$$\frac{a_\delta\hbar^2}{e^2} - \frac{8a_\delta^2\hbar^2}{e^2}(C_0 + zC + 8a_\delta)^{-1} \approx \frac{\hbar^2}{8e^2}(C_0 + zC) \tag{6.96}$$

and

$$\frac{16a_\delta^2\hbar^2 C}{e^2}(C_0 + zC + 8a_\delta)^{-2} \approx \frac{\hbar^2 C}{4e^2}\left(1 - \frac{C_0 + zC}{4a_\delta}\right) \approx \frac{\hbar^2 C}{4e^2} \qquad (6.97)$$

Introducing the approximations (6.96) and (6.97) into Eq. (6.95), we obtain the effective action of the capacitance model of Eq. (3.35), augmented by a dissipation term S_D

$$S_{\text{eff}}[\{\theta_i\}] = \int_0^{\hbar\beta} d\tau \left(\frac{\hbar^2}{8e^2}(C_0 + zC)\sum_i \dot{\theta}_i^2(\tau) - \frac{\hbar^2 C}{4e^2}\sum_{\langle ij \rangle}\dot{\theta}_i(\tau)\dot{\theta}_j(\tau)\right)$$
$$+ S_D[\{\theta_i\}] \qquad (6.98)$$

It should be remarked that this expression represents correctly the charging energy of the capacitance network of Fig. 3.4, even if the condition $C_0 \gtrsim C$, for the nearest-neighbor charging model, is not satisfied (see Section 6.7.1).

6.6.2. Limit of $8a_\delta/(C_0 + zC) \ll 1$

Next, we consider the limit of $a_\delta \ll (C_0 + zC)/8$. We also assume that $C_0 \gg zC$, so that the condition $(\bar{\lambda} \gg 1)$ for the validity of Eq. (6.95) holds. Then the expressions on the LHS of Eqs. (6.96) and (6.97) must be expanded in the small parameter $8a_\delta/(C_0 + zC)$, yielding to lowest order

$$\frac{a_\delta\hbar^2}{e^2} - \frac{8a_\delta^2\hbar^2}{e^2}(C_0 + zC + 8a_\delta)^{-1} \approx \frac{a_\delta\hbar^2}{e^2}\left(1 - \frac{8a_\delta}{C_0 + zC}\right) \approx \frac{a_\delta\hbar^2}{e^2} \qquad (6.99)$$

and

$$\frac{16a_\delta^2\hbar^2 C}{e^2}(C_0 + zC + 8a_\delta)^{-2} \approx \frac{16a_\delta^2\hbar^2 C}{e^2(C_0 + zC)^2}\left(1 - \frac{16a_\delta}{C_0 + zC}\right)$$
$$\approx \left(\frac{a_\delta\hbar^2}{e^2}\right)\frac{16a_\delta C}{(C_0 + zC)^2} \qquad (6.100)$$

We observe that the RHS of Eq. (6.100) in this regime is much smaller than $a_\delta\hbar^2/e^2$. Then the approximate effective action (6.95) takes a diagonal form

$$S_{\text{eff}}[\{\theta_i\}] \approx \frac{a_\delta\hbar^2}{e^2}\int_0^{\hbar\beta} d\tau \sum_i \dot{\theta}_i^2(\tau) + S_D[\{\theta_i\}] \qquad (6.101)$$

Hence, on decreasing the ratio $8a_\delta/(C_0 + zC)$ well below one, the nearest-neighbor charging model crosses over to an *effective* self-charging model. Physically, this follows from the fact that, owing to the large values of $(C_0 + zC)/8a_\delta$, the electrostatic energies associated with excess charges on the grains are dominated by the energies caused by the δ effect, the latter energies being inversely proportional to the parameter a_δ. The realizability of this regime has been discussed for an isolated junction in Section 5.5.4. For an array with intergrain capacitance dominated by quasiparticle tunneling, we have, using Eq. (5.199) and taking into account the Debye-screening (enhancement) factor Ω/St

$$\frac{zC}{8a_\delta} \simeq \frac{zC_{\text{ren}}}{8a_\delta} = \frac{3\pi\hbar\delta z\Omega}{64\Delta_0 R_N e^2 St} \qquad (6.102)$$

Assuming that R_N is equal to the universal critical resistance (3.2), the criterion $(C_0 + zC)/8a_\delta \gtrsim 1$ translates into an inequality for the grain surface area [see Eq. (5.204)]

$$S \lesssim \frac{0.2\pi^2\hbar^2}{mk_B T_c^{(0)}} = \bar{S}_0 \qquad (6.103)$$

We note that Eq. (6.103) results from neglecting C_0 compared to zC. Since the nearest-neighbor charging model requires $C_0 \gtrsim C$, this assumption may not be satisfied. However, including C_0 makes the inequality work with even larger value of $\bar{S}_0$. For an array of aluminum particles with $T_c^{(0)} \approx 1.2$ K, the characteristic surface area is $\bar{S}_0 \approx 0.14 \times 10^{-10}$ cm^2, which corresponds to a spherical particle of diameter 207 Å. For particles smaller than this value, the parameter a_δ should start playing a significant role in the charging action.

6.7.   Models with tunneling-dominated junction capacitance

6.7.1.   *Mean-field theory*

We now pass on to the MFA theories of the superconductor–semiconductor phase transition, in which a quasiparticle conductance is not assumed to persist all the way down to $T = 0$. Chakravarty et al. (1987) and Ferrell and Mirhashem (1988) studied models in which the geometric intergrain capacitance is dominated by the quasiparticle term ΔC. The starting point of these studies is the standard capacitance network which has been discussed in Section 6.6.1. The intergrain capacitance is equal to ΔC, where, according to Eq. (5.197)

$$\Delta C = C_{\text{ren}} = \frac{3\pi\hbar}{32\Delta_0 R_N} \qquad (6.104)$$

From Eq. (6.98) we obtain the effective action

$$S_{\text{eff}}[\{\theta_i\}] = \int_0^{\hbar\beta} d\tau \left(\frac{\hbar^2}{8e^2}(C_0 + z\Delta C) \sum_i \dot{\theta}_i^2 - \frac{\hbar^2\Delta C}{4e^2} \sum_{\langle ij \rangle} \dot{\theta}_i\dot{\theta}_j \right.$$

$$\left. + J \sum_{\langle ij \rangle} \cos\theta_{ij} \right) \qquad (6.105)$$

In Eq. (6.105) we have assumed temperatures well below the BCS transition temperature, so that the quasiparticle *conductance channel is suppressed.* The dissipative action S_D then reduces to the capacitance term proportional to ΔC and the Josephson coupling term. A further assumption made is that the grains are sufficiently small so that the self-capacitance C_0 is negligible compared to ΔC. This implies that the Coulomb matrix is long range so that the model is quite *remote* from the nearest-neighbor charging model (see Section 3.4.2). Consequently, the criterion for phase ordering is modified as will be shown here.

The superconductor–semiconductor transition, for the model discussed, can be studied with use of the mean-field theory for the off-diagonal model represented by the Hamiltonian (3.91). As $T \to 0$, the self-consistent equation (3.109) reduces to (see Section 6.8.1)

$$1 = \frac{zJ}{2U_{11}} \qquad (6.106)$$

where U_{11} is the diagonal element of the Coulomb interaction, defined by the charging Hamiltonian [see Eq. (3.91)]

$$\widehat{H}_c = 2 \sum_{\langle ij \rangle} U_{ij} \widehat{n}_i \widehat{n}_j \qquad (6.107)$$

where $\widehat{n}_i$ is the number of excess pairs on grain i. Comparing the expression (6.107) with the general form for the electrostatic energy given in Eq. (3.30), we obtain, using the relation $Q_i = 2en_i$

$$U_{ij} = e^2 \left(\widehat{C}^{-1} \right)_{ij} \qquad (6.108)$$

According to Eq. (6.105), the capacitance matrix takes the form

$$\widehat{C}_{ij} = \begin{cases} C_0 + z\Delta C, & \text{for } i = j \\ -\Delta C, & \text{for } \langle ij \rangle = \text{nearest neighbors} \end{cases} \qquad (6.109)$$

As pointed out in Section 3.4, for $C_0 \ll z\Delta C$, the parameter λ is small and the expansion (3.52) does not converge. Consequently, the lowest-order result (3.53) cannot be used for the diagonal element of the Coulomb matrix. Instead, we invert the matrix (6.109) by using the method of the Fourier transform. Confining ourselves to a two-dimensional square array, we have from Eqs. (3.42)–(3.44)

$$\left(\widehat{C}^{-1} \right)_{ii} = \widehat{C}^{-1}(\mathbf{R} = 0) = \frac{1}{N} \sum_{\mathbf{k}} \left(\widehat{C}_{\mathbf{k}} \right)^{-1}$$

$$\simeq \frac{1}{N} \sum_{\mathbf{k}} \frac{1}{C_0 + a^2 k^2 \Delta C} \qquad (6.110)$$

where we have used the long-wavelength approximation (3.42) for the summand. This is justified, since for $C_0/(z\Delta C) \ll 1$ the $\mathbf{k}$ summation is dominated by small values of the wave vector k. However, distinct from the calculation of $\widehat{C}^{-1}(\mathbf{R} \gg a)$, the evaluation of $\widehat{C}^{-1}(\mathbf{R} = 0)$ must use a proper finite upper cutoff $k_D = 2\sqrt{\pi}/a$. Thus we obtain, converting the summation in Eq. (6.110) to an integral

$$
U_{11} = e^2 \left(\widehat{C}^{-1}\right)_{11} = \frac{a^2 e^2}{2\pi} \int_0^{k_D} \frac{k \, dk}{C_0 + a^2 k^2 \Delta C}
$$
$$
= \frac{e^2}{4\pi\Delta C} \ln\left(1 + \frac{4\pi\Delta C}{C_0}\right) \tag{6.111}
$$

Inserting this result into Eq. (6.106), we obtain a criterion for the onset of global coherence

$$
1 = \frac{2\pi z J \Delta C}{e^2 \ln\left(1 + \frac{4\pi\Delta C}{C_0}\right)} \tag{6.112}
$$

Substituting for ΔC and J the expressions (6.104) and (5.147), respectively, Eq. (6.112) can be written as a condition for the critical resistance R_N^c

$$
R_N^c = \frac{1}{8}\sqrt{3\pi^3 z}\left(\frac{\hbar}{e^2}\right)\left[\ln\left(1 + \frac{3\pi^2\hbar}{8\Delta_0 R_N^c C_0}\right)\right]^{-\frac{1}{2}} \tag{6.113}
$$

This result was first obtained, using a somewhat different method (see Section 6.8), by Chakravarty et al. (1987). The criterion obtained by Ferrell and Mirhashem (1988) differs from Eq. (6.113) by the absence of the logarithmic factor. For an array with coordination number z, they find the critical resistance to be

$$
R_N^c = \frac{\sqrt{3}\pi z}{16}\left(\frac{\hbar}{e^2}\right) \tag{6.114}
$$

The logarithmic correction factor is responsible for relatively weak deviations of R_N^c from universality. For example, changing the ratio of the self-charging energy $e^2/2C_0$ to the gap Δ_0 by an order of magnitude produces a fifty percent change in R_N^c. Hence, Eq. (6.114) can be regarded only as qualitatively correct, whereas expression (6.113) is more accurate.

6.7.2.   *Onset of superconductivity and the δ effect*

We are now in a position to derive a generalized criterion for the onset of superconductivity, which takes into account the δ effect. First, we note that the expression (6.113) is assumed to hold for values of C_0 satisfying the inequality $C_0 \ll z\Delta C$. This inequality is expected to hold in ultrathin films, containing *very small* grains. For such grains, however, the electron level separation δ may be comparable to the intergrain charging energy, so that the "δ effect" plays a significant role. Then, the correct starting point is the effective action (6.23) which, for $T \to 0$, reduces to

$$S_{\text{eff}}[\{\theta_i\}] = \int_0^{\hbar\beta} d\tau \left(\frac{a_\delta \hbar^2}{e^2} \sum_i \dot{\theta}_i^2 - \frac{8a_\delta^2 \hbar^2}{e^2} \sum_{ij} (\bar{C}^{-1})_{ij} \dot{\theta}_i \dot{\theta}_j \right.$$

$$\left. + \frac{\hbar^2 \Delta C}{8e^2} \sum_{\langle ij \rangle} \dot{\theta}_{ij}^2 - J \sum_{\langle ij \rangle} \cos \theta_{ij} \right) \qquad (6.115)$$

where the matrix $\bar{C}_{ij}$ is, according to Eq. (6.19), given by

$$\bar{C}_{ij} = \begin{cases} 8a_\delta + C_0 + zC, & \text{for } i = j \\ -C, & \text{for } \langle ij \rangle = \text{nearest neighbors} \end{cases} \qquad (6.116)$$

Distinct from the model of Section 6.7.1, the geometric capacitance C is now retained in the expression for $\bar{C}_{ij}$. To derive the criterion for the onset of superconductivity, we define an effective charging energy, H_c^{eff}, in which all phase inertial terms of the action (6.115) are lumped into a single-capacitance matrix $\tilde{C}_{ij}$

$$H_c^{\text{eff}} = \frac{\hbar^2}{8e^2} \sum_{ij} \tilde{C}_{ij} \dot{\theta}_i \dot{\theta}_j \qquad (6.117)$$

Equations (6.115)–(6.117) imply

$$\tilde{C}_{ij} = (C_\delta + z\Delta C)\delta_{ij} - \Delta C \delta(\mathbf{R}_i, \mathbf{R}_j + \mathbf{h}) - C_\delta^2 (\bar{C}^{-1})_{ij} \qquad (6.118)$$

where $\mathbf{h}$ is the lattice translation vector and C_δ is the effective capacitance associated with the δ effect, defined as

$$C_\delta = 8a_\delta \qquad (6.119)$$

Applying to the model (6.115) the MFA method of Section 3.7, the condition for the onset of phase ordering at $T = 0$ is [see Eq. (3.109)]

$$1 = \frac{zJ}{2\tilde{U}_{11}} \qquad (6.120)$$

where $\tilde{U}_{ij}$ is related to the capacitance matrix $\tilde{C}_{ij}$ by

$$\tilde{U}_{ij} = e^2 (\tilde{C}^{-1})_{ij} \qquad (6.121)$$

In what follows, we use again the method of the Fourier transform to invert the matrix (6.118). Using Eq. (3.43), we obtain for a two-dimensional planar network (in the long-wavelength approximation)

$$\tilde{C}(k) = C_\delta + \Delta C a^2 k^2 - \frac{C_\delta^2}{C_\delta + C_0 + Ca^2 k^2}$$

$$= \Delta C a^2 k^2 + \frac{C_\delta (C_0 + Ca^2 k^2)}{C_\delta + C_0 + Ca^2 k^2} \qquad (6.122)$$

The second term on the RHS of this equation corresponds to capacitors C_δ and $C(k) = C_0 + Ca^2k^2$ connected in *series*. Hence, if $C_\delta \ll C(k)$, the capacitance C_δ acts as a buffer, diminishing the contribution of the geometric capacitance to $\widetilde{C}$. Assuming that the grains are of spherical shape, the geometric capacitances are proportional to the grain diameter d, whereas C_δ is proportional to d^2 [see Eq. (6.30)]. This implies that the inequality $C_\delta \ll C(k)$ may hold for ultrathin films (where d is very small).

To obtain a more complete picture of how the geometric capacitances affect the condition for R_N^c, let us consider first the case of relatively large grains, for which $C_\delta \gg C_0 + zC$. Then the expression (6.122) yields, to first order in $C(k)/C_\delta$

$$\widetilde{C}(k) \simeq \Delta C a^2 k^2 + C_0 + C a^2 k^2 \qquad (6.123)$$

This result implies [upon recalling Eqs. (6.110) and (6.111)] that $\widetilde{U}_{11}$ can be obtained from Eq. (6.111) simply by replacing ΔC by $\Delta C + C$. Thus the criterion (6.112) is modified as follows

$$1 = \frac{2\pi z J(\Delta C + C)}{e^2 \ln\left(1 + \frac{4\pi(\Delta C + C)}{C_0}\right)} \qquad (6.124)$$

From this expression, it is evident that the universality of critical resistance is spoiled, mainly because of the presence of the geometric capacitance C in the numerator. We note that the logarithmic term in the denominator perturbs the universality in a much milder way. The role of the capacitance C in Eq. (6.124) is measured by the ratio $\Delta C/C$, which can be expressed, using Eq. (6.104), as

$$\frac{\Delta C}{C} = \frac{3\pi\hbar}{32\Delta_0 R_N C} \qquad (6.125)$$

Letting $R_N = h/4e^2$, this expression implies that $\Delta C/C \simeq 1$ when the capacitance C reaches the value

$$C \simeq \frac{3e^2}{16\Delta_0} \qquad (6.126)$$

If we apply this expression to aluminum, where $\Delta_0 \simeq 1.7 \times 10^{-4}$ eV, we obtain $C \simeq 1.8 \times 10^{-16}$ F. The latter value happens to be an order of magnitude larger than the estimates of the geometric capacitance in ultrathin films, made by Jaeger et al. (1989). However, C may play a measurable role in the condition (6.124), when applied to experimental studies on granular films with larger grains (Yamada et al., 1992).

Let us now turn to the possible "buffer role" of C_δ. We assume that $C_\delta \ll C(k)$ and expand Eq. (6.122), yielding, to first order in $C_\delta/C(k)$

$$\widetilde{C}(k) \simeq a^2 k^2 \Delta C + C_\delta \qquad (6.127)$$

Using this expression, we calculate with the help of Eqs. (6.111) and (6.121) the Coulomb matrix element needed in Eq. (6.120)

$$\tilde{U}_{11} = e^2 \left(\tilde{C}^{-1} \right)_{11} = \frac{e^2}{4\pi\Delta C} \ln \left(1 + \frac{4\pi\Delta C}{C_\delta} \right) \tag{6.128}$$

Introducing this result into Eq. (6.120), the condition for the critical resistance R_N^c is obtained in the form [see Eq. (6.113)]

$$R_N^c = \frac{1}{8} \sqrt{3\pi^3 z} \frac{\hbar}{e^2} \left[\ln \left(1 + \frac{3\pi^2 \hbar}{8\Delta_0 R_N^c C_\delta} \right) \right]^{-\frac{1}{2}} \tag{6.129}$$

We see that in the limit $C_\delta \ll C_0 + zC$, the geometric capacitances disappear entirely from the criterion for R_N^c, owing to the capacitance C_δ acting as a buffer. To check the realizability of the condition $C_\delta \ll C(k)$, we make an estimate of the ratio $C_\delta/\Delta C$. This ratio gives us qualitative information on $C_\delta/C(k)$, since $C(k) \lesssim C_0 + zC$ plays a role only if comparable to ΔC. According to Eq. (6.30), C_δ can be written as

$$C_\delta = \frac{3nSe^2}{2\epsilon_F k_F} \tag{6.130}$$

where S is the surface area of the grain. Using this result and Eq. (6.104), we have, assuming $R_N = R_N^c = h/4e^2$

$$\frac{C_\delta}{\Delta C} = \frac{8\Delta_0 nS}{\epsilon_F k_F} \tag{6.131}$$

Let us consider again an ultrathin aluminum film, for which the onset of global superconductivity has been found near a nominal thickness of about 50 Å (Jaeger et al., 1989). Introducing for S the surface of a spherical grain of diameter 50 Å, Eq. (6.131) yields $C_\delta/\Delta C \simeq 0.1$, indicating that the δ effect may be important. It should be noted, however, that for Al grains of this size, the level splitting δ satisfies the condition (3.6), so that the spontaneous superconducting ordering within such grains may be inhibited. Moreover, as suggested by Eq. (3.12), the temperature width of the critical fluctuation region is of order one, indicating that strong fluctuations of the amplitude of the order parameter are present. Jaeger et al. (1989) present experimental arguments for the absence of such fluctuations and suggest that the ultrathin films contain nonspherical clusters of size larger than the film thickness (see the discussion after Eq. (3.12) for a similar cluster description). In view of this, the estimate given of $C_\delta/\Delta C \simeq 0.1$ should be revised to a larger value. Even if $C_\delta \ll C(k)$, this theory is still unable to explain the observed universality of R_N^c. This is because the expression (6.129) involves material- and structure-dependent constants Δ_0 and C_δ, yielding a weakly nonuniversal value of R_N^c. Another interesting problem is presented by the already mentioned experiments by Yamada et al. (1992), in which R_N^c is found to increase about four times, as the grain size increases from 280 to 1200 Å. A possible explanation of this observation can be made by referring to Eq. (6.124) (Šimánek, 1993). If the ratio $C/\Delta C$ approaches one, we expect a sizable increase of R_N^c from its "universal" value based on Eq. (6.113). According to Eq. (6.126), this takes place when $C \simeq 5.8 \times 10^{-17}$ F. It is possible that geometric intergrain capacitances of this order are present in the 1200 Å film, contributing to the predicted increase of R_N^c. The observation of nearly universal resistance $R_N^c \simeq 5.5k$ Ω for the 280 Å film still remains, however, unexplained in the framework of models in which dissipation is absent at $T = 0$.

We conclude this section with a conjecture about a possible explanation of the observed universal resistance in granular films. According to Sections 6.4 and 6.5, the dissipative self-charging model yields a material- and thickness-independent value of R_N^c, if a normal conductance channel persists down to $T = 0$. Now, the quasiparticle conduction mechanism contributes only if the grains are gapless superconductors. For 280 Å particles, the gaplessness seems, however, excluded since the Ginzburg fluctuation width is, according to Eq. (3.12), only about 3×10^{-2}. An interesting alternative, which allows conduction down to $T = 0$, has been proposed by Fisher et al. (1990). The Cooper pairs are represented by repulsively interacting bosons moving in a random potential that is caused by the film disorder. At the superconductor–insulator transition, the gapless Bose system exhibits a finite conductivity at $T = 0$, since the Cooper pairs are capable of ordinary diffusion. It should be interesting to develop a phenomenological model for the effective action by considering the coupling of the phase variables to the gapless excitations of this Bose system (see Section 4.3). With such an action, the onset of superconductivity could be studied using the method of Section 6.5, thus avoiding problems with the nonuniversal logarithmic terms.

6.8. Mapping to a quantum Ginzburg–Landau functional

We now review the basic steps of the coarse-graining method, which Doniach (1981) has developed to map the problem of a two-dimensional Josephson array, with charging energy, to a $(2+1)$-dimensional Ginzburg–Landau functional. Our goal is to clarify the connection between the MFA method, yielding the condition (6.106), and the method used by Chakravarty et al. (1987) for the onset of global phase coherence. The starting point is the partition function

$$Z_G = \prod_i \int \mathcal{D}\theta_i \, \exp\left(-\frac{1}{\hbar} S_{\text{eff}}[\{\theta_i\}]\right) \qquad (6.132)$$

where the effective action is given by

$$S_{\text{eff}}[\{\theta_i\}] = S_0[\{\theta_i\}] + S_1[\{\theta_i\}] \qquad (6.133)$$

The functional $S_0[\{\theta_i\}]$ represents the charging part of the action [see Eq. (6.105)] and $S_1[\{\theta_i\}]$ is the Josephson part

$$S_1[\{\theta_i\}] = -J \int_0^{\hbar\beta} d\tau \sum_{\langle ij \rangle} \cos\theta_{ij}(\tau) \qquad (6.134)$$

We assume that $T \simeq 0$, so that the dissipative action reduces to a capacitance term proportional to ΔC and a Josephson coupling term. In the method of Doniach (1981), Eq. (6.134) is rewritten as a *quadratic* functional of new variables $\psi_i(\tau)$, which are related to the original phase variables $\theta_i(\tau)$ by

$$\psi_i(\tau) = e^{i\theta_i(\tau)} \qquad (6.135)$$

Using this relation we have

$$\sum_{\langle ij \rangle} \cos\theta_{ij}(\tau) = \frac{1}{2} \sum_{\langle ij \rangle} [\psi_i^*(\tau)\psi_j(\tau) + \text{c.c.}]$$

$$= N \sum_{\mathbf{k}} A(\mathbf{k})\psi_{\mathbf{k}}^*(\tau)\psi_{\mathbf{k}}(\tau) \qquad (6.136)$$

where $\psi_{\mathbf{k}}(\tau)$ is the Fourier transform of the site variable $\psi_i(\tau)$, defined via

$$\psi_i(\tau) = \sum_{\mathbf{k}} \psi_{\mathbf{k}}(\tau)e^{i\mathbf{k}\cdot\mathbf{R}_i} \qquad (6.137)$$

and $A(\mathbf{k})$ is given by

$$A(\mathbf{k}) = \frac{1}{2} \sum_{\mathbf{h}} \cos(\mathbf{k}\cdot\mathbf{h}) \qquad (6.138)$$

where $\mathbf{h}$ is a set of z lattice vectors and the factor of $1/2$ ensures that each bond $\langle ij \rangle$ in (6.136) is counted only once. Let us now consider the part of the integrand of Eq. (6.132) involving the action S_1. Discretizing the τ variable, we can write this part, with use of Eqs. (6.134) and (6.136), as

$$Z_1[\theta] = \exp\left(-\frac{1}{\hbar}S_1[\{\theta_i\}]\right)$$

$$= \exp\left(\frac{JN\Delta\tau}{\hbar} \sum_{\mathbf{k},i} A(\mathbf{k})\psi_{\mathbf{k}}^*(\tau_i)\psi_{\mathbf{k}}(\tau_i)\right) \qquad (6.139)$$

The RHS of this equation can be transformed by means of the Stratonovich transformation, similar to that described in Section 5.2.2. Applying the identity (5.10) at each fixed pair of values $(\mathbf{k}, \tau_i)$, the functional $Z_1[\theta]$ can be written as a path integral over the complex stochastic variable $\varphi_{\mathbf{k}}(\tau)$

[see Eq. (5.19)]

$$Z_1[\theta] \sim \int \mathcal{D}^2\varphi_{\mathbf{k}}(\tau) \exp\left(-\frac{1}{\hbar}\int_0^{\hbar\beta} d\tau \sum_{\mathbf{k}}\left\{|\varphi_{\mathbf{k}}(\tau)|^2\right.\right.$$
$$\left.\left.-\sqrt{JNA(\mathbf{k})}[\psi_{\mathbf{k}}(\tau)\varphi_{\mathbf{k}}^*(\tau) + \text{c.c.}]\right\}\right) \qquad (6.140)$$

Introducing this result into Eq. (6.132) and performing the path integration over the variables $\theta_i(\tau)$, we obtain

$$Z_G \sim \prod_i \int \mathcal{D}\theta_i \exp\left(-\frac{1}{\hbar}S_0[\{\theta_i\}]\right) Z_1[\theta]$$

$$\sim \int \mathcal{D}^2\varphi_{\mathbf{k}}(\tau) \exp\left(-\frac{1}{\hbar}\int_0^{\hbar\beta} d\tau \sum_{\mathbf{k}}|\varphi_{\mathbf{k}}(\tau)|^2\right)$$

$$\times \left\langle \exp\left(-\frac{1}{\hbar}\int_0^{\hbar\beta} d\tau\, F(\tau)\right)\right\rangle_0 Z_0 \qquad (6.141)$$

where

$$F(\tau) = \sum_{\mathbf{k}} \sqrt{JNA(\mathbf{k})}[\psi_{\mathbf{k}}(\tau)\varphi_{\mathbf{k}}^*(\tau) + \text{c.c.}] \qquad (6.142)$$

On the RHS of (6.141) we have introduced the average

$$\langle\cdots\rangle_0 = \frac{1}{Z_0}\prod_i \int \mathcal{D}^2\theta_i \exp\left(-\frac{1}{\hbar}S_0[\{\theta_i\}]\right)\cdots \qquad (6.143)$$

where

$$Z_0 = \prod_i \int \mathcal{D}^2\theta_i \exp\left(-\frac{1}{\hbar}S_0[\{\theta_i\}]\right) \qquad (6.144)$$

Using the method of cumulant expansion, the expectation value on the RHS of Eq. (6.141) becomes, to lowest order in $F(\tau)$

$$\left\langle \exp\left(-\frac{1}{\hbar}\int_0^{\hbar\beta} d\tau\, F(\tau)\right)\right\rangle_0$$

$$= \left\langle 1 - \frac{1}{\hbar}\int_0^{\hbar\beta} d\tau\, F(\tau) + \frac{1}{2\hbar^2}\int_0^{\hbar\beta} d\tau_1 \int_0^{\hbar\beta} d\tau_2\, F(\tau_1)F(\tau_2) + \cdots \right\rangle_0$$

$$\approx \exp\left(\frac{1}{2\hbar^2}\int_0^{\hbar\beta} d\tau_1 \int_0^{\hbar\beta} d\tau_2\, \langle F(\tau_1)F(\tau_2)\rangle_0\right) \qquad (6.145)$$

Using Eq. (6.142), we have

$$\langle F(\tau_1)F(\tau_2)\rangle_0 = JN\sum_{\mathbf{k}} A(\mathbf{k})\,[\varphi_{\mathbf{k}}^*(\tau_1)\varphi_{\mathbf{k}}(\tau_2)$$
$$\times \langle\psi_{\mathbf{k}}^*(\tau_2)\psi_{\mathbf{k}}(\tau_1)\rangle_0 + \text{c.c.}] \qquad (6.146)$$

Next we Fourier transform the τ variable to a discrete, even Matsubara frequency $\omega_m = 2\pi m/\hbar\beta$. We define the following functions of $\mathbf{k}$ and ω_m

$$\varphi(\mathbf{k}, \omega_m) = \int_0^{\hbar\beta} d\tau \, e^{i\omega_m \tau} \varphi_{\mathbf{k}}(\tau) \tag{6.147}$$

and

$$Y(\mathbf{k}, \omega_m) = \int_0^{\hbar\beta} d\tau \, e^{i\omega_m \tau} \langle \psi_{\mathbf{k}}^*(\tau) \psi_{\mathbf{k}}(0) \rangle_0 \tag{6.148}$$

Using Eq. (6.147), we obtain for the exponent on the RHS of Eq. (6.141)

$$\frac{1}{\hbar} \int_0^{\hbar\beta} d\tau \sum_{\mathbf{k}} |\varphi(\mathbf{k}, \tau)|^2 = \frac{1}{\hbar^2\beta} \sum_{\mathbf{k}, \omega_m} |\varphi(\mathbf{k}, \omega_m)|^2 \tag{6.149}$$

Now we can Fourier transform the exponent on the RHS of Eq. (6.145), yielding, with the help of Eqs. (6.146)–(6.148)

$$\frac{1}{2\hbar^2} \int_0^{\hbar\beta} d\tau_1 \int_0^{\hbar\beta} d\tau_2 \, \langle F(\tau_1) F(\tau_2) \rangle_0$$
$$= \frac{JN}{\hbar^3\beta} \sum_{\mathbf{k}, \omega_m} A(\mathbf{k}) Y(\mathbf{k}, \omega_m) |\varphi(\mathbf{k}, \omega_m)|^2 \tag{6.150}$$

Introducing the expressions (6.149) and (6.150) into Eq. (6.141), the partition function Z_G becomes a path integral over the stochastic field $\varphi(\mathbf{k}, \omega_m)$

$$Z_G \sim \int \mathcal{D}^2 \varphi(\mathbf{k}, \omega_m) \exp\left[-\frac{1}{\hbar^2\beta} \sum_{\mathbf{k}, \omega_m} |\varphi(\mathbf{k}, \omega_m)|^2 \right.$$
$$\left. \times \left(1 - \frac{JN}{\hbar} A(\mathbf{k}) Y(\mathbf{k}, \omega_m) \right) + \cdots \right] \tag{6.151}$$

where the ellipsis stands for the quartic and higher-order terms in $\varphi(\mathbf{k}, \omega_m)$, which are generated by carrying out the cumulant expansion (6.145) beyond second order. As $T \to 0$, the discrete Matsubara frequencies ω_m become continuous, and the exponent of (6.151) can be identified with the action of a $(2+1)$-dimensional Ginzburg–Landau model [see Eq. (2.152)].

6.8.1. Onset of global phase coherence at $T = 0$

In the framework of the Ginzburg–Landau model, the global phase coherence sets in when the average order parameter $\varphi(\mathbf{k} = 0, \omega_m = 0)$ first becomes nonzero. From Eq. (6.151) we see that this takes place when the following condition holds

$$1 = \frac{JN}{\hbar} A(\mathbf{k} = 0) Y(\mathbf{k} = 0, \omega_m = 0) \tag{6.152}$$

Using Eqs. (6.137), (6.138), and (6.148), the condition (6.152) can be written in the form

$$1 = \frac{zJ}{2\hbar} \sum_{i(j)} \int_0^{\hbar\beta} d\tau \left\langle e^{i\theta_i(\tau)} e^{-i\theta_j(0)} \right\rangle_0 \tag{6.153}$$

where the summation is over all i sites with the fixed site j excluded. The condition (6.153) agrees with the self-consistency condition (3.96). The explicit evaluation of the phase correlator has been done in Section 3.7, resulting in a condition (3.109) for the transition temperature. Setting $T_c = 0$ in Eq. (3.109), the condition for the onset of phase coherence at $T = 0$ becomes

$$1 = \frac{zJ}{2U_{11}} \tag{6.154}$$

in agreement with Eq. (6.106). This clarifies the relation between the Ginzburg–Landau formulation and that based on Efetof theory for the condition of the onset of coherence at $T = 0$ (Efetof, 1980).

The mean-field instability criterion (6.154) should be compared with the MFA condition (3.20) for the transition temperature T_c^c of the classical three-dimensional xy model. Then we see that the critical value of U_{11} [satisfying condition (6.154)] is in direct correspondence with T_c^c. Clearly, U_{11} is the parameter that, at $T = 0$, drives the instability of the model (6.115). Now it is known that the *actual transition temperature* T_c of the three-dimensional xy model is *smaller* then the MFA value T_c^c. From the correspondence shown, we expect that the critical value of U_{11} will also be less than implied by the MFA criterion (6.154). This can be understood by recalling that fluctuations (neglected within MFA) tend to destroy the order so that the instability can take place with a smaller value of U_{11}. We can substantiate this conclusion in a more formal way, by considering the quartic terms in the Ginzburg–Landau expansion (6.151). Denoting the coefficient of the $|\varphi(\mathbf{k}, \omega_m)|^2$ at $\mathbf{k} = 0$, $\omega_m = 0$ by u_2, we have, using Eqs. (6.152) and (6.154),

$$u_2(\alpha) = (\alpha_c^0 - \alpha) \tag{6.155}$$

where $\alpha_c^0 = 1$ and

$$\alpha = \frac{zJ}{2U_{11}} \tag{6.156}$$

We note that Eq. (6.155) corresponds to the inverse of the propagator $g_0(\mathbf{k}, \omega) = \langle \varphi(\mathbf{k}, \omega) \varphi(-\mathbf{k}, \omega) \rangle_0$ calculated at $\mathbf{k} = 0$ and $\omega = 0$, within the Gaussian approximation. Let us consider the correction to this propagator produced by the quartic terms in Eq. (6.151). In the Hartree approximation, the renormalized propagator $g(\mathbf{k}, \omega)$ is obtained from the Dyson equation

$$g(\mathbf{k}, \omega) = g_0(\mathbf{k}, \omega) - \lambda u_4 g_0(\mathbf{k}, \omega) D g(\mathbf{k}, \omega) \tag{6.157}$$

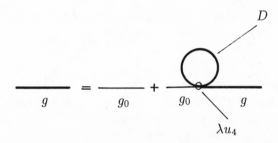

FIG. 6.2. Dyson's equation for the renormalized propagator g in the Hartree approximation. The bare propagator g_0 corresponds to the Gaussian approximation of the quantum Ginzburg–Landau functional of the variable φ. The closed-loop contribution D is defined by Eq. (6.158), u_4 is the coefficient of the quartic term in Eq. (6.151), and λ is a constant of order 1.

where λ is a constant of order one, u_4 is the coefficient of the quartic term in (6.151), and D is the closed-loop contribution (see Fig. 6.2)

$$D = \frac{1}{(2\pi)^3} \int d^2\mathbf{q} \int d\omega\, g(\mathbf{k}, \omega) \tag{6.158}$$

From Eq. (6.157) we have

$$\frac{1}{g} = \frac{1}{g_0} + \lambda u_4 D \tag{6.159}$$

Taking $\mathbf{k} = 0$, $\omega = 0$, this equation can be written

$$u_2^{\mathrm{ren}}(\alpha) = u_2(\alpha) + \lambda u_4 D(\alpha) \tag{6.160}$$

where u_2^{ren} is the renormalized coefficient of the quadratic term in (6.151). In view of Eq. (6.155), Eq. (6.160) can be written as

$$\alpha_c - \alpha = \alpha_c^0 - \alpha + \lambda u_4 D(\alpha) \tag{6.161}$$

To obtain the shift of α_c with respect to α_c^0, we let $\alpha = \alpha_c$ in this equation and obtain

$$\alpha_c = \alpha_c^0 + \lambda u_4 D(\alpha_c) \tag{6.162}$$

From Eqs. (6.158)–(6.161), it is easy to see that $D(\alpha_c)$ is a positive constant independent of α_c. Thus, Eq. (6.160) implies that $\alpha_c > \alpha_c^0 = 1$ so that the critical value of U_{11} is *smaller than its mean-field value*. As a result of this shift, the condition for the onset of global coherence, given in Eq. (6.112), is replaced by an *inequality*

$$1 < \frac{2\pi z J \Delta C}{e^2 \ln\left(1 + \frac{4\pi\Delta C}{C_0}\right)} \tag{6.163}$$

Expressing ΔC and J in terms of R_N by means of Eqs. (6.104) and (5.147), respectively, the condition (6.163) yields [see (6.113)]

$$R_N^c < \frac{1}{8}\sqrt{3\pi^3 z}\frac{\hbar}{e^2}\left[\ln\left(1 + \frac{3\pi^2\hbar}{8\Delta_0 R_N^c C_0}\right)\right]^{-\frac{1}{2}} \tag{6.164}$$

We see that the MFA theory tends to overestimate the critical resistance. So far, quantitative calculations of the quantity $\lambda u_4 D$ have not been made, to the author's knowledge. However, Ferrell and Mirhashem (1988) have calculated corrections to the MFA value of R_N^c using an expansion in descending powers of z (the number of grain nearest neighbors). In the Bethe approximation, they find a fractional correction of -4% for $z = 4$. Note that the sign of their correction agrees with the prediction of Eq. (6.164). The size of the depression seems too small to affect the MFA results in a profound manner; so, the problems with explaining the observed near-perfect universal R_N^c remain unresolved.

6.8.2. Correlation length in a 2D Josephson array

In a granular superconductor, one defines two kinds of correlation lengths, both derivable from mapping onto an effective Ginzburg–Landau model.

1. The first one describes the decay of correlations near $T_{\rm BCS}$ (the transition temperature of the material forming the grains).

2. The other one is related to the phase-locking transition.

In the following, we shall be mostly interested in quantum phase transitions, tuned by varying the parameter $zJ/2U_{11}$. We thus focus our attention on the correlation length at $T = 0$.

With regard to (1), the correlation length near $T_{\rm BCS}$ (for a Josephson-coupled array) has been considered by Deutscher et al. (1974). These authors derive a GL free-energy functional by starting from the Hamiltonian of the form

$$H = \sum_i \Delta\Omega\left(\alpha|\psi_i|^2 + \frac{\beta}{2}|\psi_i|^4\right) - J\sum_{\langle ij\rangle}\cos\theta_{ij} \tag{6.165}$$

where $\Delta\Omega$ is the volume of each grain, α and β are the GL parameters defined in Eq. (2.1), and $\psi_i = |\psi_i|e^{i\theta_i}$ is the order parameter that we assume not to vary significantly throughout the volume of the grain. This can be ensured by having the linear dimension of the grains *smaller* than the BCS correlation length $\xi_{\rm BCS}$ of the material forming the grains. Going over to a continuum approximation, the expression (6.165) yields a GL free energy that, for a 2D array, takes the form

$$F = d\int d^2r\left(\alpha|\psi|^2 + \frac{\beta}{2}|\psi|^4 + \frac{J}{2d|\psi|^2}|\nabla\psi|^2\right) \tag{6.166}$$

where d is the thickness of the sample. Approximating $|\psi|^{-2}$ in the Josephson-coupling term by a real constant ψ_0^{-2}, this expression implies the following GL differential equation for the order parameter $\psi(\mathbf{r})$

$$-\left(\frac{J}{2d\psi_0^2}\right)\nabla^2\psi(\mathbf{r}) + \alpha\psi(\mathbf{r}) + \beta|\psi(\mathbf{r})|^2\psi(\mathbf{r}) = 0 \tag{6.167}$$

Let us now introduce a dimensionless parameter $f = \psi/\psi_0$, where ψ_0 satisfies the GL equation for a homogeneous sample

$$\alpha\psi_0 + \beta\psi_0^3 = 0 \tag{6.168}$$

The function $f(\mathbf{r})$ satisfies the differential equation

$$-\frac{J}{2d\alpha\psi_0^2}\nabla^2 f(\mathbf{r}) + f(\mathbf{r}) - f^3(\mathbf{r}) = 0 \tag{6.169}$$

This equation defines a characteristic length ξ_{eff} given by

$$\xi_{\text{eff}}^2 = \frac{J}{2d\alpha\psi_0^2} \tag{6.170}$$

Since the GL parameter $\alpha = \hbar^2/2m^*\xi_{\text{BCS}}^2$, the expression (6.170) can be rewritten as

$$\xi_{\text{eff}}^2 = \left(\frac{Jm^*}{\hbar^2\psi_0^2 d}\right)\xi_{\text{BCS}}^2 \tag{6.171}$$

The ratio $\xi_{\text{eff}}/\xi_{\text{BCS}}$ is usually a small number. For instance, in an aluminum array of thickness 100 Å with a Josephson coupling $J/k_B = 10$ K, we obtain $\xi_{\text{eff}}/\xi_{\text{BCS}} \simeq 2.6 \times 10^{-3}$. Consequently, for moderate values of J, the effective correlation length ξ_{eff} is expected to be well below the lattice spacing of the array so that the grains fluctuate *independently* (for temperatures $T \gtrsim T_{\text{BCS}}$). A true thermodynamic phase transition can take place only by lowering the temperature down to $T_c \simeq zJ/2k_B$ (the transition temperature of phase locking).

   With regard to (2), the correlation length at $T = 0$ (in the presence of charging energy) can be extracted from the effective action of the $(2+1)$-dimensional GL model given in Eq. (6.151). Expanding the product $A(\mathbf{k})Y(\mathbf{k}, \omega_m = 0)$ about $\mathbf{k} = 0$, we obtain, to order k^2,

$$A(\mathbf{k})Y(\mathbf{k}, \omega_m = 0) \simeq \frac{z}{2}\left(1 - \frac{k^2 a^2}{z}\right)Y(0, \omega_m = 0) \tag{6.172}$$

where we used the expression (6.138) and the result

$$Y(\mathbf{k}, \omega_m = 0) = Y(0, \omega_m = 0) \tag{6.173}$$

which follows from the identity [see Eqs. (6.135), (6.137), and (6.148)]

$$Y(\mathbf{k}, \tau) = \sum_{i,j} e^{-i\mathbf{k}\cdot\mathbf{R}_{ij}}\left\langle e^{i\theta_i[\tau]}e^{-i\theta_j[0]}\right\rangle_0 \tag{6.174}$$

by taking into account the fact that the phase correlator is *site-diagonal* [as shown explicitly in Eq. (3.106)]. Inserting the expansion (6.172) into the expression (6.151), the second-order term of the effective GL action can be written

$$S_{\text{eff}}^{(2)}[\varphi] = \frac{1}{\hbar\beta} \sum_{\mathbf{k},\omega_m} |\varphi(\mathbf{k},\omega_m)|^2 \left[\left(1 - \frac{zJ}{2U_{11}} \right) + \left(\frac{J}{2U_{11}} \right) a^2 k^2 \right] \qquad (6.175)$$

The expression in the square brackets implies the following formula for the correlation length, valid for $zJ/2U_{11} \lesssim 1$ (incoherent phase),

$$\xi^2 = \left(\frac{a^2}{z} \right) \frac{\alpha}{1 - \alpha} \qquad (6.176)$$

where $\alpha = zJ/2U_{11}$ is to be distinguished from the GL parameter appearing in Eq. (6.170). In the coherent phase, which is realized when $\alpha > 1$, the quartic term of the GL action (6.151) must be taken into account. Expanding this term about the average order parameter, the $k = 0$ coefficient of the second-order term is changed from $(1 - \alpha)$ to $2(\alpha - 1)$. Then the correlation length in the phase coherent state is given by

$$(\xi')^2 = \frac{a^2}{z} \frac{\alpha}{2(\alpha - 1)} \qquad (6.177)$$

For large values of α ($\alpha \gg 1$), the phases of the grains are locked together, and the corresponding correlation length is, according to Eq. (6.177), of the order of $\xi' \simeq a/\sqrt{2z}$. This result is in a qualitative agreement with the correlation length of a Heisenberg ferromagnet with nearest-neighbor interactions.

7

TWO-DIMENSIONAL SUPERCONDUCTORS

7.1. Introduction

So far we have considered phase transitions in arrays that exhibit conventional order (with $\langle \cos \theta \rangle \neq 0$) in the superconducting phase. Three-dimensional arrays with or without charging energy belong to this category. For two-dimensional arrays with charging energy one must distinguish between the $T = 0$ and the finite-temperature regimes. According to Section 2.9, the partition function for $T = 0$ can be mapped onto that of a classical $(2 + 1)$-dimensional problem, which exhibits conventional ordering and a phase transition into a disordered phase upon increasing the charging energy. The $T > 0$ case is characterized, however, by diverging thermal fluctuations of the phase variable, which preclude the possibility of broken symmetry at any finite temperature [see Eq. (2.6)]. In Chapter 2 we have alluded to the fact that, in spite of this, the two-dimensional systems retain their superfluid properties as long as there are no thermally induced *free* vortices. For neutral superfluid films this has been first explained by Berezinskii (1971) and subsequently, in greater detail, by Kosterlitz and Thouless (1973). These authors envisaged a superfluid–normal phase transition involving the proliferation of thermally induced vortices. Below the critical temperature, the vortices and antivortices form bound pairs and the film remains superfluid. Above this temperature, the pairs unbind due to the screening produced by thermally induced vortices of high density. This leads to the presence of free vortices and a destruction of superfluidity. The idea of a topological phase transition in superfluids has a venerable history. Feynman (1955) conjectured a picture of the λ transition in bulk helium involving the interaction of rotons. Subsequently, Wiegel (1973) has developed a first-principles theory of Bose condensation based on the vortex-ring model. Since the two-dimensional counterparts of vortex rings are vortex–antivortex dipoles, there is a conceptual analogy between the λ transition and the vortex-unbinding transition in two dimensions. However, these ideas can be carried out rigorously only in two dimensions, where the superfluid Hamiltonian can be mapped onto a classical Coulomb gas—a problem for which a Debye-screening theory is available (Hauge and Hemmer, 1971).

In this chapter, we concentrate on the problem of vortex unbinding in two-dimensional superconductors. We consider first a two-dimensional array describing a neutral superfluid and show, using the theorem by Mermin (1967), that $\langle \cos\theta \rangle = 0$ at any finite temperature. Then we proceed to the case of a superconducting film in a transverse static magnetic field and demonstrate that it exhibits a Meissner effect as long as the phase rigidity (stiffness) is finite. Section 7.2.1 contains a derivation of the effective screening length for transverse fields.

In Section 7.2.2, we return to the model of a neutral array and consider the renormalization of the stiffness, due to the anharmonicity of the periodic Hamiltonian, using the method of SCHA. The results, first obtained by Pokrovskii and Uimin (1973), exhibit an abruptly vanishing stiffness at a temperature close to the temperature of the vortex–antivortex unbinding transition (Kosterlitz and Thouless, 1973). This transition involves a Coulomb gas of logarithmically interacting vortex pairs, the Hamiltonian of which is derived in Section 7.3.

The vortex-unbinding transition itself is described in Section 7.4. The central results are the scaling equations of Kosterlitz (1974), which are derived from the dielectric-screening model in Section 7.4.1.

In Section 7.5, we consider the vortex-unbinding transition in two-dimensional superconductors, where coupling of supercurrent to the vector potential is taking place. The modification of the vortex energy caused by this coupling is described in Section 7.5.2. The results show the significance of the effective screening length $\lambda_\perp$ (for transverse magnetic fields) for the vortex-unbinding transition. For a thin film of circular shape with radius R, the vortex energy is proportional to $\ln R$ only when $R \ll \lambda_\perp$. If, on the other hand, $R \gg \lambda_\perp$, the screening of the fields makes the vortex energy entirely independent of the film size (like in a bulk superconductor). Consequently, the vortex free energy is always dominated by the entropy term and the two-dimensional film, with $R \gg \lambda_\perp$, is never superconducting at any finite temperature (being filled with free vortices). Only samples of radius $R \ll \lambda_\perp$ can exhibit superconductivity.

This is confirmed in Section 7.5.3, where we consider the interaction energy of the vortex–antivortex pair. We show that this energy has a logarithmic dependence on the distance only if this distance is much shorter than the screening length. These results have been first derived by Pearl (1964). The possibility of a vortex-unbinding transition in thin superconducting films with $\lambda_\perp \gg R$ was first pointed out by Beasley et al. (1979). These authors have proposed a simple relation between the unbinding transition temperature in dirty films and their normal sheet resistance. In Section 7.5.4 we present a derivation of this relation, starting from the free energy of a two-dimensional Josephson array. We also point out some remarkable similarities between the propagation of Cooper pairs in Josephson-coupled arrays and that in dirty superconducting films.

The effects of quantum phase fluctuations on the vortex-unbinding transition are discussed in Section 7.6, Major impetus to these theoretical investigations came from the experiments of Hebard (1979) and Hebard and Vandenberg (1980), which showed a dramatic decrease of the vortex-unbinding transition temperature on increasing the sheet resistance through a critical value of about 13 kΩ. Section 7.6.1 describes a calculation of the renormalized transition temperature of vortex unbinding, using the SCHA (Šimánek, 1980b). In Section 7.6.2 we discuss, in some detail, phase fluctuations around the static vortex configuration, which renders the renormalization problem into an inhomogeneous one. A comparison of the SCHA theory with experiment is described in Section 7.6.3. Renormalization of the transition temperature within the MFA is considered in Section 7.6.4. The result of the paper of Maekawa et al. (1981), showing that the renormalized stiffness is proportional to $\langle \cos\theta \rangle^2_{\mathrm{MFA}}$, is confirmed using an alternative method of derivation. If the order parameter $\langle \cos\theta \rangle_{\mathrm{MFA}}$ is calculated with the method of Section 3.5 (including the 2π-antiperiodic states), then the theory of Maekawa et al. (1981) predicts a reentrant vortex-unbinding transition. Finally, in Section 7.6.5 we review some aspects of the study of the effect of quantum fluctuations based on the renormalization-group (RG) analysis. José (1984) starts from the scaling equations of Kosterlitz (1974) (see Section 7.4.1) in which one uses a scale-dependent charge renormalized by quantum phase fluctuations. We discuss in some detail the validity of the approximations used in arriving at the expression for the renormalized charge. We also discuss briefly the results of Monte Carlo simulations, which have been motivated by the RG analysis. Particularly interesting is the evidence for a new type of coherent state, into which the system enters upon lowering the temperature towards zero. In contrast to the ordinary reentrant phase transition, this low-temperature phase exhibits a reduced but *nonzero* stiffness (Jacobs et al., 1984). This finding, as well as the pronounced effects of transverse magnetic field on the reentrant transition (Jacobs et al., 1987), present a challenging unsolved problem for theory.

7.2. Superconductivity in the absence of long-range order

Let us consider a Josephson-coupled array, in two dimensions, without Coulomb interaction. As shown in Section 3.3, the Hamiltonian of this array can be written as

$$H = -J \sum_{\langle ij \rangle} \cos(\theta_i - \theta_j) \tag{7.1}$$

The absence of long-range order in this model can been established rigorously using the classical version of the Bogoliubov inequality derived by Mermin (1967). For a planar array with coordination number z and intergrain spacing a, this inequality reads

$$1 > \frac{k_B T m^2}{N} \sum_{\mathbf{k}} \left(zJa^2k^2 + |h||m| \right)^{-1} \tag{7.2}$$

where N is the number of sites in the sample. h is the auxiliary magnetic field designed to induce symmetry breaking but taken equal to zero at the end of the calculation. The quantity m is the magnetization per rotor

$$m = \frac{1}{N} \sum_n \langle \cos \theta(n) \rangle \tag{7.3}$$

In the thermodynamic limit, we obtain from Eq. (7.2), converting the two-dimensional $\mathbf{k}$ sum to an integral

$$\begin{aligned}
1 &> \frac{k_B T m^2 a^2}{2\pi} \int_0^{k_D} \frac{k \, dk}{z J a^2 k^2 + |h||m|} \\
&= \frac{k_B T m^2}{4\pi z J} \ln \frac{z J a^2 k_D^2 + |h||m|}{|h|m}
\end{aligned} \tag{7.4}$$

where k_D is the Debye cutoff. As the symmetry-breaking field h is taken to zero, the logarithmic function on the RHS of Eq. (7.4) goes to infinity, and the inequality can be satisfied only if $m = 0$, establishing the absence of long-range order.

We now consider a harmonic approximation of the Hamiltonian (7.1), which we expect to exhibit the same physics as long as $k_B T \ll J$. A continuum version of this Gaussian model is of the form

$$H = \frac{1}{2} J \int d^2 \rho \, (\boldsymbol{\nabla} \theta)^2 \tag{7.5}$$

The partition function is given by the path integral [see Eq. (A5.8)]

$$Z \sim \int \mathcal{D}\theta \, \exp \left(-\frac{1}{2} \beta J \int d^2 \rho \, (\boldsymbol{\nabla} \theta)^2 \right) \tag{7.6}$$

It is easy to show that this model is again characterized by the absence of long-range order. In fact, we can use the derivation that led to Eq. (A9.21) to obtain

$$\begin{aligned}
m &= \langle \cos \theta(\mathbf{r}) \rangle \\
&= \frac{\int \mathcal{D}\theta \, \cos \theta(\mathbf{r}) \exp \left(-\frac{1}{2} \beta J \int d^2 \rho \, (\boldsymbol{\nabla} \theta)^2 \right)}{\int \mathcal{D}\theta \, \exp \left(-\frac{1}{2} \beta J \int d^2 \rho \, (\boldsymbol{\nabla} \theta)^2 \right)} \\
&= \exp \left(-\frac{1}{2\pi \beta J} \int_0^{k_D} \frac{dk}{k} \right) = 0
\end{aligned} \tag{7.7}$$

The k integral on the RHS of this equation again diverges logarithmically at the lower limit, which implies vanishing of the order parameter m. This result hinges on the fact that the phase excitations, associated with Eq.

(7.5), are *Goldstone* bosons. In case of a charged superfluid, the supercurrent couples to the gauge field and the Goldstone boson is transformed into a *massive* vector boson. Consequently, the argument leading to the RHS of (7.7) does not hold for a charged superfluid. However, if the film is very thin (and dirty), the screening length associated with the self-induced gauge fields may be large compared to the size of the sample [see Eq. (7.28)]. Then a logarithmic quasidiverging integral is again obtained in (7.7), leading to $m \simeq 0$. In what follows, we assume that the given condition holds. We are now going to show that, in spite of the nearly vanishing order parameter, the film is still superfluid in the sense that it exhibits the Meissner effect.

Applying a static magnetic field perpendicular to the plane of the array, the Hamiltonian (7.5) is modified as follows

$$H_{\mathbf{A}} = \frac{1}{2} J \int d^2\rho \left(\boldsymbol{\nabla}\theta - \frac{e^*}{\hbar c}\mathbf{A} \right)^2 = \int d^3 r\, h(\mathbf{r}) \qquad (7.8)$$

where we define the three-dimensional Hamiltonian density

$$h(\mathbf{r}) = h(\boldsymbol{\rho}, z) = \frac{1}{2} J \delta(z) \left(\boldsymbol{\nabla}\theta - \frac{e^*}{\hbar c}\mathbf{A} \right)^2 \qquad (7.9)$$

where $\boldsymbol{\rho}$ and z are components of $\mathbf{r}$ in the plane and perpendicular to it, respectively. Using this result in Eq. (2.22), we obtain the current density

$$\mathbf{j} = -c\frac{\partial h}{\partial \mathbf{A}} = \frac{e^* J}{\hbar}\delta(z)\left(\boldsymbol{\nabla}\theta - \frac{e^*}{\hbar c}\mathbf{A} \right) \qquad (7.10)$$

To derive the Meissner effect, we need a thermally averaged current density induced by the gauge field to first order in $\mathbf{A}$. From Eqs. (7.8)–(7.10) we have

$$\begin{aligned} \langle \mathbf{j}(\mathbf{r}) \rangle &= \frac{e^* J}{\hbar}\delta(z)\frac{\int \mathcal{D}\theta \left(\boldsymbol{\nabla}\theta - \frac{e^*}{\hbar c}\mathbf{A} \right) \exp\left(-\beta \int d^3 r\, h(\mathbf{r}) \right)}{\int \mathcal{D}\theta \, \exp\left(-\beta \int d^3 r\, h(\mathbf{r}) \right)} \\ &\simeq -\frac{e^{*2} J}{c\hbar^2}\delta(z)\bigg(\mathbf{A}(\mathbf{r}) \\ &\qquad - \beta J \int d^2\rho' \, \langle \boldsymbol{\nabla}\theta(\boldsymbol{\rho}') \cdot \mathbf{A}(\boldsymbol{\rho}')\boldsymbol{\nabla}\theta(\boldsymbol{\rho}) \rangle_0 \bigg) \end{aligned} \qquad (7.11)$$

where $\langle \cdots \rangle_0$ implies thermal average evaluated with the field-free Hamiltonian $H_{\mathbf{A}=0}$. The RHS of Eq. (7.11) exhibits a manifestly gauge invariant linear dependence on $\mathbf{A}(\mathbf{r})$. This can be shown by taking the ith component of Eq. (7.11) and Fourier transforming the latter. Defining the three-dimensional transform

$$A_i(\mathbf{q}, k) = \int d^2\rho \int dz\, e^{-i(\mathbf{q}\cdot\boldsymbol{\rho}+kz)} A_i(\mathbf{r}) \qquad (7.12)$$

and its inverse

$$A_i(\mathbf{r}) = \frac{1}{(2\pi)^3} \int d^2q \int dk \, e^{i(\mathbf{q}\cdot\boldsymbol{\rho}+kz)} A_i(\mathbf{q}, k) \qquad (7.13)$$

we have

$$\int d^2\rho \int dz \, e^{-i(\mathbf{q}\cdot\boldsymbol{\rho}+kz)} \delta(z) A_i(\mathbf{r}) = \int d^2\rho \, e^{-i\mathbf{q}\cdot\boldsymbol{\rho}} A_i(\boldsymbol{\rho}) = A_i(\mathbf{q}) \qquad (7.14)$$

where $A_i(\mathbf{q})$ is the two-dimensional Fourier transform of $A_i(\mathbf{r}) = A_i(\boldsymbol{\rho})$. Expressing the phase variables $\theta(\boldsymbol{\rho})$ in terms of the two-dimensional transforms $\theta_\mathbf{q}$, we obtain for the ith component of the second term on the RHS of Eq. (7.11)

$$\beta J \int d^2\rho' \langle \nabla_j \theta(\boldsymbol{\rho}') \nabla_i \theta(\boldsymbol{\rho}) \rangle_0 A_j(\boldsymbol{\rho}')$$

$$= -\beta J \int d^2\rho' \int \frac{d^2q'}{(2\pi)^2} \int \frac{d^2q}{(2\pi)^2} q'_j q_i \langle \theta_{\mathbf{q}'} \theta_\mathbf{q} \rangle_0 e^{i(\mathbf{q}'\cdot\boldsymbol{\rho}' + \mathbf{q}\cdot\boldsymbol{\rho})} A_j(\boldsymbol{\rho}')$$

$$= \beta J \int \frac{d^2q}{(2\pi)^2} q_i q_j \int d^2\rho' \langle \theta_\mathbf{q} \theta_{-\mathbf{q}} \rangle_0 e^{i\mathbf{q}\cdot(\boldsymbol{\rho}-\boldsymbol{\rho}')} A_j(\boldsymbol{\rho}')$$

$$= \int \frac{d^2q}{(2\pi)^2} \frac{q_i q_j}{q^2} e^{i\mathbf{q}\cdot\boldsymbol{\rho}} A_j(\mathbf{q}) \qquad (7.15)$$

On the RHS of this equation we have used the result

$$\langle \theta_\mathbf{q} \theta_{-\mathbf{q}} \rangle_0 = \frac{1}{\beta J q^2} \qquad (7.16)$$

which follows from Eq. (7.6) by applying the methods of Appendix A9. Using Eqs. (7.14) and (7.15), the Fourier transform of the ith component of Eq. (7.11) can be written as

$$\langle j_i(\mathbf{q}, k) \rangle = -\frac{e^{*2}J}{c\hbar^2} \left(\delta_{ij} - \frac{q_i q_j}{q^2} \right) A_j(\mathbf{q}) \qquad (7.17)$$

This expression has the correct form dictated by the gauge invariance (Schafroth, 1951), and it yields a Meissner effect, as long as the constant J is nonzero. Thus, in keeping with the general arguments (Anderson, 1984), the existence of superfluidity in our system is intimately linked to the non-vanishing phase rigidity defined by the Hamiltonian (7.5). The expression in the angular brackets in Eq. (7.17) ensures that only the transverse component of $\mathbf{A}$ contributes to the induced current. In the special case of the London gauge ($\nabla\cdot\mathbf{A}_\mathrm{L} = 0$), we have $q_j A_{\mathrm{L}j} = 0$, so that only the first term contributes, yielding

$$\langle \mathbf{j}(\mathbf{q}, k) \rangle = -\Lambda \mathbf{A}_\mathrm{L}(\mathbf{q}) \qquad (7.18)$$

where

$$\Lambda = \frac{e^{*2}J}{c\hbar^2} \tag{7.19}$$

An inverse Fourier transform of Eq. (7.18) then yields, with the help of Eqs. (7.12)–(7.14)

$$\mathbf{J}(\mathbf{r}) = \langle \mathbf{j}(\mathbf{r}) \rangle = -\Lambda \, \delta(z) \, \mathbf{A}_L(\mathbf{r}) \tag{7.20}$$

which is the London equation for a thin superconducting film.

7.2.1. *Penetration depth for a planar array*

Equation (7.20) enables us to find the magnetic penetration depth for fields perpendicular to the plane of the array. The London gauge field $\mathbf{A}_L$ is composed of the external and induced parts

$$\mathbf{A}_L(\mathbf{r}) = \mathbf{A}_{ext}(\mathbf{r}) + \mathbf{A}_{ind}(\mathbf{r}) \tag{7.21}$$

where the induced field satisfies the Maxwell equation

$$\nabla^2 \mathbf{A}_{ind}(\mathbf{r}) = -\frac{4\pi}{c} \mathbf{J}(\mathbf{r}) \tag{7.22}$$

From Eqs. (7.20)–(7.22), we obtain

$$\left(\nabla^2 - \frac{4\pi\Lambda}{c} \delta(z) \right) \mathbf{A}_L(\mathbf{r}) = \nabla^2 \mathbf{A}_{ext}(\mathbf{r}) \tag{7.23}$$

Performing a three-dimensional Fourier transform of this equation, we obtain with use of Eqs. (7.12)–(7.14)

$$\left(q^2 + k^2 \right) \mathbf{A}_L(\mathbf{q}, k) + \frac{4\pi\Lambda}{c} \mathbf{A}_L(\mathbf{q}) = \left(q^2 + k^2 \right) \mathbf{A}_{ext}(\mathbf{q}, k) \tag{7.24}$$

To obtain the relation between the two-dimensional Fourier transforms $\mathbf{A}_L(\mathbf{q})$ and $\mathbf{A}_{ext}(\mathbf{q})$, we divide this equation by $q^2 + k^2$ and integrate over k. With the use of Eq. (7.12), we have

$$\int_{-\infty}^{\infty} dk \, \mathbf{A}_L(\mathbf{q}, k) = 2\pi \int d^2\rho \, e^{-i\mathbf{q} \cdot \boldsymbol{\rho}} \mathbf{A}_L(\boldsymbol{\rho}) = 2\pi \mathbf{A}_L(\mathbf{q}) \tag{7.25}$$

Equations (7.24) and (7.25) imply

$$\mathbf{A}_L(\mathbf{q}) = \frac{q}{q + q_c} \mathbf{A}_{ext}(\mathbf{q}) \tag{7.26}$$

where

$$q_c = \frac{2\pi\Lambda}{c} = \frac{2\pi e^{*2}J}{c^2\hbar^2} \tag{7.27}$$

The relation (7.26) confirms the Meissner property of a two-dimensional array. To be specific, let us consider a slowly varying external field for which $q \ll q_c$. Then the total field $\mathbf{A}_L(\mathbf{q})$ approaches zero as q times $\mathbf{A}_{ext}(\mathbf{q})$. Hence, the long-wavelength external field is expelled from the sample. On the other hand, for $q \gg q_c$, Eq. (7.26) implies that $\mathbf{A}_L(\mathbf{q}) \simeq \mathbf{A}_{ext}(\mathbf{q})$. Thus, the effective screening length for magnetic fields perpendicular to the film appears to be (de Gennes, 1966)

$$\lambda_\perp = \frac{1}{2q_c} = \frac{c^2\hbar^2}{4\pi e^{*2} J} \tag{7.28}$$

This expression does not show an explicit dependence on the film thickness d. However, since J is proportional to d (for $d < a$), we have from Eq. (7.28) an implicit dependence: $\lambda_\perp \sim 1/d$. Actually, we can establish a more general relation

$$\lambda_\perp = \frac{\lambda_L^2}{d} \tag{7.29}$$

where λ_L is the London penetration depth of a bulk superconductor. The two-dimensional screening length of this form has been first shown by Pearl (1964) to play a role in the interaction of vortices in a thin superconducting film. To prove (7.29) for granular arrays, let us consider a three-dimensional cubic array with intergrain distance a. By extending the width of the tunneling gaps from d to a, the Josephson coupling constant of the 3D array is increased from the two-dimensional J to

$$J^{3D} = \frac{a}{d} J \tag{7.30}$$

The Hamiltonian of the 3D array, obtained in the harmonic and continuum approximation from Eq. (7.1), is of the form

$$H = \frac{1}{2a} J^{3D} \int d^3 r \, (\boldsymbol{\nabla}\theta)^2 \tag{7.31}$$

Repeating the calculation of the induced current (using the same steps as for the two-dimensional array), we obtain from (7.31) the London equation for a bulk superconductor in the standard form

$$\mathbf{J}(\mathbf{r}) = -\frac{c}{4\pi\lambda_L^2} \mathbf{A}_L(\mathbf{r}) \tag{7.32}$$

where

$$\lambda_L^{-2} = \frac{4\pi e^{*2} J^{3D}}{ac^2\hbar^2} = \frac{4\pi e^{*2} J}{dc^2\hbar^2} \tag{7.33}$$

which proves the relation (7.29).

7.2.2. *Josephson-coupled array with renormalized stiffness*

According to Eqs. (7.18) and (7.19), the two-dimensional array represented by the Hamiltonian (7.5) remains superconducting as long as the Josephson coupling constant J, representing the stiffness, is nonzero. Physically, one expects that the superconductivity property breaks down at some finite temperature above which the Gaussian model (7.5) ceases to be valid. In Section 7.4 we present arguments for this breakdown, based on the picture of vortex–antivortex dissociation. This physical situation can be imitated by studying directly the model (7.1) with use of the SCHA approximation (see Section 3.9). The partition function of the system is given by

$$Z = \int \mathcal{D}\theta \, \exp\left(-\beta J \sum_{\langle ij\rangle}[1-\cos(\theta_{ij})]\right) = \int \mathcal{D}\theta \, \exp\left(-\frac{1}{\hbar}S[\theta]\right) \quad (7.34)$$

Following the SCHA method, we replace $S[\theta]$ by a trial harmonic action

$$S_{\text{tr}}[\theta] = \frac{1}{2}\hbar\beta K \sum_{\langle ij\rangle}\theta_{ij}^2 \quad (7.35)$$

where K is the renormalized stiffness, to be determined variationally. Adapting the derivation given in Eqs. (3.131)–(3.163) to the present classical model, we obtain

$$K = J\exp\left(-\frac{1}{2}\bar{D}_{ij}\right) \quad (7.36)$$

where $\bar{D}_{ij}$ is the bond average of the quantity D_{ij} given by

$$D_{ij} = \frac{2}{N\beta K}\sum_{\mathbf{k}}\frac{1-\cos\mathbf{k}\cdot\mathbf{R}_{ij}}{f(\mathbf{k})} \quad (7.37)$$

where $f(\mathbf{k})$ is the bond-structure factor defined in Eq. (3.155). Converting the $\mathbf{k}$ sum to an integral, we have

$$\bar{D}_{ij} = \frac{1}{z}\sum_{j(i)}D_{ij} = \frac{2k_BTa^2}{zK}\int_0^{k_D}\frac{d^2k}{(2\pi)^2} = \frac{k_BT}{2K} \quad (7.38)$$

where we have put $z = 4$, assuming a square planar array. Inserting this result into Eq. (7.36), we obtain a self-consistent equation for the relative stiffness $k_1 = K/J$ of the form

$$k_1 = \exp\left(-\frac{k_BT}{4Jk_1}\right) \quad (7.39)$$

This result was first obtained by Pokrovskii and Uimin (1973). For $T \to 0$, Eq. (7.39) yields $k_1 = 1$. On increasing T, k_1 decreases slowly until at $T = T_c$, where

$$T_c = \frac{1.47J}{k_B} \quad (7.40)$$

there is a precipitous drop of the relative stiffness to zero. The trial harmonic action (7.35) implies that, in the continuum approximation, the Hamiltonian (7.5) is replaced by

$$H_{\text{eff}} = \frac{1}{2}K \int d^2r \, (\nabla\theta)^2 \tag{7.41}$$

The vanishing of the stiffness constant K, which follows from Eq. (7.39), then implies a destruction of superconductivity for $T \geq T_c$. Hence, the anharmonic model (7.1) exhibits dramatic renormalization effects in the SCHA. Although there are doubts about the validity of this method near T_c, it is interesting that a more powerful approach (based on the vortex–antivortex dissociation) also shows a jump in the rigidity, taking place at the Kosterlitz–Thouless transition temperature, which is close to the value of T_c given by Eq. (7.40) (Nelson and Kosterlitz, 1977).

7.3. Vortices in the two-dimensional xy model

For the continuum version of the xy model, described by Hamiltonian (7.5), vortex degrees freedom can be introduced by using the complex-potential description of the planar flow (Kosterlitz, 1974). The path integral for the partition function (7.6) can be evaluated using the method of steepest descent, in which the phase variable $\theta(\mathbf{r})$ is separated as follows

$$\theta(\mathbf{r}) = \varphi(\mathbf{r}) + \theta_1(\mathbf{r}) \tag{7.42}$$

In keeping with the notation used in classic papers on this subject, we denote the coordinate in the plane by $\mathbf{r}$. In Eq. (7.42), the function $\varphi(\mathbf{r})$ is the extremal path satisfying

$$\frac{1}{2}J \int d^2r [\nabla\varphi(\mathbf{r})]^2 = \text{minimum} \tag{7.43}$$

and $\theta_1(\mathbf{r})$ represents a fluctuation about $\varphi(\mathbf{r})$. Equation (7.43) implies that $\varphi(\mathbf{r})$ satisfies Laplace's equation

$$\nabla^2\varphi = 0 \quad (\text{mod } 2\pi) \tag{7.44}$$

The fact that $\varphi(x,y)$ is a harmonic function suggests that we introduce a complex potential $f(z)$ defined in the complex plane $z = x + iy$ as

$$f(z) = \varphi(x,y) + i\psi(x,y) \tag{7.45}$$

where the real functions φ and ψ satisfy the Cauchy–Riemann conditions

$$\frac{\partial\varphi}{\partial x} = \frac{\partial\psi}{\partial y}$$
$$\frac{\partial\varphi}{\partial y} = -\frac{\partial\psi}{\partial x} \tag{7.46}$$

We can make contact with two-dimensional superfluid flow by identifying $\varphi(x,y)$ with the velocity potential. The flow velocity is then given by

$$\mathbf{v} = \kappa \boldsymbol{\nabla} \varphi \qquad (7.47)$$

where κ is a constant to be specified later [see Eq. (8.4)]. Nontrivial solutions of Eq. (7.44) are characterized by nonzero vorticity defined by the velocity circulation integral. Let us consider first a single vortex of vorticity $\bar{q}_i$, centered at $\mathbf{r}_i$. We have

$$\bar{q}_i = \frac{1}{2\pi} \oint_C \mathbf{v} \cdot d\mathbf{s} \qquad (7.48)$$

where the line integral is around a closed contour C surrounding the point $\mathbf{r}_i$. Using Eqs. (7.46) and (7.47), Eq. (7.48) can be expressed by means of Green's theorem in the following manner

$$\bar{q}_i = \frac{1}{2\pi} \oint_C (v_x dx + v_y dy) = \frac{\kappa}{2\pi} \oint_C \left(\frac{\partial \psi}{\partial y} dx - \frac{\partial \psi}{\partial x} dy \right)$$
$$= -\frac{\kappa}{2\pi} \int d^2 r \, \nabla^2 \psi(\mathbf{r}) \qquad (7.49)$$

Equations (7.44) and (7.45) imply that ψ satisfies a Laplace equation, except at $\mathbf{r}_i$, where Eq. (7.49) implies a δ-function divergence. Thus we obtain

$$\nabla^2 \psi = -2\pi q_i \, \delta(\mathbf{r} - \mathbf{r}_i) \qquad (7.50)$$

where $q_i = \bar{q}_i / \kappa = \pm 1$. The solution of this equation is a two-dimensional Green's function

$$\psi(\mathbf{r}) = -q_i \ln |\mathbf{r} - \mathbf{r}_i| \qquad (7.51)$$

We note that Eqs. (7.50) and (7.51) correspond to a pointlike vortex. A real vortex has a core of finite size. The presence of such a core can be imitated by replacing the pointlike δ function $\delta(\mathbf{r} - \mathbf{r}_i)$ by a cylindrical $\bar{\delta}$ function defined as

$$\bar{\delta}(\mathbf{r} - \mathbf{r}_i) = \frac{1}{2\pi\varepsilon_v} \delta(|\mathbf{r} - \mathbf{r}_i| - \varepsilon_v) \qquad (7.52)$$

where ε_v is the radius of the vortex core. According to Eqs. (7.45) and (7.51), the function $\psi(x,y)$ is an imaginary part of the complex potential

$$f(z) = -iq_i \ln(z - z_i) \qquad (7.53)$$

where $z_i = x_i + iy_i$. The velocity potential $\varphi(x,y)$ is given by the real part of Eq. (7.53), yielding

$$\varphi(x,y) = q_i \operatorname{Im} \ln(z - z_i) \qquad (7.54)$$

We see that, owing to the fact that $\theta(\mathbf{r})$ is an angular variable, $\varphi(x, y)$ is a multivalued function. It exhibits jumps equal to 2π upon encircling the vortex center z_i.

Let us now consider a set of $2N$ vortices and define a vortex distribution function

$$\rho(\mathbf{r}) = \sum_{i=1}^{2N} q_i \, \bar{\delta}(\mathbf{r} - \mathbf{r}_i) \qquad (7.55)$$

In view of Eqs. (7.50) and (7.55), the function $\psi(x, y)$ describing this set satisfies the equation

$$\nabla^2 \psi = -2\pi\rho(\mathbf{r}) \qquad (7.56)$$

7.3.1. *Vortex Hamiltonian*

Introducing the ansatz (7.42) into Eq. (7.5), we obtain

$$H = \frac{1}{2} J \int d^2r \left[(\boldsymbol{\nabla}\varphi)^2 + (\boldsymbol{\nabla}\theta_1)^2 + 2(\boldsymbol{\nabla}\theta_1) \cdot (\boldsymbol{\nabla}\varphi) \right] \qquad (7.57)$$

Assuming that the fluctuation $\theta_1(\mathbf{r})$ vanishes on the boundary of the film, the last term on the RHS of Eq. (7.57) can be written [using integration by parts and Eq. (7.44)]

$$\int d^2r \, (\boldsymbol{\nabla}\theta_1) \cdot (\boldsymbol{\nabla}\varphi) = -\int d^2r \, \theta_1 \nabla^2\varphi = 0 \qquad (7.58)$$

Thus the fluctuation θ_1 and the vortex degrees of freedom are uncoupled in the thermodynamics of the system. On the other hand, if θ_1 is to represent an *externally* driven supercurrent, the term given describes the current–vortex interaction responsible for the *Magnus force* (see Section 8.4.5).

Let us focus our attention on the vortex contribution, given by the first term of Eq. (7.57). Using Eqs. (7.46) and (7.47), we write this contribution as

$$H_v = \frac{1}{2} J \int d^2r \, (\boldsymbol{\nabla}\varphi)^2 = \frac{1}{2\kappa} J \int d^2r \left(v_x \frac{\partial\psi}{\partial y} - v_y \frac{\partial\psi}{\partial x} \right)$$

$$= \frac{1}{2\kappa} J \int d^2r \left[\frac{\partial}{\partial y}(v_x\psi) - \frac{\partial}{\partial x}(v_y\psi) - \psi\left(\frac{\partial v_x}{\partial y} - \frac{\partial v_y}{\partial x} \right) \right] \qquad (7.59)$$

The last term on the RHS of this equation can be expressed via the vortex distribution (7.55). In fact, using Eqs. (7.46), (7.47), and (7.56), we have

$$\frac{\partial v_x}{\partial y} - \frac{\partial v_y}{\partial x} = \kappa\nabla^2\psi = -2\pi\kappa\rho \qquad (7.60)$$

The first two terms on the RHS of Eq. (7.59) can be transformed by means of Green's theorem to a line integral

$$\frac{1}{2\kappa} J \int d^2r \left(\frac{\partial}{\partial y}(v_x\psi) - \frac{\partial}{\partial x}(v_y\psi) \right) = -\frac{1}{2\kappa} J \oint_C \psi(\mathbf{v} \cdot d\mathbf{s}) \qquad (7.61)$$

where contour C surrounds all $2N$ vortices. Using Eqs. (7.60) and (7.61), the vortex Hamiltonian (7.59) becomes

$$H_v = \pi J \int d^2r\, \psi(\mathbf{r})\rho(\mathbf{r}) - \frac{1}{2\kappa}J \oint_C \psi(\mathbf{r})(\mathbf{v}\cdot d\mathbf{s}) \tag{7.62}$$

7.3.2. Neutrality constraint

The second term of Eq. (7.62) imposes a following constraint on the vortex distribution function

$$\int d^2r\, \rho(\mathbf{r}) = 0 \tag{7.63}$$

To show this, we assume that all vortices are confined to a finite region near the center of contour C which we assume to be a circle of radius R. Assuming that R is much greater than the size of this region, the function $\psi(\mathbf{r})$ on the contour C is found from Eqs. (7.51), (7.55), and (7.56) to be

$$\psi(\mathbf{r}_C) = -\int d^2r'\, \rho(\mathbf{r}')\ln|\mathbf{r}_C - \mathbf{r}'|$$

$$\approx -\ln R \int d^2r'\, \rho(\mathbf{r}'), \qquad \text{for } R \gg |\mathbf{r}'| \tag{7.64}$$

where $\mathbf{r}_C$ is the coordinate on C that satisfies $|\mathbf{r}_C - \mathbf{r}'| \approx R$. Since the RHS of this equation is independent of $\mathbf{r}_C$, we can take $\psi(\mathbf{r}_C)$ out of the contour integral and write the second term of Eq. (7.62) as

$$-\frac{1}{2\kappa}J \oint_C \psi(\mathbf{v}\cdot d\mathbf{s}) = \pi J \ln R \left(\int d^2r\, \rho(\mathbf{r}) \right)^2 \tag{7.65}$$

where we have used, consistent with Eqs. (7.49) and (7.56),

$$\oint_C \mathbf{v}\cdot d\mathbf{s} = 2\pi\kappa \int d^2r\, \rho(\mathbf{r}) = 2\pi \sum_{i=1}^{2N} \bar{q}_i \tag{7.66}$$

We see that as long as $\sum_{i=1}^{2N} q_i \neq 0$, the expression (7.65) diverges logarithmically with the radius R of the contour of integration. The same kind of divergence is exhibited by the kinetic energy of a single vortex. In this case, we have from Eqs. (7.51), (7.52), and (7.55), assuming a vortex of vorticity q at $\mathbf{r} = 0$

$$\psi(\mathbf{r}) = -q\ln r \tag{7.67}$$

$$\rho(\mathbf{r}) = \frac{q}{2\pi\varepsilon_v}\delta(r - \varepsilon_v) \tag{7.68}$$

Using Eqs. (7.62) and (7.65)–(7.68), the energy of a single vortex is given by

$$H_v^{(1)} = -\frac{\pi J q^2}{\varepsilon_v} \int_0^\infty dr\, \delta(r - \varepsilon_v)r\ln r + \pi J q^2 \ln R$$

$$= \pi J q^2 \ln(R/\varepsilon_v) \tag{7.69}$$

The expression on the RHS of Eq. (7.69) can be also obtained directly from
the expression

$$H_v^{(1)} = \frac{1}{2} J \int d^2 r \, (\nabla \varphi)^2 \tag{7.70}$$

by substituting for $\nabla \varphi$ the gradient, whose tangential component is given
by $1/r$, and by integrating the variable r over the interval $(\varepsilon_v < r < R)$.
In view of (7.65), only such configurations are admitted in the partition
function, for which the *net vorticity (7.66) vanishes*. If the vortices are
mapped onto a system of charges, then this condition implies a *neutral*
Coulomb gas. In the vortex system, the neutrality is realized by having
two sets of N vortices of equal but opposite vorticity. Assuming that the
constraint (7.63) is satisfied, the vortex Hamiltonian follows from Eq. (7.62)
by substituting for $\psi(\mathbf{r})$ the solution of (7.56), written in the form of a
superposition of Green's functions (7.51)

$$\psi(\mathbf{r}) = - \int d^2 r' \, \rho(\mathbf{r} - \mathbf{r}') \ln |\mathbf{r} - \mathbf{r}'| \tag{7.71}$$

In this way, we obtain with use of Eq. (7.55)

$$
\begin{aligned}
H_v &= \pi J \int d^2 r \, \psi(\mathbf{r}) \rho(\mathbf{r}) \\
&= -\pi J \int d^2 r \int d^2 r' \, \rho(\mathbf{r}') \ln |\mathbf{r} - \mathbf{r}'| \rho(\mathbf{r}) \\
&= -\pi J \sum_{i,j} q_i q_j \int d^2 r \int d^2 r' \, \bar{\delta}(\mathbf{r}' - \mathbf{r}_i) \ln |\mathbf{r} - \mathbf{r}'| \bar{\delta}(\mathbf{r} - \mathbf{r}_j)
\end{aligned} \tag{7.72}
$$

Using the expression (7.52) for the $\bar{\delta}$ functions and performing the integrations on the RHS of Eq. (7.72), we have

$$H_v = -2\pi J \sum_{i<j} q_i q_j \ln |\mathbf{r}_i - \mathbf{r}_j| - \pi J \ln a \sum_i q_i^2 \tag{7.73}$$

where a is a characteristic length of the order of the core diameter. The
first term on the RHS of this equation is derived under the assumption:
$|\mathbf{r}_i - \mathbf{r}_j| \geq a$. Taking advantage of the neutrality condition, we can add to
the RHS of Eq. (7.73) a term

$$\pi J \ln a \sum_{i,j} q_i q_j = 2\pi J \ln a \sum_{i<j} q_i q_j + \pi J \ln a \sum_i q_i^2 \tag{7.74}$$

which simplifies H_v, yielding

$$H_v = -2\pi J \sum_{i<j} q_i q_j \ln \left(\frac{|\mathbf{r}_i - \mathbf{r}_j|}{a} \right) \tag{7.75}$$

where $|\mathbf{r}_i - \mathbf{r}_j| \geq a$. The Hamiltonian (7.75) shows a logarithmic behavior, typical for the two-dimensional Coulomb gas. The condition $|\mathbf{r}_i - \mathbf{r}_j| \geq a$ is essential to prevent the collapse of the gas at low temperatures (Hauge and Hemmer, 1971). It should be mentioned that vortex degrees of freedom can also be exposed for the lattice model (7.1) by applying the transformation of Villain (1975) to the path integral for the partition function (José et al., 1977).

7.4. Vortex-unbinding transition

The classical neutral Coulomb gas in two dimensions exhibits a fascinating transition from insulating to metallic state as the temperature is increased (Hauge and Hemmer, 1971; Kosterlitz and Thouless, 1973). In view of the demonstrated mapping between the two-dimensional superfluid systems and the Coulomb gas of vortices, this phase transition has important ramifications for the physics of thin granular superconducting films. The peculiar behavior of two-dimensional systems with continuous symmetry can be traced to the unusual distance dependence of the correlation function. Using Appendix A9, we can calculate the phase correlator for the Gaussian model (7.5), yielding [see Eq. (2.6)]

$$\langle \exp\{i[\theta(\mathbf{r}) - \theta(\mathbf{r}')]\}\rangle \sim |\mathbf{r} - \mathbf{r}'|^{-\frac{k_B T}{2\pi J}} \tag{7.76}$$

This result is somewhat surprising: Since long-range order is absent, one would expect an exponentially decaying correlator. Such a slow power-law decay of the phase correlator suggests that certain subtle type of phase order persists, in spite of the lacking conventional long-range order. As first pointed out by Berezinskii (1971), the low-temperature phase with a power-law decay of correlations is characterized by vortices and antivortices *bound* into *neutral* dipoles. These dipoles cause small renormalization of the stiffness constant, but as long as the vortex pairs remain bound, the stiffness is nonzero so that the system remains superfluid. This type of ordering (taking place in the vortex degrees of freedom) is usually referred to as *topological order*. Now, we may inquire about the possibility of a topological phase transition from the phase of bound pairs to that of free vortices. That the presence of a free vortex is detrimental to superconductivity is a well-known fact. The Lorentz force produced by a uniform supercurrent causes the vortex to traverse the sample in a direction perpendicular to the supercurrent. The resulting electromotive force, which appears as a result of the Faraday induction law, then leads to a finite resistance.

Of more fundamental importance is the disappearance of the Meissner effect, which is to be attributed to the vanishing of the stiffness [see Eq. (7.19)]. A qualitative argument for this effect can be presented by taking the example of a neutral superfluid, represented by Hamiltonian (7.1). The stiffness constant is a measure of the change of free energy upon applying an external phase "twist" to the ends of the sample. In the absence of a free vortex, the free energy in the twisted configuration is determined by the expression (7.5), where $(\nabla\theta)^2$ is given the phase angle $\theta(\mathbf{r})$ slowly turning from one end of the sample to the other. Then the resulting stiffness constant is given by the unrenormalized spin-wave value J [see Eq. (7.18)].

This picture is drastically changed when a free vortex is introduced in the system. The externally set-up phase twist produces a supercurrent that interacts with the vortex via *Magnus force*. The vortex driven by this force again traverses the sample, *relieving* the externally induced phase difference. Thus, free energy is lowered by the motion of vortex across the sample. This process is analogous to a motion of dislocations which tends to relieve the stress in a crystal. In this manner, the stiffness becomes renormalized to zero at a temperature for which the first vortex pair unbinds. This picture of the topological phase transition, in the two-dimensional xy model, has been worked out in considerable detail by Kosterlitz and Thouless (1973). Hence, the vortex-unbinding transition is often called the Kosterlitz–Thouless–Berezinskii (KTB) transition. It should, however, be pointed out that Stanley and Kaplan (1966) conjectured the possibility of this phase transition much earlier, based on a high-temperature expansion for the susceptibility, though the vortex mechanism was not known at that time.

If the phase correlator is calculated using the variational method of Section 7.2.2, then one obtains the expression shown on the RHS of Eq. (7.76), except that J is now replaced by the effective stiffness constant K, which satisfies the self-consistency condition (7.39). The collapse of K, predicted for $T \geq T_c$, then implies, according to Eq. (7.76), an infinitely rapid power-law decay. This suggests that, at $T = T_c$, there is a transition to a phase with a different behavior of the correlation function (exponential decay). Kosterlitz (1974) has developed a powerful RG method for this phase transition, based on the Coulomb-gas description [see Eq. (7.75)]. This method is an extension of the scaling procedure of Anderson et al. (1970) and Anderson and Yuval (1971) for the Kondo problem and the Ising model with long-range interactions decaying as $1/r^2$. The same RG equations can be obtained from the Debye-screening picture of two-dimensional neutral plasma (Young, 1978), which we describe in Section 7.4.1.

7.4.1. *Scaling equations for the dielectric-screening model*

So far we have tacitly assumed that the number of vortex pairs in the system is a given constant N. In reality, the number of thermally excited pairs in the xy model fluctuates and the thermodynamic properties must be calculated from the grand partition function Z_G

$$Z_G = 1 + \sum_{N=1}^{\infty} \frac{z^{2N}}{(N!)^2} \int \prod_{i=1}^{2N} d^2r_i \, e^{-\beta H_v(2N)} \qquad (7.77)$$

where $H_v(2N)$ is the vortex Hamiltonian of Eq. (7.75), for N positive and N negative vortices, and $z = e^{\beta\mu}$ is the fugacity. The quantity μ is the chemical potential given by the energy needed to create a vortex–antivortex pair separated by the distance of a core diameter. The scaling equations for the dielectric-screening problem are derived by introducing a scale-dependent dielectric constant $\varepsilon(r)$ defined as (Kosterlitz and Thouless, 1973)

$$F(r) = F_0(r)/\varepsilon(r) \tag{7.78}$$

where $F_0(r)$ is the "bare" attractive force between a vortex and an antivortex, measured in the absence of other vortices. $F(r)$ is the same force renormalized by screening due to the presence of other vortices. According to Eq. (7.75), the bare force is given by

$$F_0(r) = -\frac{dV_0(r)}{dr} \tag{7.79}$$

where $V_0(r)$ corresponds to an attractive potential

$$V_0(r) \simeq 2\pi J \ln(r/a) \tag{7.80}$$

where it is assumed that $r \geq a$. We note that Eq. (7.80) follows from (7.75) by assuming that the winding numbers are $q_i = \pm 1$ (since vortices with $|q_i| > 1$ are energetically unfavorable). From Eqs. (7.78)–(7.80) we see that, as a result of dielectric screening, the magnitude of the force decreases from its unscreened value $2\pi J/r$ to a screened value $2\pi J/[r\varepsilon(r)]$. The dielectric description can be pursued further by defining the polarizability $\chi(r)$ through the relation

$$\varepsilon(r) = 1 + 4\pi\chi(r) \tag{7.81}$$

where

$$\chi(r) = \int_a^r dr'\, n(r')\alpha(r') \tag{7.82}$$

The quantity $\alpha(r)$ is a polarizability of a vortex pair with separation r. The expression (7.82) represents the polarizability due to all pairs of separation less than r. It should be noted that only these pairs can reorient themselves in the field of a pair of size r and thus contribute to the screening of the force $F_0(r)$. The quantity $n(r)\,dr$ is the average number of pairs in the interval $(r, r + dr)$, and it is given by the Boltzmann distribution

$$n(r)\,dr = 2\pi r n_0 \exp\left\{-\beta[2\mu + V(r)]\right\} dr \tag{7.83}$$

where n_0 is a quantity proportional to a^{-4} and independent of β, J, μ, and r. Note that $2\mu + V(r)$ corresponds to the energy needed to create a pair at the smallest distance $r = a$ and then increasing the separation to r. The potential $V(r)$ is, according to Eqs. (7.78)–(7.80), defined as

$$-\frac{dV}{dr} = \frac{F_0(r)}{\varepsilon(r)} \tag{7.84}$$

where

$$F_0(r) = -2\pi J \frac{d\ln(r/a)}{dr} \tag{7.85}$$

Integrating Eq. (7.84), we obtain with use of Eq. (7.85)

$$V(r) = 2\pi J \int_{\ln a}^{\ln r} \frac{d\ln r}{\varepsilon(r)} \tag{7.86}$$

Let us now derive the expression for the polarizability $\alpha(r)$. The expression (7.75) implies that the magnitude q_v of the vortex charge (in Coulomb-gas representation) is

$$q_v = (2\pi J)^{\frac{1}{2}}|q_i| = (2\pi J)^{\frac{1}{2}} \tag{7.87}$$

Since the dipole moment of the pair with separation r is $q_v r$, we have for the average dipole moment induced by the applied electric field $\mathbf{E}$

$$\langle p \rangle_r = \frac{\int_0^{2\pi} d\varphi \, (q_v r \cos\varphi) \exp\left(\beta q_v r E \cos\varphi\right)}{\int_0^{2\pi} d\varphi \, \exp\left(\beta q_v r E \cos\varphi\right)} \tag{7.88}$$

where φ is the angle between the pair and the applied field. For a small electric field, the exponentials in Eq. (7.88) can be expanded, yielding, to first order in E,

$$\langle p \rangle_r \simeq \frac{1}{2}\beta q_v^2 r^2 E \tag{7.89}$$

Using Eq. (7.87), Eq. (7.89) implies that the polarizability of a pair of size r is:

$$\alpha(r) = \frac{\langle p \rangle_r}{E} = \pi\beta J r^2 \tag{7.90}$$

Introducing this result and Eq. (7.83) into Eq. (7.82), we obtain

$$\chi(r) = 2\pi^2 \beta n_0 J \bar{y}^2 \int_a^r dr' \, e^{-\beta V(r')} r'^3 \tag{7.91}$$

where

$$\bar{y} = e^{-\beta\mu} \tag{7.92}$$

is the activity for creating a vortex pair.

Equations (7.86) and (7.91) form a system of coupled integral equations. They can be converted to a system of coupled differential equations, which can be further reduced to a form of the scaling equations first derived by Kosterlitz (1974). This is done by first introducing a dimensionless scaling parameter

$$l = \ln(r/a) \tag{7.93}$$

Furthermore, one defines the scale-dependent screened charge

$$K(l) = \frac{J\beta}{\varepsilon(l)} \tag{7.94}$$

and fugacity

$$\widetilde{y}(l) = \sqrt{2n_0} r^2 e^{-\frac{1}{2}\beta V(r)} \widetilde{y} \tag{7.95}$$

The reader should avoid confusing $K(l)$ with the renormalized stiffness K of Eq. (3.145). Now we construct a differential equation for $K^{-1}(l)$. From Eq. (7.94), we have with help of Eq. (7.93)

$$\frac{dK^{-1}(l)}{dl} = \frac{1}{\beta J} \frac{d\varepsilon(l)}{dl} = \frac{1}{\beta J} r \frac{d\varepsilon(r)}{dr} \tag{7.96}$$

Using the relation (7.81), we determine $d\varepsilon(r)/dr$ by differentiating Eq. (7.91)

$$\frac{d\varepsilon(r)}{dr} = 4\pi \frac{d\chi(r)}{dr} = 8\pi^3 n_0 \beta J r^3 e^{-\beta V(r)} \widetilde{y}^2 \tag{7.97}$$

Introducing this result into Eq. (7.96) and recalling Eq. (7.95), we obtain

$$\frac{dK^{-1}(l)}{dl} = 8\pi^3 n_0 r^4 e^{-\beta V(r)} \widetilde{y}^2 = 4\pi^3 \widetilde{y}^2(l) \tag{7.98}$$

This is one of the desired scaling equations. The other one follows by differentiating Eq. (7.95) with respect to l

$$\frac{d\widetilde{y}(l)}{dl} = r \frac{d\widetilde{y}(l)}{dr} = r\sqrt{2n_0} \widetilde{y} \left(2r - \frac{1}{2} r^2 \beta \frac{dV}{dr} \right) e^{-\frac{1}{2}\beta V(r)}$$

$$= \left(2 - \frac{1}{2} r \beta \frac{dV}{dr} \right) \widetilde{y}(l) \tag{7.99}$$

From Eq. (7.86) we have with use of Eq. (7.94)

$$\frac{1}{2} r \beta \frac{dV}{dr} = \frac{\beta \pi J}{\varepsilon(r)} = \pi K(l) \tag{7.100}$$

Introducing this expression on the RHS of Eq. (7.99), we obtain the second scaling equation

$$\frac{d\widetilde{y}(l)}{dl} = [2 - \pi K(l)] \widetilde{y}(l) \tag{7.101}$$

7.4.2. Linearized scaling equations

Equations (7.98) and (7.101) have the following fixed point

$$\widetilde{y}_f = 0, \qquad 2 - \pi K_f = 0 \tag{7.102}$$

To linearize the scaling equations about this fixed point, we introduce the parameters x and y, defined as (Kosterlitz, 1974)

$$x = -(2 - \pi K) \tag{7.103}$$

and

$$y = 4\pi\widetilde{y} \tag{7.104}$$

Using Eq. (7.103), we calculate the LHS of Eq. (7.98)

$$\frac{dK^{-1}}{dl} = \frac{d}{dl}\left(\frac{\pi}{x+2}\right) = -\frac{\pi}{(x+2)^2}\frac{dx}{dl} \tag{7.105}$$

The RHS of (7.98) is expressed, by means of (7.104), as

$$4\pi^3\widetilde{y}^2(l) = \frac{\pi}{4}y^2 \tag{7.106}$$

Equations (7.105) and (7.106) allow us to transform Eq. (7.98) to the form

$$dx = -\frac{1}{4}(x+2)^2 y^2\, dl \tag{7.107}$$

Equation (7.101) becomes, using Eqs. (7.103) and (7.104)

$$dy = -xy\, dl \tag{7.108}$$

In terms of the parameters x and y, the fixed point (7.102) is expressed as

$$y_f = 0, \qquad x_f = 0 \tag{7.109}$$

The linearized scaling equations are obtained by letting $x = x_f = 0$ on the RHS of Eq. (7.107)

$$dx = -y^2\, dl \tag{7.110}$$

In what follows, we consider the solution of Eqs. (7.108) and (7.110). Eliminating dl from these equations, we obtain

$$x\, dx = y\, dy \tag{7.111}$$

Solutions that are consistent with Eq. (7.111) have the form

$$x^2 - y^2 = x_0^2 \tag{7.112}$$

7.4.3. Flow in xy space

In Fig. 7.1 we show the trajectories in the xy plane, corresponding to the solutions (7.112). Equation (7.110) implies that $dx < 0$ for all values of x and y. Hence, x always decreases on iteration consistent with the direction of the arrows in Fig. 7.1. On the other hand, Eq. (7.108) implies that y decreases in regions I and III (where $xy > 0$), whereas it increases in regions II and IV (where $xy < 0$). To discuss the critical point, we must first establish the boundary conditions of the scaling equations. These are

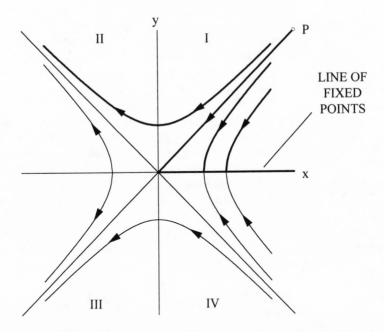

FIG. 7.1. Scaling trajectories for the dielectric-screening model of interacting vortices in a 2D superfluid. The parameters x and y are defined by Eqs. (7.103) and (7.104), respectively. The trajectories represent solutions of the Eq. (7.112), and the point P is a critical point.

given by the initial values x_i and y_i, corresponding to the unrenormalized parameters $K(l = 0)$ and $\widetilde{y}(l = 0)$. Using Eq. (7.103), we have

$$x_i = -[2 - \pi K(l = 0)] = -\left(2 - \frac{\pi J \beta}{\varepsilon(l = 0)}\right) = (\pi J \beta - 2) \qquad (7.113)$$

where we have used Eq. (7.94) and put $\varepsilon(l = 0) = 1$, which expresses the fact that there is no screening of the pair with the smallest possible size $r \sim a$, corresponding to $l = 0$ [see Eq. (7.93)]. From Eq. (7.95), we have

$$y_i = \widetilde{y}(l = 0) = \sqrt{2n_0} a^2 \bar{y} e^{-\frac{1}{2}\beta V(l=0)} = \sqrt{2n_0} a^2 e^{-\beta \mu} \qquad (7.114)$$

where we have put $V(l = 0) = V(r = a) = 0$, which follows directly from Eq. (7.86). Let us now consider the point P on the asymptote, whose coordinates are the initial values x_i, y_i, satisfying the condition

$$x_i = y_i \qquad (7.115)$$

The trajectory then flows directly into the fixed point (7.109). For all initial points to the right of P, the trajectories terminate on the line $x_f > 0$,

$y_f = 0$. This *line of fixed points* corresponds to the phase in which there are *no free* vortices (all pairs are bound). Note, that free vortices correspond to pairs with the separation $r \to \infty$, and the quantity $y_f = 4\pi\tilde{y}(l \to \infty)$ is according to Eq. (7.95) proportional to the square root of the density of such pairs [see also Eq. (7.83)]. Now consider the trajectories that start at initial points to the left of point P. Such trajectories flow to $x \to -\infty$, $y \to \infty$. This is called a *runaway* which is symptomatic of the breakdown of the linearization of the original scaling equations (7.107) and (7.108). In this strong-coupling limit, the number of dissociated pairs, proportional to $\tilde{y}^2(l \to \infty)$, diverges. Thus the segment connecting the point P to the origin provides a demarcation between the set of initial values, leading to superfluidity ($x_f > 0$, $y_f = 0$) and a set that terminates in a state of free vortices (preventing superfluidity). Hence the point P can be identified with the critical point. The transition temperature is defined by the coordinates of the critical point x_i, y_i. Inserting Eqs. (7.113) and (7.114) into (7.115), we obtain a condition for the Kosterlitz–Thouless transition temperature T_{KT}

$$\frac{\pi J}{k_B T_{KT}} - 2 = a^2 \sqrt{2n_0} \exp\left(-\frac{\mu}{k_B T_{KT}}\right) \qquad (7.116)$$

It turns out that the RHS of Eq. (7.116) is well below one. Thus, we obtain an approximate expression for the transition temperature

$$T_{KT} \simeq \frac{\pi}{2k_B} J \qquad (7.117)$$

It is interesting that this result can also be obtained from a simple thermodynamic consideration (Kosterlitz and Thouless, 1973): The free energy of a single vortex is

$$F = E - TS \qquad (7.118)$$

where the energy E is obtained from Eq. (7.69) by setting the vortex charge q equal to one. Thus

$$E = \pi J \ln(R/\varepsilon_v) \qquad (7.119)$$

Since the vortex center can be placed in $(R/\varepsilon_v)^2$ possible places, the entropy is

$$S = 2k_B \ln(R/\varepsilon_v) \qquad (7.120)$$

From Eqs. (7.118)–(7.120), we have for the free energy

$$F = (\pi J - 2k_B T) \ln(R/\varepsilon_v) \qquad (7.121)$$

The formation of isolated free vortices will, according to this result, be favorable when the temperature is high enough so that the entropy term takes over. The critical temperature thus obtained is in agreement with Eq. (7.117).

7.5. Vortex-unbinding in charged superfluids

We now consider thin superconducting films and two-dimensional Josephson arrays involving *charged* superfluid. In such systems, the vortex–antivortex interaction is modified, compared to the case of a neutral superfluid, by a coupling of the supercurrent to the gauge field. As shown by Pearl (1964), the interaction energy increases as $\ln r$ for separations $r \ll \lambda_\perp$, where $\lambda_\perp$ is the penetration depth for magnetic fields perpendicular to the film. For $r \gg \lambda_\perp$, the attractive interaction decreases in magnitude as $1/r$. This relatively slow decay is caused by less efficient magnetic screening of the thin superconducting sheet. Nevertheless, this decay is not sufficiently slow to keep the vortex–antivortex pairs with large separations from being thermally dissociated at very low temperatures. It turns out that the logarithmic law is essential for the existence of the KTB transition (Kosterlitz and Thouless, 1973) and thus superconducting films appear to be excluded, at least theoretically, as candidates for such a phase transition. Beasley et al. (1979) have pointed out that, for sufficiently dirty films, the screening length $\lambda_\perp$ can actually exceed the size of the sample. Hence, the $\ln r$ law applies for pairs within such films. This idea led to a surge of experimental activity focused on the KTB transition in superconducting films and arrays. In this section, we review the basic properties of a single vortex and vortex-pair interactions in a thin superconducting film and investigate the conditions for the existence of the KTB transition. We use the model of a two-dimensional array, described by Hamiltonian (7.9). The role of charging energy in the KTB transition will be discussed in Section 7.6.

7.5.1. *Current distribution for a single vortex*

Let us consider a planar array with a single vortex described by angular variable $\theta(\boldsymbol{\rho})$. According to Eq. (7.54), we have, assuming a vortex of unit vorticity placed at $\boldsymbol{\rho} = 0$,

$$\boldsymbol{\nabla}\theta(\boldsymbol{\rho}) = \frac{1}{\rho}\boldsymbol{\theta}_0 \tag{7.122}$$

where $\boldsymbol{\theta}_0$ is a unit vector in θ direction. For an array of finite thickness d, the Hamiltonian density $h(\boldsymbol{\rho}, z)$ is [see Eq. (7.9)]

$$h(\boldsymbol{\rho}, z) = \begin{cases} \frac{J}{2d}\left(\boldsymbol{\nabla}\theta - \frac{e^*}{\hbar c}\mathbf{A}\right)^2, & \text{for } -\frac{d}{2} < z < \frac{d}{2} \\ 0, & \text{otherwise} \end{cases} \tag{7.123}$$

which implies a current density

$$\begin{aligned} \mathbf{j}(\mathbf{r}) = \mathbf{j}(\boldsymbol{\rho}, z) &= -c\frac{\partial h}{\partial \mathbf{A}} \\ &= \begin{cases} \frac{e^* J}{\hbar d}\left(\boldsymbol{\nabla}\theta - \frac{e^*}{\hbar c}\mathbf{A}\right) = \mathbf{j}(\boldsymbol{\rho}), & \text{for } -\frac{d}{2} < z < \frac{d}{2} \\ 0, & \text{otherwise} \end{cases} \end{aligned} \tag{7.124}$$

We assume that d is so small that $\nabla\theta$ and $\mathbf{A}$ can be regarded as independent of z. It should be pointed out that $\mathbf{A}$ in Eq. (7.124) does *not* correspond to the London gauge (see Tinkham, 1975). However, the derivation of the vortex-induced current density is especially easy to do in the London gauge, by using the results of Section 7.2.1. We thus introduce, following Eq. (7.20), the two-dimensional London gauge field

$$\mathbf{A}_L(\boldsymbol{\rho}) = \mathbf{A} - \frac{\hbar c}{e^*}\nabla\theta \qquad (7.125)$$

In the absence of an external magnetic field, the field $\mathbf{A}$ on the RHS of this equation can be identified with $\mathbf{A}_{\text{ind}}$, and a comparison with Eq. (7.21) yields

$$\mathbf{A}_{\text{ext}}(\boldsymbol{\rho}) = -\frac{\hbar c}{e^*}\nabla\theta = -\frac{\hbar c}{e^*\rho}\boldsymbol{\theta}_0 \qquad (7.126)$$

where we have used Eq. (7.122). Thus the vortex itself acts as a source of external field. From Eqs. (7.124) and (7.125), the current density $\mathbf{j}(\boldsymbol{\rho})$ is given by the London equation

$$\mathbf{j}(\boldsymbol{\rho}) = -\frac{e^{*2}J}{\hbar^2 cd}\mathbf{A}_L(\boldsymbol{\rho}) \qquad (7.127)$$

To calculate $\mathbf{A}_L(\boldsymbol{\rho})$, we use Eq. (7.26), where $\mathbf{A}_{\text{ext}}(\mathbf{q})$ is the two-dimensional Fourier transform of Eq. (7.126)

$$\mathbf{A}_{\text{ext}}(\mathbf{q}) = \int d^2\rho\, e^{-i\mathbf{q}\cdot\boldsymbol{\rho}}\mathbf{A}_{\text{ext}}(\boldsymbol{\rho}) = -\frac{2\pi\hbar c\boldsymbol{\theta}_0}{e^* q} \qquad (7.128)$$

Introducing this result into Eq. (7.26), we obtain

$$\mathbf{A}_L(\mathbf{q}) = -\frac{2\pi\hbar c\boldsymbol{\theta}_0}{e^*(q + q_c)} \qquad (7.129)$$

The inverse Fourier transform of this equation yields

$$\mathbf{A}_L(\boldsymbol{\rho}) = -\frac{2\pi\hbar c\boldsymbol{\theta}_0}{e^*}\int \frac{d^2q}{(2\pi)^2}\frac{e^{i\mathbf{q}\cdot\boldsymbol{\rho}}}{q + q_c} = -\frac{\hbar c\boldsymbol{\theta}_0}{e^*}\int_0^\infty dq\,\frac{J_0(q\rho)q}{q + q_c} \qquad (7.130)$$

Explicit dependence of $\mathbf{A}_L$ and $\mathbf{j}$ on ρ can be found from this expression in two extreme limits: $\rho \ll \lambda_\perp$ and $\rho \gg \lambda_\perp$, where $\lambda_\perp$ is the effective screening length (7.28).

For $\rho \ll \lambda_\perp$, we have $q \gg q_c$ and the expressions (7.127) and (7.130) yield:

$$j(\boldsymbol{\rho}) \simeq \frac{e^* J\theta_0}{\hbar d}\int_0^\infty dq\, J_0(q\rho) = \frac{e^* J\theta_0}{\hbar d\rho}, \qquad \text{for } \rho \ll \lambda_\perp \qquad (7.131)$$

We note that the same result can be obtained directly from Eq. (7.124) by neglecting $\mathbf{A}$. Thus, for $\boldsymbol{\rho} \ll \lambda_\perp$, screening due to induced magnetic vector potential is absent, and the current is completely determined by the gradient of angular variable (in the same way as in a neutral superfluid).

For $\rho \gg \lambda_\perp$, we can approximate Eq. (7.130) as follows

$$\mathbf{A}_L(\boldsymbol{\rho}) \simeq -\frac{\hbar c \theta_0}{e^*} \int_0^{\frac{1}{\rho}} dq \frac{q}{q_c} = -\frac{\hbar c \theta_0}{2e^* q_c \rho^2}, \qquad \text{for } \rho \gg \lambda_\perp \qquad (7.132)$$

where $J_0(q\rho)$ has been replaced by a step function with an upper cutoff at $1/\rho$. Substituting this result into the Eq. (7.127), we obtain

$$\mathbf{j}(\boldsymbol{\rho}) \simeq \frac{e^* J \theta_0}{2\hbar d q_c \rho^2}, \qquad \text{for } \rho \gg \lambda_\perp \qquad (7.133)$$

We see that, even beyond the screening length $\lambda_\perp$, $\mathbf{j}(\boldsymbol{\rho})$ exhibits only a slow decrease with distance. The reason for this is the rather inefficient screening by the thin layer of superelectrons. The results (7.131) and (7.133) have been first derived by Pearl (1964).

7.5.2. Self-energy of a vortex

The energy E of a single-vortex configuration is given by the sum of the kinetic energy of Eq. (7.8) and a magnetic energy (de Gennes, 1966). Using Eqs. (7.123) and (7.124), we obtain

$$E = E_K + E_M = \frac{1}{2}\frac{\hbar^2 d^2}{e^{*2} J} \int d^2\rho\, \mathbf{j}^2(\boldsymbol{\rho}) + \frac{1}{8\pi} \int d^3r\, H^2(\mathbf{r}) \qquad (7.134)$$

where the integral for the kinetic energy is over the film area, and the magnetic energy density is integrated over the entire three-dimensional space. Using the identity (Jackson, 1975)

$$\frac{1}{8\pi} \int d^3r\, H^2(\mathbf{r}) = \frac{1}{2c^2} \int d^3r \int d^3r' \frac{\mathbf{j}(\mathbf{r}) \cdot \mathbf{j}(\mathbf{r}')}{|\mathbf{r} - \mathbf{r}'|} \qquad (7.135)$$

with $\mathbf{j}(\mathbf{r})$ given by Eq. (7.124), the expression (7.134) can be written in terms of the integrals of $\mathbf{j}(\rho)$ over the film area only

$$E = \frac{1}{2}\frac{\hbar^2 d^2}{e^{*2} J} \int d^2\rho\, \mathbf{j}^2(\boldsymbol{\rho}) + \frac{d^2}{2c^2} \int d^2\rho \int d^2\rho' \frac{\mathbf{j}(\boldsymbol{\rho}) \cdot \mathbf{j}(\boldsymbol{\rho}')}{|\boldsymbol{\rho} - \boldsymbol{\rho}'|} \qquad (7.136)$$

Introducing the two-dimensional Fourier transform $\mathbf{j}(\mathbf{q})$, we can express E in terms of Parseval's relation

$$E = \frac{1}{2}\frac{\hbar^2 d^2}{e^{*2} J} \int \frac{d^2q}{(2\pi)^2} \mathbf{j}(\mathbf{q}) \cdot \mathbf{j}(-\mathbf{q}) + \frac{d^2}{2c^2} \int \frac{d^2q}{(2\pi)^2} \mathbf{j}(\mathbf{q}) \left(\frac{2\pi}{q}\right) \cdot \mathbf{j}(-\mathbf{q}) \quad (7.137)$$

According to Eqs. (7.127) and (7.129), we have

$$\mathbf{j}(\mathbf{q}) = \frac{2\pi J e^* \theta_0}{\hbar d(q + q_c)} \qquad (7.138)$$

Introducing this result into Eq. (7.137) and combining the kinetic and magnetic contributions, we have

$$E = \pi J \int_{q_{min}}^{q_{max}} \frac{dq}{q + q_c} = \pi J \ln \left(\frac{q_{max} + q_c}{q_{min} + q_c} \right) \tag{7.139}$$

Assuming a superconducting film (array) in the form of a circle of radius R with the vortex at its center, the cutoffs are given by $q_{min} \simeq 2\pi/R$ and $q_{max} \simeq 2\pi/\xi$, where ξ is the size of the vortex core. If $R \gg \lambda_\perp$, then the following inequality is expected to hold

$$q_{max} \gg q_c \gg q_{min} \tag{7.140}$$

so that Eq. (7.137) can be approximated as

$$E \simeq \pi J \ln(q_{max}/q_c) \simeq \pi J \ln(\lambda_\perp/\xi), \qquad \text{for } R \gg \lambda_\perp \tag{7.141}$$

We see that the self-energy of the vortex tends to a constant value, *independent* of R, for $R \gg \lambda_\perp$. This is in *contrast* to the vortex self-energy of a neutral superfluid, which exhibits a logarithmic dependence on R [see Eq. (7.119)]. Consequently, the simple thermodynamic argument for vortex dissociation, which for neutral superfluid led to a simple expression for the critical temperature, cannot be carried through for a charged superfluid; for, if the vortex entropy is given by Eq. (7.120), the free energy (7.121) remains dominated by the entropy term. Consequently, a free vortex can appear in the system at a temperature $T > T_c(R)$, where $T_c(R)$ is obtained by setting $E - TS$ equal to zero, yielding

$$T_c(R) \simeq \frac{\pi J}{2k_B} \frac{\ln(\lambda_\perp/\xi)}{\ln(R/\xi)} \tag{7.142}$$

We see that $T_c(R)$ vanishes (logarithmically) with R increasing to infinity. Therefore, in principle, there is *no superconductivity* in a thin film, in the thermodynamic limit $R \to \infty$, at finite temperatures.

For $R \ll \lambda_\perp$, the inequality (7.140) is replaced by

$$q_{max} \gg q_{min} \gg q_c \tag{7.143}$$

and Eq. (7.139) can be approximated as follows

$$E \simeq \pi J \ln(q_{max}/q_{min}) = \pi J \ln(R/\xi), \qquad \text{for } R \ll \lambda_\perp \tag{7.144}$$

which agrees with the expression (7.119), as expected, because of the lack of magnetic screening in the range $\xi < \rho < R$ [see Eq. (7.131)]. Now the thermodynamic argument of Eqs. (7.118)–(7.121) can be applied exactly as in the case of a neutral superfluid. Hence, if the superconducting films can be made such that the magnetic penetration depth $\lambda_\perp$ greatly *exceeds the sample size*, the KTB transition is expected to take place. Recalling

Eq. (7.28), we see that this condition can be met in a two-dimensional array with very small Josephson coupling energy. Beasley et al. (1979) have considered thin superconducting films in the dirty limit, for which $\lambda_\perp$ can be shown proportional to the inverse mean-free path [see Eq. (7.179)]. Thus, for films with large normal sheet resistance, the condition $R \ll \lambda_\perp$ is expected to hold and superconductivity is possible for $T < T_{KT}$.

7.5.3. *Interaction between vortices*

We now consider a vortex at $\boldsymbol{\rho} = 0$ and an antivortex at $\boldsymbol{\rho} = \boldsymbol{\rho}_{12}$. The net current density associated with this pair is a superposition

$$\mathbf{j}_{tot}(\boldsymbol{\rho}) = \mathbf{j}(\boldsymbol{\rho}) - \mathbf{j}(\boldsymbol{\rho} - \boldsymbol{\rho}_{12}) = \mathbf{j}_1(\boldsymbol{\rho}) + \mathbf{j}_2(\boldsymbol{\rho}) \tag{7.145}$$

where $\mathbf{j}(\boldsymbol{\rho})$ is the current density, defined by Eq. (7.127). Introducing Eq. (7.145) into the expression (7.136), we obtain for the total energy of the system

$$E = E_1 + E_2 + E_{12} \tag{7.146}$$

where $E_1 = E_2$ is the vortex self-energy , given by Eq. (7.139), and E_{12} is the interaction energy

$$
\begin{aligned}
E_{12} = \frac{\hbar^2 d^2}{e^{*2}J} \int d^2\rho\, \mathbf{j}_1(\boldsymbol{\rho}) \cdot \mathbf{j}_2(\boldsymbol{\rho}) + \frac{d^2}{2c^2} \int d^2\rho \\
\times \int d^2\rho'\, \frac{\mathbf{j}_1(\boldsymbol{\rho}) \cdot \mathbf{j}_2(\boldsymbol{\rho}') + \mathbf{j}_2(\boldsymbol{\rho}) \cdot \mathbf{j}_1(\boldsymbol{\rho}')}{|\boldsymbol{\rho} - \boldsymbol{\rho}'|}
\end{aligned}
\tag{7.147}
$$

Using Parseval's relation, we obtain from Eq. (7.147)

$$E_{12} = \frac{\hbar^2 d^2}{e^{*2}J} \int \frac{d^2q}{(2\pi)^2} \mathbf{j}_1(\mathbf{q}) \cdot \mathbf{j}_2(-\mathbf{q}) + \frac{d^2}{c^2} \int \frac{d^2q}{(2\pi)^2} \mathbf{j}_1(\mathbf{q}) \cdot \left(\frac{2\pi}{q}\right) \mathbf{j}_2(-\mathbf{q}) \tag{7.148}$$

From Eq. (7.145), we calculate, with use of (7.138), the Fourier transform

$$\mathbf{j}_2(-\mathbf{q}) = -\int d^2\rho\, e^{i\mathbf{q}\cdot\boldsymbol{\rho}} \mathbf{j}(\boldsymbol{\rho} - \boldsymbol{\rho}_{12}) = -\frac{2\pi J e^{*2} e^{i\mathbf{q}\cdot\boldsymbol{\rho}_{12}} \boldsymbol{\theta}_0}{\hbar d(q + q_c)} \tag{7.149}$$

where the minus sign corresponds to the *antivortex* current. Introducing this result into Eq. (7.148), we have:

$$
\begin{aligned}
E_{12}(\rho_{12}) &= -J \int \frac{d^2q\, e^{i\mathbf{q}\cdot\boldsymbol{\rho}_{12}}}{(q + q_c)^2} \left(1 + \frac{q_c}{q}\right) \\
&= -2\pi J \int_{q_{min}}^{q_{max}} \frac{dq\, J_0(q\rho_{12})}{q + q_c}
\end{aligned}
\tag{7.150}
$$

where $q_{min} \simeq 2\pi/R$ and $q_{max} \simeq 2\pi/\xi$, as in Eq. (7.139). Assuming that $R \gg \lambda_\perp$, these cutoffs satisfy the inequality (7.140). In this case, we can distinguish between two regimes, showing different type of ρ_{12} dependence of the interaction energy:

1. For $\rho \ll \lambda_\perp$, we obtain from Eq. (7.150), letting $\rho_{12} = \rho$ and making a cutoff approximation [see Eq. (7.132)]

$$
\begin{aligned}
E_{12}(\rho) &\simeq -2\pi J \int_{q_{\min}}^{1/\rho} \frac{dq}{q + q_c} \\
&= -2\pi J \ln \left(\frac{1/\rho + q_c}{q_{\min} + q_c} \right) \\
&\simeq 2\pi J \ln(\rho q_c) \\
&\simeq 2\pi J \ln(\rho/\lambda_\perp), \qquad \text{for } \rho \ll \lambda_\perp
\end{aligned}
\tag{7.151}
$$

The RHS of Eq. (7.151) shows the same *logarithmic* behavior as the vortex–antivortex interaction H_v in a neutral superfluid [see Eq. (7.75)].

2. For $\rho \gg \lambda_\perp$, Eq. (7.151) yields

$$
\begin{aligned}
E_{12}(\rho) &\simeq -2\pi J \ln \left(\frac{1/\rho + q_c}{q_{\min} + q_c} \right) \\
&\simeq -2\pi J \ln \left(1 + \frac{1}{q_c \rho} \right) \\
&\simeq -\frac{4\pi J \lambda_\perp}{\rho}
\end{aligned}
\tag{7.152}
$$

Hence, the attractive energy of the pair decreases as $1/\rho$. Such a long-range decay is again due to the reduced screening power of the superconducting array, so that the interaction takes place mostly through empty space above and below the array. Nevertheless, the $1/\rho$ dependence is not slow enough to prevent pairs with large spacing from being thermally dissociated. This can be seen from the criterion for stability of vortex–antivortex pair with separation ρ at temperature T

$$
\frac{dV(\rho)}{d\rho} > \frac{4k_B T}{\rho}
\tag{7.153}
$$

which follows from Eq. (7.99). Introducing for $V(\rho)$ the interaction (7.152), Eq. (7.153) implies a stability condition

$$
k_B T < \frac{\pi J \lambda_\perp}{\rho} = k_B T(\rho)
\tag{7.154}
$$

We see that pairs with separation $\rho \gg \lambda_\perp$ become unstable with respect to thermal dissociation for a temperature $T(\rho)$, which goes to zero on increasing ρ as $1/\rho$. This confirms the absence of superconductivity in the thermodynamic limit, as previously found from free-energy considerations for a single vortex [see Eq. (7.142)].

The other experimentally interesting limit, $R \ll \lambda_\perp$, remains to be considered. Since $\rho \lesssim R$, this also implies $1/\rho \gg q_c$, so that the Eq. (7.150) can be approximated as

$$E_{12}(\rho) = -2\pi J \ln \left(\frac{1/\rho + q_c}{q_{\min} + q_c} \right)$$

$$\simeq 2\pi J \ln \left(\frac{2\pi \rho}{R} \right), \qquad \text{for } R \ll \lambda_\perp \qquad (7.155)$$

Inserting this logarithmic function into the stability criterion (7.153), we obtain

$$\frac{dV}{d\rho} = \frac{2\pi J}{\rho} > \frac{4k_B T}{\rho} \qquad (7.156)$$

which implies a critical temperature of vortex-unbinding transition

$$T_c = T_{\text{KT}} = \frac{\pi}{2k_B} J \qquad (7.157)$$

As expected, this transition temperature agrees with the Kosterlitz–Thouless–Berezinskii transition temperature in a neutral superfluid [see Eq. (7.117)].

7.5.4. *Transition temperature for disordered superconducting films*

We consider next microscopically inhomogeneous systems represented by thin disordered superconducting films. Following Beasley et al. (1979), we first derive a relationship between the vortex-unbinding transition temperature T_{KT} and the normal sheet resistance R_N of the film. The starting point is the relation

$$k_B T_{\text{KT}} = \frac{\pi \hbar^2 n_s^{2D}}{2m^*} \qquad (7.158)$$

where n_s^{2D} is the superfluid sheet density and m^* is the mass of the Cooper pair. One way to establish this relation is to consider the phase-dependent part of the Ginzburg–Landau free energy (2.1)

$$F[\theta] = \frac{\hbar^2 |\psi|^2}{2m^*} \int d^3r \left(\boldsymbol{\nabla}\theta - \frac{e^*}{\hbar c} \mathbf{A} \right)^2 \qquad (7.159)$$

where $|\psi|^2$ is assumed constant and equal to the average superfluid density n_s [see Eq. (2.23)]. To avoid possible confusion, let us point out that, in a granular array, $|\psi|^2$ is not the local superconducting density in the grain; rather, it is a superconducting density of the array that is usually smaller depending on the parameters J and d. Noting that for a thin film of thickness d, we have

$$n_s^{2D} = dn_s = d|\psi|^2 \qquad (7.160)$$

the free energy (7.159) can be written as a two-dimensional integral over the film area:

$$F[\theta] = \frac{\hbar^2 n_s^{2D}}{2m^*} \int d^2\rho \left(\boldsymbol{\nabla}\theta - \frac{e^*}{\hbar c}\mathbf{A} \right)^2 \tag{7.161}$$

On comparing this expression with Eq. (7.8), we obtain a relation between the Josephson coupling constant J and the superfluid sheet density

$$J = \frac{\hbar^2}{m^*} n_s^{2D} \tag{7.162}$$

A similar equation has been derived by Berezinskii (1971), relating the coupling constant of the planar spin model to the superfluid density of the lattice boson model. Introducing Eq. (7.162) into (7.157), we obtain Eq. (7.158). Although we have derived the latter result starting from the model of a Josephson-coupled array, it should be pointed out that this is a universal relation that is valid also in the case of a disordered superconducting film. This is to be expected, as long as the Ginzburg–Landau description used in Eq. (7.159) is valid, which implies that we are restricted to the approximation of *local* electrodynamics. This holds provided the mean-free path l is much shorter than the BCS coherence length ξ_0 (de Gennes, 1966).

The London penetration depth λ_L of a bulk superconductor is defined in terms of the Cooper pair density n_s as follows (de Gennes, 1966)

$$\lambda_L^2 = \frac{mc^2}{4\pi n e^2} = \frac{1}{2}\frac{mc^2}{4\pi n_s e^2} \tag{7.163}$$

where $n = 2n_s$ is the free-electron density. Using this relation and Eq. (7.160), the superfluid sheet density can be written as

$$n_s^{2D} = \frac{1}{2}\frac{mc^2}{4\pi e^2}\frac{d}{\lambda_L^2} \tag{7.164}$$

The ratio d/λ_L^2 on the RHS of this expression represents, according to Eq. (7.29), the inverse effective screening length $\lambda_\perp$ for magnetic fields perpendicular to the film.

Though Eq. (7.29) has been verified for a granular array, it is again a general relation. This follows by taking the London equation for a bulk superconductor [see Eq. (7.32)]

$$\mathbf{J}(\mathbf{r}) = -\frac{c}{4\pi\lambda_L^2}\mathbf{A}_L(\mathbf{r}) \tag{7.165}$$

For a very thin film, the z dependence of $\mathbf{J}(\mathbf{r})$ takes the form

$$\mathbf{J}(\mathbf{r}) = \mathbf{J}(\boldsymbol{\rho}, z) = -\frac{cd\delta(z)}{4\pi\lambda_L^2}\mathbf{A}_L(\mathbf{r}) \tag{7.166}$$

Comparing the RHS of this equation with Eq. (7.20), we find

$$\Lambda = \frac{cd}{4\pi\lambda_L^2} \tag{7.167}$$

From this relation we find, with the help of Eqs. (7.27) and (7.28)

$$\lambda_\perp = \frac{1}{2q_c} = \frac{c}{4\pi\Lambda} = \frac{\lambda_L^2}{d} \tag{7.168}$$

In view of Eq. (7.165), this implies that the relation (7.29) is true as long as we are dealing with a local London superconductor.

Introducing Eq. (7.164) into Eq. (7.158) and taking (7.29) into account, we obtain

$$k_B T_{KT} = \frac{\hbar^2 c^2}{32 e^2 \lambda_\perp} = \frac{\phi_0^2}{32\pi^2}\frac{1}{\lambda_\perp(T_{KT})} \tag{7.169}$$

where $\phi_0 = hc/2e$ is the superconducting flux quantum. The penetration depth $\lambda_\perp$ can be linked to the sheet resistance of the film by expressing the bulk pair density n_s in terms of normal conductivity σ_n as follows (Tinkham, 1975)

$$n_s(T) = \frac{\pi\sigma_n m}{2\hbar e^2}\Delta_0(T)\tanh\frac{\beta\Delta_0(T)}{2} \tag{7.170}$$

The origin of this expression is discussed in Section 7.5.5. Inserting Eq. (7.170) into the RHS of Eq. (7.163), the penetration depth $\lambda_\perp$ becomes

$$\lambda_\perp(T) = \frac{\lambda_L^2}{d} = \frac{mc^2}{8\pi e^2 d n_s(T)} = \frac{1.78\phi_0^2}{4\pi^5}\frac{e^2}{\hbar}\frac{R_N}{k_B T_{c0}}f^{-1}(T) \tag{7.171}$$

where we express the sheet resistance as $R_N = 1/d\sigma_n$. The function $f(T)$ is given by

$$f(T) = \frac{\Delta_0(T)\tanh\frac{\beta\Delta_0(T)}{2}}{\Delta_0(T=0)} \tag{7.172}$$

On the RHS of Eq. (7.171) we have also used the BCS result $\Delta_0(T = 0) = 1.76 k_B T_{c0}$, where T_{c0} is the bulk transition temperature. Introducing Eqs. (7.171) and (7.172) into Eq. (7.169), we obtain finally the central result of the work by Beasley et al. (1979)

$$\frac{T_{KT}}{T_{c0}}f^{-1}(T_{KT}) = 2.18\frac{R_0}{R_N} \tag{7.173}$$

This formula is interesting, since the same relation for T_{KT} follows for a Josephson-coupled array (Hebard, 1979). This can be seen by inserting for J in Eq. (7.157) the expression (5.146), yielding

$$k_B T_{KT} = \frac{\pi^2\hbar\Delta_0(T_{KT})}{8e^2 R_N}\tanh\frac{\beta_{KT}\Delta_0(T_{KT})}{2}$$

$$= 2.18 k_B T_{c0}\frac{R_0}{R_N}f(T_{KT}) \tag{7.174}$$

7.5.5. *Remark on Eq. (7.170)*

The coincidence of conditions (7.173) and (7.174) can be traced to the formula (7.170), which happens to exhibit strong similarities with the expression for the temperature-dependent Josephson coupling constant. It is instructive to review the basic steps that lead to the expression (7.170). Following Tinkham (1975), we define n_s through the imaginary part of the frequency-dependent conductivity $\sigma_L(\omega)$. According to London theory, we have (noting that $n = 2n_s$)

$$\operatorname{Im}\sigma_L(\omega) = \frac{2n_s e^2}{m\omega} \tag{7.175}$$

Using the theory of linear response, it is possible to calculate $\operatorname{Im}\sigma_L(\omega)$ microscopically from the Kubo formula. This involves evaluation of the current–current correlation function for a superconducting alloy, which is presented in detail by Abrikosov et al. (1965). Using the impurity-averaging methods, these authors derive an expression for the kernel $Q(\mathbf{k},\omega)$, which is related to the low-frequency conductivity by

$$Q = Q(\mathbf{k} = 0, \omega) = -i\omega\sigma(\omega) \tag{7.176}$$

which implies

$$\operatorname{Im}\sigma(\omega) = \frac{1}{\omega}Q \tag{7.177}$$

The explicit formula for Q is [see Eq. (39.28) of Abrikosov et al., 1965]

$$Q = \frac{4\pi n e^2}{\hbar m}\tau_{\mathrm{tr}} k_B T \Delta_0^2(T) \sum_{n=1}^{\infty} \frac{1}{[\hbar^2\omega_n^2 + \Delta_0^2(T)]} \tag{7.178}$$

where τ_{tr} is the transport time between collisions. In writing (7.178), we have assumed that $\tau_{\mathrm{tr}} \ll \hbar/\Delta_0$, which is tantamount to the condition $l \ll \xi_0$. The RHS of Eq. (7.178) involves the same kind of Matsubara sum as Eq. (5.145) for the Josephson coupling constant. It corresponds to the $\mathbf{k} = 0$ limit of the propagator for a free-electron pair. Performing the n summation, Eq. (7.178) yields

$$Q = \frac{\pi n e^2}{\hbar m}\tau_{\mathrm{tr}}\Delta_0(T)\tanh\frac{\beta\Delta_0(T)}{2} = \frac{\pi}{\hbar}\sigma_n\Delta_0\tanh\frac{\beta\Delta_0(T)}{2} \tag{7.179}$$

Introducing this result into Eq. (7.177) and substituting for $\operatorname{Im}\sigma(\omega)$ the RHS of Eq. (7.175), the relation (7.170) is obtained. We see that the superfluid density in a disordered superconductor is determined by the *pair propagation* in the presence of *impurity scattering*. For the Josephson array, on the other hand, it is the *pair propagation* through the *insulating barriers* that determines the superfluid density. If this propagation is viewed as a scattering of pairs by the thin localized barriers, then it is understandable why both models yield the same dependence of the vortex-unbinding transition temperature on the normal-state resistance of the sample.

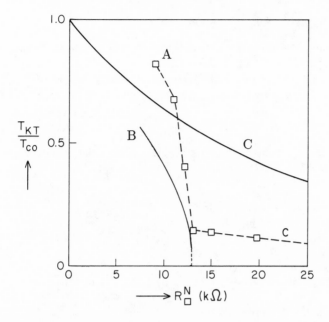

FIG. 7.2. The temperature ratio T_{KT}/T_{c0} versus the normal sheet resistance R_N for resistive transitions in lead film. Line A represents the experimental data of Hebard (1979) and Hebard and Vandenberg (1980), line B is the SCHA result obtained from Eqs. (7.180)-(7.184), and line C corresponds to the solution of Eq. (7.174).

7.6.   Effects of quantum fluctuations on T_{KT}

Comparing the relation (7.173) with the resistivity transitions, taken from a wide variety of data, Beasley et al. (1979) were able to obtain qualitative agreement. Subsequently, Hebard (1979) and Hebard and Vandenberg (1980) made a detailed study of the resistive transitions in granular lead films. They measured T_{KT} versus R_N for a sequence of transitions, obtained for the *same* 250 Å thick film by *varying the degrees of oxidation*. The results illustrated in Fig. 7.2 (curve A) show an abrupt drop of T_{KT} near $R_N \simeq 13$ kΩ, which is in strong disagreement with the smooth dependence (curve C) following from the relation (7.173). It has been suggested by Hebard (1979) that this discrepancy is due to the presence of localized electronic states. Another possible explanation attributes the sudden drop in T_{KT} to the superconductor–insulator transition caused by quantum phase fluctuations (Šimánek, 1980b). Several theoretical approaches have been developed to describe the effect of quantum fluctuations on the vortex-unbinding transition. In this section we confine ourselves to the SCHA method (Šimánek, 1980b; Šimánek and Stein, 1984), the MFA method (Maekawa et al., 1981) and the RG analysis of José (1984).

7.6.1. *Renormalization of T_{KT} within the SCHA*

Following the notation of Section 3.9, we consider the self-charging model described by the Euclidean action (3.132). In the *absence* of a *vortex*, the SCHA replaces this action by the trial action (3.133), where the renormalized stiffness is given by [see Eq. (3.145)]

$$K^{(0)} = Je^{-D_{ij}^{(0)}/2} \tag{7.180}$$

The superscript indicates that we are dealing with the *zero-vortex sector*. According to Eq. (3.137), we have

$$D_{ij}^{(0)} = \langle \theta_{ij}^2 \rangle_{\text{tr}}^{(0)} \tag{7.181}$$

where the average is taken with respect to the trial action (3.133), yielding [see Eq. (3.162)]

$$D_{ij}^{(0)} = \frac{1}{N\beta} \sum_{\mathbf{k}} \sum_n \frac{1 - \cos(\mathbf{k} \cdot \mathbf{R}_{ij})}{(\hbar^2/8U_a)\omega_n^2 + \frac{1}{2}K^{(0)}f(\mathbf{k})} \tag{7.182}$$

where $f(\mathbf{k})$ is the structure factor

$$f(\mathbf{k}) = \sum_{j(i)} [1 - \cos(\mathbf{k} \cdot \mathbf{R}_{ij})] \tag{7.183}$$

The simplest way to take into account the renormalization of T_{KT} is to replace J in Eq. (7.157) by $K^{(0)}$ of Eq. (7.180). Incorporating this replacement into Eq. (7.174), we obtain (Šimánek, 1980b)

$$\left(\frac{T_{KT}}{T_{c0}} \right)_{\text{SCHA}} = 2.18 \frac{R_0}{R_N} f(T_{KT}) \left(\frac{K^{(0)}}{J} \right) \tag{7.184}$$

The factor $K^{(0)}/J$ on the RHS of this equation represents the modification of the original expression of Beasley et al. (1979) due to charging effects. Before we describe the numerical results of calculations based on Eqs. (7.180)–(7.184), we present a derivation of the free-energy functional from which Eq. (7.184) can be justified.

7.6.2. *Free energy as a functional of vortex configuration*

First, let us point out that Eq. (7.184) is based on the assumption that the zero-point phase fluctuations are *not affected* by the presence of a vortex. This is certainly not true near the vortex center where the local intergrain coupling is strongly diminished, which leads to increased local phase fluctuations (see Fig. 7.3). Thus the renormalization of J results in an inhomogeneous problem, which can be described by separating the field $\theta_i(\tau)$ as follows (Šimánek and Stein, 1984):

$$\theta_i(\tau) = \varphi_i + \delta_i(\tau) \tag{7.185}$$

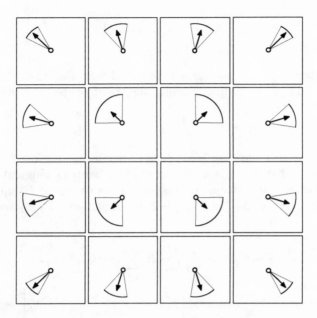

FIG. 7.3.   A portion of a square planar Josephson array, containing a vortex excitation. The circular sectors indicate the span of phase fluctuations about the static vortex configuration. The vectors correspond to the average order parameter in the directions pointed by the static vortex configuration. Note the diminished length of these vectors near the vortex center.

where φ_i is a static vortex configuration and $\delta_i(\tau)$ is a fluctuation about φ_i. Introducing the ansatz (7.185) into the Euclidean action (3.132), we obtain

$$S[\varphi, \delta] = \int_0^{\hbar\beta} d\tau \left(\frac{\hbar^2}{8U_a} \sum_i \dot{\delta}_i^2(\tau) + J \sum_{\langle ij \rangle} (1 - \cos \varphi_{ij} \cos \delta_{ij}) \right) \quad (7.186)$$

where $\varphi_{ij} = \varphi_i - \varphi_j$ and $\delta_{ij} = \delta_i - \delta_j$. We note that the angular variable φ_i satisfies the equation (minimum-action principle) that implies

$$J \sum_{j(i)} \sin \varphi_{ij} = 0 \quad (7.187)$$

where $j(i)$ denotes the nearest neighbors of site i. This allows us to disregard the $\sin \varphi_{ij} \sin \delta_{ij}$ term in the $\langle ij \rangle$ sum of Eq. (7.186). The anharmonic term $\cos \delta_{ij}$, in the latter equation, is treated within the SCHA by replacing $S[\varphi, \delta]$ by a trial action

$$S_{\mathrm{tr}}[\varphi, \delta] = \int_0^{\hbar\beta} d\tau \left(\frac{\hbar^2}{8U_a} \sum_i \dot{\delta}_i^2(\tau) + J \sum_{\langle ij \rangle} (1 - \cos \varphi_{ij}) \right)$$

$$+ \frac{1}{2} \sum_{\langle ij \rangle} K_{ij} \cos \varphi_{ij} \delta_{ij}^2(\tau) \Bigg) \tag{7.188}$$

The variational parameter K_{ij} is determined from the variational principle of Eq. (3.134), according to which the true free energy F satisfies the inequality

$$F \leq F_{\mathrm{tr}} + \frac{1}{\hbar \beta} \langle S - S_{\mathrm{tr}} \rangle_{\mathrm{tr}} = F^*[\varphi] \tag{7.189}$$

We note that this equation involves free energies for a given vortex configuration, so that $F^*[\varphi]$ is a functional of the vortex configuration $\{\varphi_i\}$. In what follows, we derive an expression for the change of F^* upon introduction of a vortex: $\Delta F^* = F^*[\varphi] - F^*[0]$. Equations (7.186) and (7.188) imply

$$\frac{1}{\hbar \beta} \langle S - S_{\mathrm{tr}} \rangle_{\mathrm{tr}} = \frac{1}{\hbar \beta} \int_0^{\hbar \beta} d\tau \sum_{\langle ij \rangle} \cos \varphi_{ij}$$

$$\times \left[J \left(1 - e^{-D_{ij}/2} \right) - \frac{1}{2} K_{ij} D_{ij} \right] \tag{7.190}$$

where

$$D_{ij} = \langle \delta_{ij}^2 \rangle_{\mathrm{tr}} \tag{7.191}$$

Following the steps outlined in Eqs. (3.136)–(3.145), the best choice of K_{ij} is given by the condition

$$\frac{1}{\hbar \beta} \left(\frac{\partial \langle S - S_{\mathrm{tr}} \rangle_{\mathrm{tr}}}{\partial D_{ij}} \right)_K = \frac{1}{2\hbar \beta} \int_0^{\hbar \beta} d\tau \sum_{\langle ij \rangle} \cos \varphi_{ij} D_{ij} \left(J e^{-D_{ij}/2} - K_{ij} \right)$$

$$= 0 \tag{7.192}$$

so that

$$K_{ij} = J e^{-D_{ij}/2} \tag{7.193}$$

Equations (7.191) and (7.193) form a pair of coupled self-consistent equations which, in contrast to Eqs. (7.180) and (7.181), yield a solution for K_{ij} lacking *translation invariance*. This is because the fluctuation $\delta_i(\tau)$ satisfies an equation with a position-dependent coefficient $K_{ij} \cos \varphi_{ij}$. This is seen from Eq. (7.188), which implies the following Euler–Lagrange equation for $\delta_i(\tau)$

$$\frac{\hbar^2}{4U_a} \ddot{\delta}_i(\tau) - \sum_{j(i)} K_{ij} \cos \varphi_{ij} \delta_{ij}(\tau) = 0 \tag{7.194}$$

Equation (7.194) describes displacements of an acoustic phonon scattered by the "potential well" (produced by softening of the coupling constants around the vortex center). The behavior of D_{ij} as a function of distance from the vortex center can be discussed qualitatively for the case of weakly renormalized stiffness constant $(K_{ij} \simeq J)$. Then the spatial dependence of the local coupling constant in Eq. (7.194) is controlled mainly by the function $\cos \varphi_{ij}$. At a lattice site (x_i, y_i), we have a vortex angle φ_i given by

$$\varphi_i = \arctan \frac{y_i}{x_i} \tag{7.195}$$

For a nearest-neighbor pair $\langle ij \rangle$ (on the x axis), Eq. (7.195) implies

$$\varphi_{ij} \simeq \frac{a}{r_{ij}}, \qquad \text{for } r_{ij} \gg a \tag{7.196}$$

where a is the lattice spacing and r_{ij} is the distance from the pair to the vortex center. Thus the local coupling constant exhibits a following dependence on r_{ij}

$$K_{ij}^{\text{loc}} = K_{ij} \cos \varphi_{ij} \simeq J \left(1 - \frac{a^2}{r_{ij}^2} \right) \qquad \text{for } r_{ij} \gg a \tag{7.197}$$

Noting that D_{ij} is dominated by short-wavelength contributions, $k \lesssim k_D$ [see Eq. (3.166)], we use the Einstein model and obtain from Eq. (7.197) the zero-point value of local D_{ij}

$$D_{ij} \sim \sqrt{\frac{U_a}{K_{ij}^{\text{loc}}}} \simeq \sqrt{\frac{U_a}{J(1 - a^2/r_{ij}^2)}} \tag{7.198}$$

Introducing this result into Eq. (7.193), we obtain, expanding the square root,

$$K_{ij} \sim J \exp \left[-\frac{1}{2} \sqrt{\frac{U_a}{J}} \left(1 + \frac{a^2}{2 r_{ij}^2} \right) \right] = K^{(0)} \exp \left(-\frac{1}{4} \sqrt{\frac{U_a}{J}} \frac{a^2}{r_{ij}^2} \right) \tag{7.199}$$

where $K^{(0)}$ is the homogeneous stiffness constant of Eq. (7.180). Since U_a/J is of order one, Eq. (7.199) indicates that the inhomogeneity of K_{ij} decays as we move away from the vortex center, the characteristic decay length being of order a. Thus we may replace in Eq. (7.188) the inhomogeneously renormalized stiffness K_{ij} by $K^{(0)}$, except for a region of radius a near the vortex center. From Eqs. (7.189) and (7.190) we obtain, for a given vortex configuration $\{\varphi_i\}$,

$$F^*[\varphi] \simeq \frac{1}{\hbar \beta} \int_0^{\hbar \beta} d\tau \sum_{\langle ij \rangle} \cos \varphi_{ij} \left[J \left(1 - e^{-D_{ij}^{(0)}/2} \right) - \frac{1}{2} K^{(0)} D_{ij}^{(0)} \right]$$
$$- \frac{1}{\beta} \ln \int \mathcal{D}\delta \, \exp \left(-\frac{1}{\hbar} S_{\text{tr}}^{(0)}[\varphi, \delta] \right) \tag{7.200}$$

where $S_{\text{tr}}^{(0)}[\varphi, \delta]$ follows from Eq. (7.188) by making the replacement $K_{ij} \to K^{(0)}$, yielding

$$S_{\text{tr}}^{(0)}[\varphi, \delta] = \int_0^{\hbar\beta} d\tau \left(\frac{\hbar^2}{8U_a} \sum_i \dot{\delta}_i^2(\tau) + J \sum_{\langle ij \rangle} (1 - \cos \varphi_{ij}) \right.$$

$$\left. + \frac{1}{2} \sum_{\langle ij \rangle} K^{(0)} \cos \varphi_{ij} \delta_{ij}^2(\tau) \right) \qquad (7.201)$$

We note that the last expression on the RHS of this equation overestimates the coupling constant K_{ij} in the neighborhood of the vortex center. This error can be partly compensated by increasing the effective vortex core radius. Introducing the expression (7.201) into (7.200), the free-energy functional $\Delta F^*[\varphi]$ becomes

$$\Delta F^*[\varphi] = F^*[\varphi] - F^*[\varphi = 0]$$

$$= \frac{1}{\hbar\beta} \int_0^{\hbar\beta} d\tau \sum_{\langle ij \rangle} (1 - \cos \varphi_{ij}) K^{(0)} \left(1 + \frac{1}{2} D_{ij}^{(0)} \right)$$

$$- \frac{1}{\beta} \ln \left(\frac{\int \mathcal{D}\delta \, \exp\left(-\frac{1}{\hbar} S_{\text{tr}}^{(\delta)}[\varphi, \delta] \right)}{\int \mathcal{D}\delta \, \exp\left(-\frac{1}{\hbar} S_{\text{tr}}^{(\delta)}[\varphi = 0, \delta] \right)} \right) \qquad (7.202)$$

where $S_{\text{tr}}^{(\delta)}[\varphi, \delta]$ is the part of expression (7.201) involving the variable δ_i only

$$S_{\text{tr}}^{(\delta)}[\varphi, \delta] = \int_0^{\hbar\beta} d\tau \left(\frac{\hbar^2}{8U_a} \sum_i \dot{\delta}_i^2(\tau) + \frac{1}{2} \sum_{\langle ij \rangle} K^{(0)} \cos \varphi_{ij} \delta_{ij}^2(\tau) \right) \qquad (7.203)$$

For a slowly varying vortex angle ϕ_i, we can use on the RHS of this equation the expansion: $\cos \phi_{ij} \simeq 1 - \phi_{ij}^2/2$. Then the second term on the RHS of Eq. (7.202) can be written

$$- \frac{1}{\beta} \ln \left(\frac{\int \mathcal{D}\delta \, \exp\left(-\frac{1}{\hbar} S_{\text{tr}}^{(\delta)}[\varphi, \delta] \right)}{\int \mathcal{D}\delta \, \exp\left(-\frac{1}{\hbar} S_{\text{tr}}^{(\delta)}[\varphi = 0, \delta] \right)} \right) = -\frac{1}{\beta} \ln \langle \exp A \rangle_0 \qquad (7.204)$$

where

$$A = \frac{1}{4\hbar} \int_0^{\hbar\beta} d\tau \sum_{\langle ij \rangle} K^{(0)} \varphi_{ij}^2 \delta_{ij}^2(\tau) \qquad (7.205)$$

The expectation value on the RHS of Eq. (7.204) is to be evaluated with use of the action (for a homogeneous superconductor)

$$S_{\text{tr}}^{(\delta)}[\varphi = 0, \delta] = \int_0^{\hbar\beta} d\tau \left(\frac{\hbar^2}{8U_a} \sum_i \dot{\delta}_i^2(\tau) + \frac{1}{2} \sum_{\langle ij \rangle} K^{(0)} \delta_{ij}^2 \right) \qquad (7.206)$$

When the sample size $R \to \infty$, the following limit holds (Šimánek and Stein, 1984)

$$\langle \exp A \rangle_0 \xrightarrow[R \to \infty]{} \exp \langle A \rangle_0 \qquad (7.207)$$

This can be understood qualitatively by noting that for large R the summation in Eq. (7.205) is dominated by terms with large distances r_{ij} from the vortex center, for which $\varphi_{ij}^2 \sim r_{ij}^{-2}$ is small [see Eq. (7.196)]. Note that for each summand in Eq. (7.205) that involves a small coefficient of $\delta_{ij}^2(\tau)$, we have an equality

$$\left\langle \exp \left(\frac{K^{(0)} \varphi_{ij}^2}{4\hbar} \int_0^{\hbar \beta} d\tau \, \delta_{ij}^2(\tau) \right) \right\rangle_0 = \exp \left(\frac{\beta K^{(0)} \varphi_{ij}^2}{4} D_{ij}^{(0)} \right) \qquad (7.208)$$

Introducing the RHS of Eq. (7.203) into (7.204) and using Eq. (7.207), the increment of free energy due to a vortex configuration φ_i is obtained from Eq. (7.202) in the form

$$\Delta F^*[\varphi] = \frac{1}{\hbar \beta} \int_0^{\hbar \beta} d\tau \sum_{\langle ij \rangle} (1 - \cos \varphi_{ij}) K^{(0)} \left(1 + \frac{1}{2} D_{ij}^{(0)} \right)$$

$$- \frac{1}{4} \sum_{\langle ij \rangle} K^{(0)} \varphi_{ij}^2 D_{ij}^{(0)}$$

$$\simeq \frac{1}{2} \sum_{\langle ij \rangle} K^{(0)} \varphi_{ij}^2 \qquad (7.209)$$

where we have used the slow-variation approximation for $\cos \varphi_{ij}$. The RHS of this equation becomes, in the continuum approximation

$$\Delta F^*[\varphi] \simeq \frac{1}{2} K^{(0)} \int d^2 \rho \, (\boldsymbol{\nabla} \varphi)^2 \qquad (7.210)$$

Equation (7.210) is of the same form as the vortex kinetic energy on the LHS of expression (7.43) resulting from the Hamiltonian for a Josephson-coupled array in the absence of charging energy. In view of the approximation (7.200), which neglects large enhancements of $\langle \delta_{ij}^2 \rangle$ near the vortex center, the expression (7.210) is valid except near the vortex center. A vortex core, with a radius on the order of the lattice spacing a, can take care of this correction. Equation (7.210) implies that, as long as the vortex–antivortex spacing is much larger than the lattice spacing, the interaction is logarithmic, with a coupling constant given by $K^{(0)}$. Moreover, if the screening length $\lambda_\perp$ is much larger than the sample size R, the gauge field $\mathbf{A}$ can be neglected so that the free energy of the form (7.210) is justified. Thus, the expression (7.184) is verified for the case of weak renormalization [see Eq. (7.197)]. For strong renormalization, the spatial dependence of the local coupling constant K_{ij}^{loc} is faster than implied by the RHS of

Eq. (7.197). In the next stage of iteration, we may replace K_{ij} by the RHS of Eq. (7.199), leading to

$$K_{ij}^{\text{loc}} \sim K^{(0)} \exp\left(-\frac{1}{4}\sqrt{\frac{U_a}{J}}\frac{a^2}{r_{ij}^2}\right)\left(1 - \frac{a^2}{r_{ij}^2}\right) \qquad (7.211)$$

This leads to a potential well (for the phase fluctuation) varying more rapidly with r_{ij} [see Eq. (7.194)]. Thus, we may expect that the decay of the inhomogeneity of K_{ij} with the distance r_{ij} also becomes faster than suggested by Eq. (7.198), providing a justification of the approximation (7.200) also in the strong-coupling limit. In the immediate vicinity of the transition, where $K^{(0)}$ drops abruptly to zero, the SCHA approximation is, however, considered unreliable. Maekawa et al. (1981) showed, using the MFA method, that T_{KT} is in this region proportional to $\langle\cos\delta\rangle_{\text{MFA}}^2$ that, in contrast to $K^{(0)}$ vanishes continuously at the critical value of α (see Section 7.6.4).

7.6.3. Comparison of SCHA theory with experiment

The relative stiffness, $k_1 = K^{(0)}/J$, appearing on the RHS of Eq. (7.184) is, in general, a function of temperature. If $k_B T_{\text{KT}} \ll (K^{(0)}U_a)^{\frac{1}{2}}$, then the parameter k_1 can be evaluated using the $T = 0$ approximation (Šimánek, 1980b). Since $k_B T_{\text{KT}} \simeq \pi K^{(0)}/2$, the $T = 0$ approximation is justified when $k_1 \ll z/\alpha$, where α is given by Eqs. (3.76) and (5.147)

$$\alpha(T = 0) = \frac{zJ(T = 0)}{2U_a} = \frac{\pi R_0 \Delta_0(0)}{2R_N U_a} \qquad (7.212)$$

where $z = 4$ has been used, assuming a square planar array. The relative stiffness k_1 is determined by the solution of the self-consistent SCHA equation (3.167), which involves α as a parameter. The results show that, as α decreases towards the critical value $\alpha_c \simeq 1.33$, k_1 decreases and drops abruptly from $k_1 = 0.13$ to zero at α_c. Hence, the condition $k_1 \ll z/\alpha$ is well satisfied for values of α several times the critical value α_c. To complete the link between the SCHA theory and the experiment, we need to know the dependence of the charging energy on R_N. In our original treatment (Šimánek, 1980b) it has been assumed, on intuitive grounds, that $U_a \sim R_N$. The rationale behind this assumption goes as follows: The superconducting film is represented by a regular array of Josephson-coupled granular clusters of size l_c. Thus, it is reasonable to assume that the charging energy $U_a \sim l_c^{-1}$. The variation of l_c with the sheet resistance R_N can be deduced, from Fig. 3 of the paper by Hebard (1979), to be $l_c^{-1} \sim R_N$. In a model with an off-diagonal Coulomb interaction, the proportionality of the diagonal charging energy to the sheet resistance follows more directly from the renormalization of the intergrain capacitance [see Eq. (6.111)]. Parenthetically, we suggest that it should be interesting to extend the present SCHA method to such a model. Using the assumption $U_a \sim R_N$, Eq. (7.212) implies

$$\alpha(T = 0) \simeq \frac{225}{R_N^2} \qquad (7.213)$$

where R_N is given in kΩ. The numerical constant on the RHS of this equation results from matching the experimental critical sheet resistance $R_N^c \simeq 13$ kΩ with the theoretical value $\alpha^c = 1.33$. Using the relation (7.213), the dependence of k_1 on α can be translated into the $k_1(R_N)$ dependence. According to Eqs. (7.180)–(7.182), the $k_1(\alpha)$ dependence is obtained by solving numerically Eq. (3.167). Inserting the resulting $k_1(R_N)$ function on the RHS of Eq. (7.184) and solving the resulting equation for T_{KT}, we obtain the curve B of Fig. 7.2. Curve B is in qualitative agreement with the experimental points (curve A). The experimental tail in the region of $R_N \gtrsim 13$ kΩ is not reproduced by the SCHA theory, which leads to a *vanishing* phase rigidity above R_N^c. This tail, however, may be an experimental artifact due to the method that determines T_{KT} indirectly on the basis of higher-temperature data (Hebard, private communication). If the background above R_N^c is subtracted from the experimental data, then the SCHA curve represents a reasonable fit.

7.6.4. *Renormalization of* T_{KT} *within MFA*

We now describe a calculation of the transition temperature T_{KT}, which employs the MFA in the evaluation of the renormalized phase rigidity. One motivation for such a treatment is the breakdown of the SCHA approximation at the critical value of the parameter α. Moreover, as Maekawa et al. (1981) first demonstrated, the MFA approach leads to an interesting prediction of a reentrant KTB transition. These authors have used a somewhat ad hoc, though correct, procedure based on the averaging of Josephson current. We present here a rederivation of their result, using an MFA version of the variational principle of Section 7.6.2. We start from Eq. (7.189) with S given by Eq. (7.186). The trial action has the MFA form

$$S_{tr}[\varphi, \delta] = \int_0^{\hbar\beta} d\tau \left(\frac{\hbar^2}{8U_a} \sum_i \dot{\delta}_i^2(\tau) \right.$$

$$\left. + J \sum_{\langle ij \rangle} [1 - \gamma \cos \varphi_{ij} (\cos \delta_i + \cos \delta_j)] \right) \qquad (7.214)$$

where γ is a variational parameter to be determined from Eq. (7.189) by requiring that $\delta F^*[\varphi] = 0$ when $\gamma \to \gamma + \delta\gamma$. For simplicity, we assume at the outset a position-independent γ. This assumption can be justified using steps similar to the SCHA Eqs. (7.195)–(7.199). Our goal now is to find the functional $\Delta F^*[\varphi]$ describing the increment of the free energy due to the introduction of vorticity. From Eqs. (7.186) and (7.214), we have for $\varphi_i \neq 0$

$$
\frac{1}{\hbar\beta}\langle S[\varphi,\delta] - S_{\text{tr}}[\varphi,\delta]\rangle_{\text{tr}}
$$

$$
= \frac{J}{\hbar\beta}\int_0^{\hbar\beta} d\tau \sum_{\langle ij\rangle} \cos\varphi_{ij}\langle\gamma(\cos\delta_i + \cos\delta_j) - \cos\delta_{ij}\rangle_{\text{tr}} \qquad (7.215)
$$

According to Eq. (7.189), we have using Eq. (7.215)

$$
\Delta F^*[\varphi] = \frac{J}{\hbar\beta}\int_0^{\hbar\beta} d\tau \sum_{\langle ij\rangle}(\cos\varphi_{ij} - 1)\langle\gamma(\cos\delta_i + \cos\delta_j) - \cos\delta_{ij}\rangle_{\text{tr}}
$$

$$
- \frac{1}{\beta}\ln\left(\frac{\int \mathcal{D}\delta \exp\left(-\frac{1}{\hbar}S_{\text{tr}}[\varphi,\delta]\right)}{\int \mathcal{D}\delta \exp\left(-\frac{1}{\hbar}S_{\text{tr}}[\varphi=0,\delta]\right)}\right) \qquad (7.216)
$$

To bring this expression to its final form, we need to find the condition for the best choice of γ. The variational principle $\delta F^*[\varphi] = 0$ can be written as [compare with Eq. (3.139)]

$$
\left(\frac{\partial F^*}{\partial\gamma}\right)_\Delta + \left(\frac{\partial F^*}{\partial\Delta}\right)_\gamma \frac{\partial\Delta}{\partial\gamma} = 0 \qquad (7.217)
$$

where

$$
\Delta = \langle\cos\delta_i\rangle_{\text{tr}} = \langle\cos\delta_j\rangle_{\text{tr}} \qquad (7.218)
$$

We note that Eq. (7.218) is an approximation that disregards the nonequivalence of lattice sites (due to the presence of a vortex). This approximation is justified, except for the immediate vicinity of the vortex center. Following the analogy to the SCHA case, we now show that the first term on the LHS of Eq. (7.217) vanishes. From Eq. (7.189) we have, for a given configuration $\{\varphi_i\}$

$$
\left(\frac{\partial F^*}{\partial\gamma}\right)_\Delta = \frac{\partial F_{\text{tr}}}{\partial\gamma} + \frac{1}{\hbar\beta}\frac{\partial}{\partial\gamma}\langle S[\varphi,\delta] - S_{\text{tr}}[\varphi,\delta]\rangle_{\text{tr}} \qquad (7.219)
$$

Using Eqs. (7.214) and (7.218), the first term on the RHS of this equation can be written

$$
\frac{\partial F_{\text{tr}}}{\partial\gamma} = -\frac{J}{\hbar\beta}\sum_{\langle ij\rangle}\int_0^{\hbar\beta} d\tau \cos\varphi_{ij}\langle\cos\delta_i + \cos\delta_j\rangle_{\text{tr}}
$$

$$
= -2J\Delta\sum_{\langle ij\rangle}\cos\varphi_{ij} \qquad (7.220)
$$

From Eq. (7.215) we obtain

$$
\frac{1}{\hbar\beta}\frac{\partial}{\partial\gamma}\langle S[\varphi,\delta] - S_{\text{tr}}[\varphi,\delta]\rangle_{\text{tr},\Delta} = 2J\Delta\sum_{\langle ij\rangle}\cos\varphi_{ij} \qquad (7.221)
$$

Introducing Eqs. (7.220) and (7.221) into the RHS of Eq. (7.219), we obtain

$$\left(\frac{\partial F^*}{\partial \gamma}\right)_\Delta = 0 \tag{7.222}$$

Since F_{tr} does not involve the parameter Δ explicitly, we have $(\partial F_{\mathrm{tr}}/\partial \Delta)_\gamma = 0$ so that

$$
\begin{aligned}
\left(\frac{\partial F^*}{\partial \Delta}\right)_\gamma &= \frac{1}{\hbar\beta}\frac{\partial}{\partial \Delta}\langle S[\varphi,\delta] - S_{\mathrm{tr}}[\varphi,\delta]\rangle_{\mathrm{tr},\gamma} \\
&= J\sum_{\langle ij\rangle}\cos\varphi_{ij}\frac{\partial}{\partial \Delta}\left[\gamma(\langle\cos\delta_i\rangle_{\mathrm{tr}} + \langle\cos\delta_j\rangle_{\mathrm{tr}}) - \langle\cos\delta_{ij}\rangle_{\mathrm{tr}}\right] \\
&= J\sum_{\langle ij\rangle}\cos\varphi_{ij}\frac{\partial}{\partial \Delta}\left(2\gamma\Delta - \Delta^2\right) \\
&= 2J\sum_{\langle ij\rangle}\cos\varphi_{ij}(\gamma - \Delta)
\end{aligned}
\tag{7.223}
$$

where Eqs. (7.215) and (7.218) and the MFA result $\langle\cos\delta_{ij}\rangle_{\mathrm{tr}} = \Delta^2$ have been used [see Eq. (3.61)]. Introducing Eqs. (7.222) and (7.223) into (7.217), we obtain, as expected

$$\gamma = \Delta \tag{7.224}$$

Using this result and Eq. (7.218), the first term on the RHS of Eq. (7.216) becomes

$$\frac{J}{\hbar\beta}\int_0^{\hbar\beta} d\tau \sum_{\langle ij\rangle}(\cos\varphi_{ij} - 1)\langle\gamma(\cos\delta_i + \cos\delta_j) - \cos\delta_{ij}\rangle_{\mathrm{tr}}$$

$$= -J\Delta^2\sum_{\langle ij\rangle}(1 - \cos\varphi_{ij}) \simeq -\frac{1}{2}J\Delta^2\sum_{\langle ij\rangle}\varphi_{ij}^2 \tag{7.225}$$

The second term can be written, expanding the $\cos\varphi_{ij}$ term in $S_{\mathrm{tr}}[\varphi,\delta]$

$$-\frac{1}{\beta}\ln\left\langle\exp\left(-\frac{J\gamma}{2\hbar}\int_0^{\hbar\beta}d\tau\sum_{\langle ij\rangle}\varphi_{ij}^2(\cos\delta_i + \cos\delta_j)\right)\right\rangle_{\mathrm{tr},\varphi=0}$$

$$\approx -\frac{1}{\beta}\ln\left[\exp\left(-\beta J\Delta^2\sum_{\langle ij\rangle}\varphi_{ij}^2\right)\right] = J\Delta^2\sum_{\langle ij\rangle}\varphi_{ij}^2 \tag{7.226}$$

where we have approximated the trace of the exponential function by an exponential function of the trace of the exponent [see Eq. (7.207)]. Using Eqs. (7.225) and (7.226), Eq. (7.216) yields for the functional $\Delta F^*[\varphi]$

$$\Delta F^*[\varphi] = \frac{1}{2}J^{\mathrm{ren}}\sum_{\langle ij\rangle}\varphi_{ij}^2 \tag{7.227}$$

where

$$J^{\text{ren}} = J\Delta^2(T) = J\langle\cos\delta\rangle^2_{\text{MFA}} \qquad (7.228)$$

The expectation value $\langle\cos\delta\rangle_{\text{MFA}}$ in Eq. (7.228) is to be evaluated with respect to the Hamiltonian (3.62) by means of the expression (3.64). Equation (7.227) implies that the renormalized vortex-unbinding transition temperature, within the MFA, follows by replacing the bare coupling constant J in Eq. (7.157) by the expression (7.228), yielding

$$T_{\text{KT}} = \frac{\pi J}{2k_B}\Delta^2(T_{\text{KT}}) \qquad (7.229)$$

As a result of this replacement, Eq. (7.174) is modified to [see (7.184)]

$$k_B T_{\text{KT}} = 2.18 k_B T_{c0}\frac{R_0}{R_N}f(T_{\text{KT}})\Delta^2(T_{\text{KT}}) \qquad (7.230)$$

This result agrees with the work of Maekawa et al. (1981). As first pointed out by these authors, the expression (7.230) predicts a possibility of reentrant KTB transition. This can be seen by examining the T dependence of the product $f(T)\Delta^2(T)$. The function $f(T)$ decreases monotonically from one to zero as T/T_{c0} increases from zero to one. The function $\Delta^2(T)$, however, exhibits a nonmonotonic T dependence, if calculated from Eq. (3.64) with the 4π-periodic states included. For values of α in the interval $0.78 < \alpha < 1$, the function $\Delta(T)$ has a form shown in Fig. 2 of Maekawa et al. (1981). The graphical solution of Eq. (7.230) for a value of α in this interval is sketched in Fig. 7.4. We see that, owing to the nonmonotonic behavior of $\Delta(T_{\text{KT}})$, there are two solutions for T_{KT}. On decreasing the temperature, a transition to bound vortex–antivortex pairs first takes place at T_{KT}. Further lowering of the temperature leads to vortex unbinding at T^*_{KT}, which persist down to $T = 0$. The latter effect is due to the loss of phase rigidity, produced by quantum zero-point phase fluctuations in the self-charging model. As shown in Section 6.6.2, this model appears justified for very small grains (via the δ effect). In this case, the reentrant KTB transition, predicted by Maekawa et al. (1981), is a possibility worth of further experimental and theoretical pursuit.

7.6.5. *Renormalization-group analysis and Monte Carlo work*

José (1984) described a method of studying the effects of quantum phase fluctuations on the KTB scenario, by using a semiclassical approximation for the renormalized stiffness. We now describe derivations of his renormalization-group scaling equations, using the techniques and notation developed in previous parts of this book. The starting point is again the self-charging model described by the Euclidean action (3.132). Separating the phase $\theta_i(\tau)$ according to Eq. (7.185), the partition function is

$$0.78 < \alpha < 1$$

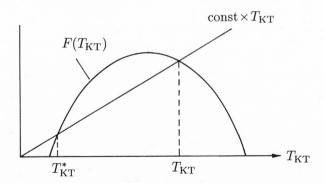

FIG. 7.4. A sketch illustrating the graphical solution of Eq. (7.230). The function $F(T_{KT}) = f(T_{KT})\Delta^2(T_{KT})$, and the constant multiplying T_{KT} is equal to $R_N/2.18R_0$.

written with use of Eq. (7.186) as

$$Z \sim \int \mathcal{D}\varphi \int \mathcal{D}\delta \, \exp\left(-\frac{1}{\hbar}S[\varphi,\delta]\right)$$

$$= \int \mathcal{D}\varphi \, \exp\left(-\beta J \sum_{\langle ij \rangle}(1 - \cos\varphi_{ij})\right) Z_\delta[\varphi] \qquad (7.231)$$

where $Z_\delta[\varphi]$ is a functional of the vortex phase φ, obtained from Eq. (7.186) by expanding the $\cos\varphi_{ij}$ and $\cos\delta_{ij}$ terms (assuming slowly varying φ and small fluctuations δ_i). To the lowest order in δ_{ij}, we obtain

$$Z_\delta[\varphi] \simeq \int \mathcal{D}\delta \, \exp\left(-\frac{1}{\hbar}S_0[\delta] + \frac{J}{4\hbar}\int_0^{\hbar\beta} d\tau \sum_{\langle ij \rangle}\varphi_{ij}^2 \delta_{ij}^2(\tau)\right)$$

$$\simeq Z_0\left(1 + \frac{J\beta}{4}\sum_{\langle ij \rangle}\varphi_{ij}^2 \langle \delta_{ij}^2 \rangle_0\right)$$

$$\approx Z_0 \exp\left(\frac{J\beta}{4}\sum_{\langle ij \rangle}\varphi_{ij}^2 \langle \delta_{ij}^2 \rangle_0\right) \qquad (7.232)$$

where

$$S_0[\delta] = \int_0^{\hbar\beta} d\tau \left(\frac{\hbar^2}{8U_a}\sum_i \dot{\delta}_i^2 + \frac{1}{2}J\sum_{\langle ij \rangle}\delta_{ij}^2\right) \qquad (7.233)$$

and

$$Z_0 = \int \mathcal{D}\delta \, \exp -\frac{1}{\hbar} S_0[\delta] \qquad (7.234)$$

The thermal average $\langle \delta_{ij}^2 \rangle_0$ is given by

$$\langle \delta_{ij}^2 \rangle_0 = \frac{1}{Z_0} \int \mathcal{D}\delta \, \exp\left(-\frac{1}{\hbar} S_0[\delta] \right) \delta_{ij}^2 \qquad (7.235)$$

It is convenient to define, in analogy to Eq. (3.163), the bond-averaged quantity

$$\overline{\langle \delta_{ij}^2 \rangle}_0 = \frac{1}{z} \sum_{j(i)} \langle \delta_{ij}^2 \rangle_0 \qquad (7.236)$$

Replacing $\langle \delta_{ij}^2 \rangle_0$ in Eq. (7.232) by the bond-averaged value (7.236) and inserting the result into Eq. (7.231), we obtain

$$Z \approx \int \mathcal{D}\varphi \, \exp\left(-\frac{1}{2}\beta \bar{J} \sum_{\langle ij \rangle} \varphi_{ij}^2 \right) \qquad (7.237)$$

where $\bar{J}$ is the renormalized coupling constant

$$\bar{J} \simeq J\left(1 - \frac{1}{2}\overline{\langle \delta_{ij}^2 \rangle}_0 \right) \qquad (7.238)$$

We note that this result is valid in the weak-coupling limit $\overline{\langle \delta_{ij}^2 \rangle}_0 \ll 1$. It could have been obtained by simply taking the lowest-order term in the expansion of the exponential on the RHS of the SCHA equation (3.145). We now consider the path integral (7.235), from which the renormalized coupling constant can be derived in the form used by José (1984). This integral is of the same kind as that used in Section 3.9 for the evaluation of $\bar{D}_{ij}$. We only need to replace in Eq. (3.163) the constant K by J. In this way we obtain

$$\overline{\langle \delta_{ij}^2 \rangle}_0 = \frac{1}{zN\beta} \sum_{\mathbf{k}} \sum_n \frac{f(\mathbf{k})}{\frac{\hbar^2}{8U_a}[\omega_n^2 + \omega^2(\mathbf{k})]} \qquad (7.239)$$

where $f(\mathbf{k})$ is given by Eq. (3.164) and $\omega(\mathbf{k})$ is defined as

$$\omega(\mathbf{k}) = \frac{2}{\hbar} \sqrt{U_a J f(\mathbf{k})} \qquad (7.240)$$

Performing the sum over the even Matsubara frequencies, we have

$$\frac{1}{\beta} \sum_{n=-\infty}^{\infty} \frac{1}{\omega_n^2 + \omega^2(\mathbf{k})} = \frac{\hbar}{2\omega(\mathbf{k})} \coth\left(\frac{\hbar\omega(\mathbf{k})}{2k_B T} \right) \qquad (7.241)$$

From Eqs. (7.239)–(7.241) we have, for a square planar array, using the Debye approximation [see Eq. (3.166)]

$$\overline{\langle \delta_{ij}^2 \rangle}_0 = \frac{a^3}{4\pi} \sqrt{\frac{U_a}{J}} \int_0^{k_D} dk\, k^2 \coth\left(\beta \sqrt{U_a J} ka\right) \tag{7.242}$$

where a is the lattice spacing and $k_D = 2\sqrt{\pi}/a$. The semiclassical approximation follows from this equation by expanding the coth function and keeping only the two lowest-order terms

$$\overline{\langle \delta_{ij}^2 \rangle}_0 \simeq \frac{a^3}{4\pi} \sqrt{\frac{U_a}{J}} \int_0^{k_D} dk\, k^2 \left(\frac{1}{\beta\sqrt{U_a J}ka} + \frac{1}{3}\beta\sqrt{U_a J}ka\right)$$

$$= \frac{1}{2\beta J} + \frac{\pi}{3} U_a \beta \tag{7.243}$$

The first term on the RHS of this equation corresponds to the purely classical limit, whereas the second term represents the lowest-order *quantum correction*. Introducing this result into Eq. (7.238), we have for the renormalized coupling constant, in the semiclassical approximation

$$\bar{J} = J\left(1 - \frac{1}{4\beta J} - \frac{\pi}{6}U_a\beta\right) \tag{7.244}$$

The starting point of the RG analysis of José (1984) is the scaling equations (7.98) and (7.101), in which the scale-dependent charge $K(l)$ is replaced by $\bar{K}(l)$, representing a scale-dependent charge *renormalized by quantum fluctuations*. According to Eq. (7.94), we have for the initial values of $K(l)$ and $\bar{K}(l)$

$$\beta J = K(l = 0) \tag{7.245}$$

and

$$\beta \bar{J} = \bar{K}(l = 0) \tag{7.246}$$

Equations (7.244)–(7.246) imply

$$\bar{K}(l = 0) = K(l = 0)\left(1 - \frac{1}{4K(l=0)} - \frac{\pi}{6}\frac{U_a}{J}K(l=0)\right) \tag{7.247}$$

The second term on the RHS of this equation yields an irrelevant constant that can be discarded. Introducing the parameter x_u defined as

$$x_u = \frac{\pi}{6}\frac{U_a}{J} = \frac{\pi}{3\alpha} \tag{7.248}$$

the scale-dependent form of Eq. (7.247) can be written

$$\bar{K}(l) = K(l)[1 - K(l)x_u] \tag{7.249}$$

The RG recursion formulas of José (1984), obtained by introducing the RHS of this equation into Eqs. (7.98) and (7.101), have the form

$$\frac{dK}{dl} = 4\pi^3 \tilde{y}^2(l) \frac{K^2(1 - Kx_u)^2}{2Kx_u - 1} \tag{7.250}$$

$$\frac{d\tilde{y}}{dl} = [2 - \pi K(1 - Kx_u)]\tilde{y} \tag{7.251}$$

These equations have fixed points that follow from

$$\tilde{y}_f = 0 \tag{7.252}$$

$$2 - \pi K_f(1 - K_f x_u) = 0 \tag{7.253}$$

The two roots, $K_f^{(1,2)}$, obtained by solving the quadratic equation (7.253) for K_f, are

$$K_f^{(1)} = K_{\mathrm{KT}} = \frac{1}{2x_u} \left(1 - \sqrt{1 - \frac{8x_u}{\pi}} \right) \tag{7.254}$$

and

$$K_f^{(2)} = K_2 = \frac{1}{2x_u} \left(1 + \sqrt{1 - \frac{8x_u}{\pi}} \right) \tag{7.255}$$

The physical meaning of K_{KT} becomes apparent by assuming a small charging energy ($x_u \ll 1$) and expanding K_{KT}^{-1} in x_u. From Eq. (7.254), we have

$$K_{\mathrm{KT}}^{-1} = \frac{2x_u}{1 - \left(1 - \frac{4x_u}{\pi} - \frac{8x_u^2}{\pi^2} + \cdots \right)} \simeq \frac{\pi}{2} \left(1 - \frac{2x_u}{\pi} \right) \tag{7.256}$$

For $x_u = 0$, the RHS of this equation yields the fixed point of the classical planar model [see Eq. (7.102)]. Thus K_{KT} may be identified with this fixed point, renormalized by the quantum fluctuations. Note that Eq. (7.256) predicts for the renormalized vortex-unbinding temperature (to first order in x_u)

$$T_{\mathrm{KT}} \simeq \frac{\pi}{2} J \left(1 - \frac{2x_u}{\pi} \right) \tag{7.257}$$

The inverse of K_2, to order x_u, is from Eq. (7.255)

$$K_2^{-1} = \frac{2x_u}{1 + \left(1 - \frac{4x_u}{\pi} - \frac{8x_u^2}{\pi^2} + \cdots \right)} \simeq 2x_u \tag{7.258}$$

Hence, K_2^{-1} vanishes when $x_u = 0$ as expected, since there is no other fixed point than K_{KT} in the classical limit. Equations (7.254) and (7.255) imply that the fixed points K_{KT} and K_2 merge into each other at the critical value $x_u^c = \pi/8$. The RG flow diagram indicates that in the region of temperatures $JK_2^{-1} < k_B T < JK_{\mathrm{KT}}^{-1}$ the system is in a phase-coherent state.

On increasing x_u, this region shrinks until at x_u^c, it disappears completely. Then a phase-incoherent state is stable at all temperatures (José, 1984).

At this point, it is instructive to make a contact with the MFA of Maekawa et al. (1981). Their theory also yields two transition temperatures obtained by solving Eq. (7.229) for T_{KT}. We note that this equation is obtained from the simple formula $k_B T_{KT} = \pi J^{\text{ren}}/2$. The same formula can be used to derive Eqs. (7.253)–(7.255); one only needs to substitute Eq. (7.244) for J^{ren}. This substitution yields

$$k_B T_{KT} = \frac{\pi}{2}\bar{J}(T_{KT}) = \frac{\pi}{2}J\left(1 - \frac{k_B T_{KT}}{4J} - \frac{\pi U_a}{6k_B T_{KT}}\right) \qquad (7.259)$$

If the second term on the RHS of this equation is dropped, we obtain [multiplying through by $2/\pi k_B T_{KT}$ and using Eqs. (7.245) and (7.248)]

$$2 = \pi K_{KT}(1 - K_{KT}x_u) \qquad (7.260)$$

where $K_{KT} = J/k_B T_{KT}$. As expected, Eq. (7.260) coincides with the fixed-point condition (7.253). Hence, we see that the double roots for the fixed points $K_f^{(1)}$ and $K_f^{(2)}$ have the same origin as the solutions for T_{KT} shown in Fig. 7.4. Any function of T_{KT} that yields two intersections with the straight line passing through the origin is a possible candidate for reentrance. In the approximations of José (1984), this function is given by

$$F(T_{KT}) \simeq 1 - x_u\frac{J}{k_B T_{KT}} \qquad (7.261)$$

This function yields double roots for T_{KT}, owing to the kneelike shape in the region of low T_{KT}. There is, however, a problem with the approximation used to obtain this result. If the expansion of the coth function in (7.242) is carried out to include the three lowest-order terms, Eq. (7.249) is augmented by a term $\sim K^2 x_u^2$, which enters with a *positive* sign. Consequently, the function (7.261) is modified so that there is no rising part that would allow double roots. This is confirmed by a numerical calculation of the integral (7.242), which results in $\overline{\langle \delta_{ij}^2 \rangle}_0$ monotonically increasing with temperature. Thus the expression on the RHS of Eq. (7.238) is, in contrast to the approximation (7.261), a monotonically *decreasing* function of temperatures that can never lead to a reentrant KTB transition. José (1984) suggested that reentrance can still be present, if one carries out the spin-wave expansion (7.232) to the next (fourth) order in δ_{ij}. Expanding the $\cos \delta_{ij}$ term in Eq. (7.186) to this order, the partition function Z_δ defined in Eq. (7.231) can be evaluated with the result

$$Z_\delta[\varphi] \simeq Z_0\left(1 + \frac{1}{4}J\beta\sum_{\langle ij \rangle}\varphi_{ij}^2\langle\delta_{ij}^2\rangle_0 - \frac{J\beta}{48}\sum_{\langle ij \rangle}\varphi_{ij}^2\langle\delta_{ij}^4\rangle_0\right) \qquad (7.262)$$

For Gaussian averages we have

$$\langle \delta_{ij}^4 \rangle_0 = 3 \left(\langle \delta_{ij}^2 \rangle_0 \right)^2 \tag{7.263}$$

Introducing this result into Eq. (7.262), Eq. (7.238) is replaced by

$$\bar{J} \simeq J \left[1 - \frac{1}{2} \overline{\langle \delta_{ij}^2 \rangle_0} + \frac{1}{8} \left(\overline{\langle \delta_{ij}^2 \rangle_0} \right)^2 \right] \tag{7.264}$$

Since the last term on the RHS of this equation increases with temperature, the entire expression in the brackets will have the T dependence necessary to obtain double roots from the equation $k_B T_{KT} = \pi \bar{J}(T_{KT})/2$. At the same time, we see that the RHS of Eq. (7.261) coincides with the first three terms of the expansion of

$$\bar{J} \simeq J \exp \left(-\frac{1}{2} \overline{\langle \delta_{ij}^2 \rangle_0} \right) \tag{7.265}$$

Since the RHS of this equation is a monotonically decreasing function of temperature, conclusions about reentrance, based on the approximation (7.264), may not be justified.

So far, the existence of double roots for T_{KT} seems fairly well established in the MFA approach of Maekawa et al. (1981). Thus it should be interesting to study the RG flow diagram associated with the scaling equations involving the MFA-renormalized stiffness of Eq. (7.228). This should be especially feasible for $\alpha \simeq 1$, since then we have $|\Delta(T)| \ll 1$, so that expansion in the small parameter q can be used in calculating $\Delta^2(T)$ [see Eqs. (3.64) and (3.65)].

Additional insight into the reentrant transition of the self-charging model has emerged from the nonperturbative quantum Monte Carlo analysis (Jacobs et al., 1984). In this work, a dramatic decrease of the helicity modulus and specific heat has been observed on lowering the temperature towards zero. This provides a clear signal for the existence of a reentrant phase transition. Moreover, the Monte Carlo data exhibited strong fluctuations, preventing reliable determination of the helicity modulus and specific heat in the temperature region around this transition. These fluctuations are believed to be associated with the metastability that is symptomatic of a first order transition. The data also indicate that the helicity modulus tends to a finite value at $T = 0$. This suggests that, unlike what would be expected in a standard reentrant scenario, the low-temperature phase represents a new type of coherent state with a reduced but *nonzero phase rigidity*. The Monte Carlo study was subsequently extended to include the effect of an external transverse magnetic field (Jacobs et al., 1987). The field appears to modify the low-temperature (reentrant) transition quite strongly. The authors have concentrated on the case of a fully frustrated network ($f = 1/2$) and observed that the reentrant transition takes place at a temperature about five times larger than the corresponding field-free value ($f = 0$). Moreover, the discontinuity of the helicity modulus at this temperature is about thirty times the $f = 0$ value. At present, a complete theoretical understanding of these interesting results is lacking. Heuristically, it has been suggested by Jacobs et al. (1987) that the reentrant transition is produced by proliferation of bound vortex pairs induced by zero-point phase fluctuations.

8

DYNAMICS AND QUANTUM TUNNELING OF VORTICES

8.1. Introduction

Within the past decade there has been a resurgence of interest in the dynamics of superfluid and superconducting vortices. One source of motivation for this topic may be attributed to the discovery of the quantum Hall effect (von Klitzing et al., 1980), which led to the suggestion that elementary excitations in 2D systems may obey nonstandard statistics (Wilczek, 1982; Halperin, 1984). Haldane and Wu (1985) pointed out a powerful analogy between the dynamics of a vortex in a 2D superfluid film and the electron motion in a cyclotron orbit. Led by this analogy, Haldane and Wu (1985) have determined the geometric phase change of the superfluid wave function as the vortex is transported around a closed path. These results have implications for a possible fractional statistics of vortices in 2D superfluids and superconductors. Moreover, the calculated geometric phase can be linked to the pseudo-Lorentz force acting on a moving vortex in a stationary fluid. This force as well as the force on a vortex in a moving fluid (Magnus force) are important ingredients in the vortex dynamics of massless vortices. In superconductors, the magnitude of the pseudo-Lorentz force can be diagnosed via the Hall effect. In fact, if this magnitude is the same as that in an uncharged homogeneous superfluid, then there is a *perfect* Hall effect. Experimentally, however, a very small Hall angle is seen in superconductors (Tinkham, 1975). The explanation for this is still an open question, though some interesting ideas come from considering the geometric phase for a vortex transported around a closed path in a *boson lattice* superconductor (Fisher, 1991). It turns out that the calculated phase factor is 2π times the number of bosons encircled, implying that the vortex sees each boson effectively attached to one unit of a fictitious magnetic flux. Similarly, a charged boson encircling the vortex picks up a phase 2π (Aharonov and Bohm, 1959). This is the essence of the vortex–charge duality transformation, in which the boson density is replaced by the curl of a fictitious gauge field felt by the vortices (Fisher and Lee, 1989). This concept has been used by Fisher (1990) to study the zero-temperature melting of the vortex-glass phase in a disordered superconducting film, taking place as the applied magnetic field is increased towards a certain critical value.

In the liquid phase, the field-induced vortices Bose condense, an effect that is dual to the Cooper pair condensation in the superconducting phase.

The other reason for increased interest in vortex physics is the discovery of superconducting oxides with unexpectedly high transition temperatures (Bednorz and Müller, 1986). The observation of anomalously high decay rates of the magnetization in Ba-La-Cu-O and Sr-La-Cu-O at millikelvin temperatures indicates the possibility of a flux motion by quantum tunneling (Mota et al., 1987, 1988). In general, this process is enabled by the fact that the vortex is endowed with Newtonian dynamics associated with the presence of a nonzero inertial mass. There is also a dissipation, the origin of which is believed to be the ohmic losses, taking place in the core of a moving vortex. Consequently, the description of quantum vortex tunneling bears strong similarity with that of the MQT in a resistively shunted Josephson junction discussed in Chapter 4. However, the vortex physics in a disordered system becomes more complicated because of a possible localization due to the combined effect of random pinning potential and the intervortex interactions.

Blatter et al. (1991) have published a theory of quantum tunneling rates of pinned vortices in bulk superconductors, which seems to compare favorably with the observed magnetization decay rates in high-T_c oxides. This theory is based on a previous work of Glazman and Fogel (1984), in which the quantum creep velocity of an overdamped vortex is assumed proportional to the tunneling rate, calculated in the strong-dissipation limit [see Eq. (4.149)]. More recently, the same assumption was made to interpret the anomalous temperature dependence of the resistance, observed in ultrathin superconducting films (Liu et al., 1992). This interpretation as well as the work of Blatter et al. (1991) raises fundamental questions regarding the existence of a vortex-glass phase, which is capable of surviving the "zero-point melting" caused by quantum tunneling.

Another related question is about the role of *dissipation* in the vortex mobility at $T = 0$. The work of Schmid (1983) and of Fisher and Zwerger (1985) may offer some insight into this problem. These authors consider the mobility of a particle that diffuses quantum mechanically along a periodic array of potential wells in the presence of dissipation. As the dissipation reaches a certain critical value, the linear mobility drops to zero. Mooij and Schön (1992) have suggested that a similar localization transition may apply to a vortex in granular arrays.

Before discussing applications of this concept to actual experiments, let us emphasize that if the potential barrier is tilted (by a sufficiently strong transport current) the ensuing *nonlinear* mobility may be *nonzero* even in the limit of strong dissipation. This is substantiated by calculations of nonlinear mobility in the limit of weak corrugation (Fisher and Zwerger, 1985). For a strong pinning potential, with subsequent minima differing in energy by an amount larger than the zero-point energy of well vibrations, the absence of dissipative localization is suggested by the following arguments.

As mentioned in Section 6.1, the mechanism of dissipative localization in a periodic potential is a logarithmic attractive interaction between kinks and antikinks. These correspond to hopping paths between subsequent minima of the potential in forward and backward directions, respectively. In the case of the tilted potential, the instanton path corresponding to a classical motion in an inverted potential is qualitatively different from that in the periodic potential. There are isolated bounces describing quantum down-hill tunneling. As long as the energy difference between subsequent potential minima is larger that the zero-point vibrational energy in the well, there are no quantum tunneling processes in the up-hill direction. Since the particle is unable to return to the initial position (after tunneling in the down-hill direction), dissipative localization does not take place. In Section 8.8.4 we point out that the logarithmic time decay of magnetization in bulk superconductors presumably involves quantum vortex transport in the *presence* of strong barrier tilting. In view of this, dissipative localization of vortices in high-T_c superconductors may be disregarded, though the estimates based on parameters used by Blatter et al. (1991) show that the dissipative constant is well above the critical value. The situation is less clear for ultrathin films, where a similar estimate indicates a dissipation of the same order as the critical value. The resistance data, however, involve a vortex mobility which is nearly linear and thus it may be, in principle, influenced by dissipative localization effects.

The chapter starts with a review of the classical hydrodynamic theory of vortex motion in 2D fluids. In Section 8.2 we derive the Hamilton equations of motion for the vortex center. Section 8.3 describes the canonical quantization procedure for the vortex and derives the analogy to the 2D electron in a magnetic field. In Section 8.4 we review the method of quantum adiabatic phase (Berry, 1984) and consider its application to a vortex in a superfluid film. The geometric phase is derived using an alternative method of derivation based on the time-dependent nonlinear Schrödinger equation for the macroscopic wave function of the superfluid. We also discuss the interpretation of the geometric phase in terms of the pseudo-Lorentz field and present a derivation of the Magnus force. In Section 8.5, these forces are reexamined with the view of an application to vortices in real superconducting materials. In particular, we discuss the Hall resistance measurements in light of the vortex dynamics in superconductors.

Section 8.6 is concerned with the origin of the inertial mass of vortices in a TDGL model. Starting from a hypothetical model involving neutral Cooper pairs, we identify two contributions to the vortex mass: The first one stems from the phase fluctuations, and the other one is the core contribution, first derived by Suhl (1965). Interestingly, the sum of these masses satisfies an energy–mass relation, characteristic of the Lorentz invariance of the TDGL model. In a real superconductor, the phase fluctuations couple to the gauge fields, a problem that is considered in Section 8.6.5. Eliminating the scalar potential from the partition function, we derive an effective action that is a functional of the phase variable only. This enables us to derive the electromagnetic vortex mass in the form derived by Suhl (1965). It turns out that this form corresponds to imposing a *phase-pinning constraint* from the outset. Physically, this is tantamount to assuming that the electrostatic-energy fluctuations dominate over the vortex kinetic energy associated with the phase fluctuations of a neutral model. In the context of granular superconductivity, this limit corresponds to discarding the δ effect (see Section 6.3). In Section 8.6.7 we estimate the correction to the electromagnetic mass caused by relaxing the phase-pinning constraint. We also show how a more careful treatment based on elimination of the scalar potential is needed to resolve the following paradox:

> Noting that the electromagnetic mass is inversely proportional to the electrostatic energy generated by vortex translation, we may consider a hypothetical limit of infinite electromagnetic mass obtained, for instance, by letting the dielectric constant go to infinity, or by simply taking the carrier charge to zero. The latter case corresponds to a neutral superfluid for which the vortices, in the *classical limit, are massless.* Actually, the vortex in a *neutral TDGL fluid* carries a *finite* mass, generated by the phase fluctuations. For a charged TDGL superfluid, the correct vortex mass is a *reduced* mass composed of the electromagnetic and phase-inertial contributions (the latter is due to the finite compressibility of the superfluid). Thus the actual vortex mass remains finite even in the limit of a diverging electromagnetic mass. This "reduced-mass effect" disappears as soon as the phase-pinning constraint is invoked.

The origin of the vortex mass in a granular array is discussed in Section 8.7. Starting from the phase-dependent action functional (6.115) enables us to incorporate properly the contribution of the δ effect to the inertial mass. We also discuss the effect of spin-wave fluctuations about the vortex configuration on the effective action. This effect is particularly pronounced when the action is of a Lorentz-invariant form, yielding an *acoustic* spin-wave spectrum. In this case, the action can be expressed in the form of a magnetostatic energy of a system of currents flowing along the trajectories of the vortex centers (Popov, 1983). Eckern and Schmid (1989) derived a dissipative contribution of the acoustic spin waves to the effective action starting from this magnetostatic analogy. Section 8.7 also contains a derivation of the vortex dissipation mechanism due to ohmic losses. We start from the effective action (6.66), which has been *microscopically* established, and by going over to the continuum approximation, we obtain a vortex-viscosity coefficient of the form, previously derived phenomenologically by Bardeen and Stephen (1965) for a bulk superconductor. Our derivation raises questions about the correct value of the sheet resistance to be used in the expression for the vortex viscosity. The problem is related to the blocking of the quasiparticle-conductance channel at temperatures well below the bulk superconducting transition temperature (see Chapter 6). Since the usual models of a vortex in granular superconductors assume that the grains have a *position-independent magnitude* of the order parameter, the sheet resistance entering the viscosity coefficient should be the *subgap resistance*, implying a vanishing of the viscosity as $T \to 0$. This is in contrast with the model of Bardeen and Stephen (1965), in which a normal (gapless) vortex core is assumed. Perhaps a more realistic model allowing for an analogous depression of the order parameter near the vortex center would yield an enhanced low-temperature viscosity in a granular array. The section is concluded by a review of the classical dynamics of a massive vortex in a dissipative Josephson-junction array. An equation of motion for the vortex center in a 2D array in the presence of applied current is derived in a form suitable for the analysis of the quantum vortex tunneling data, recently reported by van der Zant et al. (1991).

Section 8.8 discusses three experimental studies involving a possible quantum tunneling of vortices. In Section 8.8.1 we review the work of van der Zant et al. (1991) investigating the temperature dependence of the resistance in underdamped arrays. This work attempts to make a critical comparison of the data with the MQT theory (see Chapter 4). Unfortunately, our poor knowledge of the vortex structure near its center introduces large uncertainties of the estimates of the vortex mass and viscosity. In spite of this, the data suggest that MQT processes are present. Next, we discuss the recent study of the resistive transition in ultrathin films (Liu et al., 1992). The linear T dependence of the log of the resistance observed by these authors is derived. We also include a discussion of the dissipative localization and its effect on the variable-range hopping mechanism of vortex transport. Section 8.8.4 deals with the theoretical aspects of quantum flux creep in bulk superconductors. We review the path-integral method used by Blatter et al. (1991) to estimate the vortex-tunneling rates in the limit of strong dissipation. The method of dimensional estimates (of the effective bounce action) is introduced via a simple example of single-particle tunneling through a 1D potential barrier. Numerical estimates of the characteristic tunneling time for a high-T_c superconductor indicate that the vortex–core bound state may not be able to follow adiabatically the vortex motion during the tunneling event (Šimánek, 1992). This raises questions about the validity of the picture in which the vortex tunnels with its core rigidly attached to the line. Finally, an orientational estimate is made of the dissipation constant α characterizing the possibility of the dissipative localization of the line. It is found that $\alpha \approx 15$, indicating vanishing of the linear vortex mobility.

8.2. Classical dynamics of vortices in 2D superfluids

Let us first outline the standard phenomenological approach to the dynamics of point vortices in an ideal 2D incompressible fluid. As first shown by Kirchhoff [1883], the *rectangular position coordinates* of the vortex constitute a pair of *conjugate variables*, which allow the classical hydrodynamics to be written in a *Hamiltonian form*. Let us first consider a single vortex, centered at $\mathbf{r}_j$, in an unbounded fluid and determine the concomitant fluid-velocity field $\mathbf{v}(\mathbf{r})$. Using Eqs. (7.45)–(7.47), we obtain

$$
\begin{aligned}
v_x &= \kappa \frac{\partial \psi}{\partial y} \\[2mm]
v_y &= -\kappa \frac{\partial \psi}{\partial x}
\end{aligned}
\tag{8.1}
$$

The function $\psi(x, y)$ is, according to Eq. (7.51),

$$
\psi(x, y) = -q_j \ln |\mathbf{r} - \mathbf{r}_j|
\tag{8.2}
$$

where $q_j = \pm 1$ and the constant κ is determined by the quantization of circulation (Fetter and Walecka, 1971). For a superfluid consisting of bosons, each of mass m^*, we have, using Eq. (7.47), a circulation

$$
m^* \oint_C \mathbf{v}_s \cdot d\mathbf{s} = \oint_C \mathbf{p} \cdot d\mathbf{s} = \kappa m^* \oint_C \boldsymbol{\nabla} \varphi \cdot d\mathbf{s}
\tag{8.3}
$$

where C is a contour surrounding the vortex center. The integral on the RHS of this equation is equal to 2π times an integer [see Eq. (7.54)]. Moreover, applying the Bohr–Sommerfeld quantization condition to the LHS of (8.3), we obtain

$$
\kappa = \frac{\hbar}{m^*}
\tag{8.4}
$$

From Eqs. (8.1) and (8.2), we have

$$
\mathbf{v}(\mathbf{r}) = q_j \kappa \mathbf{z} \times \boldsymbol{\nabla}_{\mathbf{r}} \ln |\mathbf{r} - \mathbf{r}_j|
\tag{8.5}
$$

where $\mathbf{z}$ is a unit vector perpendicular to the plane of flow. Let us now insert a second vortex at $\mathbf{r}_i \neq \mathbf{r}_j$ and determine the fluid velocity at its center, induced by the presence of the vortex at $\mathbf{r}_j$. According to Eq. (8.5), this velocity is

$$
\mathbf{v}(\mathbf{r}_i) = q_j \kappa \mathbf{z} \times \boldsymbol{\nabla}_{\mathbf{r}_i} \ln |\mathbf{r}_i - \mathbf{r}_j|
\tag{8.6}
$$

This result can be generalized to a system of vortices with circulation q_j, centered at $\mathbf{r}_j \neq \mathbf{r}_i$

$$
\mathbf{v}(\mathbf{r}_i) = \mathbf{z} \times \sum_{j \neq i} \kappa q_j \boldsymbol{\nabla}_{\mathbf{r}_i} \ln |\mathbf{r}_i - \mathbf{r}_j|
\tag{8.7}
$$

The final step of the derivation of Hamilton's equations rests on setting $\mathbf{v}(\mathbf{r}_i)$ of Eq. (8.7) equal to the velocity of the center of the ith vortex. In other words, the vortex center is assumed to move at the *local fluid velocity* at its center. This is a consequence of the basic theorem of classical hydrodynamics (Lamb, 1945; Fetter et al., 1966), discussed next.

8.2.1. *Basic theorem of the hydrodynamics of incompressible fluid*

We start from the Euler equation for an incompressible fluid (Landau and Lifshitz, 1975).

$$\frac{\partial \mathbf{v}}{\partial t} - \mathbf{v} \times (\boldsymbol{\nabla} \times \mathbf{v}) = -\boldsymbol{\nabla}\left(\frac{p}{\rho} + \frac{1}{2}v^2 + U\right) \tag{8.8}$$

where p, ρ, and U are the pressure, density, and the potential energy density, respectively. Introducing the vorticity $\boldsymbol{\omega}$

$$\boldsymbol{\omega} = \boldsymbol{\nabla} \times \mathbf{v} \tag{8.9}$$

and taking the curl of Eq. (8.8), we obtain

$$\frac{\partial \boldsymbol{\omega}}{\partial t} = \boldsymbol{\nabla} \times (\mathbf{v} \times \boldsymbol{\omega}) \tag{8.10}$$

Using a vector identity, the RHS of this equation can be written as

$$\boldsymbol{\nabla} \times (\mathbf{v} \times \boldsymbol{\omega}) = -(\mathbf{v} \cdot \boldsymbol{\nabla})\boldsymbol{\omega} + (\boldsymbol{\omega} \cdot \boldsymbol{\nabla})\mathbf{v} + \mathbf{v}(\boldsymbol{\nabla} \cdot \boldsymbol{\omega}) - \boldsymbol{\omega}(\boldsymbol{\nabla} \cdot \mathbf{v}) \tag{8.11}$$

From Eq. (8.9), we see that $\boldsymbol{\nabla} \cdot \boldsymbol{\omega} = 0$. Moreover, for an *incompressible* fluid, we have $\boldsymbol{\nabla} \cdot \mathbf{v} = 0$. Thus, the last two terms on the RHS of Eq. (8.11) vanish, and we obtain with use of Eq. (8.10)

$$\frac{\partial \boldsymbol{\omega}}{\partial t} = -(\mathbf{v} \cdot \boldsymbol{\nabla})\boldsymbol{\omega} + (\boldsymbol{\omega} \cdot \boldsymbol{\nabla})\mathbf{v} \tag{8.12}$$

Let us now apply this result to a moving vortex centered at $\mathbf{r}_i(t)$. From Eqs. (7.55) and (7.60), we have

$$\boldsymbol{\omega}(\mathbf{r}, t) = 2\pi \mathbf{z} q_i \kappa \, \bar{\delta}(\mathbf{r} - \mathbf{r}_i(t)) \tag{8.13}$$

yielding

$$\frac{\partial \boldsymbol{\omega}}{\partial t} = -2\pi \mathbf{z} q_i \kappa (\dot{\mathbf{r}}_i \cdot \boldsymbol{\nabla}) \bar{\delta}(\mathbf{r} - \mathbf{r}_i(t)) \tag{8.14}$$

where $\bar{\delta}$ is a "cylindrical δ function," which takes into account the finite size of the vortex core [see Eq. (7.52)]. The second term on the RHS of Eq. (8.12) becomes with the use of Eq. (8.13)

$$(\boldsymbol{\omega} \cdot \boldsymbol{\nabla})\mathbf{v} = 2\pi q_i \kappa \, \bar{\delta}(\mathbf{r} - \mathbf{r}_i(t)) \frac{\partial \mathbf{v}}{\partial z} = 0 \tag{8.15}$$

where we have used the fact that the fluid velocity $\mathbf{v}$ for 2D flow is *independent of z*. Using Eq. (8.13), the first term on the RHS of Eq. (8.12) can also be expressed via $\bar{\delta}$

$$(\mathbf{v} \cdot \boldsymbol{\nabla})\omega = 2\pi \mathbf{z} q_i \kappa (\mathbf{v} \cdot \boldsymbol{\nabla})\bar{\delta}(\mathbf{r} - \mathbf{r}_i(t)) \tag{8.16}$$

Inserting the results (8.14)–(8.16) into Eq. (8.12), we obtain

$$2\pi \mathbf{z} q_i \kappa [(\dot{\mathbf{r}}_i - \mathbf{v}) \cdot \boldsymbol{\nabla}]\bar{\delta}(\mathbf{r} - \mathbf{r}_i(t)) = 0 \tag{8.17}$$

Noting that $\boldsymbol{\nabla}\bar{\delta}$ is nonzero only within the vortex core, Eq. (8.15) implies

$$\dot{\mathbf{r}}_i = \mathbf{v}(\mathbf{r}_i) \tag{8.18}$$

Equation (8.18) completes the proof of the basic hydrodynamic theorem.

8.2.2. *Hamilton equations of motion*

Introducing Eq. (8.18) into (8.7), we obtain the following equation of motion for the vortex center $\mathbf{r}_i$

$$\dot{\mathbf{r}}_i = \mathbf{z} \times \sum_{j \neq i} q_j \kappa \boldsymbol{\nabla}_{\mathbf{r}_i} \ln |\mathbf{r}_i - \mathbf{r}_j| \tag{8.19}$$

The RHS of this equation can be expressed in terms of the Hamiltonian H_v of a system of vortices derived in Section 7.3.1. Multiplying Eq. (8.19) by $2\pi J q_i / \kappa$ and using Eq. (7.75), we obtain

$$\frac{2\pi J q_i}{\kappa} \dot{x}_i = -2\pi J \sum_{j \neq i} q_i q_j \frac{\partial}{\partial y_i} \ln |\mathbf{r}_i - \mathbf{r}_j| = \frac{\partial H_v}{\partial y_i} \tag{8.20}$$

and

$$\frac{2\pi J q_i}{\kappa} \dot{y}_i = -\frac{\partial H_v}{\partial x_i} \tag{8.21}$$

These equations can be brought to a more familiar form (Fetter, 1967), by expressing the Josephson-coupling constant J via the superfluid sheet density n_s^{2D}. Using Eqs. (7.162) and (8.4), we have

$$\frac{2\pi J q_i}{\kappa} = 2\pi \kappa q_i n_s^{2D} m^* = \gamma_i \rho_s \tag{8.22}$$

where we introduced the quantized circulation [see Eq. (8.4)]

$$\gamma_i = 2\pi \kappa q_i = \pm \frac{h}{m^*} \tag{8.23}$$

and the *mass* sheet density of the superfluid

$$\rho_s = n_s^{2D} m^* \tag{8.24}$$

Using Eq. (8.22), the pair of Eqs. (8.20) and (8.21) takes the form (Fetter, 1967; McCauley, 1979)

$$\rho_s \gamma_i \dot{x}_i = \frac{\partial H}{\partial y_i}, \qquad \rho_s \gamma_i \dot{y}_i = -\frac{\partial H}{\partial x_i} \tag{8.25}$$

Defining the new variables

$$Q_i = (\rho_s \gamma_i)^{\frac{1}{2}} x_i$$
$$P_i = (\rho_s \gamma_i)^{\frac{1}{2}} y_i \tag{8.26}$$

Eqs. (8.25) become a set of true Hamilton's equations (Fetter, 1967)

$$\dot{Q}_i = \frac{\partial H}{\partial P_i}, \qquad -\dot{P}_i = \frac{\partial H}{\partial Q_i} \tag{8.27}$$

Equations (8.26) and (8.27) confirm that the rectangular vortex coordinates constitute a pair of canonically conjugate variables (Kirchhoff, 1883).

8.3.   Quantization of the vortex system—2D electron analogy

Following the standard canonical quantization procedure, the vortex system is quantized by interpreting Q_i and P_i as noncommuting operators satisfying the following commutation relations (Fetter, 1967)

$$[Q_i, P_i] = i\hbar \delta_{ij} \tag{8.28}$$

Using Eqs. (8.23), (8.24), and (8.26), the commutator (8.28) can be expressed in terms of rectangular coordinates of the vortex center

$$[x_i, y_i] = \frac{i\hbar \delta_{ij}}{\rho_s \gamma_i} = \frac{iq_i \delta_{ij}}{2\pi n_s^{2D}} \tag{8.29}$$

where $q_i = \pm 1$. Starting from Eq. (8.29), Haldane and Wu (1985) deduced an interesting analogy between the *vortex* and the *2D electron in a magnetic field*. Introducing a characteristic length

$$l = \left(2\pi n_s^{2D}\right)^{-\frac{1}{2}} \tag{8.30}$$

and letting $q_i = 1$, Eq. (8.29) becomes

$$[x_i, y_i] = il^2 \delta_{ij} \tag{8.31}$$

This commutation relation is formally identical to that for the *guiding center* of a 2D electron moving in a cyclotron orbit. To show this, we consider the electronic Hamiltonian of the form

$$H_e = \frac{1}{2m} \left(\mathbf{p} - \frac{e}{c}\mathbf{A}\right)^2 \tag{8.32}$$

For a static magnetic field $\mathbf{H} \parallel \mathbf{z}$, a convenient choice of the vector potential $\mathbf{A}$ is (Coulomb gauge)

$$\mathbf{A} = \frac{1}{2}(\mathbf{H} \times \mathbf{r}) \tag{8.33}$$

Using Eq. (8.33), the components of the kinematic momenta Π_x and Π_y are

$$\Pi_x = mv_x = p_x - \frac{eA_x}{c} = p_x + \frac{eH}{2c}y \tag{8.34}$$

$$\Pi_y = mv_y = p_y - \frac{eH}{2c}x \tag{8.35}$$

There are two constants of motion, X and Y, given by

$$X = x + \frac{c}{eH}\Pi_y \tag{8.36}$$

$$Y = y - \frac{c}{eH}\Pi_x \tag{8.37}$$

To show that X is a constant of motion, let us consider the commutator $[X, H_e]$. Using Eqs. (8.32) and (8.36), we have

$$[X, H_e] = \frac{1}{2m}[x, \Pi_x^2] + \frac{c}{2meH}[\Pi_y, \Pi_x^2] \tag{8.38}$$

The first term on the RHS of this equation is evaluated with use of the commutator $[x, p_x] = i\hbar$, yielding

$$\frac{1}{2m}[x, \Pi_x^2] = \frac{i\hbar}{m}\Pi_x \tag{8.39}$$

This contribution is exactly canceled by the second term on the RHS of Eq. (8.38), since the latter is found to be

$$\frac{c}{2meH}[\Pi_y, \Pi_x^2] = -\frac{c\Pi_x}{meH}[\Pi_x, \Pi_y] = -\frac{i\hbar}{m}\Pi_x \tag{8.40}$$

Thus we find that the RHS of Eq. (8.38) vanishes, implying that the variable X is a constant of the motion. A similar proof shows that Y is also a constant of the motion. The variables X and Y are called *guiding centers* of an electron orbit. They cannot take definite values simultaneously, since they *do not commute*. Rather, they satisfy the relation

$$[X, Y] = -\frac{i\hbar c}{eH} \tag{8.41}$$

which can be easily verified by substituting for X and Y Eqs. (8.36) and (8.37) and using Eqs. (8.34) and (8.35). For the sake of comparison with the commutator (8.31), it is useful to express the RHS of Eq. (8.41) in terms of the *magnetic length* l_H defined as

$$l_H = \left(\frac{\hbar c}{eH}\right)^{\frac{1}{2}} \qquad (8.42)$$

Further insight into the meaning of l_H can be gained by considering the magnetic flux passing through the circle of radius l_H. Using Eq. (8.42), we obtain

$$\phi = \pi l_H^2 H = \frac{hc}{2e} = \phi_0 \qquad (8.43)$$

Thus we see that exactly *one flux quantum* passes through this circle. Equations (8.41) and (8.42) imply

$$[X, Y] = -i l_H^2 \qquad (8.44)$$

which should be compared with the vortex commutator (8.31). From Eq. (8.30), it follows that $2\pi l^2$ is the surface area per superfluid particle. On the other hand, Eq. (8.43) implies that $2\pi l_H^2$ is the surface area per magnetic flux quantum hc/e. This *particle flux correspondence* is central to the conjecture made by Haldane and Wu (1985) about the phase change of the superfluid wave function as a vortex is *transported around a closed path*. This phase change is another example of the *geometrical phase factor* found by Berry (1984) in his study of the quantum adiabatic theorem. For an electron slowly transported around a closed contour threaded by a magnetic flux, the geometric phase is given by the *number of flux quanta enclosed by the contour* (Aharonov and Bohm, 1959). This fact, in conjunction with the correspondence between the flux and the superfluid particle, suggests that the Berry phase for a vortex is given by the *number of superfluid particles* enclosed by the contour. The next section is devoted to a proof of this conjecture and to its physical implications for vortex dynamics.

8.4.   Berry's phase and forces on vortices

8.4.1.   *Quantum adiabatic phases—a review*

For the sake of completeness, we begin with a short review of Berry (1984) analysis. Let $H[\mathbf{R}(t)]$ be a Hamiltonian depending on a parameter $\mathbf{R} = (X, Y, Z, \ldots)$. The wave function of the system $\psi(t)$ evolves according to the Schrödinger equation

$$i\hbar\dot{\psi}(t) = H[\mathbf{R}(t)]\psi(t) \qquad (8.45)$$

For each $\mathbf{R}$, we introduce a complete set of energy eigenstates, $u_n(\mathbf{R})$, satisfying

$$H[\mathbf{R}]u_n(\mathbf{R}) = E_n(\mathbf{R})u_n(\mathbf{R}) \qquad (8.46)$$

Let us now consider the time evolution of the wave function $\psi(t)$ between $t = 0$ and T, subject to the initial condition

$$\psi(t = 0) = u_n[\mathbf{R}(t = 0)] \tag{8.47}$$

We assume that $H[\mathbf{R}]$ depends on time only through $\mathbf{R}(t)$ and that, during the time evolution, $u_n[\mathbf{R}(t)]$ remains *nondegenerate*. Following the notation of Schiff (1968), the wave function $\psi(t)$ is written as

$$\psi(t) = \sum_m a_m(t) u_m[\mathbf{R}(t)] \exp\left(-\frac{i}{\hbar} \int_0^t dt' \, E_m[\mathbf{R}(t')] + i\gamma_m(t)\right) \tag{8.48}$$

If we assume that the time dependence of the parameter $\mathbf{R}(t)$ is *slow*, then Eq. (8.45) can be solved, with the ansatz (8.48), in the *adiabatic limit*, which amounts to setting $a_m(t) = \delta_{m,n}$ (Schiff, 1968). Thus the wave function at time t is

$$\psi(t) = \exp\left(-\frac{i}{\hbar} \int_0^t dt' \, E_n[\mathbf{R}(t')] + i\gamma_n(t)\right) u_n[\mathbf{R}(t)] \tag{8.49}$$

The first term in the exponent of Eq. (8.49) is the standard dynamical phase factor. The second term is the geometrical phase factor discovered by Berry (1984). An explicit form for γ_n can be found by inserting (8.49) into (8.45). With the help of Eq. (8.46), we obtain

$$\dot{\gamma}_n(t) u_n[\mathbf{R}(t)] = i \dot{u}_n[\mathbf{R}(t)] \tag{8.50}$$

which implies, using orthonormality of the set $u_n(\mathbf{R})$

$$\dot{\gamma}_n(t) = \left\langle u_n[\mathbf{R}(t)] \left| i \frac{\partial}{\partial t} \right| u_n[\mathbf{R}(t)] \right\rangle \tag{8.51}$$

Integrating this equation from 0 to T, the geometric phase at $t = T$ is given by (Berry, 1984)

$$\begin{aligned}
\gamma_n(T) &= \int_0^T dt \left\langle u_n[\mathbf{R}(t)] \left| i \frac{\partial}{\partial t} \right| u_n[\mathbf{R}(t)] \right\rangle \\
&= \int_0^T dt \, \frac{\partial \mathbf{R}}{\partial t} \left\langle u_n[\mathbf{R}] \left| i \frac{\partial}{\partial \mathbf{R}} \right| u_n[\mathbf{R}] \right\rangle \\
&= \int_{\mathbf{R}(0)}^{\mathbf{R}(T)} d\mathbf{R} \cdot \left\langle u_n[\mathbf{R}] \left| i \frac{\partial}{\partial \mathbf{R}} \right| u_n[\mathbf{R}] \right\rangle
\end{aligned} \tag{8.52}$$

where the second equality stems from the fact that u_n depends on t through the time dependence of $\mathbf{R}(t)$. Actually, Eq. (8.51) was also derived by Schiff (1968). However, he uses a gauge transformation to eliminate the phases γ_n from the problem. It is interesting to review his argument and see the conditions under which it fails, as it highlights the meaning of the geometric phase. First, we note that since the overall phase of $u_n[\mathbf{R}]$ is arbitrary, we can make an $\mathbf{R}$-dependent gauge transformation (Schiff, 1968)

$$u_n[\mathbf{R}] \to e^{i\xi_n(\mathbf{R})}u_n[\mathbf{R}] \tag{8.53}$$

Under this transformation, the matrix element on the RHS of Eq. (8.52) is changed as follows

$$\mathbf{A}_n = \left\langle u_n[\mathbf{R}] \left| i\frac{\partial}{\partial \mathbf{R}} \right| u_n[\mathbf{R}] \right\rangle \to \mathbf{A}_n - \frac{\partial \xi_n}{\partial \mathbf{R}} \tag{8.54}$$

From Eqs. (8.52) and (8.54), we see that the corresponding change of γ_n is

$$\gamma_n(T) = \int_{\mathbf{R}(0)}^{\mathbf{R}(T)} d\mathbf{R} \cdot \mathbf{A}_n \to \gamma_n(T) - \int_{\mathbf{R}(0)}^{\mathbf{R}(T)} d\mathbf{R} \cdot \frac{\partial \xi_n}{\partial \mathbf{R}}$$
$$= \gamma_n(T) - \xi_n[\mathbf{R}(T)] + \xi_n[\mathbf{R}(0)] \tag{8.55}$$

If the gauge transformation (8.53) is chosen such that

$$\xi_n[\mathbf{R}(T)] - \xi_n[\mathbf{R}(0)] = \gamma_n(T) \tag{8.56}$$

then the RHS of (8.55) vanishes, so that $\gamma_n(T)$ is eliminated. In this way, Schiff (1968) arrives at the expression (8.49) with $\gamma_n(t)$ set equal to zero. However, this procedure *fails* when $\mathbf{R}(T) = \mathbf{R}(0)$ (Berry, 1984). In this case, we have from Eq. (8.52)

$$\Delta\gamma_n(T) = \int_{\mathbf{R}(0)}^{\mathbf{R}(T)=\mathbf{R}(0)} d\mathbf{R} \cdot \mathbf{A}_n = \oint_C d\mathbf{R} \cdot \mathbf{A}_n \tag{8.57}$$

where C is a closed loop traversed counterclockwise. Since the gauge transformation (8.53) must be *single valued* on C, we have

$$\xi_n[\mathbf{R}(T)] = \xi_n[\mathbf{R}(0)] + 2\pi m \tag{8.58}$$

where m is an integer. Using this result in Eq. (8.55), we see that under the gauge transformation (8.53)

$$\gamma_n(T) \to \gamma_n(T) - 2\pi m \tag{8.59}$$

so that $\gamma_n(T)$ is *gauge invariant* (mod 2π) and cannot be eliminated. Therefore, the full expression (8.49), including the geometric phase (8.57), must be used when $\mathbf{R}(T) = \mathbf{R}(0)$.

8.4.2.   *Berry phase for vortex in a superfluid film*

Haldane and Wu (1985) have evaluated the geometric phase (8.57), using the wave function of an N-body system of interacting bosons containing vortex excitation (Feynman, 1955). We use here an alternative approach in which $\psi(t)$ of Eq. (8.45) is identified with the macroscopic wave function of the condensate. This wave function satisfies the nonlinear Schrödinger equation

$$i\hbar\dot{\psi} = -\left(\frac{\hbar^2}{2m^*}\nabla^2 + \mu\right)\psi + g|\psi|^2\psi \tag{8.60}$$

where μ is the chemical potential of the interacting Bose system and g is the amplitude of the interparticle repulsive δ-function potential. If the RHS of Eq. (8.60) is set equal to zero, then we obtain the Gross–Pitaevskii equation for the *stationary* condensate wave function with $E_0 = 0$ (Gross, 1961; Pitaevskii, 1961). This equation allows a vortex solution of the form (Fetter and Walecka, 1971)

$$u[\mathbf{R}] = \psi_\infty f(|\mathbf{r} - \mathbf{R}|)e^{i\varphi(\mathbf{r}-\mathbf{R})} \tag{8.61}$$

where ψ_∞ is the condensate amplitude at infinite distance from the vortex center $\mathbf{R}$. The function f obeys a nonlinear differential equation and must be found numerically. A good approximation for $f(r)$ is (Fetter, 1971)

$$\begin{aligned} f(r) &\propto \tfrac{r}{\xi}, & r &\to 0 \\ f(r) &\sim 1 - \tfrac{\xi^2}{2r^2}, & r &\to \infty \end{aligned} \tag{8.62}$$

where ξ is the coherence length. For a slowly varying position of the vortex center $\mathbf{R}(t)$, the time evolution of $\psi(t)$ can be treated within the adiabatic approximation of Section 8.4.1. In other words, if the wave function at $t = 0$ is given by $u[\mathbf{R}(t = 0)]$, it *remains* in this state also at a later time $t = T$. According to Eqs. (8.52) and (8.57), the geometric phase $\Delta\gamma_0$ for a vortex transported around the contour C is

$$\Delta\gamma_0 = i\oint_C d\mathbf{R} \cdot \left\langle u[\mathbf{R}] \left| \frac{\partial}{\partial\mathbf{R}} \right| u[\mathbf{R}] \right\rangle \tag{8.63}$$

We note that the subscript zero on $\Delta\gamma_0$ refers to the case where there are *no other vortices inside* C. Using the condensate wave function (8.61), the matrix element on the RHS of Eq. (8.63) becomes

$$\begin{aligned} \left\langle u[\mathbf{R}] \left| \frac{\partial}{\partial\mathbf{R}} \right| u[\mathbf{R}] \right\rangle &= |\psi_\infty|^2 \int d^2r\, f(|\mathbf{r} - \mathbf{R}|)e^{-i\varphi(\mathbf{r}-\mathbf{R})} \\ &\quad \times \frac{\partial}{\partial\mathbf{R}}\left[f(\mathbf{r} - \mathbf{R})e^{i\varphi(\mathbf{r}-\mathbf{R})} \right] \\ &= i|\psi_\infty|^2 \int d^2r\, f^2(|\mathbf{r} - \mathbf{R}|)\frac{\partial}{\partial\mathbf{R}}\varphi(\mathbf{r} - \mathbf{R}) \end{aligned} \tag{8.64}$$

The last equality of Eq. (8.64) has been obtained by taking into account the $\mathbf{R}$ independence of the normalization of the wave function (8.61), implying

$$\int d^2r\, f\frac{\partial f}{\partial\mathbf{R}} = \frac{1}{2}\frac{\partial}{\partial\mathbf{R}}\int d^2r\, f^2 = 0 \tag{8.65}$$

To evaluate the RHS of (8.64), we use Eqs. (7.47) and (8.5), yielding for a vortex centered at $\mathbf{R}$

$$\frac{\partial}{\partial \mathbf{R}} \varphi(\mathbf{r} - \mathbf{R}) = -\frac{1}{\kappa} \mathbf{v}(\mathbf{r}) = -q\mathbf{z} \times \boldsymbol{\nabla}_r \ln(\mathbf{r} - \mathbf{R}) \qquad (8.66)$$

where $q = \pm 1$ is a dimensionless vortex charge. Introducing this result into Eq. (8.64) and using the latter in Eq. (8.63), we obtain the change of phase

$$\Delta \gamma_0 = q|\psi_\infty|^2 \int d^2r \oint_C d\mathbf{R} \cdot [\mathbf{z} \times \boldsymbol{\nabla}_r \ln(\mathbf{r} - \mathbf{R})] f^2(|\mathbf{r} - \mathbf{R}|) \qquad (8.67)$$

This result is similar to the Berry phase derived by Haldane and Wu (1985), the main difference being that they consider *multivortex configurations* so that the function f^2 is replaced by an expectation value of the density operator in the presence of *many* vortices. In view of Eq. (8.62), we write

$$|\psi_\infty|^2 f^2(|\mathbf{r} - \mathbf{R}|) = \rho_0 + \delta\rho(|\mathbf{r} - \mathbf{R}|) \qquad (8.68)$$

where ρ_0 is the *undepleted* superfluid sheet density and $\delta\rho$ represents the *depletion* in the vortex core. Introducing Eq. (8.68) into (8.67), we have, using Eq. (8.66)

$$\Delta \gamma_0 = -\rho_0 \int d^2r \oint_C d\mathbf{R} \cdot \frac{\partial\varphi(\mathbf{r} - \mathbf{R})}{\partial \mathbf{R}}$$

$$+ q \int d^2r\, \delta\rho(|\mathbf{r} - \mathbf{R}|) \oint_C (d\mathbf{R} \times \mathbf{z}) \cdot \boldsymbol{\nabla}_r \ln(\mathbf{r} - \mathbf{R}) \qquad (8.69)$$

The second term on the RHS of this equation vanishes, since

$$\int d^2r\, \delta\rho(|\mathbf{r} - \mathbf{R}|) \boldsymbol{\nabla}_r \ln(\mathbf{r} - \mathbf{R}) = \int d^2r\, \delta\rho(|\mathbf{r} - \mathbf{R}|) \frac{\mathbf{r} - \mathbf{R}}{|\mathbf{r} - \mathbf{R}|^2} = 0 \quad (8.70)$$

where the second equality follows from the *rotational symmetry* of $\delta\rho$ (about the vortex center). In the first term on the RHS of Eq. (8.69), we consider the integral

$$I(\mathbf{r}) = \oint_C d\mathbf{R} \cdot \frac{\partial\varphi(\mathbf{r} - \mathbf{R})}{\partial \mathbf{R}} \qquad (8.71)$$

It is convenient to evaluate this integral in the two-dimensional R_x, R_y plane. Noting that $\varphi(\mathbf{r} - \mathbf{R}) = \varphi(\mathbf{R} - \mathbf{r}) + 2\pi(\text{integer})$ and using Stokes theorem, we have

$$I(\mathbf{r}) = \oint_C d\mathbf{R} \cdot \frac{\partial\varphi(\mathbf{R} - \mathbf{r})}{\partial \mathbf{R}}$$

$$= -\int_{A_C} d^2R \nabla^2 \psi(\mathbf{R} - \mathbf{r})$$

$$= 2\pi q \int_{A_C} d^2R\, \delta(\mathbf{R} - \mathbf{r}) \qquad (8.72)$$

The last two equalities were obtained using Eqs. (7.46)–(7.50). Since A_C is the area bounded by the contour C, we obtain from (8.72)

$$I(\mathbf{r}) = \begin{cases} 2\pi q, & \text{for } \mathbf{r} \text{ inside } C \\ 0, & \text{otherwise} \end{cases} \tag{8.73}$$

Introducing this result into (8.69) and using (8.70), the geometric phase is

$$\Delta\gamma_0 = -2\pi q \rho_0 A_C \tag{8.74}$$

Hence, $\Delta\gamma_0$ is $-2\pi q$ times the mean number of superfluid particles within the contour. This confirms the conjecture made using the vortex–electron analogy at the end of Section 8.3. If a vortex is transported around a contour containing other vortices, then one must take into account the depletion of superfluid density caused by the latter. For an *incompressible* fluid, the following expression should be used for the geometric phase (Haldane and Wu, 1985)

$$\Delta\gamma = -2\pi q(\rho_0 A_C - N_v \pi \xi^2 \rho_0) \tag{8.75}$$

where N_v is the number of enclosed vortices. We remark that the sign of the expressions for $\Delta\gamma_0$ and $\Delta\gamma$ given in (8.74) and (8.75) differs from that given by Haldane and Wu (1985). We arrive at the negative sign also by evaluating the Berry phase, using the effective Euclidean action for the boson system.

8.4.3. *Fractional statistics of vortices*

The depletion term on the RHS of Eq. (8.75) makes the vortices behave like objects with "fractional statistics." This can be seen as follows. Let us consider a state with two vortices at $\mathbf{R}_a$ and $\mathbf{R}_b$. Now we adiabatically move $\mathbf{R}_a$ around a loop C. If this loop does not contain $\mathbf{R}_b$, the wave function of the state is changed by the factor $\Delta\gamma_0$ given in Eq. (8.74). On the other hand, if C contains $\mathbf{R}_b$, the corresponding geometric phase must be determined from Eq. (8.75) with $N_v = 1$. Thus the extra phase change, due to the presence of a vortex at $\mathbf{R}_b$ *inside* C, is (letting $q = +1$)

$$\Delta\gamma_b = \Delta\gamma - \Delta\gamma_0 = 2\pi\nu \tag{8.76}$$

where

$$\nu = \pi\xi^2 \rho_0 \tag{8.77}$$

is the superfluid (number) deficit.

The statistics of vortices is determined through the operation of the *exchange* of two vortices. The effect of this exchange on the wave function of the system is defined by

$$\psi(\mathbf{R}_a, \mathbf{R}_b) = e^{-i\theta}\psi(\mathbf{R}_b, \mathbf{R}_a), \qquad 0 \le \theta \le 2\pi \tag{8.78}$$

where θ is the *statistical phase*. The exchange of two vortices is associated with a phase change

$$\theta = \frac{1}{2}\Delta\gamma_b = \pi\nu \tag{8.79}$$

where the factor $1/2$ in the first equality follows from the fact that the exchange amounts to moving one of the vortices, say the ath, around a half circle enclosing the bth vortex. The latter vortex is moved on half loop that remains inside the path of the ath vortex, thus making sure that the vortices are never brought on top of each other. In this way, only the ath vortex picks up half of a Berry phase. From Eqs. (8.78) and (8.79) we conclude that vortices obey fractional statistics with a statistical phase $\theta = \pi^2\xi^2\rho_0$. Unfortunately, there is a problem with applying this result to real superfluid films, since they do not satisfy the condition of *incompressibility*. Haldane and Wu (1985) have determined the asymptotic superfluid density $\rho(r)$ of a *compressible* ^{4}He film in the presence of a vortex

$$\rho(r) = \rho(\infty)\left[1 - \left(\frac{\xi}{r}\right)^2\right] \tag{8.80}$$

Integrating this expression over the area of the loop C, we obtain a depletion that, due to the presence of the $(\xi/r)^2$ term, depends logarithmically on the radius of the contour. Thus, in view of Eqs. (8.77) and (8.79), there is *no finite* value of θ as the radius goes to infinity. To our knowledge, a similar investigation of the Cooper pair depletion (and a possible fractional statistics) has not been made for thin superconducting films. For granular arrays, the difficulty lies in our poor knowledge of the structure of the vortex core. A qualitative estimate of the Cooper pair deficit ν can be made from Eq. (8.77), by substituting for ρ_0 the sheet density n_s^{2D} given by Eq. (7.162) and letting the core radius ξ equal to the lattice spacing a of the array [see Eq. (7.199)]. In this way, we obtain for the superfluid deficit

$$\nu \approx \pi a^2 n_s^{2D} = \pi a^2 Jm^*/\hbar^2 \tag{8.81}$$

The RHS of this equation suggests that ν is a continuously variable parameter, allowing the vortices to behave like objects with fractional statistics. However, a conclusive statement about vortex statistics would require a more careful analysis of the behaviour of the superfluid density near the vortex core.

8.4.4. *Pseudo-Lorentz force in a neutral superfluid*

An interesting interpretation can be given to the Berry phase (8.74) by comparing it with the phase factor induced by adiabatic transport of an electric charge q_e around a magnetic flux (Aharonov and Bohm, 1959). For this phase factor, we write

$$\Delta\gamma_{AB} = \frac{q_e}{\hbar c}\Phi = \frac{q_e}{\hbar c}\oint_C d\mathbf{R}\cdot\mathbf{A}(\mathbf{R}) \tag{8.82}$$

This expression motivates us to express Eq. (8.69) in the form

$$\Delta\gamma_0 = \frac{q}{\hbar c} \oint_C d\mathbf{R} \cdot \mathbf{A}_f(\mathbf{R}) \tag{8.83}$$

where $\mathbf{A}_f(\mathbf{R})$ is a fictitious gauge field. On comparing the RHS of this expression with the first term on the RHS of Eq. (8.69), we have

$$\mathbf{A}_f(\mathbf{R}) = -\frac{\hbar c \rho_0}{q} \int d^2r\, \frac{\partial\varphi(\mathbf{r}-\mathbf{R})}{\partial\mathbf{R}} \tag{8.84}$$

Using Eq. (8.66), we have from Eq. (8.84)

$$\mathbf{A}_f(\mathbf{R}) = -\hbar c \rho_0 \int d^2r \left(\mathbf{z} \times \frac{\mathbf{R}-\mathbf{r}}{|\mathbf{R}-\mathbf{r}|^2}\right) \tag{8.85}$$

For further interpretation of the fictitious gauge field (8.84), it is convenient to replace the continuum superfluid by a discrete one, composed of pointlike particles

$$\rho(\mathbf{r}) = \sum_i \delta(\mathbf{r} - \mathbf{r}_i) \tag{8.86}$$

This amounts to replacing (8.85) by its discretized version

$$\mathbf{A}_f(\mathbf{R}) = \sum_i \mathbf{A}_f(\mathbf{R} - \mathbf{r}_i) \tag{8.87}$$

where

$$\mathbf{A}_f(\mathbf{R} - \mathbf{r}_i) = -\hbar c \left(\mathbf{z} \times \frac{\mathbf{R}-\mathbf{r}_i}{|\mathbf{R}-\mathbf{r}_i|^2}\right) \tag{8.88}$$

$\mathbf{A}_f(\mathbf{R} - \mathbf{r}_i)$ is a fictitious gauge field generated by the ith superfluid particle. Associated with it is a pseudomagnetic field $\mathbf{H}_f = \boldsymbol{\nabla} \times \mathbf{A}_f$. The z component of this field is, according to Eqs. (8.5), (8.13) and (8.88),

$$H_{f,z}(\mathbf{R} - \mathbf{r}_i) = -\frac{\hbar c}{q}\mathrm{curl}_z\left(\frac{\partial\varphi}{\partial\mathbf{R}}\right) = -2\pi\hbar c\, \bar{\delta}(\mathbf{R} - \mathbf{r}_i) \tag{8.89}$$

This result indicates that the pseudomagnetic field is in the form of a flux *line*, carrying a single flux quantum hc, attached to the particle. Obviously, such a localized field does not lead to a Lorentz force on the vortex, unless the particle crosses the vortex path. Since the net pseudofield is a superposition from all particles in the plane, we have, using Eqs. (8.86)–(8.89) and going to the continuum limit:

$$H_{f,z} = -\sum_i H_{f,z}(\mathbf{R} - \mathbf{r}_i) = -hc\sum_i \bar{\delta}(\mathbf{R} - \mathbf{r}_i) \to -hc\rho_0 \tag{8.90}$$

where ρ_0 is the superfluid sheet density equal to n_s^{2D} in an undepleted superfluid. Equation (8.90) implies that in an unbounded fluid with a spatially independent density ρ_0, the pseudofield is a constant vector

$$\mathbf{H}_f = -\mathbf{z}hc\rho_0 \tag{8.91}$$

The pseudo-Lorentz force, acting on a vortex moving with velocity $\mathbf{v}_v$, is then

$$\mathbf{F}_L = \frac{q}{c}(\mathbf{v}_v \times \mathbf{H}_f) = qh\rho_0(\mathbf{z} \times \mathbf{v}_v) \tag{8.92}$$

It should be pointed out that this result holds only for a translationally invariant fluid, as implied by the assumed continuum description [see Eq. (8.90)].

8.4.5. Magnus force and equation of motion for vortex in a neutral superfluid

Consider now a vortex (which is at rest) in the presence of a uniform superfluid flow with velocity $\mathbf{v}_s$. As a result of the interaction of the vorticity with the flow, there is a force on the vortex called the *Magnus force*

$$\mathbf{F}_M = -qh\rho_0(\mathbf{z} \times \mathbf{v}_s) \tag{8.93}$$

To derive this expression, we start from the continuum approximation (7.5) for the kinetic energy of a neutral superfluid

$$H = \frac{1}{2}J \int d^2\rho(\boldsymbol{\nabla}\theta)^2 \tag{8.94}$$

The phase variable θ is split as follows

$$\theta(\boldsymbol{\rho}) = \varphi(\boldsymbol{\rho}) + \theta_1(\boldsymbol{\rho}) \tag{8.95}$$

where $\varphi(\boldsymbol{\rho})$ and $\theta_1(\boldsymbol{\rho})$ describe the velocity potential of the vortex and the uniform flow, respectively. Using Eqs. (7.47) and (8.95), the Hamiltonian (8.94) can be written as

$$H = \frac{J}{2\kappa^2} \int d^2\rho \left(\kappa\boldsymbol{\nabla}\varphi + \mathbf{v}_s\right)^2 \tag{8.96}$$

The Magnus force is given by

$$\mathbf{F}_M = -\frac{\partial}{\partial \mathbf{R}} H_{\text{int}}(\mathbf{R}) \tag{8.97}$$

where $H_{\text{int}}(\mathbf{R})$ is the *interaction* part of (8.96), which is a function of the vortex center $\mathbf{R}$

$$H_{\text{int}}(\mathbf{R}) = \kappa\rho_m \int d^2\rho\,\mathbf{v}_s \cdot \boldsymbol{\nabla}\varphi(\boldsymbol{\rho} - \mathbf{R}) \tag{8.98}$$

where $\rho_m = n_s^{2D} m^*$ is the mass sheet superfluid density [see Eq. (7.162)]. Noting that

$$\frac{\partial}{\partial \mathbf{R}} \varphi(\boldsymbol{\rho} - \mathbf{R}) = -\boldsymbol{\nabla}\varphi(\boldsymbol{\rho} - \mathbf{R}) \tag{8.99}$$

and using a vector identity, we obtain from Eqs. (8.97) and (8.98)

$$\mathbf{F}_M = \kappa\rho_m \int d^2\rho \left[\mathbf{v}_s \times (\boldsymbol{\nabla}\times\boldsymbol{\nabla}\varphi) + (\mathbf{v}_s \cdot \boldsymbol{\nabla})\boldsymbol{\nabla}\varphi\right] \tag{8.100}$$

The second term on the RHS of Eq. (8.100) can be shown to vanish by using Eq. (7.54). The first term can be written, using Eq. (8.13), as

$$\mathbf{F}_M = 2\pi\kappa\rho_m q \int d^2\rho \left[\mathbf{v}_s \times \mathbf{z}\,\bar{\delta}(\boldsymbol{\rho} - \mathbf{R})\right] = \frac{qh\rho_m}{m^*}(\mathbf{v}_s \times \mathbf{z}) \tag{8.101}$$

where $\mathbf{v}_s = \mathbf{v}_s(\mathbf{R})$ is the superfluid velocity at the vortex center. Since $\rho_m/m^* = n_s^{2D} = \rho_0$, Eq. (8.101) coincides with (8.93).

We can now proceed to set up the equation of motion for a vortex. Since the vortex is massless, it moves with a velocity such that the total force on it is zero

$$\mathbf{f} + \mathbf{F}_M + \mathbf{F}_L = 0 \tag{8.102}$$

where $\mathbf{f}$ stands for forces other than $\mathbf{F}_M$ and $\mathbf{F}_L$. For example, $\mathbf{f}$ can represent a dissipative (drag) force

$$\mathbf{f}_D = -\eta\mathbf{v}_v \tag{8.103}$$

Introducing Eqs. (8.92) and (8.93) into (8.102), we have

$$\mathbf{f} = qh\rho_0[(\mathbf{v}_v - \mathbf{v}_s) \times \mathbf{z}] \tag{8.104}$$

This equation is known in the physics of vortex lines in helium II as the Magnus formula (Wilks, 1967). In the case of $\mathbf{f} = 0$, Eq. (8.104) implies that the pseudo-Lorentz force is *exactly canceled* by the Magnus force, so that $\mathbf{v}_v = \mathbf{v}_s$ (see Fig. 8.1). Thus we recover the basic theorem of classical hydrodynamics, according to which the vortex in an ideal Euclidean fluid moves at the local superfluid velocity [see Eq. (8.18)]. It turns out that in superconductors this behavior is not seen. Rather, vortices are typically observed to move *parallel* to the *Magnus force* with only a small component along the direction of transport current (Hagen et al., 1990). Possible explanations of this behavior are discussed next.

8.5. Dynamics of vortices in 2D superconductors

Let us now proceed to the case of charged superfluids and study the forces acting on a moving vortex in the presence of a transport supercurrent. In order to derive the analog of the Magnus force, we first need to establish the

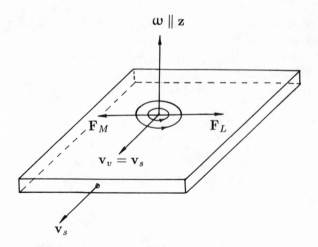

FIG. 8.1. Motion of a vortex in 2D ideal Eulerian fluid. The vorticity vector $\boldsymbol{\omega}$ is along the z axis, and the superfluid velocity is $\mathbf{v}_s$. The vortex center moves at the velocity $\mathbf{v}_v = \mathbf{v}_s$, and the Magnus force exactly cancels the pseudo-Lorentz force.

differential equation for vortex-induced supercurrent $\mathbf{j}_v$. From Eq. (7.124), we have with use of Eqs. (7.160) and (7.162)

$$
\mathbf{j}_v(\mathbf{r}) = \mathbf{j}(\boldsymbol{\rho}, z) =
\begin{cases}
\frac{e^* \hbar |\psi|^2}{m^*} \left(\boldsymbol{\nabla}\varphi - \frac{e^*}{\hbar c} \mathbf{A}_v \right), & \text{for } -\frac{d}{2} < z < \frac{d}{2} \\
0, & \text{otherwise}
\end{cases}
\tag{8.105}
$$

where $\varphi(\boldsymbol{\rho})$ is the angular variable of the vortex given by Eq. (7.54). Taking the curl of this equation and expressing the curl of $\boldsymbol{\nabla}\varphi$ by means of Eq. (8.13), we obtain

$$
\frac{m^* c}{e^{*2} |\psi|^2} (\boldsymbol{\nabla} \times \mathbf{j}_v) + \mathbf{H}_v = \frac{e^* q \mathbf{z}}{|e^*|} \phi_0 \bar{\delta}(\boldsymbol{\rho} - \mathbf{R})
\tag{8.106}
$$

where $\mathbf{H}_v = \boldsymbol{\nabla} \times \mathbf{A}_v$, $\mathbf{R}$ is the vortex-center coordinate, and $\phi_0 = hc/2e$. If we introduce the London depth λ_L in a bulk superconductor, then Eq. (8.106) coincides with the standard form also valid for straight flux lines in 3D superconductors [see Eq. (5-10) of Tinkham, 1975].

8.5.1.   *Analog of Magnus force*

First, it should be made clear that this force is defined as that experienced by a *static* vortex due to the interaction with the *transport current*. Some authors define the Magnus force as that acting on a moving vortex (see Fazio et al., 1992). We again use the formula (8.97) to calculate $\mathbf{F}_M$

but with the interaction energy determined from Eq. (7.136), which also includes the kinetic energy of the supercurrents and the magnetic-field energy. Introducing into Eq. (7.136) for $\mathbf{j}$ the superposition of transport and vortex currents, $\mathbf{j}_s + \mathbf{j}_v$, we obtain the interaction energy

$$H_{\text{int}}(\mathbf{R}) = \frac{\hbar^2 d^2}{e^{*2} J} \int d^2\rho \, \mathbf{j}_s(\boldsymbol{\rho}) \cdot \mathbf{j}_v(\boldsymbol{\rho} - \mathbf{R}) + \frac{d}{c} \int d^2\rho \, \mathbf{j}_s(\boldsymbol{\rho}) \cdot \mathbf{A}_v(\boldsymbol{\rho} - \mathbf{R}) \quad (8.107)$$

where

$$\mathbf{A}_v(\boldsymbol{\rho} - \mathbf{R}) = \frac{d}{c} \int d^2\rho' \, \frac{\mathbf{j}_v(\boldsymbol{\rho}' - \mathbf{R})}{|\boldsymbol{\rho} - \boldsymbol{\rho}'|} \quad (8.108)$$

is the vortex gauge field. Introducing this expression into Eq. (8.97) and using steps similar to those in Eqs. (8.98)–(8.100), we obtain

$$\mathbf{F}_M = \frac{\hbar^2 d^2}{e^{*2} J} \int d^2\rho \, \mathbf{j}_s \times (\boldsymbol{\nabla} \times \mathbf{j}_v) + \frac{d}{c} \int d^2\rho \, \mathbf{j}_s \times (\boldsymbol{\nabla} \times \mathbf{A}_v) \quad (8.109)$$

Using Eq. (8.106), the first term on the RHS of this equation can be written

$$\frac{\hbar^2 d^2}{e^{*2} J} \int d^2\rho \, \mathbf{j}_s \times (\boldsymbol{\nabla} \times \mathbf{j}_v) \propto \int d^2\rho \, \mathbf{j}_s$$
$$\times \left(\frac{e^* q\mathbf{z}}{|e^*|} \phi_0 \, \bar{\delta}(\boldsymbol{\rho} - \mathbf{R}) - \mathbf{H}_v(\boldsymbol{\rho} - \mathbf{R}) \right) \quad (8.110)$$

It is easy to see that the two vectors in the large parentheses on the RHS of this equation point in the *same* direction (for *any* sign of q or e^*). Moreover, if $\mathbf{j}_s$ varies slowly on the scale of spatial variations of $\mathbf{H}_v$, then the vectors contribute to the integral with the same magnitude, implying that the first term on the RHS of Eq. (8.109) exactly *vanishes*. For slowly varying $\mathbf{j}_s$, the second term yields

$$\mathbf{F}_M = \frac{d}{c} \mathbf{j}_s(\mathbf{R}) \times \boldsymbol{\phi} \quad (8.111)$$

where $\boldsymbol{\phi}$ is a vector of magnitude ϕ_0, pointing in the direction of $\mathbf{H}_v$. We remark that this expression reduces precisely to Eq. (8.93) upon putting $\mathbf{j}_s = e^* \rho_0 \mathbf{v}_s / d$. This is to be expected, since Eq. (8.111) is, in principle, valid for charges of any magnitude (including $e^* \to 0$). Thus the concept of the Magnus force seems well established for both the neutral and charged superfluids. We note that the expression (8.111) is commonly referred to as the Lorentz force (Tinkham, 1975).

8.5.2. *Pseudo-Lorentz force in homogeneous materials*

The question of pseudo-Lorentz forces in superconductors remains open and of considerable interest at the present time (Fisher, 1991). Let us first consider the case of a pure, homogeneous material. It has been shown by Nozieres and Vinen (1966) that the equation of motion for the vortex is given by the Magnus formula (8.104), just like in a neutral superfluid. We note that in a translationally invariant superfluid, the pseudo-Lorentz force can be obtained from the Magnus force by applying the principle of Galilean invariance. In other words, the force on a vortex moving with the velocity $\mathbf{v}_v$ should be equal to the Magnus force produced by fluid moving with the velocity $-\mathbf{v}_v$. It has been often argued that, in superconductors, the positive background of charge creates a preferred frame of reference, which spoils the Galilean invariance. In fact, this argument has been advanced by some authors as an explanation for the absence of pseudo-Lorentz force in superconductors. Careful derivation of the vortex equation of motion by Nozieres and Vinen (1966) has proved that this argument is fallacious since the uniform charge of the background does not affect the local equation of motion of the superfluid.

We can substantiate this conclusion by applying the concept of continuity of the force $\mathbf{F}_L$ with respect to the changes of the carrier charge e^*. If the effect of the background charge is to cancel $\mathbf{F}_L$, then upon endowing the neutral superfluid with an infinitesimal charge, the pseudo-Lorentz force should drop precipitously to zero, violating the principle of continuity. de Gennes and Matricon (1964) have also suggested that the motion of flux lines in ideal type II superconductors satisfies the same equation of motion as the vortex in a neutral superfluid. Thus we may conclude that the expression (8.92) is applicable to homogeneous superconductors as well. This is interesting, since this formula has been derived from the Berry phase of the condensate wave function. For a charged superfluid, the Schrödinger equation (8.60) is modified by replacing $(\hbar/i)\mathbf{\nabla}$ by a gauge-invariant combination $(\hbar/i)\mathbf{\nabla} - (e^*/c)\mathbf{A}$. But we can choose a gauge such that the vortex solution of this equation is still of the form (8.61). Consequently, the Berry phase for a vortex transported adiabatically around a closed contour in a 2D homogeneous charged superfluid is still given by the expression (8.69) from which the pseudo-Lorentz force follows in the form of (8.92), as in a neutral superfluid.

8.5.3. *Vortex dynamics and Hall resistance in superconductors*

The conclusion of the previous section is that in a pure, homogeneous superconductor, the equation of motion of the vortex takes the form of (8.104). Moreover, if the friction force vanishes, then this equation implies that $\mathbf{v}_v = \mathbf{v}_s$, so that the vortex moves at the superfluid velocity at its core. An immediate consequence of this fact is the appearance of a perfect Hall effect in a superconductor in the mixed state. This can be shown as follows. If a transport current $\mathbf{j}_s$ is present, then the vortices move at the velocity

$$\mathbf{v}_v = \mathbf{v}_s = \frac{\mathbf{j}_s}{ne} \qquad (8.112)$$

For a square sample of size L, a single vortex induces an electromotive force of magnitude $v_v \phi_0 / Lc$ upon traversing the sample. This leads to an electric field, $v_v \phi_0 / L^2 c$, directed perpendicular to $\mathbf{j}_s$. Since ϕ_0 / L^2 is just H, the vector of this induced field can be written

$$\mathbf{E} = -\frac{1}{c}\mathbf{v}_v \times \mathbf{H} = -\frac{1}{nec}\mathbf{j}_s \times \mathbf{H} \tag{8.113}$$

where we have used Eq. (8.112). Using the standard definition of the Hall resistance R_H, Eq. (8.113) implies

$$R_H = -\frac{1}{nec} \tag{8.114}$$

We thus see that a homogeneous superconductor exhibits a perfect Hall effect. Moreover, if the charge carriers are of the same sign in the super-conducting and normal metal, the sign of the electric field (8.113) is the same as that obtained from the Hall voltage measured in the normal metal. Most experimental data show a Hall resistance much smaller than expected on the basis of Eq. (8.114). From the point of view of vortex dynamics, of special interest are measurements yielding a Hall voltage of opposite sign to that in the normal metal. Hagen et al. (1990) have observed an unexpected sign reversal in epitaxial $YBa_2Cu_3O_7$ thin films at temperatures just below the superconducting transition temperature. They have proposed an explanation based on the equation of motion (8.104), in which one uses for $\mathbf{f}$ a general drag force of the form (Hall and Vinen, 1956)

$$\mathbf{f} = -\eta \mathbf{v}_v - \eta' \mathbf{z} \times \mathbf{v}_v \tag{8.115}$$

If the phenomenological damping constant η' satisfies the condition $\eta' > ne/c$, then Eq. (8.104) yields a vortex velocity of opposite sign than $\mathbf{v}_s$. In conjunction with Eq. (8.113), this then implies a Hall voltage of anomalous sign.

Fisher (1991) has recently discussed the Hall effect for disordered superconducting films exhibiting a superconductor–insulator transition induced by a transverse magnetic field. In a previous work, Fisher et al. (1990) have modeled such films by repulsively interacting bosons moving in a random potential (see Section 6.7). This model has been shown to exhibit not only a universal resistance at the transition ($T = 0$), but also a universal Hall resistance of a magnitude comparable to the longitudinal resistance. Recent experiments by Hebard and Paalanen (1990) have, however, shown a very small Hall effect, with a Hall resistance three orders of magnitude below the value of the longitudinal resistance. The data also appear to lack the predicted universality. The near vanishing of the Hall resistivity indicates that the vortices move essentially perpendicular to the transport current $\mathbf{j}_s$. Such behavior is expected if the equation of motion for the vortex is of the form

$$-\eta \mathbf{v}_v = qh\rho_0 (\mathbf{z} \times \mathbf{v}_s) \tag{8.116}$$

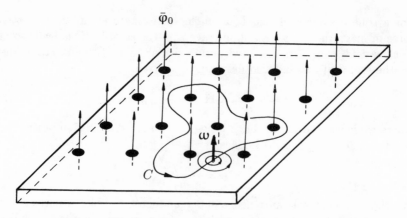

FIG. 8.2. Adiabatic transport of a vortex, characterized by vorticity ω, around contour C in the plane of a 2D lattice boson model. Each boson is shown to carry a unit of fictitious magnetic flux $\bar{\varphi}_0 = hc$ [see Eq. (8.90)].

Recalling Eqs. (8.102) and (8.103), Eq. (8.116) corresponds to taking the pseudo-Lorentz force to be exactly equal to zero. As mentioned in Section 8.5.2, it is the Galilean invariance of the homogeneous media that is essential for the existence of a fully developed pseudo-Lorentz force capable of exactly balancing the Magnus force. If the translational invariance is broken, then this simple balance disappears. Fisher (1991) proposed a following microscopic argument to show that the pseudo-Lorentz force will vanish in a periodic lattice. He considers bosons hopping on a square lattice and evaluates the Berry phase under the assumption that the boson density is exactly *one* boson per site. Without repeating his derivation (which uses a path-integral formulation) we may directly recall our Eq. (8.74), which shows that when a vortex moves through such lattice, enclosing a loop, it picks up a phase factor $\Delta\gamma_0 = -2\pi q n_B$, where n_B is an integer number of sites enclosed. Since $e^{i\Delta\gamma_0} = 1$, there is really no change in the wave function, so that one may argue that the vortices move in a zero fictitious magnetic field. An alternative way to view the same physics is to recall Eq. (8.89), which shows that every boson carries one unit of localized fictitious magnetic flux. Since the bosons reside most of the time at the sites, the vortex manages to avoid the flux lines as it moves between the sites (see Fig. 8.2). Consequently, a nearly zero pseudo-Lorentz force ensues for strong boson localization. On the other hand, a delocalization of boson charge may lead to an increase of this force. In the context of Josephson-junction arrays, this point of view has been advanced recently by Fazio et al. (1992). Clearly, if there is less than one boson per site, then the phase $\Delta\gamma_0$ will be less than 2π times an integer and a nonzero pseudo-Lorentz force $\mathbf{F}_L$ is again expected. Fisher (1991) argues that the sign of $\mathbf{F}_L$ in this case is opposite to that of Eq. (8.92). This could be understood by imagining that it is the boson *holes* carrying opposite flux that contribute in a partially occupied site. If the fraction of holes per site is α, then the vortex equation of motion is obtained by augmenting the LHS of Eq. (8.116) by a pseudo-Lorentz force $\mathbf{F}_L = -\alpha q h \rho_0 (\mathbf{z} \times \mathbf{v}_v)$, yielding

$$- \eta \mathbf{v}_v - \alpha q h \rho_0 (\mathbf{z} \times \mathbf{v}_v) = q h \rho_0 (\mathbf{z} \times \mathbf{v}_s) \qquad (8.117)$$

Assuming $\mathbf{v}_s \parallel \mathbf{x}$, the component $v_{v,x}$ can be obtained from (8.117) in the form

$$v_{v,x} = -\frac{\alpha (q h \rho_0)^2}{\eta^2 + (q h \rho_0)^2} v_{s,x} \qquad (8.118)$$

showing that the vortex velocity has an upstream component, leading to a Hall resistivity of a sign that is opposite to that of Eq. (8.114). In this context, it is useful to call attention to Eq. (8.76), showing a similar sign reversal of the Berry phase corresponding to the deficit of superfluid particles. On the other hand, an excess average number of bosons (over the value of one per site) would lead to a pseudo-Lorentz force $\alpha q h \rho_0 (\mathbf{z} \times \mathbf{v}_v)$. If this expression is used in the equation of motion, then a vortex velocity is obtained that is of the same sign as $v_{s,x}$, so that the Hall resistivity now has the same sign as that in Eq. (8.114). These considerations suggest that a system exhibiting *particle–hole symmetry* should have a *zero Hall resistance*. This can be confirmed by using general symmetry arguments (Fisher, 1991). These ideas can be applied to granular superconductors as well. For instance, the Hamiltonian of the off-diagonal charging model (3.91) is invariant under the transformation $\widehat{n}_i \to -\widehat{n}_i$, $\theta_i \to -\theta_i$. In the presence of an external magnetic field $\mathbf{H}$, this transformation must be accompanied by $\mathbf{H} \to -\mathbf{H}$ [see Eq. (2.7)], and the general argument of Fisher (1991) is applicable. We may thus conclude that in granular superconductors the Hall resistance should vanish. The same conclusion can be reached from the analysis of the Berry phase. Since an integer number of Cooper pairs is assumed to reside at each grain, the vortex picks up a phase -2π times an integer similar to the boson model given, with an average occupation of one per site.

8.6. Massive vortices in the TDGL model

First estimates of the inertial mass of flux lines in superconductors were made by Suhl (1965). Starting from the TDGL model [see our Eq. (2.7)], he has identified two main contributions to the vortex mass:

1. The core kinetic energy, caused by the spatial variation of $|\psi|$ near the vortex center. For a moving vortex, this variation contributes to the kinetic energy in the Lagrangian by a term proportional to $(\partial |\psi| / \partial t)^2$ [see Eq. (2.24)]. Thus, a core contribu-

tion to the inertial mass is obtained that, for a vortex line of unit length, is of the form (Suhl, 1965)

$$\mu_{\text{core}} = \frac{\pi \hbar^2 \psi_\infty^2}{4 m v_F^2} \qquad (8.119)$$

where ψ_∞ is the magnitude of the condensate wave function in a bulk superconductor. In ordinary (low-T_c) superconductors, μ_{core} is of order 10^3 electron masses per cm of the line. Estimates made for high-T_c materials show a remarkably large value of μ_{core} (10^6 times larger than that in low-T_c superconductors).

2. The electromagnetic mass originating from the energy of the electric field induced by the moving vortex. It is given by the expression (Suhl, 1965)

$$\mu_{\text{em}} = \frac{1}{3} \kappa^2 \left(\frac{v_F}{c} \right)^2 \mu_{\text{core}} \qquad (8.120)$$

where $\kappa = \lambda_L / \xi$ (λ_L is the London penetration depth).

Owing to the smallness of the factor $(v_F/c)^2$, the ratio $\mu_{\text{em}}/\mu_{\text{core}}$ in bulk superconductors is usually much smaller than one. On the other hand, in granular superconductors, the electromagnetic contribution to vortex mass is very important (Šimánek, 1983). This is partly because the core contribution cannot be identified unless $|\psi|$ exhibits significant spatial variations near the vortex core. Since most theoretical models of granular arrays neglect these variations, the usual assumption made is that $\mu_{\text{core}} \simeq 0$. The derivation of the electromagnetic mass in bulk superconductors and granular arrays involves a subtle but significant step that deserves further analysis. It is related to the competition between the charging energy and the energy of the δ effect, discussed at some length in Sections 5.5 and 6.2. There we saw that disregarding the δ effect is tantamount to assuming that the time derivative of the phase variable θ is *pinned* to the local potential V, thus satisfying the Josephson relation

$$\hbar \dot{\theta}(\mathbf{r}, \tau) = -2eV(\mathbf{r}, \tau) \qquad (8.121)$$

Using this relation in Eq. (2.42), we obtain a condition that the local *charge density vanishes*. We note that the same condition is implicit in the derivation of μ_{em} by Suhl (1965). By assuming a perfect electrostatic screening, he sets the charge density on the vortex line equal to zero, thus forcing a perfect pinning of the local phase variable to the potential. In the context of granular arrays, we have shown that the Josephson relation (8.121) may be violated under certain conditions. Specifically, if the energy associated with the δ effect is comparable with the charging energy, the effective action is modified compared to the form resulting from assuming Eq. (8.121) at the outset. In Sections 5.5 and 6.2 we have derived this modified action by integrating out the stochastic electrostatic potentials $V_i(t)$. We note that the vortex mass associated with phase fluctuations will depend on the precise form of the functional for the action. Thus, it is meaningful to ask:

To what extent is the expression for vortex mass affected by a similar violation of relation (8.121) in bulk superconductors?

In what follows, we present a systematic analysis of the contributions to the vortex mass, based on the Lagrangian density (2.7).

8.6.1. Vortex in a neutral superfluid

We first consider a hypothetical TDGL model describing Cooper pairs formed by neutral fermions at $T = 0$. The corresponding Lagrangian density, given by Eq. (2.27), contributes to the vortex kinetic energy by

$$E_K = E_K^{(1)} + E_K^{(2)} = \frac{\gamma}{2m^*}\left(\frac{\hbar}{c}\right)^2 \int d^3r \left[\left(\frac{\partial\theta}{\partial t}\right)^2 |\psi|^2 + \left(\frac{\partial|\psi|}{\partial t}\right)^2\right] \quad (8.122)$$

where we have put, following Eq. (2.23), $n_0 = |\psi|^2$. For a vortex line, exhibiting space–time fluctuations, we assume

$$\psi(\mathbf{r}, t) = |\psi(\mathbf{r} - \mathbf{u})| \exp[i\theta(\mathbf{r} - \mathbf{u})] \quad (8.123)$$

where $\mathbf{u} = \mathbf{u}(z, t)$ is the two-dimensional displacement field, describing an instantaneous trajectory of the vortex line, and $\mathbf{r}$ is the radius vector in the xy plane. The phase variable is given by [see Eq. (7.54)]

$$\theta(\mathbf{r}) = \arctan\frac{y}{x} \quad (8.124)$$

Following Suhl (1965), the $\mathbf{r}$ dependence of $|\psi(\mathbf{r})|$ is approximated by

$$|\psi(\mathbf{r})| = \psi_\infty f(r) \quad (8.125)$$

where ψ_∞ is the magnitude of the order parameter in the bulk and the function $f(r)$ is given by

$$f(r) = \begin{cases} r/\xi, & \text{for } r < \xi \\ 1, & \text{for } r > \xi \end{cases} \quad (8.126)$$

8.6.2. Vortex mass due to θ fluctuations

Let us first evaluate the contribution to the kinetic energy, stemming from the phase fluctuations. From Eq. (8.124), we have

$$\frac{\partial\theta}{\partial t} = -\left(\dot{u}_x\frac{\partial\theta}{\partial x} + \dot{u}_y\frac{\partial\theta}{\partial y}\right) = \frac{1}{x^2 + y^2}(\dot{u}_x y - \dot{u}_y x) \quad (8.127)$$

Introducing Eqs. (8.125)–(8.127) into the first term on the RHS of Eq. (8.122), we have for a circular slab of radius R and width L

$$
\begin{aligned}
E_K(1) &= \frac{\hbar^2 \gamma \psi_\infty^2}{2m^* c^2} \int_{-L/2}^{L/2} dz \int_0^R dr \, \frac{f^2(r)}{r} \int_0^{2\pi} d\theta \left(\dot{u}_x^2 \sin^2 \theta + \dot{u}_y^2 \cos^2 \theta \right) \\
&= \frac{\pi \hbar^2 \gamma \psi_\infty^2}{2m^* c^2} \left[\frac{1}{2} + \ln(R/\xi) \right] \int_{-L/2}^{L/2} dz \left(\frac{\partial u(z,t)}{\partial t} \right)^2
\end{aligned}
\tag{8.128}
$$

Now we introduce the inertial mass μ_θ per unit length of the vortex line, by writing the kinetic energy as

$$
E_K^{(1)} = \frac{1}{2} \int_{-L/2}^{L/2} dz \, \mu_\theta \left(\frac{\partial u}{\partial t} \right)^2
\tag{8.129}
$$

Comparing this expression with the RHS of Eq. (8.128), we obtain for the θ contribution to the vortex mass

$$
\mu_\theta \simeq \frac{\pi \hbar^2 \gamma \psi_\infty^2}{m^* c^2} \left[\ln(R/\xi) + \frac{1}{2} \right]
\tag{8.130}
$$

8.6.3. Lorentz invariance and energy–mass relation

The expression (8.130) is reminiscent of the energy of a single vortex, derived in Eq. (7.69). Since the TDGL model is invariant under Lorentz transformation, it is meaningful to ask if the vortex mass satisfies the energy–mass relation

$$
\varepsilon_\theta = \mu_\theta c_{\text{eff}}^2
\tag{8.131}
$$

where ε_θ is the vortex rest energy (per unit length) due to the spatial variations of the phase variable and the effective speed c_{eff} is given by [see Eq. (2.7)]

$$
c_{\text{eff}}^2 = \frac{v_F^2}{3}
\tag{8.132}
$$

To verify this relation, we evaluate with the use of Eqs. (2.27), (8.5), (8.125), and (8.126) the potential energy of the stationary vortex associated with spatial variations of θ

$$
\begin{aligned}
E_v^{(1)} &= \frac{\hbar^2}{2m^*} \int_{-L/2}^{L/2} dz \int d^2 r \, (\boldsymbol{\nabla}\theta)^2 |\psi(\mathbf{r})|^2 = \frac{\pi \hbar^2 \psi_\infty^2 L}{m^*} \int_0^R dr \, \frac{f^2(r)}{r} \\
&= \frac{\pi \hbar^2 \psi_\infty^2 L}{m^*} \left[\frac{1}{2} + \ln(R/\xi) \right] = \varepsilon_\theta L
\end{aligned}
\tag{8.133}
$$

From Eqs. (8.130) and (8.133), we see that the relation (8.131) is indeed satisfied.

8.6.4. *Derivation of the core contribution to vortex mass*

The core contribution μ_{core}, first derived by Suhl (1965), follows from the second term on the RHS of Eq. (8.122). Equation (8.125) implies

$$\frac{\partial |\psi|}{\partial t} = -\left(\dot{u}_x \frac{\partial |\psi|}{\partial x} + \dot{u}_y \frac{\partial |\psi|}{\partial y} \right) = -\psi_\infty \left(\dot{u}_x \frac{x}{r} + \dot{u}_y \frac{y}{r} \right) \frac{\partial f}{\partial r} \qquad (8.134)$$

Using this expression and Eq. (8.126), we have from Eq. (8.122)

$$
\begin{aligned}
E_K^{(2)} &= \frac{\gamma \psi_\infty^2}{2m^*} \left(\frac{\hbar}{c} \right)^2 \int_{-L/2}^{L/2} dz \int_0^R dr \left(\frac{\partial f}{\partial r} \right)^2 r \int_0^{2\pi} d\theta \left(\dot{u}_x^2 \cos^2 \theta + \dot{u}_y^2 \sin^2 \theta \right) \\
&= \frac{\pi \gamma \psi_\infty^2 \hbar^2}{4m^* c^2} \int_{-L/2}^{L/2} dz \left(\frac{\partial \mathbf{u}(z,t)}{\partial t} \right)^2
\end{aligned}
\qquad (8.135)
$$

By writing $E_K^{(2)}$ in a form similar to Eq. (8.129), the RHS of Eq. (8.135) yields the following formula for the core mass

$$\bar{\mu}_{\text{core}} = \frac{\pi \hbar^2 \gamma \psi_\infty^2}{2m^* c^2} = \frac{3\pi \hbar^2 \psi_\infty^2}{4m v_F^2} \qquad (8.136)$$

where we have used the expression $\gamma = 3c^2/v_F^2$. The RHS of Eq. (8.136) differs from the expression (8.119) (quoted by Suhl, 1965) by a factor of 3. This difference can be traced to the form of the Lagrangian density used to derive this result [it misses the factor of 3 appearing in the first term of Eq. (7.7)]. Associated with the spatial variation of $|\psi(\mathbf{r})|$ there is a potential energy that, for a stationary vortex, is given by [see Eq. (2.27)]

$$E_v^{(2)} = \frac{\hbar^2}{2m^*} \int_{-L/2}^{L/2} dz \int d^2r \left[\boldsymbol{\nabla} |\psi(\mathbf{r})| \right]^2 = \varepsilon_{\text{core}} L \qquad (8.137)$$

We now show that μ_{core} and $\varepsilon_{\text{core}}$ also satisfy the energy–mass relation

$$\varepsilon_{\text{core}} = \bar{\mu}_{\text{core}} c_{\text{eff}}^2 = \frac{1}{3} \bar{\mu}_{\text{core}} v_F^2 \qquad (8.138)$$

To verify this relation, we calculate, using Eqs. (8.125) and (8.126), the integral

$$\int d^2r \left[\boldsymbol{\nabla} |\psi(\mathbf{r})| \right]^2 = 2\pi \psi_\infty^2 \int_0^R dr\, r \left(\frac{df(r)}{dr} \right)^2 = \pi \psi_\infty^2 \qquad (8.139)$$

Introducing this result into Eq. (8.137), we obtain for the vortex–core energy per unit length of a vortex line the expression

$$\varepsilon_{\text{core}} = \frac{\pi \psi_\infty^2 \hbar^2}{2m^*} \qquad (8.140)$$

This result and Eq. (8.136) show that the energy–mass relation (8.138) is satisfied.

Hence, the total inertial mass of a vortex line in a neutral superfluid, satisfying a wavelike TDGL equation, is, according to Eqs. (8.130) and (8.136)

$$\mu_{\text{neutral}} = \mu_\theta + \mu_{\text{core}} = \frac{\pi\hbar^2\gamma\psi_\infty^2}{m^*c^2}[\ln(R/\xi) + 1] \qquad (8.141)$$

One unusual aspect of this result is that the core contribution is *exceeded* by the mass, generated by *phase fluctuations*. This is due to the wavelike dynamics of the TDGL model, which allows phase fluctuations to exist even in the absence of charging energy. In the context of granular superconductivity, this limit corresponds to a dominance of the δ effect (see Section 6.2). As pointed out in Chapter 2, the Lagrangian (2.7) appears to be confirmed microscopically, as far as the phase variable is concerned. Thus, we expect that the θ contribution to the vortex mass is correctly represented by Eq. (8.130). On the other hand, the TDGL equation, derived microscopically at $T = 0$ by Abrahams and Tsuneto (1966) for the magnitude of the order parameter, is somewhat different from the equation of motion based on Lagrangian (2.7). Duan and Leggett (1992) have pointed out that, in a microscopically correct Lagrangian, the energy term corresponding to the (space or time) variation of $|\psi|$ is only 1/3 times the value given in Eq. (2.7). This implies that the core mass and core energy are just 1/3 of the values predicted by Eqs. (8.136) and (8.140), respectively. Incidentally, this then leads to a core vortex mass given by the Eq. (8.119). Duan and Leggett (1992) have also found that the velocity of the phase fluctuations c_{eff}, involved in the energy–mass relation (8.131), can be considerably enhanced over the value $v_F/\sqrt{3}$ by the *interactions* between the quasiparticles. In the low-frequency, collision-dominated regime, c_{eff} becomes equal to the velocity of the first sound. The velocity of the amplitude variations remains unaffected by the Fermi liquid effects, since it is not associated with density-induced polarization fluctuations (Leggett, 1975).

8.6.5.   *Vortex in a TDGL model of a superconductor*

To study the vortex dynamics in a superconductor, one must start from the full expression (2.7), which takes into account the coupling of the supercurrents to the gauge fields. It turns out that the core contribution to the vortex mass is not affected in any essential way by this coupling, and thus we focus our attention on the electromagnetic mass and its modifications due to possible deviations from the Josephson relation (8.121). The method we follow here is an extension of that introduced in Section 6.2. We start from the grand partition function for the superconductor in the presence of gauge fields

$$Z_G \rightarrow \int \mathcal{D}^2\psi(\mathbf{r}, \tau) \int \mathcal{D}V(\mathbf{r}, \tau) \int \mathcal{D}\mathbf{A}(\mathbf{r}, \tau) \exp\left(-\frac{1}{\hbar}S[\psi, V, \mathbf{A}]\right) \quad (8.142)$$

where the Euclidean action S is written in terms of the Lagrangian density (2.7)

$$S[\psi, V, \mathbf{A}] = \int d^3r \int d\tau (\mathcal{L}_m[\psi, V, \mathbf{A}] + \mathcal{L}_{em}[V, \mathbf{A}]) \tag{8.143}$$

We note that Eq. (8.142) could be derived from first principles, by performing a Stratonovich transformation in a similar way as Eqs. (6.12) and (6.13) have been obtained from the Hamiltonian (6.2). By using directly the Lagrangian density (2.7) in the expression (8.143), we are bypassing the lengthy procedure of averaging over the fermionic variables. In fact, this procedure is already contained in the derivation of (2.7) (Abrahams and Tsuneto, 1966). Our goal now is to express the partition function (8.142) as a path integral over the phase variable alone. The resulting effective action is then a functional of the $\dot{\theta}(\mathbf{r}, \tau)$ variable, from which the contribution to the vortex inertial mass can be obtained. According to Eq. (2.24), the matter Lagrangian density $\mathcal{L}_m$ (in the imaginary-time formulation) is taken in the form

$$\mathcal{L}_m[\psi, V\mathbf{A}] = \frac{\gamma}{2m^*}\left(\frac{\hbar}{c}\dot{\theta} + \frac{e^*}{c}V\right)^2 |\psi|^2 + \bar{\mathcal{L}}_m \tag{8.144}$$

where $\bar{\mathcal{L}}_m$ contains all the remaining terms in (2.24) that do not involve $\dot{\theta}$. We note that $V(\mathbf{r}, \tau)$ is now a random Stratonovich potential of a similar nature to that appearing in Eq. (5.169). The electromagnetic Lagrangian density is approximated by

$$\mathcal{L}_{em}[V, \mathbf{A}] \simeq \frac{1}{8\pi}[\boldsymbol{\nabla}V(\mathbf{r}, \tau)]^2 \tag{8.145}$$

This corresponds to a total disregard of the magnetic vector potential $\mathbf{A}(\mathbf{r}, \tau)$. One might wonder how can this approximation be justified for a moving vortex, when $\partial \mathbf{A}/\partial \tau$ is induced, contributing to the electric field. As will be shown, the electromagnetic energy is dominated by electric fields whose wavelength is of the order of the vortex–core size [see Eq. (8.157)]. In what follows we will be assuming that the London penetration depth λ_L is much greater than the core size. In such case, Eqs. (7.124) and (7.131) show that the vector potential induced by the vortex is much less than $(\hbar c/e^*)\boldsymbol{\nabla}\theta$ and thus it can be eliminated from the dynamics. We note that a similar argument is used to justify the neglect of $\mathbf{A}$ in the effective action for granular superconductors. In view of Eqs. (8.144) and (8.145), the part of Eq. (8.142) contributing to the inertial mass of a vortex is the path integral

$$Z[\theta] \rightarrow \int \mathcal{D}V(\mathbf{r}, \tau) \exp\left\{-\frac{1}{\hbar}\int d^3r \int d\tau\right.$$

$$\left. \times \left[\frac{\gamma}{2m^*}\left(\frac{\hbar}{c}\dot{\theta} + \frac{e^*}{c}V\right)^2 |\psi|^2 + \frac{1}{8\pi}(\boldsymbol{\nabla}V)^2\right]\right\}$$

$$= \int \mathcal{D}V(\mathbf{r}, \tau) \exp\left(-\frac{1}{\hbar} S[\theta, V]\right) \tag{8.146}$$

An additional approximation, made in (8.146), is that fluctuations of $|\psi|$ are neglected. If the expression (8.125) is inserted for $|\psi|^2$, the integrand in the exponent of Eq. (8.140) loses its translational invariance, causing complications in the evaluation of the path integral over $V(\mathbf{r}, \tau)$. For this reason, we let $|\psi|^2 = \psi_\infty^2$, and make a correction of this in the final integral for vortex energy by introducing a lower-cutoff length of the order of the core size [see Eq. (8.157)]. To perform the path integral (8.146), we use the method of Fourier transforms (see Section 5.5.1). Thus we define

$$V(\mathbf{r}, \tau) = \sum_{\mathbf{k}, n} v(\mathbf{k}, n) e^{i(\mathbf{k} \cdot \mathbf{r} - \omega_n \tau)} \tag{8.147}$$

$$\theta(\mathbf{r}, \tau) = \sum_{\mathbf{k}, n} \theta(\mathbf{k}, n) e^{i(\mathbf{k} \cdot \mathbf{r} - \omega_n \tau)} \tag{8.148}$$

Introducing these expansions into Eq. (8.146), the action $S[\theta, V]$ takes the form

$$
\begin{aligned}
S[\theta, V] = {} & \frac{\hbar^2 \psi_\infty^2 \gamma}{2m^* c^2} (\hbar\beta\Omega) \sum_{\mathbf{k}, n} \omega_n^2 |\theta(\mathbf{k}, n)|^2 \\
& + \hbar\beta\Omega \sum_{\mathbf{k}, n} \left(\frac{1}{8\pi} k^2 + \frac{\psi_\infty^2 \gamma e^{*2}}{2m^* c^2}\right) |v(\mathbf{k}, n)|^2 \\
& + i(\hbar\beta\Omega) \frac{\hbar e^* \psi_\infty^2 \gamma}{m^* c^2} \sum_{\mathbf{k}, n} \omega_n \left[\theta^i(\mathbf{k}, n) v^r(\mathbf{k}, n)\right. \\
& \left. - \theta^r(\mathbf{k}, n) v^i(\mathbf{k}, n)\right]
\end{aligned}
\tag{8.149}
$$

where θ^r, θ^i and v^r, v^i are the real and imaginary parts of $\theta(\mathbf{k}, n)$ and $v(\mathbf{k}, n)$, respectively. Using Eqs. (8.147)–(8.149), the path integral (8.146) reduces to a multiple Gaussian integral over the Fourier components $v^r(\mathbf{k}, n)$ and $v^i(\mathbf{k}, n)$ [see Eq. (5.188)]. Defining the effective action $S_{\text{eff}}[\theta]$ by

$$Z[\theta] = \exp\left(-\frac{S_{\text{eff}}[\theta]}{\hbar}\right) \tag{8.150}$$

we thus obtain

$$S_{\text{eff}}[\theta] = \frac{(\hbar\Omega\beta)\hbar^2 \psi_\infty^2 \gamma}{2m^* c^2} \sum_{\mathbf{k}, n} \frac{k^2 \omega_n^2 |\theta(\mathbf{k}, n)|^2}{\left(k^2 + \frac{4\pi \psi_\infty^2 \gamma e^{*2}}{m^* c^2}\right)} \tag{8.151}$$

Starting from this expression, we now derive the electromagnetic mass of the vortex and its modifications due to the deviations from the Josephson relation (8.121).

8.6.6. *Electromagnetic mass of a vortex*

Let us assume that for the important range of k values in the sum (8.151), the following inequality holds

$$\frac{4\pi\psi_\infty^2 \gamma e^{*2}}{m^* c^2} = k_c^2 \gg k^2 \tag{8.152}$$

Then the expression (8.151) simplifies to

$$S_{\text{eff}}^{(0)}[\theta] \simeq (\hbar\beta\Omega)\frac{\hbar^2}{8\pi e^{*2}} \sum_{\mathbf{k},n} k^2 \omega_n^2 |\theta(\mathbf{k},n)|^2$$

$$= \frac{\hbar^2}{8\pi e^{*2}} \int_\Omega d^3 r \int_0^{\hbar\beta} d\tau \left[\boldsymbol{\nabla}\dot\theta(\mathbf{r},\tau)\right]^2 \tag{8.153}$$

where the RHS follows by inverting the Fourier sum to an Euclidean space–time integral. The RHS of Eq. (8.153) corresponds to the following expression for the kinetic energy caused by *real-time* variations of the phase variable

$$E_K = \frac{\hbar^2}{8\pi e^{*2}} \int d^3 r \left[\boldsymbol{\nabla}\dot\theta(\mathbf{r},t)\right]^2 \tag{8.154}$$

Let us now evaluate the integral (8.154) for a moving vortex line, for which the phase variable is given by the expression (8.124). Using Eq. (8.5), we have in cylindrical coordinates

$$\boldsymbol{\nabla}\theta(\mathbf{r}-\mathbf{u}) = q\left(\mathbf{z}\times\frac{\mathbf{r}-\mathbf{u}}{|\mathbf{r}-\mathbf{u}|}\right) = \frac{q}{|\mathbf{r}-\mathbf{u}|}\boldsymbol{\theta}_0(\mathbf{r}-\mathbf{u}) \tag{8.155}$$

where $\boldsymbol{\theta}_0$ is a unit vector pointing in the azimuthal direction. With the use of this result, we calculate

$$\boldsymbol{\nabla}\dot\theta(\mathbf{r}-\mathbf{u}) = -(\dot{\mathbf{u}}\cdot\boldsymbol{\nabla})\boldsymbol{\nabla}\theta(\mathbf{r}-\mathbf{u}) = \frac{q}{|\mathbf{r}-\mathbf{u}|^2}\left(\dot{u}_r\boldsymbol{\theta}_0 + \dot{u}_\theta\frac{(\mathbf{r}-\mathbf{u})}{|\mathbf{r}-\mathbf{u}|}\right) \tag{8.156}$$

Introducing the RHS of this equation into (8.154), we obtain

$$E_K = \frac{\hbar^2}{8\pi e^{*2}} \int_{-L/2}^{L/2} dz \int d^2 r \left(\frac{\partial\mathbf{u}}{\partial t}\right)^2 \frac{1}{|\mathbf{r}-\mathbf{u}|^4}$$

$$\simeq \frac{\hbar^2}{4e^{*2}} \int_{-L/2}^{L/2} dz \left(\frac{\partial\mathbf{u}}{\partial t}\right)^2 \int_\xi^\infty \frac{dr}{r^3}$$

$$= \frac{\hbar^2}{8e^{*2}\xi^2} \int_{-L/2}^{L/2} dz \left(\frac{\partial\mathbf{u}}{\partial t}\right)^2 \tag{8.157}$$

where we have taken advantage of the independence of the $\int d^2 r$ integral of the position of the vortex center $\mathbf{u}$. We also have introduced a lower limit ξ into the radial integral, justified by the fact that the supercurrents (proportional to $\nabla\theta$) must vanish within the normal vortex core of this size. Comparing the RHS of Eq. (8.157) with a mass-defining equation [see Eq. (8.129)], the electromagnetic mass of a unit length of the vortex line is obtained in the form

$$\mu_{em} = \frac{\hbar^2}{4e^{*2}\xi^2} \tag{8.158}$$

Using this result and Eq. (8.136), we calculate the ratio of the electromagnetic to the core mass

$$\frac{\mu_{em}}{\bar{\mu}_{core}} = \frac{2}{3}\left(\frac{v_F}{c}\right)^2 \frac{mc^2}{4\pi ne^2\xi^2} = \frac{2}{3}\left(\frac{\lambda_L}{\xi}\right)^2\left(\frac{v_F}{c}\right)^2 \tag{8.159}$$

where we have used the relation $\psi_\infty^2 = n/2$ and the definition of London penetration depth (7.163).

8.6.7. Mass correction from relaxing the phase-pinning constraint

First, let us point out that the electromagnetic mass (8.158) can be obtained directly from the electric-field energy density (8.145) by invoking the Josephson relation (8.121). This can be seen clearly from the expression (8.154). We now estimate the correction to the vortex mass produced by relaxing the relation (8.121). Expanding the denominator in (8.151) in a small parameter k^2/k_c^2, we have to first order

$$
\begin{aligned}
S_{\text{eff}}[\theta] &\simeq (\hbar\beta\Omega)\frac{\hbar^2}{8\pi e^{*2}}\sum_{\mathbf{k},n}k^2\omega_n^2\left(1-\frac{k^2}{k_c^2}\right)|\theta(\mathbf{k},n)|^2 \\
&= S_{\text{eff}}^{(0)}[\theta] - \frac{\hbar^2}{8\pi e^{*2}k_c^2}\int_\Omega d^3r\int_0^{\hbar\beta}d\tau\left(\nabla^2\dot{\theta}\right)^2 \\
&= S_{\text{eff}}^{(0)}[\theta] - \delta S_{\text{eff}}[\theta]
\end{aligned}
\tag{8.160}
$$

To evaluate the integral on the RHS of this equation, we take the divergence of Eq. (8.156), yielding

$$\nabla^2\dot{\theta} = -\frac{\dot{u}_\theta q}{|\mathbf{r}-\mathbf{u}|^3} \tag{8.161}$$

The correction to the effective action $S_{\text{eff}}^{(0)}[\theta]$ is then obtained from Eqs. (8.160) and (8.161) in the form

$$
\begin{aligned}
\delta S_{\text{eff}}[\theta] &= -\frac{\hbar^2}{8\pi e^{*2}k_c^2}\int_0^{\hbar\beta}d\tau\int_{-L/2}^{L/2}dz\left(\frac{\partial u}{\partial t}\right)^2\int_\xi^\infty\frac{r\,dr}{r^6}\int_0^{2\pi}d\theta\,\sin^2\theta \\
&= -\frac{\hbar^2}{32e^{*2}k_c^2\xi^4}\int_0^{\hbar\beta}d\tau\int_{-L/2}^{L/2}dz\left(\frac{\partial u}{\partial t}\right)^2
\end{aligned}
\tag{8.162}
$$

The RHS of this equation implies a mass correction, which can be written, using Eq. (8.158) as

$$\delta\mu = -\frac{1}{12}\left(\frac{v_F}{c}\right)^2\left(\frac{\lambda_L}{\xi}\right)^2\mu_{em} \qquad (8.163)$$

Interestingly, the factor multiplying μ_{em} on the RHS of this relation has the same form as that appearing on the RHS of Eq. (8.120). As pointed out, this factor is usually well below one in bulk materials. Even in high-T_c superconductors, where large values of order 10^2–10^3 are expected for the ratio λ_L/ξ, the Fermi velocity is so small that the factor $(\kappa v_F/c)^2/3$ is only about 10^{-2}. It should be noted, however, that the ratio $|\delta\mu/\mu_{em}|$ can be possibly enhanced by taking into account the dielectric screening of the local electric fields near the vortex core. The polarization of the electronic, weakly bound states, can contribute to the *short-wavelength* dielectric constant (Landau, 1950). Parenthetically, we remark that in London theory the $q \to 0$, $\omega \to 0$ limit of the dielectric function diverges (Tinkham, 1975). For an orientational assessment of the role of dielectric screening, we start from the following electromagnetic Lagrangian density [see Eq. (8.145)]

$$\mathcal{L}_{em}[V, \mathbf{A}] \simeq \frac{\varepsilon}{8\pi}[\boldsymbol{\nabla} V(\mathbf{r}, \tau)]^2 \qquad (8.164)$$

where ε is an effective dielectric constant, representing the screening on a length scale of order ξ. Using this expression, the effective action (8.151) is replaced by

$$S_{\text{eff}}[\theta] = \frac{(\hbar\Omega\beta)\hbar^2\psi_\infty^2\gamma}{2m^*c^2}\sum_{\mathbf{k},n}\frac{\varepsilon k^2\omega_n^2|\theta(\mathbf{k},n)|^2}{\left(\varepsilon k^2 + \frac{4\pi\psi_\infty^2\gamma e^{*2}}{m^*c^2}\right)} \qquad (8.165)$$

If the wave vector k satisfies the inequality

$$k^2 \ll k_c^2/\varepsilon \qquad (8.166)$$

where k_c is defined in Eq. (8.152), the summand in Eq. (8.165) can be expanded in a similar way as in Eq. (8.160), yielding

$$\delta\mu_\varepsilon = -\frac{\varepsilon}{12}\left(\frac{v_F}{c}\right)^2\left(\frac{\lambda_L}{\xi}\right)^2\mu_{em,\varepsilon} \qquad (8.167)$$

where $\mu_{em,\varepsilon}$ is the electromagnetic mass in the presence of dielectric screening

$$\mu_{em,\varepsilon} = \frac{\hbar^2\varepsilon}{4e^{*2}\xi^2} \qquad (8.168)$$

Equation (8.167) suggests that, dependent on the size of ε, the mass correction $\delta\mu_\varepsilon$ can now be made relatively more significant. For a high-T_c

superconductor, a value of $\varepsilon \simeq 10^2$ would be required to make the ratio $\delta\mu_\varepsilon/\mu_{em,\varepsilon}$ of order one. In this context, it should be noted that a large static dielectric constant has been measured on the nonsuperconducting tetragonal phase of $YBa_2Cu_3O_x$ (Testardi et al., 1988). At $T \simeq 4.3$ K, a value of $\varepsilon \simeq 30$ has been observed in these samples. At this moment, it may be premature to make firm conclusions about the numerical importance of the mass correction $\delta\mu_\varepsilon$. Nevertheless, the theoretical expression (8.165) helps us to answer the following question of principle. Let us consider a hypothetical situation of $\varepsilon \to \infty$ (or $e^* \to 0$). Then, the electromagnetic mass (8.168) goes to infinity and so does the total vortex mass $\mu_{core} + \mu_{em,\varepsilon}$. Since the limit $\varepsilon \to \infty$ corresponds to the *absence* of Coulomb charging energy, the infinite total mass is an absurd result. The resolution of this paradox is contained in the expression (8.165), which in the limit $\varepsilon \to \infty$ goes to

$$S_{\text{eff}}[\theta] \xrightarrow[\varepsilon \to \infty]{} \frac{\gamma}{2m^*}\left(\frac{\hbar}{c}\right)^2 \psi_\infty^2 \int_\Omega d^3r \int_0^{\hbar\beta} d\tau \left(\frac{\partial\theta}{\partial\tau}\right)^2 \qquad (8.169)$$

The expression on the RHS of this equation corresponds to the kinetic energy $E_K^{(1)}$ of the neutral superfluid, defined in Eq. (8.122). The latter equation then leads to the vortex mass (8.130), which is *finite*. The physical reason behind this is that the effective mass resulting from (8.165) behaves in an analogous way as the effective capacitance of the self-charging model in the presence of the δ effects. From Eq. (6.25) we see that the capacitance C_0 (of electrostatic origin) is connected in series with the fictitious capacitance proportional to a_δ. As $C_0 \to \infty$, the effective capacitance of this combination is given by the fictitious capacitance. Similarly, the contribution of the diverging electromagnetic vortex mass is blocked by the kinetic energy term $E_K^{(1)}$ of the neutral model, and it is this term that dominates in the limit of $\varepsilon \to \infty$. In a sense, the total vortex mass behaves like a reduced mass composed of the electromagnetic mass and that due to the kinetic energy term of the neutral model. Though the expression (8.165) may be mostly of academic interest for known bulk superconductors, estimates made in Section 8.7 show that it can be of numerical significance for granular superconductors.

8.7.   Massive vortices in granular superconductors

Since vortex excitations in granular superconductors involve only the phase of the order parameter, the appropriate effective action is that of Eq. (6.23). Moreover, if we confine ourselves to temperatures close to zero and assume a nonzero gap in the quasiparticle spectrum, then we can start directly from the action of Eq. (6.115). In general, the inverse-capacitance matrix $(\bar{C}^{-1})_{ij}$ appearing in this action poses formal complications in further analysis. Equation (6.116) implies that $\bar{C}_{ij}$ is of short range. Then, recalling Section 3.4.3, we see that $(\bar{C}^{-1})_{ij}$ has a range that varies depending on the ratio

$(8a_\delta + C_0)/C$. If this ratio is much less than one, then the range of $(\bar{C}^{-1})_{ij}$ is much greater than the lattice spacing of the array. In the other limit, the inverse capacitance matrix is of short range, implying that all phase-inertial terms in (6.115) involve products $\dot{\theta}_i \dot{\theta}_j$, where i and j are at most separated by one lattice spacing. This simplifies considerably the calculation of the inertial mass of a vortex. In what follows, we consider the extreme limit $C = 0$, in which case $(\bar{C}^{-1})_{ij}$ is a diagonal matrix (see Section 6.7.1). We also confine ourselves to two-dimensional arrays.

8.7.1. Inertial mass of a vortex in an array with vanishing intergrain geometric capacitance

For $C = 0$, Eq. (6.116) reduces to

$$\bar{C}_{ij} = (8a_\delta + C_0)\delta_{ij} \tag{8.170}$$

which implies

$$(\bar{C}^{-1})_{ij} = (8a_\delta + C_0)^{-1}\delta_{ij} \tag{8.171}$$

Inserting Eq. (8.171) into (6.115), the effective action takes the form

$$S_{\text{eff}}[\{\theta_i\}] = \int_0^{\hbar\beta} d\tau \left(\frac{\hbar^2}{8e^2} C_{\text{eff}} \sum_i \dot{\theta}_i^2 + \frac{\hbar^2 \Delta C}{8e^2} \sum_{\langle ij \rangle} \dot{\theta}_{ij}^2 \right.$$

$$\left. + J \sum_{\langle ij \rangle} \cos\theta_{ij} \right) \tag{8.172}$$

where ΔC is given by the expression (6.104) and C_{eff} is an effective capacitance resulting from capacitors C_0 and $C_\delta = 8a_\delta$, connected in series

$$C_{\text{eff}} = \frac{8a_\delta C_0}{8a_\delta + C_0} \tag{8.173}$$

Except for an immediate neighborhood of the vortex center, the phase variable exhibits smooth spatial variations. This allows us to replace Eq. (8.172) by a continuum approximation [see Eq. (7.5)]. Let us first focus our attention on the first term on the RHS of Eq. (8.172), which we denote $S_1[\{\theta_i\}]$. Using Eqs. (5.184) and (8.173), we have for a sample of thickness d

$$S_1[\{\theta_i\}] = \frac{\hbar^2}{8e^2} \frac{C_{\text{eff}}}{\Delta\Omega} \int_0^{\hbar\beta} d\tau \sum_i \dot{\theta}_i^2 \Delta\Omega \rightarrow g \int_0^{\hbar\beta} d\tau \int d^2\rho \, \dot{\theta}^2(\boldsymbol{\rho}, \tau)$$

$$= S_1[\theta] \tag{8.174}$$

where $\Delta\Omega$ is the volume of the primitive cell of the array and the coefficient g is given by

$$g = \frac{\hbar^2 C_{\text{eff}} d}{8e^2 \Delta\Omega} = \frac{\frac{3n\hbar^2 C_0 \Omega_{\text{eff}} d}{16\epsilon_F \Delta\Omega}}{C_0 + \frac{3n\Omega_{\text{eff}} e^2}{2\epsilon_F}} \tag{8.175}$$

where the second equality follows from Eq. (8.173). The effective volume $\Omega_{eff} \simeq S k_F^{-1}$, where S is the surface area of the grain and k_F is the Fermi wave vector [see Eq. (2.103)]. From the discussion presented in Section 6.3 it follows that the contribution of the geometric capacitance C_0 to C_{eff} is diminished by the presence of C_δ. In the extreme limit $\Omega_{eff} \to 0$, the expression on the RHS of (8.175) goes to

$$g \xrightarrow[\Omega_{eff} \to 0]{} \frac{3n\hbar^2 \Omega_{eff} d}{16\epsilon_F \Delta\Omega} \simeq \frac{3n\hbar^2 d}{16\epsilon_F} \qquad (8.176)$$

where we assume $\Omega_{eff}/\Delta\Omega \to 1$ and $C_0/\Omega_{eff} \to \infty$ as $\Omega_{eff} \to 0$. This assumption is justified for arrays where C_0 scales with the *linear* dimension of the superconducting regions. Introducing the RHS of Eq. (8.176) into (8.174), we obtain an effective action $S_1[\theta]$ corresponding to a kinetic energy $E_K^{(1)}$ of the neutral superfluid [see Eq. (8.122)].

Next, we consider the second term on the RHS of Eq. (8.172), which we denote as $S_2[\{\theta_i\}]$. In the continuum approximation, this term becomes

$$S_2[\{\theta_i\}] = \frac{\hbar^2}{8e^2} \Delta C \int_0^{\hbar\beta} d\tau \sum_{\langle ij \rangle} \dot{\theta}_{ij}^2 \to \frac{\hbar^2 \Delta C}{8e^2} \int_0^{\hbar\beta} d\tau \int d^2\rho \, (\nabla \dot{\theta})^2$$

$$= S_2[\theta] \qquad (8.177)$$

We notice immediately that the RHS of this equation has the form of the electromagnetic action of a bulk superconductor (8.153). It is interesting to compare (for a film of thickness d) the coefficient of the latter expression, $\hbar^2 d/8\pi e^{*2}$, with that on the RHS of Eq. (8.177). Using Eq. (6.104), we thus evaluate the ratio

$$r = \frac{\frac{\hbar^2 \Delta C}{8e^2}}{\frac{\hbar^2 d}{8\pi e^{*2}}} = \frac{3\pi^2 \hbar}{8\Delta_0 R_N d} = \frac{3\pi^2}{8}\left(\frac{R_0}{R_N}\right)\left(\frac{e^2}{d\Delta_0}\right) \qquad (8.178)$$

where $R_0 = \hbar/e^2$. For a 100 Å thick aluminum film, the ratio $e^2/d\Delta_0 \simeq 8.4 \times 10^2$. Taking $R_0/R_N \simeq 1$, the ratio (8.178) becomes $r \simeq 3.1 \times 10^3$. This result implies that the inertial mass of a vortex in granular superconductors receives a contribution from the intergrain capacitance, which is large compared to the electromagnetic vortex mass in bulk materials.

To extract from Eqs. (8.174) and (8.177) the expression for vortex kinetic energy, we introduce the ansatz

$$\theta(\boldsymbol{\rho}, \tau) = \theta(\boldsymbol{\rho} - \mathbf{u}(\tau)) \qquad (8.179)$$

where $\mathbf{u}(\tau)$ is a two-dimensional vector of the vortex center and $\boldsymbol{\rho}$ is a radius vector in the xy plane. The phase variable θ has the same form as that previously used for continuous superfluids [see Eq. (8.124)]. Using Eqs. (8.174) and (8.175), we see that $S_1[\theta] + S_2[\theta]$ generates, in the continuum approximation, a kinetic energy

$$E_K = g \int d^2\rho\, \dot{\theta}^2 + \frac{\hbar^2 \Delta C}{8e^2} \int d^2\rho\, (\boldsymbol{\nabla}\dot{\theta})^2 \qquad (8.180)$$

where the dot denotes a real-time derivative. Inserting the ansatz (8.179), we obtain from Eq. (8.180) [see Eqs. (8.127) and (8.157)]

$$E_K = g \int_{l_c}^{R} \frac{d\rho}{\rho} \int_{0}^{2\pi} d\theta\, (\dot{u}_x^2 \sin^2\theta + \dot{u}_y^2 \cos^2\theta) + \frac{\pi\hbar^2 \Delta C}{4e^2} \int_{l_c}^{R} \frac{d\rho}{\rho^3} (\dot{\mathbf{u}})^2 \quad (8.181)$$

where a lower cutoff has been imposed on the ρ integrals, assuming a finite core of size l_c. Performing the integrations, the vortex kinetic energy follows from Eq. (8.181) in the form

$$E_K = \frac{1}{2}(m + M)\left(\frac{\partial \mathbf{u}}{\partial t}\right)^2 \qquad (8.182)$$

where

$$m = 2\pi g \ln(R/l_c) \qquad (8.183)$$

and

$$M = \frac{\pi\hbar^2 \Delta C}{4e^2 l_c^2} \qquad (8.184)$$

In our previous work, concerned with estimates of M, we have used for the lower cutoff the value $l_c \simeq a$, where a is the lattice spacing (Šimánek, 1983). This is consistent with Eq. (8.177), which assumes that each grain is associated with an order parameter of fixed magnitude and a smoothly varying phase (without underlying quantum fluctuations). We note that a correlation length ξ' of order a also follows from Eq. (8.177) when the parameter $\alpha = zJ/QU_{11}$ is much greater than 1. However, as this equation shows, ξ' increases as the parameter α approaches the critical value $\alpha_c = 1$. Hence, it seems more accurate to set $l_c = \xi'$, which depends on α. Calculations taking into account the discreteness of the granular array show a position-dependent vortex mass. For example, in a square planar array, the mass M exhibits about 20% spatial modulation (Šimánek, 1983).

One may object that the results (8.182)–(8.184) are based on ansatz (8.179) for θ, which is too simple since it disregards the spin-wave fluctuations about the vortex configuration [see Eq. (7.185)]. Let us examine this first in the limit $C_{\text{eff}} = 0$. Then the continuum version of the effective action (8.172) can be written, with use of Eq. (8.177), as

$$S_{\text{eff}}^{(1)}[\theta] = \frac{\hbar^2 \Delta C}{8e^2} \int_{0}^{\hbar\beta} d\tau \int d^2\rho \left[\left(\boldsymbol{\nabla}\dot{\theta}\right)^2 + \omega_p^2 (\boldsymbol{\nabla}\theta)^2\right] \qquad (8.185)$$

where

$$\omega_p^2 = \frac{4Je^2}{\hbar^2 \Delta C} \qquad (8.186)$$

The expression (8.186) defines the Josephson plasma frequency of the junction, formed by a pair of nearest-neighbor grains in the array. From the effective action (8.185), we see that (for θ corresponding to small oscillations) the spectrum is that of an *optical* spin-wave mode. We thus expect that, as long as the vortex excitations involve frequencies well below ω_p, the spin-wave fluctuations would have little effect on the effective vortex action (Eckern and Schmid, 1989). In this case, the ansatz (8.179) seems to capture the essential physics rather well.

Unfortunately, there seems to be a problem with this argument when the parameter α is close to the critical value $\alpha_c = 1$. In this case, the constant J in Eq. (8.172) should be replaced by its renormalized value J^{ren}. This value has been derived, in the mean-field approximation, by Maekawa et al. (1981). It is given in Eq. (7.228), which shows that J^{ren} drops to zero as $\alpha \to \alpha_c$. Thus, in the vicinity of the critical point, the plasma mode goes *soft*, and it may not be possible to isolate the spin-wave fluctuations from the vortex excitation any more.

Next, let us consider the case of $\Delta C = 0$, in the limit $\Omega_{\text{eff}} \to 0$. That implies an effective action [see Eq. (8.177)]

$$S_{\text{eff}}^{(2)}[\theta] = g \int_0^{\hbar\beta} d\tau \int d^2\rho \left(\dot{\theta}^2 + \frac{v_F^2}{3}(\boldsymbol{\nabla}\theta)^2 \right) \tag{8.187}$$

To obtain the second term on the RHS of this expression, we have used the relations (7.160) and (7.162). We see that the action (8.187) corresponds to *acoustic modes* propagating with velocity $\bar{c} = v_F/\sqrt{3}$. In the other limit of $C_0 \ll 3n\Omega_{\text{eff}}e^2/2\epsilon_F$, the acoustic phasons propagate at a velocity $\bar{c} \simeq a\sqrt{4Je^2/\hbar^2 C_0}$, where a is the lattice spacing of the array. Now it is clearly impossible to separate the vortex excitations from the spin waves. In fact, the kinetic energy associated with vortex translations can be converted into acoustic spin waves, acting as a mechanism for vortex dissipation (Eckern and Schmid, 1989). This is discussed next, using the elegant formulation of the effective action by Popov (1983).

8.7.2. *Effective action for vortex (magnetic analogy)*

Following Popov (1983), we define a (2+1)-dimensional vector $\widetilde{\mathbf{x}} = (x, y, \bar{c}\tau)$, where $\bar{c}$ is the speed of the acoustic phason, and the corresponding gradient is $\widetilde{\boldsymbol{\nabla}} = (\partial/\partial x, \partial/\partial y, \partial/\bar{c}\,\partial\tau)$. Then the action $S_{\text{eff}}^{(2)}[\theta]$ of Eq. (8.187) can be written as

$$\begin{aligned} S_{\text{eff}}^{(2)}[\theta] &= \frac{J}{2\bar{c}} \int d\tau\,\bar{c} \int d^2r \left[(\boldsymbol{\nabla}\theta)^2 + \frac{1}{\bar{c}^2}\left(\frac{\partial\theta}{\partial\tau}\right)^2 \right] \\ &= \frac{J}{2\bar{c}} \int d^3x \left(\widetilde{\boldsymbol{\nabla}}\theta \right)^2 = \frac{J}{2\bar{c}} \int d^3x \left(\widetilde{\mathbf{h}} \right)^2 \end{aligned} \tag{8.188}$$

where we have introduced a fictitious auxiliary magnetic field $\widetilde{\mathbf{h}}$

$$\widetilde{\mathbf{h}} = \widetilde{\boldsymbol{\nabla}}\theta \tag{8.189}$$

In Eq. (8.188), the two-dimensional vector in the xy plane is denoted by $\mathbf{r}$ (consistent with the notation of Chapter 7, which we find convenient to use in the following derivation). According to Eq. (8.188), the extremal paths satisfies the equation [see Eq. (7.44)]

$$\widetilde{\nabla}^2\theta = 0 \quad (\text{mod } 2\pi) \tag{8.190}$$

so that $\widetilde{\mathbf{h}}$ is a solenoidal field satisfying

$$\widetilde{\boldsymbol{\nabla}} \cdot \widetilde{\mathbf{h}} = 0 \tag{8.191}$$

In analogy to magnetostatics (in three spatial dimensions), the constraint (8.191) is satisfied by setting

$$\widetilde{\mathbf{h}} = \widetilde{\boldsymbol{\nabla}} \times \widetilde{\mathbf{A}} \tag{8.192}$$

where

$$\widetilde{\mathbf{A}}(\widetilde{\mathbf{x}}) = \frac{1}{\bar{c}} \int d^3x' \, \frac{\widetilde{\mathbf{J}}(\widetilde{\mathbf{x}}')}{|\widetilde{\mathbf{x}} - \widetilde{\mathbf{x}}'|} \tag{8.193}$$

$\widetilde{\mathbf{J}}(\widetilde{\mathbf{x}}')$ is an auxiliary $(2+1)$-dimensional current density. Taking the curl of Eq. (8.192), we obtain, with the use of Eq. (8.193),

$$\widetilde{\boldsymbol{\nabla}}\times\widetilde{\mathbf{h}} = \widetilde{\boldsymbol{\nabla}}\times\left(\widetilde{\boldsymbol{\nabla}}\times\int d^3x'\,\frac{\widetilde{\mathbf{J}}(\widetilde{\mathbf{x}}')}{\bar{c}|\widetilde{\mathbf{x}} - \widetilde{\mathbf{x}}'|}\right) = -\frac{1}{\bar{c}}\int d^3x'\,\widetilde{\nabla}^2\frac{\widetilde{\mathbf{J}}(\widetilde{\mathbf{x}}')}{|\widetilde{\mathbf{x}} - \widetilde{\mathbf{x}}'|} \tag{8.194}$$

where we have used the vector identity

$$\widetilde{\boldsymbol{\nabla}}\times(\widetilde{\boldsymbol{\nabla}}\times\widetilde{\mathbf{A}}) = \widetilde{\boldsymbol{\nabla}}(\widetilde{\boldsymbol{\nabla}}\cdot\widetilde{\mathbf{A}}) - \widetilde{\nabla}^2\widetilde{\mathbf{A}} \tag{8.195}$$

and imposed a continuity equation for the conserved current $\widetilde{\mathbf{J}}$

$$\widetilde{\boldsymbol{\nabla}}\cdot\widetilde{\mathbf{J}} = 0 \tag{8.196}$$

Noting that $\widetilde{\nabla}^2$ on the RHS of Eq. (8.194) operates on the $\widetilde{\mathbf{x}}$ variable, we evaluate

$$\widetilde{\nabla}^2\frac{1}{|\widetilde{\mathbf{x}} - \widetilde{\mathbf{x}}'|} = -4\pi\,\delta(\widetilde{\mathbf{x}} - \widetilde{\mathbf{x}}') \tag{8.197}$$

Inserting this result into the RHS of Eq. (8.194), we obtain

$$\widetilde{\boldsymbol{\nabla}}\times\widetilde{\mathbf{h}} = \frac{4\pi}{\bar{c}}\widetilde{\mathbf{J}}(\widetilde{\mathbf{x}}) \tag{8.198}$$

Starting from this relation, we can clarify the physical meaning of the auxiliary current density $\widetilde{\mathbf{J}}(\widetilde{\mathbf{x}})$. Let us consider first the τ component of Eq. (8.198). Using Eq. (8.189), we have

$$\left(\widetilde{\boldsymbol{\nabla}} \times \widetilde{\mathbf{h}}\right)_\tau = \left(\widetilde{\boldsymbol{\nabla}} \times \widetilde{\boldsymbol{\nabla}}\theta\right)_\tau = \frac{1}{\kappa}\left(\frac{\partial v_y}{\partial x} - \frac{\partial v_x}{\partial y}\right) = 2\pi\rho(\mathbf{r}) \qquad (8.199)$$

where $\rho(\mathbf{r})$ is the vortex-distribution function [see Eq. (7.55)]

$$\rho(\mathbf{r}, \tau) = \sum_i q_i\, \bar{\delta}(\mathbf{r} - \mathbf{r}_i(\tau)) \qquad (8.200)$$

Equations (8.198)–(8.200) imply that the τ component of $\widetilde{\mathbf{J}}$ is

$$J_\tau = \widetilde{\mathbf{J}}_\tau(\mathbf{r}) = \frac{\bar{c}}{2}\rho(\mathbf{r}, \tau) \qquad (8.201)$$

The x, y components of $\widetilde{\mathbf{J}}$ can be found easily by inserting Eq. (8.201) into the continuity equation (8.196)

$$\boldsymbol{\nabla}\cdot\mathbf{J} = \frac{\partial J_x}{\partial x} + \frac{\partial J_y}{\partial y} = -\frac{1}{\bar{c}}\frac{\partial J_\tau}{\partial \tau} = -\frac{1}{2}\frac{\partial \rho}{\partial \tau} \qquad (8.202)$$

The specific form of the τ dependence of $\rho(\mathbf{r}, \tau)$, given by Eq. (8.200), implies that

$$\frac{\partial \rho}{\partial \tau} = -\sum_i q_i\dot{\mathbf{r}}_i \cdot \boldsymbol{\nabla}\bar{\delta}(\mathbf{r} - \mathbf{r}_i) = -\boldsymbol{\nabla}\cdot\sum_i q_i\dot{\mathbf{r}}_i\, \bar{\delta}(\mathbf{r} - \mathbf{r}_i) \qquad (8.203)$$

Comparing this result with Eq. (8.202), we see that the two-component vector field $\mathbf{J} = (J_x, J_y)$ is given by

$$\mathbf{J} = \frac{1}{2}\sum_i q_i\dot{\mathbf{r}}_i\, \bar{\delta}(\mathbf{r} - \mathbf{r}_i(\tau)) \qquad (8.204)$$

This relation shows that $\mathbf{J}$ is a fictitious 2D current density carried by vortices moving along the trajectories $\mathbf{r}_i(\tau)$. In view of the relations (8.191)–(8.193) and (8.198), there is a complete analogy with magnetostatics so that the action (8.188) can be rewritten in terms of the current densities $\widetilde{\mathbf{J}}(\widetilde{\mathbf{x}})$ as [see Eq. (7.135)]

$$S_{\mathrm{eff}}^{(2)}[\theta] = \frac{J}{2\bar{c}}\int d^3x\,\left[\widetilde{\mathbf{h}}(\widetilde{\mathbf{x}})\right]^2 = \frac{2\pi J}{\bar{c}^3}\int d^3x\int d^3x'\,\frac{\widetilde{\mathbf{J}}(\widetilde{\mathbf{x}})\cdot\widetilde{\mathbf{J}}(\widetilde{\mathbf{x}}')}{|\widetilde{\mathbf{x}} - \widetilde{\mathbf{x}}'|} \qquad (8.205)$$

The expression on the RHS of this equation has the form of the magnetostatic energy of a system of tubular currents flowing along the trajectories $\mathbf{r}_i(\tau)$ of the vortex centers (see Fig. 8.3). It should be noted that Popov

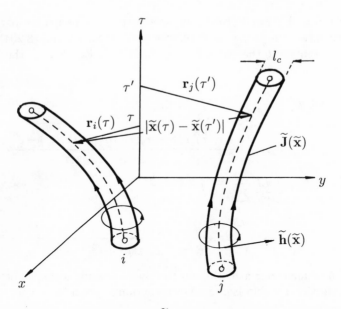

FIG. 8.3. Auxiliary tubular currents $\widetilde{\mathbf{J}}(\widetilde{\mathbf{x}})$ flowing along the vortex trajectories $\mathbf{r}_i(\tau)$ and $\mathbf{r}_j(\tau)$ in $(2+1)$-dimensional space $(x, y, c\tau)$. The diameter of the current tubes l_c is given by the diameter of the vortex core.

(1983) actually derives this action starting from the Hamiltonian of interacting system of *bosons*. By eliminating the fast density fluctuations from the partition function, he arrives at an effective action for slow phase fluctuations that has a form similar to Eq. (8.188). Another related approach is to use the coherent-state representation for the partition function of a 2D system of interacting bosons and apply the mathematical manipulations developed recently by Lee and Fisher (1991) in their derivation of the vortex–charge dual representation. There are terms in their Lagrangian corresponding to the coupling of the vortex current $\widetilde{\mathbf{J}}$ to the boson density fluctuations. By eliminating these fluctuations, an effective action of the form (8.205) ensues. Thus, vortex dynamics of Newtonian character is not confined just to superconducting Fermi systems. This is not surprising, since the phase-inertial terms, which in Eq. (8.188) stem from the fermion TDGL dynamics, are also present in the effective action of interacting Bose systems (Fisher and Grinstein, 1988).

8.7.3. *Vortex dissipation caused by coupling to acoustic spin waves*

This mechanism of vortex dissipation can be extracted from the effective action (8.205), since this action involves the interaction between vortex currents that is *nonlocal in the time variable* [compare Eq. (4.31)]. Follow-

ing Eckern and Schmid (1989), we now derive the dissipative part of the
effective action. Inserting the explicit formulas (8.201) and (8.204) for the
vortex currents into the expression on the RHS of Eq. (8.205), the effective
action is written as follows

$$S_{\text{eff}}^{(2)}[\theta] = \frac{\pi J}{2\bar{c}} \sum_{i,j} q_i q_j \int_0^{\hbar\beta} d\tau \int_0^{\hbar\beta} d\tau' \int d^2r \int d^2r'$$

$$\times \frac{\bar{\delta}(\mathbf{r} - \mathbf{r}_i(\tau))\,\bar{\delta}(\mathbf{r}' - \mathbf{r}_j(\tau'))\big[\dot{\mathbf{r}}_i(\tau)\cdot\dot{\mathbf{r}}_j(\tau') + \bar{c}^2\big]}{(|\mathbf{r}-\mathbf{r}'|^2 + \bar{c}^2|\tau-\tau'|^2)^{\frac{1}{2}}}$$

$$= \frac{\pi J}{2\bar{c}} \sum_{i,j} q_i q_j \int_0^{\hbar\beta} d\tau$$

$$\times \int_0^{\hbar\beta} d\tau' \frac{\dot{\mathbf{r}}_i(\tau)\cdot\dot{\mathbf{r}}_j(\tau') + \bar{c}^2}{[|\mathbf{r}_i(\tau) - \mathbf{r}_j(\tau')|^2 + \bar{c}^2|\tau-\tau'|^2]^{\frac{1}{2}}} \quad (8.206)$$

Let us now focus our attention on the case involving a single-vortex excitation. Setting $i = j$ in Eq. (8.206), its action is given by

$$S_{\text{eff}}^{(i)}[\mathbf{r}_i(\tau)] = \frac{\pi J}{2\bar{c}} \int_0^{\hbar\beta} d\tau \int_0^{\hbar\beta} d\tau' \frac{\dot{\mathbf{r}}_i(\tau)\cdot\dot{\mathbf{r}}_i(\tau') + \bar{c}^2}{[|\mathbf{r}_i(\tau) - \mathbf{r}_i(\tau')|^2 + \bar{c}^2|\tau-\tau'|^2]^{\frac{1}{2}}} \quad (8.207)$$

We assume that the vortex center moves slowly, so that

$$|\mathbf{r}_i(\tau) - \mathbf{r}_i(\tau')| \ll \bar{c}|\tau - \tau'| \quad (8.208)$$

Then the denominator on the RHS of Eq. (8.207) can be expanded as
follows:

$$S_{\text{eff}}^{(i)}[\mathbf{r}_i(\tau)] = \frac{\pi J}{2\bar{c}} \int_0^{\hbar\beta} d\tau \int_0^{\hbar\beta} d\tau' \frac{\dot{\mathbf{r}}_i(\tau)\cdot\dot{\mathbf{r}}_i(\tau') + \bar{c}^2}{\bar{c}|\tau-\tau'|}$$

$$\times \left(1 - \frac{|\mathbf{r}_i(\tau) - \mathbf{r}_i(\tau')|^2}{2\bar{c}^2|\tau-\tau'|^2} + \cdots\right) \quad (8.209)$$

The dissipative contribution to the effective action is generated by the
zeroth-order term of this expansion, yielding

$$S_D[\mathbf{r}_i(\tau)] = \frac{\pi J}{2\bar{c}^2} \int_0^{\hbar\beta} d\tau \int_0^{\hbar\beta} d\tau' \,\dot{\mathbf{r}}_i(\tau)\frac{1}{|\tau-\tau'|}\cdot\dot{\mathbf{r}}_i(\tau') \quad (8.210)$$

Introducing the Fourier expansions

$$\mathbf{r}_i(\tau) = \sum_n \mathbf{r}_n e^{-i\omega_n\tau} \quad (8.211)$$

and

$$f(\tau) = \frac{1}{|\tau|} = \sum_n f_n e^{-i\omega_n\tau} \quad (8.212)$$

the expression (8.210) is transformed to

$$S_D[\mathbf{r}_i(\tau)] = \frac{\pi J(\hbar\beta)^2}{2\bar{c}^2} \sum_n \omega_n^2 \mathbf{r}_n f_n \cdot \mathbf{r}_{-n} \qquad (8.213)$$

In the limit $T \to 0$, the discrete Matsubara frequencies, $\omega_n = 2\pi n/\hbar\beta$, form a continuous spectrum, and the expression (8.213) goes over to a real-frequency integral

$$S_D[\mathbf{r}_i(\tau)] \xrightarrow[T \to 0]{} \frac{J}{4\bar{c}^2} \int_{-\infty}^{\infty} d\omega\, \omega^2 |\mathbf{r}(\omega)|^2 f(\omega) \qquad (8.214)$$

where $\mathbf{r}(\omega)$ is the (real-time) Fourier integral of $\mathbf{r}(t)$ and

$$f(\omega) = \int_{-\infty}^{\infty} dt\, \frac{e^{i\omega t}}{|t|} = 2\ln(1/|\omega|) \qquad (8.215)$$

Using this result, the dissipative action for a vortex, in the $T = 0$ limit, is, according to Eq. (8.214)

$$S_D[\mathbf{r}_i]_{T=0} = \frac{1}{2} \int_{-\infty}^{\infty} d\omega\, |\omega|\eta(\omega)|\mathbf{r}(\omega)|^2 \qquad (8.216)$$

where

$$\eta(\omega) = \frac{J|\omega|}{\bar{c}^2} \ln(1/|\omega|) \qquad (8.217)$$

The way Eq. (8.216) is written was suggested by the form of the dissipative action for ohmic friction [see Eq. (4.40)]. However, there is an important distinction between the Eqs. (8.216) and (4.40). As the expression (8.217) shows, $\eta(\omega) \to 0$ as $\omega \to 0$, whereas η remains *constant* for ohmic dissipation. For this reason, the friction (8.217) is called *subohmic*. As pointed out by Eckern and Schmid (1989), its main role is to produce logarithmic corrections to the free acceleration of a vortex by external current.

8.7.4. *Vortex dissipation due to ohmic losses*

In bulk superconductors, this mechanism was first considered by Bardeen and Stephen (1965). It involves the dissipation of the electric fields generated, via the Faraday effect, by the moving vortex. The central assumption of their model is that the vortex contains a core of radius ξ that behaves as a normal metal, in the sense that its quasiparticle excitation spectrum is *gapless*. Caroli et al. (1964) have found that the core bound states exhibit at $T \ll T_c$ a spectrum with an effective gap Δ_0^2/ϵ_F. Thus, for ordinary (low-T_c) superconductors in which the effective gap is negligible, the assumption of a normal core is justified. The electric fields induced within the core can then be dissipated by ordinary Joule effect. In high-T_c superconductors, however, the effective gap is no longer negligible, and it can, in fact, be much larger than $k_B T$. Unless there is some other mechanism providing gaplessness at low temperatures, the vortex dissipation due to ohmic losses should be *blocked* in high-T_c materials. An analogous situation is encountered in dissipative Josephson-junction arrays: In Eq. (5.155) we have seen that the kernel $\alpha(\tau)$ assumes an ohmiclike form only when the superconducting gap Δ_0 vanishes.

In what follows, we shall derive an effective action for the ohmic dissipation of a vortex in a regular 2D array. From general arguments of Caldeira and Leggett (1981), the vortex dissipative action is expected to take the form [see Eq. (4.43)]

$$S_D[\mathbf{u}(\tau)] = \frac{\eta}{2\pi} \int_0^{\hbar\beta} d\tau \int_0^{\hbar\beta} d\tau' \left(\frac{\mathbf{u}(\tau) - \mathbf{u}(\tau')}{\tau - \tau'} \right)^2 \qquad (8.218)$$

where $\mathbf{u}(\tau)$ is the 2D vortex-displacement field and η is the viscosity coefficient, proportional to the junction conductance. We start from the effective action (6.66) for a planar array and use a continuum approximation

$$\begin{aligned}
S_D[\theta] &\simeq \frac{1}{8} \int_0^{\hbar\beta} d\tau \int_0^{\hbar\beta} d\tau' \sum_{\langle ij \rangle} \alpha(\tau - \tau')[\theta_{ij}(\tau) - \theta_{ij}(\tau')]^2 \\
&\rightarrow \frac{1}{8} \int_0^{\hbar\beta} d\tau \int_0^{\hbar\beta} d\tau' \\
&\qquad \times \int d^2\rho \left[\boldsymbol{\nabla}\theta(\boldsymbol{\rho},\tau) - \boldsymbol{\nabla}\theta(\boldsymbol{\rho},\tau')\right]^2 \alpha(\tau - \tau')
\end{aligned} \qquad (8.219)$$

where $\alpha(\tau)$ is given by Eq. (6.67). Using the ansatz (8.179) and assuming a slowly moving vortex, we perform an expansion in the displacements $\mathbf{u}$, yielding

$$\boldsymbol{\nabla}[\theta(\boldsymbol{\rho},\tau) - \theta(\boldsymbol{\rho},\tau')] \simeq -\boldsymbol{\nabla}\{[\mathbf{u}(\tau) - \mathbf{u}(\tau')] \cdot \boldsymbol{\nabla}\theta\} \qquad (8.220)$$

The expression on the RHS of this equation is evaluated using the explicit form for θ given in Eq. (8.124), yielding

$$\{\boldsymbol{\nabla}[\theta(\boldsymbol{\rho},\tau) - \theta(\boldsymbol{\rho},\tau')]\}^2 \simeq \frac{1}{\rho^4}[\mathbf{u}(\tau) - \mathbf{u}(\tau')]^2 \qquad (8.221)$$

Inserting this result on the RHS of Eq. (8.219), we obtain

$$S_D[\mathbf{u}(\tau)] \simeq \frac{\pi}{4} \int_0^{\hbar\beta} d\tau \int_0^{\hbar\beta} d\tau' \int_{l_c}^\infty \frac{d\rho}{\rho^3}[\mathbf{u}(\tau) - \mathbf{u}(\tau')]^2 \alpha(\tau - \tau') \qquad (8.222)$$

where a cutoff of the order of the (granular) coherence length has been imposed on radial integration, similar to Eq. (8.181). Using Eq. (6.67) and comparing the expression on the RHS of Eq. (8.222) with the general formula (8.218), we obtain the viscosity coefficient η for a vortex in a gapless

2D array

$$\eta = \frac{\pi\hbar^2}{8e^2 R_N l_c^2} \tag{8.223}$$

It is interesting to compare this result with the viscosity derived previously by Bardeen and Stephen (1965). We thus consider a film of thickness d and express the junction resistance in terms of the bulk resistivity ρ_n as follows

$$R_N = \frac{\rho_n}{d} \tag{8.224}$$

Introducing this relation into Eq. (8.223), we obtain for a vortex of length d

$$\eta = \frac{\pi\hbar^2 d}{8e^2 \rho_n l_c^2} \tag{8.225}$$

This expression has the form derived by Bardeen and Stephen (1965), and it becomes identical with it, if we set $l_c = \xi/2$. This is expected since in both cases we are dealing with the same general origin of dissipation, namely decay of the electric fields in the region of the vortex core.

8.7.5. Dynamics of massive vortices in granular superconductors

We are now in a position to write down the equation of motion for a vortex in a dissipative array of Josephson junctions. We confine ourselves to the case $m = 0$. Then the inertial mass of the vortex is equal to M [see Eq. (8.184)]. We also assume that there is a nonzero ohmic dissipation described by the effective action (8.218). In Section 8.5.3, we have presented both experimental and theoretical arguments that the pseudo-Lorentz force in superconductors is very small. Thus, we take $\mathbf{F}_L = 0$ in the vortex equation of motion. Consequently, the action of the ith vortex in the presence of other vortices is

$$S_{\mathrm{eff}}[\mathbf{u}] = \int_0^{\hbar\beta} d\tau \left[\frac{1}{2} M(\dot{\mathbf{u}}_i)^2 + H_v \right]$$
$$+ \frac{1}{2} \int_0^{\hbar\beta} d\tau \int_0^{\hbar\beta} d\tau' \, q(\tau) K(\tau - \tau') q(\tau') \tag{8.226}$$

where we have chosen to write the dissipative action in a form given in Eq. (4.126). We assume that the screening length $\lambda_\perp$ exceeds the size of the sample, so that the interaction between the vortices is logarithmic and is described by the Hamiltonian H_v of Eq. (7.75). Applying the principle of minimum action, Eq. (8.226) leads to the following equation of motion for the ith vortex [see Eq. (4.129)]

$$M \frac{d^2 \mathbf{u}_i}{d\tau^2} - \frac{\partial H_v}{\partial \mathbf{u}_i} - \int_0^{\hbar\beta} d\tau' \, K(\tau - \tau') \mathbf{u}_i(\tau') = 0 \tag{8.227}$$

where the kernel $K(\tau - \tau')$ is given by Eqs. (4.127) and (4.128). Using Eqs. (7.75) and (8.5), the force exerted on the ith vortex by other vortices is

$$\mathbf{F}_v = -\frac{\partial H_v}{\partial \mathbf{u}_i} = -\frac{2\pi J}{\kappa} \sum_{i<j} q_i \mathbf{z} \times \mathbf{v}_j(\mathbf{r}_i) \qquad (8.228)$$

where $\mathbf{v}_j(\mathbf{r}_i)$ denotes the fluid velocity at the center of the ith vortex due to the jth vortex. Using Eqs. (7.162) and (8.4), we have $2\pi J/\kappa = h\rho_0$, where $\rho_0 = n_s^{2D}$ is the 2D superfluid sheet density. Recalling Eqs. (8.93) and (8.111), the expression on the RHS of Eq. (8.228) can then be cast into the form of a Magnus force produced by the total current $\mathbf{j}_v$ of all the other vortices. If there is also an external current, represented by density $\mathbf{j}_s$, the net force acting on the ith vortex is given by the sum

$$\mathbf{F}_{\text{tot}} = \mathbf{F}_v + \mathbf{F}_M = \frac{d}{c}(\mathbf{j}_v + \mathbf{j}_s) \times \boldsymbol{\phi} \qquad (8.229)$$

where $\boldsymbol{\phi}$ is a vector of magnitude ϕ_0 pointing along the z axis and the current densities are to be evaluated at $\mathbf{u}_i$. Using Eqs. (8.227)–(8.229), the classical equation of motion for the vortex becomes (in the real-time domain)

$$M\frac{d^2\mathbf{u}_i}{dt^2} + \eta\frac{d\mathbf{u}_i}{dt} = \frac{d}{c}[\mathbf{j}_v(\mathbf{u}_i) + \mathbf{j}_s(\mathbf{u}_i)] \times \boldsymbol{\phi} \qquad (8.230)$$

Notice that for a single massless vortex this equation is in agreement with Eq. (8.116). By working in a continuum approximation, we have missed the *periodic pinning* force that the vortex experiences in a granular array. This force is given by the gradient of a periodic potential, whose amplitude is determined by the barrier the vortex must overcome to move between two energy minima. This barrier has been estimated numerically by Lobb et al. (1983), for a square planar array, to be about 0.2 J. For a fixed value of $y = y_0$, the pinning potential can thus be represented by

$$V(u_x, u_{y_0}) = -0.1J \cos\left(\frac{2\pi u_x}{a}\right) \qquad (8.231)$$

where a is the lattice spacing of the array. Assuming that the tunneling gaps between the superconductors are much smaller than a, the junction current I is related to j_s by

$$I = j_s da \qquad (8.232)$$

Using this relation and Eq. (4.9), the magnitude of the Magnus force can be expressed as

$$F_M = \frac{j_s \phi_0 d}{c} = \frac{2\pi I \hbar}{ae^*} = \frac{2\pi J}{a}\left(\frac{I}{I_c}\right) \qquad (8.233)$$

where I_c is the critical current of the junction. From Eq. (8.229), it follows that the Magnus force points along the x axis if j_s flows along the y axis.

Let us consider the equation of motion (along the x axis) of a *single* vortex. In this case, the first term on the RHS of Eq. (8.230) *vanishes*. If we add to the second term the periodic force given by $-\partial V/\partial u_x$, we obtain with the use of Eqs. (8.231)–(8.233)

$$M\ddot{u}_x + \eta\dot{u}_x = \frac{2\pi J}{a}\left[\frac{I}{I_c} - 0.1\sin\left(\frac{2\pi u_x}{a}\right)\right] \qquad (8.234)$$

It is convenient, for formal reasons, to introduce a dimensionless vortex displacement, x, measured in units of $a/2\pi$. Thus, Eq. (8.234) is written, letting $u_x = xa/2\pi$

$$M_v\frac{d^2x}{dt^2} + \eta_v\frac{dx}{dt} = J\left(\frac{I}{I_c} - 0.1\sin x\right) \qquad (8.235)$$

where

$$M_v = M\left(\frac{a}{2\pi}\right)^2 \qquad (8.236)$$

and

$$\eta_v = \eta\left(\frac{a}{2\pi}\right)^2 \qquad (8.237)$$

We note that M_v is connected with the vortex velocity normalized to $a/2\pi$. If we let l_c, in Eqs. (8.184) and (8.223), equal the lattice spacing a, then the quantities M_v and η_v become independent of a. Only the intergrain capacitance ΔC and the junction resistance R_N enter the expressions (8.236) and (8.237), respectively. This formulation of vortex dynamics in 2D arrays has been used recently by van der Zant et al. (1991) to interpret their data on quantum vortex tunneling. Since the Josephson-coupling energy in the arrays studied by these authors is dominant over the charging energy, the assumption $l_c \simeq a$ seems justified [see Eq. (6.177)].

Finally, let us point out that Eq. (8.235) and the Eq. (4.10) describing a single junction are very similar, the main difference being the reduced amplitude of the periodic potential felt by the vortex in the array. This similarity enables us to use the formalism of Chapter 4 to discuss tunneling of quantum vortices (see Section 8.8).

8.8. Quantum tunneling of vortices

Processes leading to the motion of vortices in superconductors have been subjects of interest because of the finite electrical resistance they cause even below the transition temperature. The classical dynamics of a vortex is, in general, of the form given by Eq. (8.235). Though this equation corresponds to the case of a granular superconductor array, an equation of a similar form is applicable to the motion of flux lines (or flux bundles) in bulk superconductors with pinning centers. We note that these flux lines are often assumed to form a quasiperiodic array. Applying the results of Chapter

4 to Eq. (8.235), we see that the rate of the MQT of a vortex is influenced in an important way by dissipation. For an undamped vortex, the rate of escape out of a metastable minimum of the potential is mainly determined by the ratio $V_0/\hbar\omega_0$, where V_0 is the size of the barrier to be crossed and ω_0 is the well frequency [see Eq. (4.110)]. Voss and Webb (1981) studied planar square arrays of Nb junctions, for which $V_0/\hbar\omega_0 \simeq 0.07$, indicating that tunneling should be important. In fact, such MQT processes could explain the observed resistance tails extending down to the lowest temperatures. More recent study of the dynamics of vortices in underdamped arrays has been made by van der Zant et al. (1991). For the high-resistance array, these authors find a zero-bias resistance, which decreases with decreasing temperature but *levels off* at 35 mK, indicating the possibility of the vortex MQT. These experiments are discussed in Section 8.8.1.

Another experiment, involving a possible MQT of vortices in a 2D superconductor, is the measurement of the temperature dependence of the resistance in ultrathin films of Pb, Al, and Bi (Liu et al., 1992). At low temperatures and small magnetic fields, the sheet resistance exhibits a rather unusual T dependence, which the authors cited attribute to the MQT of an overdamped vortex. This interpretation is reviewed in Section 8.8.2. In Section 8.8.3 we discuss some new ingredients of the theory of MQT in disordered superconducting films. In particular, we consider the possibility of vortex localization due to dissipation and its implications for variable-range hopping in vortex transport.

An interesting manifestation of the MQT of a vortex is the anomalous decay rate of the magnetization, observed in the millikelvin region in organic and high-T_c superconductors (Mota et al., 1987, 1988). These authors have observed logarithmic time decays of the magnetization in Sr-La-Cu-O and Ba-La-Cu-O specimens in sintered and powder form. For temperatures in the range $2 < T < 10$ K, the rate is approximately linear in temperature (in keeping with the Arrhenius law), but for $T < 1$ K a *much weaker T dependence* is observed. A similar departure from the Arrhenius law has also been found in a single crystal of $YBa_2Cu_3O_{7-x}$ (Fruchter et al., 1991). Blatter et al. (1991) have offered an interpretation of these anomalies in terms of the MQT of vortices and find agreement with the experiments of Mota et al. (1988) and Fruchter et al. (1991) in the limit of strong dissipation. The theory of Blatter et al. (1991) is the subject of Section 8.8.4.

The interpretation of the low-temperature anomalies of the resistance or magnetization decay rates may be complicated by the presence of the noise. First, there is the possibility that an *extrinsic* noise establishes an effective temperature that is higher than the measured temperature of the thermal bath. This effect has been scrutinized by Liu et al. (1992), who have been able to rule it out as a source of the observed effects.

The other possible complication, which is of more intrinsic nature, is the barrier noise (Šimánek, 1989a, b). In a granular superconductor, there are time-dependent barrier fluctuations resulting from the normal-conductance noise [see Eq. (5.147)]. Rogers and Buhrman (1984, 1985) have measured the spectral density of this noise in Nb-Nb_2O_5-$PbBi$ submicrometer junctions and found it nearly *temperature independent* for $T \lesssim 15$ K. Presumably, trapping processes in the barrier involving quantum ionic tunneling are responsible for this kind of conductance noise. This may lead to an enhancement of the vortex-escape rate over the value predicted by the Arrhenius law. This enhancement is more pronounced at lower temperatures so that deviations from this law are qualitatively similar to those caused by MQT. We note that a large conductance noise has been observed in a polycrystalline sample of $YBa_2Cu_3O_7$, which gives support to the barrier noise mechanism for the anomalous magnetization decay rates in high-T_c superconductors. In a single crystal, pinning of flux lines to microscopic defects can take place (Dolan et al., 1989). These defects may contain localized electron traps with a fluctuating occupation (Eliashberg, 1988). Since the vortex core is only 10–20 Å large, the pinning potential may fluctuate strongly, owing to jumps in occupation. Incorporating this barrier fluctuation into the Langevin equation for an overdamped vortex, we have calculated the rate of escape due to thermal activation (Šimánek, 1989b). The resulting logarithmic decay of the magnetization (obtained with a T independent quantum spectral density of the barrier noise) shows a deviation from the Arrhenius T dependence that is more complicated than the observed one. It should, however, be pointed out that these calculations use a *perturbation* expansion in the ratio of spectral intensities of the barrier and the thermal noise, and this expansion breaks down as the temperature is lowered into the region, where the departure from linear dependence becomes significant. It is possible that a *nonperturbative* calculation of the decay rate would yield a T dependence closer to the experimentally observed form.

8.8.1. *Possible MQT of a vortex in underdamped arrays*

Van der Zant et al. (1991) studied the vortex motion in high-quality 2D all-aluminum Josephson-junction arrays. Of particular interest to us are their measurements on the high-resistance samples, where vortex MQT processes have a greater chance to take place. Vortices are introduced by applying a small magnetic field perpendicular to the array. As mentioned, the leveling off of the zero-bias resistance on lowering the temperature is regarded as a possible signature of the MQT. This resistance has been analyzed in terms of the equation (van der Zant et al., 1991)

$$R \approx \frac{L\phi_0 f}{ic\langle \Delta t \rangle} \qquad (8.238)$$

where L is the length of the sample (measured in the number of unit cells), f is the magnetic flux per unit cell divided by ϕ_0, i is the applied transport current, and Δt is the average time for the vortex to cross the whole array.

At low current levels, Δt is limited by the escape rate Γ out of the periodic potential. Then $\Delta t \approx W/\Gamma$, where W is the number of unit cells across the width of the sample. We note that the RHS of Eq. (8.238) is just the induced voltage (calculated from the Faraday law) divided by i. In this sense there is a similarity with the standard theory of flux creep (Tinkham, 1975). This theory assumes that bundles of flux lines are jumping between adjacent pinning sites due to thermal activation. Then the passage time of the vortex across the sample is also limited by the jump rate. If the pinning (barrier) energy is small compared to $k_B T$, the passage time will be limited by vortex viscosity η. In fact, in this limit we can calculate Δt directly from the phenomenological equation of motion (8.116). The resistance obtained in this way is what is usually called the *flux flow resistance*. Halperin and Nelson (1979) have used this approach to calculate the contribution of the unbound vortices, above the Kosterlitz–Thouless transition temperature, to the resistance of superconducting films.

We focus our attention on sample B, characterized by a high normal sheet resistance $R_N \approx 20$ kΩ. We note that the actual damping resistance to be used in the expression for the vortex-viscosity coefficient η is not known. The standard model for a vortex in the planar array assumes a rigid order parameter with a constant magnitude even near the vortex center. Thus, in contrast with the vortex in bulk superconductors, this model does not allow for a normal core. Consequently, a *subgap resistance*, usually much larger than R_N, should be used for the effective damping resistance R_e of a vortex in the arrays (van der Zant et al., 1991). Actually, this conclusion may be a result of the deficiency of the standard vortex model in planar arrays. A modification of the standard model, allowing for an enhanced vortex dissipation, is suggested by recalling the quantum phase fluctuations about the equilibrium vortex configuration. In Section 7.6.2, we have shown that these fluctuations are enhanced near the vortex center as a result of the reduced intergrain coupling (see Fig. 7.3). According to Eq. (7.193), this effect is also responsible for a reduction of the local stiffness constant. We may envisage the possibility that this constant vanishes in the immediate vicinity of the vortex center, thus preventing coherent transfer of Cooper pairs between the grains. An incoherent transfer of pairs may, however, still be possible, leading to a dissipation. The mechanism of this dissipation is similar to that taking place in the boson Hubbard model studied by Cha et al. (1991). We note that the homogeneous boson model is closely related to the self-charging model of a Josephson-coupled array introduced in Chapter 3. As pointed out by Cha et al. (1991), the source of the dissipation in the homogeneous boson model is the production of particle–hole boson pairs with a continuous energy spectrum. To obtain a finite zero-frequency conductivity in the homogeneous boson model, it is essential that the ratio $\alpha = zJ/2U_a$ is tuned to a critical value $\alpha_c = 1$. If $\alpha > 1$, the system is phase coherent (superconducting) and dissipationless. On the other hand, if $\alpha < 1$, the system is a bosonic insulator with a

Coulomb gap. Although the arrays studied by van der Zant et al. (1991) are in a phase-coherent regime ($\alpha > 1$), the vicinity of the vortex center may nevertheless exhibit a finite pair conductance due to the enhanced local phase fluctuations. Consequently, the effective resistance R_e could be significantly smaller than the subgap resistance. If $R_e \simeq R_N$ is used [as in Eq. (8.223)], then the hysteretic parameter of the vortex is given by (van der Zant et al., 1991):

$$\beta = \frac{0.4\pi c I_c R_N^2 C}{\phi_0} \tag{8.239}$$

where I_c is the junction critical current. This parameter is related to the damping coefficient κ of Eq. (4.59) by $\beta = \kappa^{-2}$. Letting $R_N \approx 20.6$ kΩ and $I_c \approx 1.5 \times 10^{-8}$ A, the parameter β assumes a large value of 6×10^2, which indicates that the vortex is highly *underdamped*. As a first approximation, let us neglect dissipation and calculate the MQT rate from the expression (4.110). For this equation, we need to know the size of the barrier V_0 and the well frequency ω_0. Sample B is characterized by $V_0/k_B = 70$ mK, and ω_0 is calculated from Eqs. (4.12) and (8.235), yielding

$$\omega_0^2 = \frac{V''}{M} = \frac{0.1J}{M_v} = \frac{V_0}{2M_v} \tag{8.240}$$

where M_v is the vortex mass given by Eq. (8.236). Van der Zant et al. (1991) use the following expression for M_v

$$M_v = \frac{\hbar C}{8e^2} \tag{8.241}$$

We note that this yields a mass that is 2π times the value that would result if $l_c \simeq a$ is used in Eq. (8.184) with ΔC replaced by the geometric junction capacitance C. In view of the uncertainties regarding the structure of the vortex core, this discrepancy is of no major concern; estimates of M_v with only an order of magnitude accuracy are possible. Unfortunately, this leads to large uncertainties of the tunneling rate (owing to the exponential dependence of Γ on $\sqrt{M_v}$), which makes it difficult to make final conclusions about the contribution of the MQT processes to the low-temperature resistance. The geometric capacitance C has been estimated to be $C = 160$ fF for sample B. Using this value, we calculate from Eqs. (8.240) and (8.241) $\hbar\omega_0/k_B = 57$ mK. Inserting this value and the barrier size $V_0 = 70$ mK into the expression (4.110), we obtain for the dissipationless MQT rate a value $\Gamma(\eta = 0) = \Gamma(0) = 3.9 \times 10^9$ s^{-1}. Next, let us estimate the effect of the dissipation on this rate. Since the vortex is *underdamped*, we can use the *weak-dissipation* limit, given in Eq. (4.139). Using the expression (8.223) with $l_c \approx a$, we obtain $\eta = (\pi\hbar/8a^2)(R_0/R_N)$. As in the estimate of the hysteretic parameter β given previously, we use $R_N = 20.6$ kΩ. For the tunneling distance we use $q_1 \approx a$. In this way, we obtain

$$\frac{\Gamma(\eta)}{\Gamma(0)} \simeq \exp\left(-\frac{\eta q_1^2}{\hbar}\right) \approx 0.92 \tag{8.242}$$

Actually, the expression used for η by van der Zant et al. (1991) is 4π times this value, producing a stronger depression of the rate [with a ratio $\Gamma(\eta)/\Gamma(0) \approx 0.37$]. We are now in a position to make a numerical estimate of the expression (8.238) for the case where the average time to traverse sample is given by MQT. Substituting into this formula $\Delta t = W/\Gamma(\eta)$, we obtain for the resistance

$$R(\eta) = \frac{L\phi_0 f}{icW}\Gamma(\eta) \qquad (8.243)$$

Using the experimental values $L = 300$, $W = 100$, $f = 0.15$, and $i = 14$ nA, this equation leads to $R(\eta) \approx 92.5 \, \Omega$, if the dissipatively reduced rate $\Gamma(\eta) = 0.37\Gamma(0)$ is used. On the other hand, for an undamped vortex, we use the MQT rate $\Gamma(0)$ and obtain $R(\eta = 0) \approx 250 \, \Omega$, a value 7 times the measured resistance. These discrepancies could possibly be removed if we took for l_c a value smaller than a. We note that a correlation length $\xi' = a/\sqrt{2z}$ has been obtained from Eq. (6.177), in the limit of large $\alpha = zJ/2U_{11}$. Alternatively, if there is some additional mechanism enhancing the mass of the vortex, the calculated resistance could be brought down to the observed value. For instance, if the magnitude of the order parameter is allowed to have spatial variations, then a core contribution to the vortex inertial mass is expected to set in [see Eq. (8.119)]. Another possible way the MQT rate may be reduced is via an increase of the dissipation [see Eq. (8.242)]. From their data on the flux flow resistance, van der Zant et al. (1991) infer that vortices experience a damping stronger than that corresponding to $R_e = R_N$. This indicates that some other damping mechanism may be present in addition to the ohmic loss. The nonadiabatic response of core states to the vortex motion may lead to such a mechanism (see Section 8.8.4).

The importance of the MQT, relative to the thermally activated decay, can be estimated by considering the ratio $\Gamma(0)/\Gamma_{th}$, where $\Gamma(0)$ is the MQT rate for an undamped vortex and Γ_{th} is the rate of thermal decay given by Eq. (4.14). Using this equation and Eq. (4.110), we obtain

$$\frac{\Gamma(0)}{\Gamma_{th}} \approx 44.2 \exp\left[-V_0\left(\frac{7}{\hbar\omega_0} - \frac{1}{k_BT}\right)\right] \qquad (8.244)$$

Inserting into this equation the parameters V_0 and $\hbar\omega_0$, representing sample B, we calculate the temperature T_0 [at which $\Gamma(0)/\Gamma_{th} \approx 1$] to be $T_0 \approx 15$mK. This result is in qualitative agreement with the leveling off of the resistance at 35 mK. The analysis given here of the experimental data of van der Zant et al. (1991) shows that MQT processes are *not excluded* as a possible cause of the low-temperature resistance. It should be pointed out, however, that barrier fluctuations of quantum origin are not excluded either. If the MQT is the dominant decay process, then the barrier noise tends to increase its rate as well (Šimánek, 1990). There is also a modification of the thermal decay rate, which contributes to the leveling of the zero-bias resistance. To confirm the existence of the barrier noise, one should measure the normal resistance noise and the T dependence of its spectral intensity on planar arrays.

8.8.2. Vortex creep and resistivity transition in disordered superconducting films

Vortex creep in disordered films is an interesting problem, since it has a bearing on such fundamental questions like Anderson localization in 2D systems, and variable-range hopping of vortices. We begin with a discussion of the recent study of the resistive transition in ultrathin films (Liu et al., 1992). The central result of this work is the unusual temperature dependence of the measured sheet resistance, which is of the form

$$R \approx R_0 \exp(T/T_0) \qquad (8.245)$$

where R_0 and T_0 are constants. It is easy to see that this result is in disagreement with the resistance resulting from thermally activated flux creep. In this case, the vortex drifts with a creep velocity that is proportional to the net jump rate r in the direction of the Magnus force (Tinkham, 1975)

$$r = r_+ - r_- \qquad (8.246)$$

where r_+ and r_- are the "down-hill" and "up-hill" jump rates, respectively. In the limit of small applied current, the difference between the up-hill and the down-hill barriers is small compared to the thermal energy, and the net jump rate of thermally activated creep can be written as

$$r = \omega_0 e^{-V_0/k_B T}\left(e^{\Delta V/k_B T} - e^{-\Delta V/k_B T}\right) \approx 2\omega_0 e^{-V_0/k_B T}\left(\frac{\Delta V}{k_B T}\right) \quad (8.247)$$

where ω_0 is the attempt frequency, V_0 is the activation energy of the order of the vortex-pinning energy and ΔV is the "barrier tilt," due to the Magnus driving force. From Eqs. (8.246) and (8.247), it follows that the temperature dependence of the vortex-creep velocity and the resulting resistance is dominated by the factor $\exp(-V_0/k_B T)$. Thus we expect, for thermally activated vortex creep, the following T dependence of the resistance

$$R(T) \propto \exp\left(-\frac{V_0(T)}{k_B T}\right) \qquad (8.248)$$

The T dependence of the pinning energy $V_0(T)$ is estimated as follows. First, we note that the pinning energy is of the order of the self-energy of the vortex. For a 2D array, the self-energy is given by Eq. (7.141). Relating the Josephson-coupling constant of the array to the properties of the disordered film by Eqs. (7.157) and (7.174), we obtain

$$V_0(T) \approx \pi J \approx 4.36 k_B T_{c0}\left(\frac{R_0}{R_N}\right)\frac{\Delta_0(T)}{\Delta_0(T=0)}\tanh\frac{\beta\Delta_0(T)}{2} \qquad (8.249)$$

where the first equality is obtained from Eq. (7.141) by neglecting the weakly T dependent logarithmic factor. We note that the RHS of Eq. (8.249) is proportional to the superfluid density $n_s(T)$ of Eq. (7.170). For $T \lesssim T_{c0}$, where T_{c0} is the (global) transition temperature we thus have $V_0(T) \propto n_s(T) \propto T_{c0} - T$. This implies, in conjunction with Eq. (8.248),

$$R(T) \propto \exp(-T_0/T) \qquad (8.250)$$

The resistance thus follows the Arrhenius law, which disagrees with the experimental data at low temperatures. Liu et al. (1992) propose that the unusual T dependence of measured resistance is due to the *quantum vortex creep*. The basic assumption of their quantitative analysis is that the vortex is *overdamped*. We note that quantum tunneling of overdamped vortices in films was first considered by Glazman and Fogel (1984). If an overdamped vortex with a viscosity η tunnels a distance q_1, the tunneling probability can be calculated from Eq. (4.149). An important assumption made by Glazman and Fogel (1984) is that the viscosity η of a vortex in a superconducting film is given by the ohmic losses described by Eq. (8.225). Letting $l_c \approx \xi(T)/2$ in Eq. (8.223), we obtain the ohmic viscosity η to be used in Eq. (4.149), yielding for the rate

$$\Gamma(\eta) \propto \exp\left(-\frac{\eta q_1^2}{\hbar}\right) = \exp\left(-\frac{\pi \hbar q_1^2}{2e^2 R_N \xi^2(T)}\right) \qquad (8.251)$$

The temperature dependence of the coherence length $\xi(T)$ turns out to be essential for the explanation of the data. It is given by the expression (Tinkham, 1975)

$$\xi^2(T) = \frac{\xi^2(0)T_{c0}}{T_{c0} - T} = \frac{0.72\xi_0 l T_{c0}}{T_{c0} - T} \qquad (8.252)$$

where l is the mean-free path and $\xi_0 = 0.18\hbar v_F/k_B T_{c0}$. For the normal sheet resistance R_N, Liu et al. (1992) use the formula

$$R_N = \frac{2\pi R_0 \hbar}{m v_F l} \qquad (8.253)$$

where $R_0 = \hbar/e^2$. This result is appropriate for ultrathin films, since it is derived under the assumption that the electron motion takes place in the xy plane of the film, with the z component of the momentum quantized in units of 2π times the inverse thickness of the film. Introducing Eqs. (8.252) and (8.253) into Eq. (8.251), we obtain

$$\ln \Gamma(\eta) \approx 1.9\left(\frac{q_1^2 m k_B}{\hbar^2}\right)(T - T_{c0}) \qquad (8.254)$$

In a disordered film, the tunneling distance is a random variable. To obtain the resistance R, Liu et al. (1992) replace q_1^2 by the configurational average

$\langle q_1^2 \rangle$. This is presumably possible, provided the distribution of the tunneling length is not exponentially wide. Since the film resistance R resulting from quantum vortex creep is proportional to the vortex mobility [see Eq. (8.238)], we have from Eq. (8.254)

$$\ln R \approx 1.9 \langle q_1^2 \rangle \left(\frac{m k_B}{\hbar^2} \right) (T - T_{c0}) \qquad (8.255)$$

We see that this result is in agreement with the T dependence of the resistance quoted in Eq. (8.245). Moreover, the constant T_0 in this equation is given by the inverse of the coefficient of the $(T - T_{c0})$ term in Eq. (8.255). This implies that the constant T_0 is independent of the normal sheet resistance of the film, a fact that is confirmed by the data on Pb films. We note that the derivation of Eq. (8.255) is based on Eq. (4.149), valid only for T below but *close* to the temperature (A15.13) (above which thermal activation dominates the rate). As pointed out in Section 4.8.2, the exponent in Eq. (8.251) *increases* 1.42 times, as T decreases to zero (Riseborough et al., 1985). This implies that the RHS of Eq. (8.255) gets multiplied by the same T dependent factor. Hence the slope of the resistance versus T curves near $T = 0$ is 1.42 times the slope near the crossover temperature (A15.13). We note that the crossover temperature has been estimated to be about 1.1 K for the measured Pb film (Liu et al., 1992). It is interesting that this modification of the linear T dependence of the resistance appears to agree with the data for the Pb film in the highest applied field. A full confirmation of this effect may require an extension of the measurements into the region of lower temperatures.

8.8.3.   *Dissipative localization and variable-range hopping of vortices*

The derivation provided of Eq. (8.255) implies that the vortex retains its mobility all the way down to $T = 0$. This is difficult to understand, in view of the possible localization effects expected in a 2D disordered system. In the absence of dissipation, one would assume that *weak localization* due to scale-dependent *quantum interference* should dominate the vortex mobility (Lee and Ramakrishnan, 1985). However, this picture is profoundly modified by the effects of strong vortex dissipation. It is known that dissipation tends to suppress intersite tunneling and leads to a destruction of quantum coherence (see Chapter 4). Since a theory of quantum diffusion in 2D disordered systems with dissipation remains to be worked out, we resort to the case of quantum Brownian motion in a periodic potential (Fisher and Zwerger, 1985). These authors find that the linear mobility of the particle is determined in a crucial way by the magnitude of the dimensionless dissipation constant α defined as

$$\alpha = \frac{\eta a^2}{2\pi\hbar} \qquad (8.256)$$

where η and a are the viscosity and the period of the potential [as defined, for instance, by Eq. (8.234)]. If $\alpha > 1$, then the *particle (vortex) mobility is zero*. This result is related to the dissipatively induced localization transition originally predicted for a symmetric double-well potential by Bray and Moore (1982) and extended to a periodic potential by Schmid (1983). An important quantity characterizing this transition for a periodic potential is the localization length ξ_{loc}. Applying a simple variational approach (Fisher and Zwerger, 1985) to a strongly overdamped vortex, we find

$$\xi_{\text{loc}}^2 \approx \frac{2a^2}{\alpha} \ln \frac{\alpha \Lambda}{2\pi M} \qquad (8.257)$$

where M is the vortex mass and Λ is a high-frequency cutoff inversely proportional to the vortex mass. In the limit of large α, the RHS of this equation varies roughly as $2a^2/\alpha$, showing that the localization length is (within logarithmic accuracy)

$$\xi_{\text{loc}} \approx a\sqrt{\frac{2}{\alpha}} = \sqrt{\frac{4\pi\hbar}{\eta}} \qquad (8.258)$$

Hence, for $\alpha > 1$, the vortex remains localized near one minimum of the periodic potential. We believe that this localization mechanism is also working for vortices moving in a *nonperiodic potential*. This is substantiated by recalling Eq. (4.149), which shows that the amplitude of tunneling decays with the tunneling distance with a characteristic decay length

$$L_{\text{tunnel}} \approx \sqrt{\frac{\hbar}{\eta}} \qquad (8.259)$$

which agrees within a numerical factor with the η dependence shown on the RHS of Eq. (8.258). If the vortex viscosity is expressed using Eq. (8.223) (where $l_c = \xi(T)/2$), then Eq. (8.259) yields a localization length

$$\xi_{\text{loc}}(T) \approx \xi(T)\sqrt{\frac{2R_N}{\pi R_0}} \qquad (8.260)$$

where $R_0 = \hbar/e^2$ and $\xi(T)$ is the superconducting coherence length. If R_N/R_0 is of order one, then the vortex is localized within a region of a size of its *own core*. Any quantum interference effects necessary for the 2D weak localization are cut off beyond $\xi(T)$. The important question that remains to be answered is that of the quantum mobility of the vortex under the combined effect of the dissipative localization and that produced by the disorder (of the randomly distributed pinning centers). The fact that at $T = 0$ and for $\alpha > 1$ the linear mobility in a periodic potential is found to vanish indicates that it remains zero also for a vortex in the presence of disorder. Let us now estimate the constant α for the films, studied by

Liu et al. (1992). Since the same viscosity η appears in Eqs. (8.251) and (8.256), we can relate α to the constant T_0 defined by Eq. (8.245)

$$\alpha \approx \frac{1}{2\pi T_0}(T - T_{c0}) \qquad (8.261)$$

where we set the square of the period of the potential in Eq. (8.256) equal to the mean square of the jump length $\langle q_1^2 \rangle$. The data shown in Fig. 1 of the work by Liu et al. (1992) show that for a lead film (with the transition temperature $T_{c0} = 0.1$ K) the T dependence $R(T) = \exp(T/T_0)$ is well displayed for temperatures such that $T - T_{c0} \simeq 1$ K, whereas $T_0 = 0.1$ K. This yields, using Eq. (8.261), a value of $\alpha \simeq 1.6$, which implies dissipative localization, though the uncertainty of the numerical constant in Eq. (8.261) makes this prediction somewhat inconclusive.

Let us now consider the variable-range hopping (VRH) of vortices in a disordered film. First, we confine ourselves to the case of zero dissipation. At $T = 0$, the vortices induced by the applied magnetic presumably freeze into a *vortex-glass phase*, corresponding to a complete vortex localization due to pinning and repulsive vortex–vortex interactions (Fisher, 1990). It should be pointed out, however, that a vortex-delocalization transition has been proposed to take place at $T = 0$ when the magnetic field increases towards a certain critical value (Fisher, 1990; Hebard and Paalanen, 1990). In what follows, we assume that the applied field is below this critical value. Then vortex creep is blocked at $T = 0$ and the relevant mode of vortex transport at *low* T is the VRH, conceived in analogy with the well-known Fermi glass example (Mott and Davies, 1971). Let us review the arguments leading to the T dependence of the electrical resistance, predicted by vortex VRH (Fisher et al., 1991). Assuming that the vortices are *exponentially* localized with a localization length a_v, the rate of tunneling through a distance r is

$$\Gamma \sim \omega_0 \exp\left(-\frac{r}{a_v} - \frac{U_r}{k_B T}\right) \qquad (8.262)$$

where U_r is a typical energy cost for a vortex to move the distance r. Arguing that vortex translation is equivalent to changing a *twist in the boundary conditions*, Fisher et al. (1991) find that U_r is of the form

$$U_r = K\left(\frac{r}{l}\right)^\theta \qquad (8.263)$$

where K is the vortex-pinning energy, l is the intervortex spacing, and $\theta \approx -1/2$. Note that U_r depends on the distance of the vortex jump quite differently than the energy level spacing for the VRH in Fermi glass (Mott and Davies, 1971). Introducing Eq. (8.263) into Eq. (8.262) and optimizing the rate with respect to r, we obtain the distance r_0 (yielding the most probable tunneling rate)

$$r_0 = \left(\frac{k_B T}{a_v K |\theta| l^{|\theta|}}\right)^{-\frac{1}{1+|\theta|}} \qquad (8.264)$$

Using this result in Eq. (8.262), the optimal jump rate takes the form

$$\Gamma^{\text{opt}} \sim \omega_0 \exp\left[-2\left(\frac{T_0}{T}\right)^p\right] \tag{8.265}$$

where $p = (1 + |\theta|)^{-1}$ and $T_0 = \frac{1}{2}(K/k_B)(l/a_v)^{|\theta|}$. Since $|\theta| \approx 1/2$, we have $p \approx 2/3$, implying that the log of vortex mobility varies as $1/T^{2/3}$. The same T dependence is expected for the electrical resistivity of the superconducting film. The data on ultrathin films, however, cannot be explained by this result (Liu et al., 1992). One reason why VRH of vortices is not seen may be the short localization length resulting from dissipation. The normal sheet resistance R_N of the studied films is such that the factor $\sqrt{2R_N/\pi R_0}$ is of order 1, so that the localization length, predicted by Eq. (8.260), is approximately equal to the superconducting coherence length. Consequently, the temperature at which the classical Arrhenius mobility [proportional to the jump rate (8.247)] crosses over to the VRH mobility may be inaccessibly low. This crossover temperature can be estimated by equating the arguments of the exponentials in Eqs. (8.247) and (8.265). Setting $V_0 = K$, this gives (Fisher et al., 1991)

$$T^* \simeq \left(\frac{a_v}{l}\right)K \tag{8.266}$$

For a dissipatively localized vortex, we have $a_v \simeq \xi(T)$. Noting that $l = \sqrt{\phi_0/B}$, the ratio $(a_v/l) \ll 1$ (if the applied field B is small), which causes the crossover temperature to be well below K/k_B. Liu et al. (1992) assume that the energy levels accessible to vortices on pinning sites are distributed over a range smaller than $k_B T$. In such a case, Eq. (8.266) implies that $T^* \ll T$, so that if VRH is present it would start dominate the transport only at temperatures well below 0.1K.

There is still a possibility that the classical Arrhenius mobility becomes modified by the barrier noise (Šimánek, 1989a, b). From Eq. (8.249), we see that time-dependent fluctuations of R_N lead to fluctuations of the pinning energy, which facilitate vortex creep owing to the exponential dependence of the hopping rate on V_0. A numerical estimate of this effect could be made, once the spectral intensity of the normal resistance noise has been measured.

8.8.4.   *Quantum flux creep in bulk superconductors*

Motivated by the anomalous low-temperature results for magnetic relaxation in high-T_c superconductors, Blatter et al. (1991) developed a theory of quantum vortex creep. The main concern of this theory is the study of an elementary process of vortex tunneling between neighboring metastable minima of the pinning potential. The question of how to relate the calculated relaxation rates to the logarithmic time decay of the magnetization is not dealt with in detail. It is simply assumed that the magnetic relaxation rate due to quantum creep is given by the reciprocal logarithm of the tunneling rate, a result known to hold for thermally activated creep (Tinkham, 1975). The derivation of thermally-induced logarithmic time decay of the magnetization involves the assumption that the energy difference between subsequent potential minima is large compared to the thermal energy. If we adopt the same derivation to the quantum flux creep, then this energy difference needs to exceed the zero-point vibrational energy (in the potential well) in order that a logarithmic time decay of the magnetization ensues. Consequently, the observed quantum flux creep involves the nonlinear vortex mobility in the presence of a strongly tilted pinning potential. As discussed in Section 8.1, the nonlinear mobility is not subject to the dissipative localization that would be expected for overdamped vortices in the limit vanishing tilt. In this section we confine ourselves to the case of weak applied magnetic fields B, such that the mean spacing between the vortices $(\phi_0/B)^{\frac{1}{2}}$ is larger than the superconducting coherence length ξ. Since Blatter et al. (1991) assume that the adjacent metastable states are ξ apart, the weak-field limit corresponds to the situation where individual vortices *tunnel independently* of each other. We consider only the limit of strong dissipation, since it leads to an encouraging agreement with the magnetic relaxation data.

To calculate the tunneling rate, we employ the path-integral method, introduced in Section 4.8. As Eq. (4.104) shows, the rate is given by the imaginary part of K_B, the bounce contribution to the transition amplitude. This contribution is proportional to $\exp(-S_B/\hbar)$, where S_B is the bounce action, which for a single particle is given by Eq. (4.86). In flux creep, the object that tunnels has a stringlike form (a flux line), which calls for an extension of the path-integral approach to *quantum fields* (see Appendix A2.2). Thus, the effective one-body transition amplitude (4.123) is to be replaced by its field-theoretic analog (A2.23). This implies that the tunneling rate is given by

$$\exp(-S_{\text{eff}}/\hbar) \qquad (8.267)$$

where S_{eff} is the effective Euclidean action evaluated with the space- and time-dependent saddle point (bounce) solution for the displacement of the vortex line. An analytic solution to this problem requires solving complicated nonlinear partial differential equations for the displacement. Blatter et al. (1991) circumvent this difficulty by applying an approximate, but efficient and physically transparent, method of *dimensional estimates* (Feigel'man et al., 1989). In this method, the effective bounce action in Eq. (8.267) is approximated by

$$S_{\text{eff}} \approx t_c U_c \qquad (8.268)$$

where t_c is the characteristic tunneling time and U_c is the barrier height of the pinning potential.

Let us write down the functional of the effective action for a vortex line in the presence of the dissipation. For the line displacement given by a 2D field $\mathbf{u}(z, \tau)$, we have

$$S_{\text{eff}}[\mathbf{u}] = \int_0^{\hbar\beta} d\tau \, L[\mathbf{u}(z, \tau), \dot{\mathbf{u}}(z, \tau)] + S_D[\mathbf{u}] \tag{8.269}$$

where L is the Euclidean Lagrangian given by [see Eq. (A2.21)]

$$L = \int dz \left[\frac{1}{2} M \left(\frac{\partial \mathbf{u}}{\partial \tau} \right)^2 + \frac{1}{2} \varepsilon_l \left(\frac{\partial \mathbf{u}}{\partial z} \right)^2 + U_{\text{pin}}(z, \mathbf{u}) \right] \tag{8.270}$$

and $S_D[\mathbf{u}]$ is the dissipative action functional

$$S_D[\mathbf{u}] = \frac{\eta}{2\pi} \int_0^{\hbar\beta} d\tau \int_0^{\hbar\beta} d\tau' \int dz \left(\frac{\mathbf{u}(z, \tau) - \mathbf{u}(z, \tau')}{\tau - \tau'} \right)^2 \tag{8.271}$$

The first term on the RHS of Eq. (8.270) represents the kinetic energy of a moving vortex line. As shown in Section 8.6.6, the mass M is dominated by the core contribution written in Eq. (8.136). The second term on the RHS of Eq. (8.270) corresponds to the elastic energy of the line. Thus, ε_l is the self-energy (core energy) per unit length of the line. The last term is the pinning potential, which depends on the concrete form of the mechanism of line pinning. We note that Eq. (8.271) is a direct generalization of the dissipative action (8.218) to the z dependent displacement field.

The method of dimensional estimates is based on the fact that the bounce action satisfies a kind of Euclidean version of the virial theorem. The validity of such a theorem is not surprising, since the *saddle point solution* of the action corresponds to the *classical motion* (Feynman and Hibbs, 1965). To illustrate this theorem, let us consider the case of a single particle of mass M moving in a potential $V(x)$. Then the Euclidean action is given by Eq. (A1.23), and its saddle point solution $x = \bar{x}$ satisfies the equation

$$- M\ddot{\bar{x}} + \frac{\partial V}{\partial \bar{x}} = 0 \tag{8.272}$$

The boundary conditions relevant for the bounce trajectory are [see Eq. (4.75)]

$$\bar{x}(0) = \bar{x}(\hbar\beta) = 0 \tag{8.273}$$

Using these equations, we now apply integration by parts to the time integral of the kinetic energy, yielding

$$\frac{1}{2} M \int_0^{\hbar\beta} d\tau \, \dot{\bar{x}}^2 = \frac{1}{2} M \dot{\bar{x}}\bar{x} \Big|_0^{\hbar\beta} - \frac{1}{2} M \int_0^{\hbar\beta} d\tau \, \ddot{\bar{x}}\bar{x} = -\frac{1}{2} \int_0^{\hbar\beta} d\tau \, \bar{x} \frac{\partial V}{\partial \bar{x}} \tag{8.274}$$

The RHS of this equation has the form of the *virial of Clausius*, known in the context of the virial theorem of classical mechanics. Approximating

$V(\bar{x})$ by a quadratic function of $\bar{x}$, Eq. (8.274) can be written as

$$\frac{1}{2}M\int_0^{\hbar\beta}d\tau\,\dot{\bar{x}}^2 \approx \int_0^{\hbar\beta}d\tau\,|V(\bar{x})| \tag{8.275}$$

To see how this identity is used to estimate the tunneling time, we consider tunneling out of a metastable state of the potential sketched in Fig. 4.4. Then the bounce trajectory $\bar{x}(\tau) = q_B(\tau)$ corresponds to a classical motion in the inverted potential of Fig. 4.5. Since the particle moves a distance q_1 in a time t_c, its average velocity is $v = q_1/t_c$ and the time integral of the kinetic energy, contributed by the complete trajectory ($q_0 \to q_1 \to q_0$), is approximately given by

$$\frac{1}{2}M\int_0^{\hbar\beta}d\tau\,\dot{\bar{x}}^2 \approx Mt_c\left(\frac{q_1}{t_c}\right)^2 \tag{8.276}$$

From the form of the potential barrier of Fig. 4.4 [see also Eq. (4.100)], it follows that the time integral of the potential energy is roughly t_cV_0. Introducing this result and Eq. (8.276) into Eq. (8.275), we obtain

$$Mt_c\left(\frac{q_1}{t_c}\right)^2 \approx t_cV_0 \tag{8.277}$$

Using the approximation $V_0 \approx \frac{1}{2}M\omega_0^2q_1^2$, where ω_0 is the frequency of oscillation about the quasiequilibrium position q_0, the Eq. (8.277) yields a tunneling time $t_c \approx \sqrt{2}/\omega_0$. We note that a similar result follows from the analytic solution for the bounce trajectory. This can be seen by recalling Eq. (A11.13), which shows that the rise time of $q_B(\tau)$ on going from q_0 to q_1 is $2/\omega_0$. Now we consider a generalization of the identity (8.277) to the case of a particle moving in the presence of ohmic dissipation. The corresponding effective action is then given by Eq. (4.43). Using Eq. (4.40), we see that the dissipation can be accounted for by replacing the mass M by a frequency-dependent mass

$$M(\omega) = M\left(1 + \frac{\eta}{M|\omega|}\right) = ME(\omega) \tag{8.278}$$

where $E(\omega)$ defines the (dissipatively induced) mass-enhancement factor. Introducing a characteristic frequency $\omega_c = 2\pi/t_c$ into the Eq. (8.277) and making the replacement $M \to M(\omega_c)$, we obtain a quadratic equation for ω_c

$$\omega_c^2 + \frac{\eta}{M}\omega_c = \frac{1}{M}V_0\left(\frac{2\pi}{q_1}\right)^2 \tag{8.279}$$

which is solved by

$$\omega_c = \frac{\eta}{2M}\left[(1+\nu)^{\frac{1}{2}} - 1\right] \tag{8.280}$$

where

$$\nu = \frac{4MV_0}{\eta^2}\left(\frac{2\pi}{q_1}\right)^2 \tag{8.281}$$

Using the result (8.280) in Eq. (8.278), the mass-enhancement factor, evaluated at $\omega = \omega_c$, becomes

$$E(\omega_c) = 1 + \frac{\eta}{M|\omega_c|} = 1 + \frac{2}{\left[(1+\nu)^{\frac{1}{2}} - 1\right]} \tag{8.282}$$

We note that, for a strongly overdamped vortex, we have $\nu \ll 1$, which implies that $E(\omega_c) \simeq \eta/M|\omega_c| \gg 1$. Then the dissipative term on the LHS of Eq. (8.279) dominates over the inertial term, yielding for the tunneling time

$$t_c = \frac{2\pi}{\omega_c} \simeq \frac{\eta q_1^2}{2\pi V_0} \approx \frac{\eta}{\pi M\omega_0^2} \tag{8.283}$$

where the third equality follows by letting $V_0 \approx \frac{1}{2}M\omega_0^2 q_1^2$. The RHS of this equation again compares favorably with the τ dependence of the bounce trajectory obtained, in the overdamped limit, in Eq. (4.142). Indeed, the argument of the exponential in the latter equation shows a characteristic time $t_c = \hbar\beta b$. Using Eq. (4.143) for b, we obtain $t_c \simeq \eta/M\omega_0^2$, in a qualitative agreement with the dimensional estimate (8.283).

Let us now apply the Eqs. (8.280)–(8.283) to an overdamped vortex line described by the effective action (8.269). In distinction to the single-particle case, there are now two contributions to V_0: one due to the elastic energy, and the other one from the pinning potential. Minimizing the sum of these energies, we obtain a solution for $\mathbf{u}(z)$ that satisfies the relation

$$\frac{1}{2}\varepsilon_l \int dz \left(\frac{\partial \mathbf{u}}{\partial z}\right)^2 = \int dz\, U_{\text{pin}}(z, \mathbf{u}) \tag{8.284}$$

Following Blatter et al. (1991), we consider now a segment of a pinned vortex line of typical length L_c. Assuming that the line tunnels to the adjacent metastable state separated by a distance ξ from the initial position, the associated elastic energy is $\frac{1}{2}L_c\varepsilon_l(\xi/L_c)^2$. Using this result in Eq. (8.284) yields

$$\frac{1}{2}L_c\varepsilon_l\left(\frac{\xi}{L_c}\right)^2 = U_c \tag{8.285}$$

where U_c is the barrier height due to pinning. We can express U_c in terms of a measurable quantity: the critical current density j_c. In the presence of applied current of density j, the effective pinning potential U_{eff} is augmented by a term due to the Magnus force. Using Eq. (8.11), we have

$$U_{\text{eff}}(x) = U_{\text{pin}}(x) - \frac{j\phi_0 l_c x}{c} \tag{8.286}$$

where $U_{\text{pin}}(x)$ is the pinning potential in the absence of applied current and x is the transverse displacement of the line (averaged over z). When j reaches the critical current density j_c, the barrier of $U_{\text{eff}}(x)$ (between the neighboring metastable states) disappears. Thus, letting $x \simeq \xi/2$ and $U_{\text{pin}}(\xi/2) = U_c$, the Eq. (8.286) yields for $j = j_c$

$$U_c = \frac{j_c \phi_0 L_c \xi}{2c} \tag{8.287}$$

The hitherto unknown length L_c can also be expressed in terms of experimentally accessible quantities. This can be done by first expressing, using Eq. (8.287), L_c via U_c and then relating U_c to ε_l by means of Eq. (8.285). In this way, we obtain

$$L_c = \frac{2cU_c}{j_c\phi_0\xi} = \frac{L_c\varepsilon_l}{j_c\phi_0\xi}\left(\frac{\xi}{L_c}\right)^2 \tag{8.288}$$

which implies

$$L_c = \xi\sqrt{\frac{j_0}{j_c}} \tag{8.289}$$

where

$$j_0 = \frac{\varepsilon_l c}{\xi\phi_0} \tag{8.290}$$

We note that j_0 is the depairing current defined by balancing the Magnus force against the restoring force of the elastic deformation. The latter force corresponds to a displacement $x = \xi$ of a line segment of length ξ.

Next, let us consider the characteristic tunneling time t_c for the vortex segment of length L_c. Since the tunneling length is equal to ξ and the effective barrier is [see Eq. (8.285)]

$$V_0 \simeq 2U_c \simeq L_c\varepsilon_l\left(\frac{\xi}{L_c}\right)^2 \tag{8.291}$$

we obtain from Eq. (8.283) a tunneling time

$$t_c \simeq \frac{\eta\xi^2}{4\pi U_c} \approx \frac{\eta L_c}{\varepsilon_l} \tag{8.292}$$

We are now in a position to estimate the effective action S_{eff}. Before doing so, let us remind ourselves of the physical content of the simple estimate of S_{eff} given by Eq. (8.268) According to Eqs. (8.269)–(8.271), there are actually four contributions to the total effective action. In view of the relation (8.284), the contributions of the elastic and pinning energy are of the same magnitude. Moreover, the contribution to the action due to the kinetic energy and the dissipative action can be related to the net potential contribution by a relation similar to Eq. (8.279). Thus, within a numerical factor, the net action is given by the *pinning contribution alone*. Since the particle spends, during the bounce, a time of order t_c under the barrier of height U_c, the pinning contribution is of order $t_c U_c$, in agreement with Eq. (8.268). Using Eqs. (8.287) and (8.292), the RHS of Eq. (8.268) can be estimated as follows

$$S_{\text{eff}} \approx t_c U_c \approx \eta \left(\frac{j_c}{j_0} \right) L_c^2 \approx \eta_0 \left(\frac{j_c}{j_0} \right) L_c^3 \qquad (8.293)$$

where η_0 is the viscosity coefficient per unit length of the vortex line, given by [see Eq. (8.225)]

$$\eta_0 = \frac{\pi \hbar^2}{2 e^2 \rho_n \xi^2} \qquad (8.294)$$

Expressing L_c on the RHS of Eq. (8.293) by means of the relation (8.289), we obtain using Eq. (8.294)

$$S_{\text{eff}} \approx \frac{\hbar^2 \xi}{e^2 \rho_n} \left(\frac{j_0}{j_c} \right)^{\frac{1}{2}} \qquad (8.295)$$

An effective bounce action of this form has been derived by Blatter et al. (1991).

Now it is important to verify the overdamping condition ($\nu \ll 1$), which has been used to derive this result. A compact expression for ν follows from Eq. (8.281) by relating V_0 to ε_l. Using Eqs. (8.291) and (8.285), we have

$$V_0 = 2 U_c = L_c \varepsilon_l \left(\frac{\xi}{L_c} \right)^2 \qquad (8.296)$$

Letting $\eta = L_c \eta_0$, $M = L_c \mu_{\text{core}}$, and $q_1 = \xi$, Eq. (8.281) yields, with the help of Eq. (8.296),

$$\nu = \frac{16 \pi^2 \mu_{\text{core}} \varepsilon_l}{\eta_0^2 L_c^2} \qquad (8.297)$$

The core energy (self-energy) of the line can be eliminated from the RHS of this expression by using the energy–mass relation (8.138)

$$\varepsilon_l = \varepsilon_{\text{core}} = \frac{1}{3} \mu_{\text{core}} v_F^2 \qquad (8.298)$$

Furthermore, we can express L_c with use of Eq. (8.289). In this way, we obtain from Eqs. (8.297) and (8.298)

$$\nu = \frac{16 \pi^2}{3} \frac{\mu_{\text{core}}^2 v_F^2}{\eta_0^2 \xi^2} \left(\frac{j_c}{j_0} \right) \qquad (8.299)$$

Blatter et al. (1991) have proposed a following simplification of the core contribution to the vortex mass (derived by Suhl, 1965)

$$\mu_{\text{core}} \approx \left(\frac{2}{\pi^3} \right) m^* k_F \qquad (8.300)$$

where m^* is the effective mass of the charge carrier ($m^* \approx 10m_e$ in high-T_c materials). Using this expression and approximating η_0 by $\eta \simeq \hbar^2 \times (e^2\rho_n\xi^2)^{-1}$, Eq. (8.299) yields

$$\nu = \left(\frac{64}{3\pi^4}\right)\left(\frac{e^2\rho_n}{\hbar\xi}\right)^2 (k_F\xi)^4 \left(\frac{j_c}{j_0}\right) \qquad (8.301)$$

A numerical estimate of ν for high-T_c superconductors has been made by Blatter et al. (1991), using $\rho_n \approx 100 \ \mu\Omega\,\mathrm{cm}$, $k_F \approx 0.5 \times 10^8 \ \mathrm{cm}^{-1}$, $\xi \approx 15 \times 10^{-8}$ cm, and $j_c/j_0 \approx 10^{-2}$. With these values, Eq. (8.301) yields $\nu \simeq 0.2$, indicating that overdamping condition is satisfied. However, it should be pointed out that large uncertainties are brought into this estimate by our poor knowledge of some of the entering parameters. For instance, using $m^* \approx 10m_e$ and $k_F \approx 0.5 \times 10^8 \ \mathrm{cm}^{-1}$, the simplified expression (8.300) yields $\mu_{\mathrm{core}} \approx 0.32 \times 10^8 m_e/\mathrm{cm}$. On the other hand, starting from Eq. (8.119) and using parameters appropriate for the Y-Ba-Cu-O sample, we find for μ_{core} a value that is 7.2 times larger implying an underdamped vortex (Šimánek, 1991). In this work we also propose another contribution to vortex mass, which comes from the time-dependent strains induced by a moving vortex line in a deformable superconductor. This contribution turns out to be three orders of magnitude smaller than μ_{core} (Coffey, 1993; Duan and Šimánek, 1993). Moreover, for a vortex moving faster than the speed of sound this polaronic cloud is expected to be shaken off (Duan and Leggett, 1992). Let us verify if this is taking place in real life. For a vortex line tunneling through distance ξ in a time t_c, the transit velocity is $v_c \approx \xi/t_c$. Using the expression (8.292), we have

$$v_c \approx \frac{\varepsilon_l\xi}{L_c^2\eta} \qquad (8.302)$$

An orientational estimate of this quantity, for a Y-Ba-Cu-O sample, yields $v_c \approx 5\times10^5$ cm/sec which should be compared with the sound velocity $v_s \approx 10^5$ cm/sec (Šimánek, 1992). Thus, the shake-off criterion seems satisfied, though not convincingly, in the overdamped limit. There is yet another complication arising from the nonadiabatic response of the core states to the motion of the vortex line. It turns out that (for the same sample) the transit velocity (8.302) is an order of magnitude larger than the critical adiabatic velocity v_a given by (Šimánek, 1992)

$$v_a \approx \frac{0.35\Delta_\infty}{\hbar k_F^2\xi} \qquad (8.303)$$

Hence, nonadiabatic transitions among core states take place as the vortex line moves in an elementary tunnel process. This implies that the concept of a normal core anchored to the moving line is not applicable to tunneling. Even for a static vortex line, there is a problem with the normal core concept in materials, where the energy $\Delta\epsilon_c = \Delta_\infty^2/\epsilon_F$ is large compared to the thermal energy k_BT. Caroli et al. (1964) solved the Bogoliubov equations (Bogoliubov et al., 1959) for the gap function of a static vortex and found that the lowest-energy bound states have a spacing of order $\Delta\epsilon_c$. In high-T_c materials, a value of $\Delta\epsilon_c \approx 10k_B$ K is possible, indicating that the core conductivity is blocked at the measured temperature. In other words, the normal core resistivity in the expression (8.294) should be replaced by a much larger *subgap* resistivity.

Finally, let us make an orientational estimate of the constant α, characterizing the possibility of a dissipatively induced localization transition of the vortex. Introducing the tunneling distance $a = \xi$ and using the expressions (8.288) and (8.294), Eq. (8.256) yields

$$\alpha = \frac{\eta\xi^2}{2\pi\hbar} = \frac{\hbar\xi}{4e^2\rho_n}\sqrt{\frac{j_0}{j_c}} \qquad (8.304)$$

Taking $\rho_n \approx 100\ \mu\Omega\,\text{cm}$, $\xi \approx 15$ Å, and $j_0/j_c \approx 10^2$, the RHS of this equation yields $\alpha = 15$. Thus, the vortices appear to be *dissipatively localized* in high-T_c materials. It should be noted, however, that such a localization pertains only to the linear vortex mobility measured in the limit of vanishing transport current. The logarithmic time decay of the magnetization, however, involves vortex transport through strongly tilted potential barriers. Dissipative localization of vortices is, therefore, not involved in these measurements.

I have recently received a preprint by Duan (1993) in which he calculates the electromagnetic mass of a vortex from the wavelike TDGL model without imposing perfect screening. The method used in his derivation differs from that used in Section 8.6.5 in that the potential is eliminated using the Poisson equation. His results are nevertheless consistent with our work. Specifically, his expression for the electromagnetic mass can be obtained directly from Eq. (8.151), if ξ^{-1} is used in the upper cutoff of the k summation.

A1

PARTITION FUNCTION FOR ONE-BODY MOTION

To introduce the path-integral method of Feynman, we start with the simplest case of a single mechanical particle in one-dimensional space. The coordinate of the particle x is an extended variable defined for $x \in [-\infty, \infty]$. Moreover, the motion is assumed to be unconstrained. Hence, if x is to represent the phase variable θ in a superconductor or a Josephson junction, then the points $\theta = 0$ and 2π are considered as *distinguishable*. The opposite case, for which these points are physically indistinguishable, leads to a model of the particle on a circle, and the corresponding path integration is discussed in Appendix A3.

Consider the partition function, Z, defined as

$$Z = \mathrm{Tr}\left(e^{-\beta \widehat{H}}\right) = \sum_n \langle n|e^{-\beta \widehat{H}}|n\rangle \qquad (A1.1)$$

where $\beta = 1/k_B T$ and the Hamiltonian operator $\widehat{H}$ is

$$\widehat{H} = -\frac{\hbar^2}{2m}\frac{d^2}{dx^2} + V(x) = \widehat{K} + V \qquad (A1.2)$$

where m is the mass of the particle and $\widehat{K}$ and $V(x)$ are the kinetic and potential energy operators, respectively. In Eq. (A1.1), the states $|n\rangle$ are the energy eigenstates. To express Z as a Feynman path integral, we introduce the coordinate representation in which $|x\rangle$ is the coordinate eigenstate with the eigenvalue x. We have $\langle x|x'\rangle = \delta(x - x')$, so that

$$\langle n|x\rangle = \int_{-\infty}^{\infty} dx'\, \psi_n^*(x')\, \delta(x - x') = \psi_n^*(x) \qquad (A1.3)$$

and

$$\langle x'|n\rangle = \int_{-\infty}^{\infty} dx\, \delta(x - x')\psi_n(x) = \psi_n(x') \qquad (A1.4)$$

Using Eqs. (A1.3) and (A1.4) and the completeness relation, we have

$$\sum_n \psi_n^*(x)\psi_n(x') = \sum_n \langle x'|n\rangle\langle n|x\rangle = \delta(x - x') \qquad (A1.5)$$

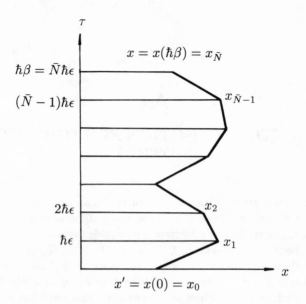

FIG. A1.1. A path of a particle traveling from x' to x in imaginary time, sliced into segments of duration $\hbar\epsilon$.

Next we insert a complete set of coordinate eigenstates into Eq. (A1.1), perform the summation over n, and use Eq. (A1.5)

$$Z = \sum_n \int_{-\infty}^{\infty} dx \int_{-\infty}^{\infty} dx' \, \langle n|x\rangle\langle x|e^{-\beta\widehat{H}}|x'\rangle\langle x'|n\rangle$$

$$= \int_{-\infty}^{\infty} dx \, \langle x|e^{-\beta\widehat{H}}|x\rangle = \int_{-\infty}^{\infty} dx \, \rho(x, x) \qquad \text{(A1.6)}$$

where we have defined the density matrix in the coordinate representation

$$\rho(x, x') = \langle x|e^{-\beta\widehat{H}}|x'\rangle \qquad \text{(A1.7)}$$

Following Feynman's "time-slicing" procedure (Feynman, 1972), we divide the interval $[0, \beta]$ into $\bar{N}$ divisions of length ϵ such that (see Fig. A1.1)

$$\bar{N}\epsilon = \beta \qquad \text{(A1.8)}$$

Inserting a complete set of coordinate eigenstates $|x_n\rangle = |x(\tau_n)\rangle$ at each point $\tau_n = n\hbar\epsilon$, we obtain from Eq. (A1.1)

$$\rho(x, x') = \langle x|e^{-\epsilon\widehat{H}}e^{-\epsilon\widehat{H}}\cdots e^{-\epsilon\widehat{H}}|x'\rangle$$

$$= \int_{-\infty}^{\infty} dx_1 \int_{-\infty}^{\infty} dx_2 \cdots \int_{-\infty}^{\infty} dx_{\bar{N}-1} \, \langle x|e^{-\epsilon\widehat{H}}|x_{\bar{N}-1}\rangle$$

$$\times \langle x_{\bar{N}-1}|e^{-\epsilon\widehat{H}}|x_{\bar{N}-2}\rangle \cdots \langle x_1|e^{-\epsilon\widehat{H}}|x'\rangle \qquad \text{(A1.9)}$$

To evaluate the matrix element $\langle x_{n+1}|e^{-\epsilon\widehat{H}}|x_n\rangle$, we take advantage of the smallness of the interval ϵ, which allows us to factorize the exponential operator $e^{-\epsilon\widehat{H}}$ as follows [see Eq. (2.136)]

$$e^{-\epsilon\widehat{H}} = e^{-\epsilon(\widehat{K}+V)} = 1 - \epsilon\widehat{K} - \epsilon V + O(\epsilon^2)$$

$$\simeq e^{-\epsilon\widehat{K}}e^{-\epsilon V} \tag{A1.10}$$

Using this approximation, we obtain

$$\langle x_{n+1}|e^{-\epsilon\widehat{H}}|x_n\rangle \simeq \langle x_{n+1}|e^{-\epsilon\widehat{K}}e^{-\epsilon V}|x_n\rangle$$

$$= \int_{-\infty}^{\infty} dy \,\langle x_{n+1}|e^{-\epsilon\widehat{K}}|y\rangle\langle y|e^{-\epsilon V}|x_n\rangle \tag{A1.11}$$

The matrix element of $e^{-\epsilon\widehat{K}}$ is explicitly given by

$$\langle x_{n+1}|e^{-\epsilon\widehat{K}}|y\rangle = \int_{-\infty}^{\infty} dx\,\delta(x - x_{n+1})\exp\left(\frac{\epsilon\hbar^2}{2m}\frac{\partial^2}{\partial x^2}\right)\delta(x - y) \tag{A1.12}$$

The differentiation of $\delta(x-y)$ in Eq. (A1.12) is executed using the integral representation

$$\delta(x - y) = \int_{-\infty}^{\infty}\frac{dk}{2\pi}e^{ik(x-y)} \tag{A1.13}$$

Then we may write

$$\exp\left(\frac{\epsilon\hbar^2}{2m}\frac{\partial^2}{\partial x^2}\right)\delta(x - y) = \int_{-\infty}^{\infty}\frac{dk}{2\pi}\left(1 + \frac{\epsilon\hbar^2}{2m}\frac{\partial^2}{\partial x^2} + \cdots\right)e^{ik(x-y)}$$

$$= \int_{-\infty}^{\infty}\frac{dk}{2\pi}\left(1 - \frac{\epsilon\hbar^2}{2m}k^2 + \cdots\right)e^{ik(x-y)}$$

$$= \int_{-\infty}^{\infty}\frac{dk}{2\pi}e^{-\epsilon\frac{\hbar^2 k^2}{2m}}e^{ik(x-y)} \tag{A1.14}$$

Introducing this result into Eq. (A1.12) and changing the order of the integrations, we have

$$\langle x_{n+1}|e^{-\epsilon\widehat{K}}|y\rangle = \int_{-\infty}^{\infty}\frac{dk}{2\pi}\int_{-\infty}^{\infty} dx\,\delta(x - x_{n+1})e^{-\epsilon\frac{\hbar^2 k^2}{2m}}e^{ik(x-y)}$$

$$= \int_{-\infty}^{\infty}\frac{dk}{2\pi}e^{-\epsilon\frac{\hbar^2 k^2}{2m}}e^{ik(x_{n+1}-y)}$$

$$= \left(\frac{m}{2\pi\epsilon\hbar^2}\right)^{\frac{1}{2}}\exp\left(-\frac{m}{2\epsilon\hbar^2}(x_{n+1} - y)^2\right) \tag{A1.15}$$

The matrix element of $e^{-\epsilon V}$ is

$$\langle y|e^{-\epsilon V}|x_n\rangle = \int_{-\infty}^{\infty} dx\,\delta(x - y)e^{-\epsilon V(x)}\,\delta(x - x_n)$$

$$= \delta(x_n - y)e^{-\epsilon V(x_n)} \tag{A1.16}$$

Using Eqs. (A1.15) and (A1.16) in Eq. (A1.11) and performing the integration over y, we obtain the desired expression for the matrix element of $e^{-\epsilon \widehat{H}}$

$$
\langle x_{n+1}|e^{-\epsilon \widehat{H}}|x_n\rangle = \left(\frac{m}{2\pi\epsilon\hbar^2}\right)^{\frac{1}{2}} \exp\left(-\frac{m}{2\epsilon\hbar^2}(x_{n+1}-x_n)^2 - \epsilon V(x_n)\right)
$$
$$
= \left(\frac{m}{2\pi\epsilon\hbar^2}\right)^{\frac{1}{2}} \exp(-\epsilon L_n) \tag{A1.17}
$$

where

$$
L_n = \frac{m}{2}\left(\frac{x_{n+1}-x_n}{\hbar\epsilon}\right)^2 + V(x_n) \tag{A1.18}
$$

Using this result in Eq. (A1.9), we obtain

$$
\rho(x,x') = \int_{-\infty}^{\infty} dx_1 \cdots \int_{-\infty}^{\infty} dx_{\bar{N}-1} \left(\frac{m}{2\pi\epsilon\hbar^2}\right)^{\frac{\bar{N}}{2}} \exp\left(-\epsilon \sum_{n=0}^{\bar{N}-1} L_n\right) \tag{A1.19}
$$

In the limit $\epsilon \to 0$, the n sum in Eq. (A1.19) becomes an integral over the continuous variable $\tau \in [0,\hbar\beta]$ (which has the dimension of time)

$$
\epsilon \sum_{n=0}^{\bar{N}-1} L_n \xrightarrow[\epsilon\to 0]{} \frac{1}{\hbar}\int_0^{\hbar\beta} d\tau\, L[x(\tau),\dot{x}(\tau)] \tag{A1.20}
$$

where

$$
L[x(\tau),\dot{x}(\tau)] = \frac{m}{2}\left(\frac{dx}{d\tau}\right)^2 + V(x,\tau) \tag{A1.21}
$$

The expression (A1.19) defines the path integral

$$
\rho(x,x') = \int_{x(0)=x'}^{x(\hbar\beta)=x} \mathcal{D}x(\tau)\, e^{-\frac{1}{\hbar}S[x]} \tag{A1.22}
$$

where

$$
S[x] = \int_0^{\hbar\beta} d\tau\, L[x(\tau),\dot{x}(\tau)] \tag{A1.23}
$$

The path integration in Eq. (A1.22) is over all paths satisfying the boundary conditions (see Fig. A1.1)

$$
\begin{aligned}
x(\tau=0) &= x_0 = x' \\
x(\tau=\hbar\beta) &= x_{\bar{N}} = x
\end{aligned} \tag{A1.24}
$$

The quantity $S[x]$ is what is often called the *Euclidean action*. We note, in this context, that a d-dimensional space combined with an imaginary time $\tau = it$ has the same geometrical properties as the $(d+1)$-dimensional

Euclidean space (Kleinert, 1990). That the variable τ has the meaning of an imaginary time can be seen by considering the $T = 0$ limit, in which the expression (A1.22) reduces to the quantum-mechanical (vacuum) transition amplitude given by the path integration in real time t [see Eq. (A10.2)]

$$\langle x(t)|x'(t')\rangle = \int \mathcal{D}x(t) \, \exp\left(\frac{i}{\hbar} \int_{t'}^{t} dt \, L^{(r)}(x,\dot{x})\right) \qquad \text{(A1.25)}$$

where $L^{(r)}$ is the real-time Lagrangian. Thus the expression (A1.22) may be regarded as an analytic continuation of (A1.25) to the imaginary time

$$\hbar\beta \to i(t - t') \qquad \text{(A1.26)}$$

Introducing the expression (A1.22) into Eq. (A1.6), we obtain the path-integral formulation for the partition function

$$Z = \int_{-\infty}^{\infty} dx \int_{x(0)=x}^{x(\hbar\beta)=x} \mathcal{D}x(\tau) \, e^{-\frac{1}{\hbar}S[x]} \qquad \text{(A1.27)}$$

A2

PARTITION FUNCTION FOR N-BODY SYSTEMS AND FIELDS

A2.1. N-body system

The result (A1.27) can be easily generalized to a many-body system that is described by the Hamiltonian

$$\widehat{H} = -\frac{\hbar^2}{2m} \sum_{i=1}^{N} \frac{\partial^2}{(\partial x^{(i)})^2} + V(x^{(1)}, x^{(2)}, \ldots, x^{(N)}) \qquad (A2.1)$$

In Eq. (A2.1), N is used for the number of particles, to be distinguished from $\bar{N}$ (number of time slices). The energy eigenstates $|n\rangle$ are the many-body energy eigenstates, which are functions of the set of particle coordinates $\{x\} = x^{(1)}, x^{(2)}, \ldots, x^{(N)}$. The coordinate eigenstates $|x\rangle$ must now be replaced by the product states

$$|x\rangle \to |\{x\}\rangle = |x^{(1)}, x^{(2)}, \ldots, x^{(N)}\rangle = \prod_{i=1}^{N} |x^{(i)}\rangle \qquad (A2.2)$$

where $\langle x'^{(i)}|x^{(i)}\rangle = \delta(x^{(i)} - x'^{(i)})$. Applying this extension to Eqs. (A1.3)–(A1.6), we obtain a generalization of Eq. (A1.5) of the form

$$\sum_n \langle \{x'\}|n\rangle\langle n|\{x\}\rangle = \prod_{i=1}^{N} \delta(x^{(i)} - x'^{(i)}) \qquad (A2.3)$$

Inserting the complete set of coordinate eigenstates (A2.2) into the expression (A1.1), we obtain with the use of the Eq. (A2.3)

$$Z = \left(\prod_{i=1}^{N} \int_{-\infty}^{\infty} dx^{(i)} \right) \rho(\{x\}, \{x\}) \qquad (A2.4)$$

where

$$\rho(\{x\}, \{x'\}) = \langle \{x\}|e^{-\beta\widehat{H}}|\{x'\}\rangle \qquad (A2.5)$$

To reduce the expression (A2.5) to a Feynman path integral, we again divide the interval $[0, \beta]$ into $\bar{N}$ slices such that $\bar{N}\epsilon = \beta$. Inserting the complete sets of coordinate eigenstates (A2.2) at each time $\tau_n = n\hbar\epsilon$, we end up with an expression of the same structure as Eq. (A1.9), except for the x_n integrations being replaced by multiple integrals over the coordinates of N particles

$$\int_{-\infty}^{\infty} dx_n \rightarrow \prod_{i=1}^{N} \int_{-\infty}^{\infty} dx_n^{(i)} \tag{A2.6}$$

Then the generalization of Eq. (A1.11) for the many-body system is

$$\langle\{x_{n+1}\}|e^{-\epsilon\widehat{H}}|\{x_n\}\rangle$$
$$= \left(\prod_{i=1}^{N} \int_{-\infty}^{\infty} dy^{(i)}\right) \langle\{x_{n+1}\}|e^{-\epsilon\widehat{K}}|\{y\}\rangle\langle\{y\}|e^{-\epsilon\widehat{V}}|\{x_n\}\rangle \tag{A2.7}$$

Following Eq. (A1.15), the matrix element of $e^{-\epsilon\widehat{K}}$ becomes a product of one-dimensional integrals

$$\langle\{x_{n+1}\}|e^{-\epsilon\widehat{K}}|\{y\}\rangle$$
$$= \prod_{i=1}^{N}\left[\int_{-\infty}^{\infty} \frac{dk^{(i)}}{2\pi} \int_{-\infty}^{\infty} dx^{(i)}\, \delta(x^{(i)} - x_{n+1}^{(i)})\right.$$
$$\left.\times \exp\left(-\frac{\epsilon\hbar^2}{2m}\left(k^{(i)}\right)^2\right) \exp\left[ik^{(i)}\left(x^{(i)} - y^{(i)}\right)\right]\right]$$
$$= \prod_{i=1}^{N}\left[\int_{-\infty}^{\infty} \frac{dk^{(i)}}{2\pi}\right.$$
$$\left.\times \exp\left(-\frac{\epsilon\hbar^2}{2m}\left(k^{(i)}\right)^2 + ik^{(i)}\left(x_{n+1}^{(i)} - y^{(i)}\right)\right)\right]$$
$$= \left(\frac{m}{2\pi\epsilon\hbar^2}\right)^{\frac{N}{2}} \prod_{i=1}^{N} \exp\left(-\frac{m}{2\epsilon\hbar^2}\left(x_{n+1}^{(i)} - y^{(i)}\right)^2\right) \tag{A2.8}$$

Using the eigenstates (A2.2), we calculate the matrix element of $e^{-\epsilon V}$ in Eq. (A2.7)

$$\langle\{y\}|e^{-\epsilon V}|\{x_n\}\rangle = \left(\prod_{i=1}^{N} \int_{-\infty}^{\infty} dx^{(i)}\right) \prod_{j=1}^{N} \delta(x^{(j)} - y^{(j)})$$
$$\times \exp\left[-\epsilon V(\{x\})\right] \prod_{k=1}^{N} \delta(x^{(k)} - x_n^{(k)})$$
$$= \left(\prod_{i=1}^{N} \delta(x_n^{(i)} - y^{(i)})\right) \exp\left[-\epsilon V(\{x_n\})\right] \tag{A2.9}$$

Introducing Eqs. (A2.8) and (A2.9) into Eq. (A2.7), we have

$$\langle\{x_{n+1}\}|e^{-\epsilon\widehat{H}}|\{x_n\}\rangle$$

$$= \left(\frac{m}{2\pi\epsilon\hbar^2}\right)^{\frac{N}{2}} \left(\prod_{i=1}^{N}\int_{-\infty}^{\infty} dy^{(i)}\right)$$

$$\times \left[\prod_{j=1}^{N}\exp\left(-\frac{m}{2\epsilon\hbar^2}(x_{n+1}^{(j)} - y^{(j)})\right)^2\right]$$

$$\times \prod_{k=1}^{N}\delta(x_n^{(k)} - y^{(k)})\exp\left[-\epsilon V(\{x_n\})\right]$$

$$= \left(\frac{m}{2\pi\epsilon\hbar^2}\right)^{\frac{N}{2}} \exp\left[-\epsilon\left(\sum_{i=1}^{N} K_n^{(i)} + V(\{x_n\})\right)\right] \qquad (A2.10)$$

where

$$K_n^{(i)} = \frac{m}{2}\left(\frac{x_{n+1}^{(i)} - x_n^{(i)}}{\hbar\epsilon}\right)^2 \qquad (A2.11)$$

is the kinetic energy of the ith particle at time τ_n. The density matrix (A2.5) can be written, following the one-body version (A1.9) and using Eq. (A2.10), as

$$\rho(\{x\}, \{x'\}) = \left(\prod_{i=1}^{N}\int_{-\infty}^{\infty} dx_1^{(i)}\right) \cdots \left(\prod_{j=1}^{N}\int_{-\infty}^{\infty} dx_{\bar{N}-1}^{(j)}\right)$$

$$\times \langle\{x\}|e^{-\epsilon\widehat{H}}|\{x_{\bar{N}-1}\}\rangle \cdots \langle\{x_1\}|e^{-\epsilon\widehat{H}}|\{x'\}\rangle$$

$$= \left(\frac{m}{2\pi\epsilon\hbar^2}\right)^{\frac{N\bar{N}}{2}} \left(\prod_{i=1}^{N}\int_{-\infty}^{\infty} dx_1^{(i)}\right) \cdots \left(\prod_{j=1}^{N}\int_{-\infty}^{\infty} dx_{\bar{N}-1}^{(j)}\right)$$

$$\times \exp\left(-\epsilon\sum_{n=0}^{\bar{N}-1} L_n\right) \qquad (A2.12)$$

where

$$L_n = \frac{m}{2}\sum_{i=1}^{N}\left(\frac{x_{n+1}^{(i)} - x_n^{(i)}}{\hbar\epsilon}\right)^2 + V(\{x_n\}) \qquad (A2.13)$$

is the Euclidean Lagrangian of the N-body system at time τ_n. Taking the limit $\epsilon \to 0$, the n summation in (A2.12) goes over to an integral over the continuous variable $\tau \in [0, \hbar\beta]$ and the density matrix becomes a "multiple-path" integral

$$\rho(\{x\}, \{x'\}) = \int_{\{x(0)\}=\{x'\}}^{\{x(\hbar\beta)\}=\{x\}} \left(\prod_{i=1}^{N}\mathcal{D}x^{(i)}(\tau)\right) e^{-\frac{1}{\hbar}S[\{x\}]} \qquad (A2.14)$$

where

$$S[\{x\}] = \int_0^{\hbar\beta} d\tau \, L[\{x(\tau)\}, \{\dot{x}(\tau)\}] \tag{A2.15}$$

The Lagrangian in Eq. (A2.15) is, from Eq. (A2.13), given by

$$L[\{x(\tau)\}, \{\dot{x}(\tau)\}] = \frac{m}{2} \sum_{i=1}^{N} \left(\frac{dx^{(i)}}{d\tau} \right)^2 + V(\{x(\tau)\}) \tag{A2.16}$$

The path integration in Eq. (A2.14) is done over the sets of paths satisfying the boundary conditions

$$\begin{aligned}
\{x(\tau = 0)\} &\equiv \{x^{(1)}(0), x^{(2)}(0), \ldots, x^{(N)}(0)\} \\
&= \{x'^{(1)}, x'^{(2)}, \ldots, x'^{(N)}\} \\
\{x(\tau = \hbar\beta)\} &\equiv \{x^{(1)}(\hbar\beta), x^{(2)}(\hbar\beta), \ldots, x^{(N)}(\hbar\beta)\} \\
&= \{x^{(1)}, x^{(2)}, \ldots, x^{(N)}\}
\end{aligned} \tag{A2.17}$$

Using the expression (A2.14) in Eq. (A2.4), we obtain the partition function of the N-body system

$$\begin{aligned}
Z = &\left(\prod_{i=1}^{N} \int_{-\infty}^{\infty} dx^{(i)} \right) \int_{\{x(0)\}=\{x\}}^{\{x(\hbar\beta)\}=\{x\}} \left(\prod_{j=1}^{N} \mathcal{D}x^{(j)}(\tau) \right) \\
&\times \exp\left(-\frac{1}{\hbar} \int_0^{\hbar\beta} d\tau \, L[\{x(\tau)\}, \{\dot{x}(\tau)\}] \right)
\end{aligned} \tag{A2.18}$$

A2.2.   Quantum fields

The expression (A2.18) allows us to find the partition function of a quantized field. The simplest prototype of a scalar field is the displacement of a continuous elastic string, which we denote by $\psi(x, \tau)$, where $x \in [0, L]$ is the position along the string of length L and $\tau \in [0, \hbar\beta]$ is the imaginary-time variable. To make a contact with the N-body problem presented, we first discretize the string by dividing it into N segments, each of length Δx. Then the displacements $\psi(x^{(i)}, \tau)$ form an enumerable set that can be mapped onto the set $\{x(\tau)\}$ of the N-body system. Thus the partition function for the quantized scalar field is obtained by making, in Eqs. (A2.13)–(A2.18), the replacement

$$\{x(\tau)\} \rightarrow \{\psi(x, \tau)\} \tag{A2.19}$$

After the limit $\Delta x \rightarrow 0$ is taken, Eq. (A2.14) yields

$$\rho(\{\psi\}, \{\psi'\}) = \int_{\{\psi(0)\}=\{\psi'\}}^{\{\psi(\hbar\beta)\}=\{\psi\}} \mathcal{D}\psi(x, \tau) \exp\left(-\frac{1}{\hbar} S[\psi] \right) \tag{A2.20}$$

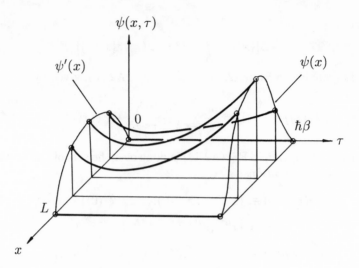

FIG. A2.1.   A set of paths, representing the displacement of a string (pinned at $x = 0$ and L) and satisfying the boundary conditions (A2.23).

where $S[\psi]$ is the Euclidean action functional

$$S[\psi] = \int_0^{\hbar\beta} d\tau \, L[\psi(x,\tau), \dot\psi(x,\tau)] \qquad (A2.21)$$

and $L[\psi, \dot\psi]$ is the Euclidean ("imaginary-time") Lagrangian of the string, which, according to Eq. (A2.16), is given by

$$L[\psi, \dot\psi] = \frac{1}{2}\rho \int_0^L dx \left[\left(\frac{\partial \psi(x,\tau)}{\partial \tau}\right)^2 + c^2 \left(\frac{\partial \psi(x,\tau)}{\partial x}\right)^2 \right] \qquad (A2.22)$$

The paths involved in the expression (A2.20) satisfy the boundary conditions

$$\{\psi(0)\} = \psi(x, \tau = 0) = \{\psi'\} = \psi'(x)$$
$$\{\psi(\hbar\beta)\} = \psi(x, \tau = \hbar\beta) = \{\psi\} = \psi(x) \qquad (A2.23)$$

Figure A2.1 shows one set of paths $\psi(x,\tau)$ forming a surface attached at $\tau = 0$ and $\hbar\beta$ to the fixed curves $\psi'(x)$ and $\psi(x)$, respectively. Integration over all possible such surfaces is to be done in Eq. (A2.20). The partition function is obtained, according to Eq. (A2.18), by integrating Eq. (A2.20) over all possible "boundary functions," yielding

$$Z = \int_{\psi(x,0)=\psi(x,\hbar\beta)} \mathcal{D}\psi(x,\tau) \exp\left(-\frac{1}{\hbar}S[\psi]\right) \qquad (A2.24)$$

where the path integral in Eq. (A2.24) is over all the functions $\psi(x,\tau)$ satisfying the condition $\psi(x,0) = \psi(x,\hbar\beta)$.

A3

PARTITION FUNCTION FOR A PARTICLE ON A CIRCLE

The particle on a circle provides us with the simplest example of path integrals with a *topological constraint* (Kleinert, 1990). In this case the particle coordinate is the angle $\theta \in [0, 2\pi]$, and the fact that the points $\theta = 0$ and 2π are *indistinguishable* is responsible for the topological constraint. The partition function Z is given by the expression (A1.1), into which we insert the angle eigenstates $|\theta\rangle$ and $|\theta'\rangle$

$$Z = \sum_n \int_0^{2\pi} d\theta \int_0^{2\pi} d\theta' \, \langle n|\theta\rangle\langle\theta|e^{-\beta\widehat{H}}|\theta'\rangle\langle\theta'|n\rangle \tag{A3.1}$$

Similar to Eqs. (A1.3)–(A1.5), we have

$$\langle n|\theta\rangle = \int_0^{2\pi} d\theta'' \, \delta(\theta - \theta'')\psi_n^*(\theta'') = \psi_n^*(\theta)$$

$$\langle\theta'|n\rangle = \int_0^{2\pi} d\theta \, \delta(\theta - \theta')\psi_n(\theta) = \psi_n(\theta') \tag{A3.2}$$

and

$$\sum_n \psi_n^*(\theta)\psi_n(\theta') = \sum_n \langle\theta'|n\rangle\langle n|\theta\rangle = \delta(\theta - \theta') \tag{A3.3}$$

Performing the n summation in Eq. (A3.1) and using Eq. (A3.3), the partition function becomes

$$Z = \int_0^{2\pi} d\theta \, \langle\theta|e^{-\beta\widehat{H}}|\theta\rangle = \int_0^{2\pi} d\theta \, \rho(\theta, \theta) \tag{A3.4}$$

The Hamiltonian $\widehat{H}$ is that of the planar rotator in an external potential $V(\theta)$

$$\widehat{H} = -\frac{\hbar^2}{2I}\frac{\partial^2}{\partial\theta^2} + V(\theta) = -\frac{\alpha}{2}\frac{\partial^2}{\partial\theta^2} + V(\theta) = \widehat{K} + V \tag{A3.5}$$

where $\alpha = \hbar^2/I$ and $I = ma^2$ is the moment of inertia of the particle of mass m constrained to move along a circular path of radius a. The potential $V(\theta)$ is a *periodic* function of the angle

$$V(\theta + 2\pi n) = V(\theta) \tag{A3.6}$$

where n is an integer. Note that (A3.6) must hold if $\theta = 0$ and 2π are *indistinguishable*.

To express the density matrix $\rho(\theta, \theta')$ as a path integral, we divide, as before, the interval $[0, \beta]$ into $\bar{N}$ divisions and obtain in analogy to Eq. (A1.9)

$$\rho(\theta, \theta') = \langle \theta | e^{-\epsilon \widehat{H}} \cdots e^{-\epsilon \widehat{H}} | \theta' \rangle$$

$$= \int_0^{2\pi} d\theta_1 \int_0^{2\pi} d\theta_2 \cdots \int_0^{2\pi} d\theta_{\bar{N}-1} \, \langle \theta | e^{-\epsilon \widehat{H}} | \theta_{\bar{N}-1} \rangle$$

$$\times \langle \theta_{\bar{N}-1} | e^{-\epsilon \widehat{H}} | \theta_{\bar{N}-2} \rangle \cdots \langle \theta_1 | e^{-\epsilon \widehat{H}} | \theta' \rangle \qquad (A3.7)$$

The matrix element of $e^{-\epsilon \widehat{H}}$ can be factorized, for small ϵ, yielding [see Eq. (A1.11)]

$$\langle \theta_{n+1} | e^{-\epsilon \widehat{H}} | \theta_n \rangle \simeq \langle \theta_{n+1} | e^{-\epsilon \widehat{K}} e^{-\epsilon V} | \theta_n \rangle$$

$$= \int_0^{2\pi} d\theta'_n \, \langle \theta_{n+1} | e^{-\epsilon \widehat{K}} | \theta'_n \rangle \langle \theta'_n | e^{-\epsilon V} | \theta_n \rangle \qquad (A3.8)$$

It is often convenient to work with the angle variable θ defined over an extended range: $\theta \in [-\infty, \infty]$. This requires one, first of all, to make a *periodic extension* of the angle eigenstates as follows

$$\langle \theta | \theta'_n \rangle = \delta(\theta - \theta'_n) \rightarrow \langle \theta | \theta'_n \rangle_p = \sum_{l_n=-\infty}^{\infty} \delta(\theta - \theta'_n + 2\pi l_n) \qquad (A3.9)$$

where l_n are integers. The RHS of Eq. (A3.9) is a "train" of δ functions, each of which can be replaced by its integral representation, yielding

$$\langle \theta | \theta'_n \rangle_p = \sum_{l_n=-\infty}^{\infty} \int_{-\infty}^{\infty} \frac{dk_n}{2\pi} e^{ik_n(\theta - \theta'_n + 2\pi l_n)} \qquad (A3.10)$$

Denoting the eigenstate of momentum $p = \hbar k$ by $|k\rangle$, we obtain using Eq. (A3.10)

$$\langle k | \theta'_n \rangle_p = \int_{-\infty}^{\infty} d\theta \, e^{-ik\theta} \langle \theta | \theta'_n \rangle_p$$

$$= \sum_{l_n=-\infty}^{\infty} e^{ik(-\theta'_n + 2\pi l_n)} \qquad (A3.11)$$

Operating on the state $|\theta'_n\rangle_p$ with the operator $e^{-\epsilon \widehat{K}}$ and using Eq. (A3.11), we have

$$e^{-\epsilon \widehat{K}} | \theta'_n \rangle_p = \int_{-\infty}^{\infty} \frac{dk_n}{2\pi} e^{-\epsilon \widehat{K}} | k_n \rangle \langle k_n | \theta'_n \rangle_p$$

$$= \sum_{l_n=-\infty}^{\infty} \int_{-\infty}^{\infty} \frac{dk_n}{2\pi} |k_n\rangle$$

$$\times \exp\left(-\frac{\alpha\epsilon}{2}k_n^2 + ik_n(-\theta_n' + 2\pi l_n)\right) \qquad (A3.12)$$

From Eqs. (A3.11) and (A3.12), the matrix element of the operator $e^{-\epsilon\widehat{K}}$ is obtained in the form

$$\langle\theta_{n+1}|e^{-\epsilon\widehat{K}}|\theta_n'\rangle_p$$

$$= \sum_{l_n=-\infty}^{\infty} \sum_{\tilde{l}_n=-\infty}^{\infty} \int_{-\infty}^{\infty} \frac{dk_n}{2\pi}$$

$$\times \exp\left(-\frac{\alpha\epsilon}{2}k_n^2 + ik_n\left[\theta_{n+1} - \theta_n' + 2\pi(l_n - \tilde{l}_n)\right]\right) \qquad (A3.13)$$

Using the identity (obtained by setting $l_n - \tilde{l}_n = \bar{l}_n = $ integer)

$$\sum_{l_n=-\infty}^{\infty} \sum_{\tilde{l}_n=-\infty}^{\infty} e^{2\pi i k_n(l_n - \tilde{l}_n)} = \sum_{\bar{l}_n=-\infty}^{\infty} e^{2\pi i k_n \bar{l}_n} \qquad (A3.14)$$

we have from Eq. (A3.13) (changing $\bar{l}_n$ to l_n)

$$\langle\theta_{n+1}|e^{-\epsilon\widehat{K}}|\theta_n'\rangle_p = \sum_{l_n=-\infty}^{\infty} \int_{-\infty}^{\infty} \frac{dk_n}{2\pi}$$

$$\times \exp\left(-\frac{\alpha\epsilon}{2}k_n^2 + ik_n(\theta_{n+1} - \theta_n' + 2\pi l_n)\right) \quad (A3.15)$$

Using Eq. (A3.9), the matrix element of the operator $e^{-\epsilon V}$ between the periodic angle eigenstates is

$$\langle\theta_n'|e^{-\epsilon V}|\theta_n\rangle_p = \int_0^{2\pi} d\theta \sum_{l_n=-\infty}^{\infty} \sum_{\tilde{l}_n=-\infty}^{\infty} \delta(\theta - \theta_n' + 2\pi l_n)$$

$$\times e^{-\epsilon V(\theta)} \delta(\theta - \theta_n + 2\pi\tilde{l}_n)$$

$$= \sum_{l_n=-\infty}^{\infty} \sum_{\tilde{l}_n=-\infty}^{\infty} \delta(\theta_n - \theta_n' + 2\pi(l_n - \tilde{l}_n))e^{-\epsilon V(\theta_n - 2\pi\tilde{l}_n)}$$

$$= \sum_{\bar{l}_n} \delta(\theta_n - \theta_n' + 2\pi\bar{l}_n)e^{-\epsilon V(\theta_n)} \qquad (A3.16)$$

where we used the periodicity property of $V(\theta)$ [see Eq. (A3.6)] and the identity

$$\sum_{l_n=-\infty}^{\infty} \sum_{\tilde{l}_n=-\infty}^{\infty} \delta(\theta_n - \theta_n' + 2\pi(l_n - \tilde{l}_n)) = \sum_{\bar{l}_n} \delta(\theta_n - \theta_n' + 2\pi\bar{l}_n) \quad (A3.17)$$

obtained by setting $l_n - \tilde{l}_n = \bar{l}_n = $ integer. Using the matrix elements (A3.15) and (A3.16) in Eq. (A3.8), we obtain

$$\langle \theta_{n+1} | e^{-\epsilon \widehat{H}} | \theta_n \rangle_p = \int_0^{2\pi} d\theta'_n \sum_{l_n=-\infty}^{\infty} \sum_{\bar{l}_n=-\infty}^{\infty} \int_{-\infty}^{\infty} \frac{dk_n}{2\pi}$$

$$\times \exp\left(-\frac{\alpha\epsilon}{2} k_n^2 + ik_n(\theta_{n+1} - \theta'_n + 2\pi l_n)\right)$$

$$\times \delta(\theta_n - \theta'_n + 2\pi \bar{l}_n) e^{-\epsilon V(\theta_n)} \qquad \text{(A3.18)}$$

Integrating over θ'_n and using the identity (A3.14), Eq. (A3.18) yields

$$\langle \theta_{n+1} | e^{-\epsilon \widehat{H}} | \theta_n \rangle_p$$

$$= \sum_{l_n=-\infty}^{\infty} \int_{-\infty}^{\infty} \frac{dk_n}{2\pi}$$

$$\times \exp\left(-\frac{\alpha\epsilon}{2} k_n^2 + ik_n(\theta_{n+1} - \theta_n + 2\pi l_n) - \epsilon V(\theta_n)\right) \quad \text{(A3.19)}$$

The integration over k_n is done by means of the formula (valid for complex numbers a and b)

$$\int_{-\infty}^{\infty} dx\, e^{-ax^2 + bx} = \left(\frac{\pi}{a}\right)^{\frac{1}{2}} e^{\frac{b^2}{4a}} \qquad \text{(A3.20)}$$

Then

$$\langle \theta_{n+1} | e^{-\epsilon \widehat{H}} | \theta_n \rangle = (2\pi\alpha\epsilon)^{-\frac{1}{2}} \sum_{l_n=-\infty}^{\infty}$$

$$\times \exp\left(-\frac{1}{2\alpha\epsilon}(\theta_{n+1} - \theta_n + 2\pi l_n)^2 - \epsilon V(\theta_n)\right) \quad \text{(A3.21)}$$

Introducing the expression (A3.21) into Eq. (A3.7), the density matrix becomes

$$\rho(\theta, \theta') = (2\pi\alpha\epsilon)^{-\frac{\bar{N}}{2}} \int_0^{2\pi} d\theta_1 \int_0^{2\pi} d\theta_2 \cdots \int_0^{2\pi} d\theta_{\bar{N}-1}$$

$$\times \sum_{l_{\bar{N}-1}=-\infty}^{\infty} \sum_{l_{\bar{N}-2}=-\infty}^{\infty} \cdots \sum_{l_0=-\infty}^{\infty}$$

$$\times \exp\left(-\frac{1}{2\alpha\epsilon}(\theta_{\bar{N}} - \theta_{\bar{N}-1} + 2\pi l_{\bar{N}-1})^2 - \epsilon V(\theta_{\bar{N}-1})\right) \cdots$$

$$\times \exp\left(-\frac{1}{2\alpha\epsilon}(\theta_1 - \theta_0 + 2\pi l_0)^2 - \epsilon V(\theta_0)\right) \qquad \text{(A3.22)}$$

The θ_n integral of the matrix element (A3.21) is

$$I_n = (2\pi\alpha\epsilon)^{-\frac{1}{2}} \sum_{l_n=-\infty}^{\infty} \int_0^{2\pi} d\theta_n$$

$$\times \exp\left(-\frac{1}{2\alpha\epsilon}(\theta_{n+1} - \theta_n + 2\pi l_n)^2 - \epsilon V(\theta_n)\right) \qquad \text{(A3.23)}$$

The *infinite* sum over l_n in Eq. (A3.23) can be absorbed into the θ_n integral, if this integral is taken over an *infinite* range $[-\infty, \infty]$. Defining the "extended" variable $\bar{\theta}_n \in [-\infty, \infty]$ by

$$\bar{\theta}_n = \theta_n - 2\pi l_n \qquad \text{(A3.24)}$$

and using the periodicity of $V(\theta_n)$, we can rewrite Eq. (A3.23) as

$$I_n = (2\pi\alpha\epsilon)^{-\frac{1}{2}} \sum_{l_n=-\infty}^{\infty} \int_0^{2\pi} d\bar{\theta}_n \, \exp\left(-\frac{1}{2\alpha\epsilon}(\theta_{n+1} - \bar{\theta}_n)^2 - \epsilon V(\bar{\theta}_n)\right)$$

$$= (2\pi\alpha\epsilon)^{-\frac{1}{2}} \int_{-\infty}^{\infty} d\bar{\theta}_n \, \exp\left(-\frac{1}{2\alpha\epsilon}(\theta_{n+1} - \bar{\theta}_n)^2 - \epsilon V(\bar{\theta}_n)\right) \qquad \text{(A3.25)}$$

The replacement of (A3.23) by the RHS of Eq. (A3.25) can be made for $1 \leq n \leq \bar{N} - 1$. We note that the integration over angle $\theta_0 = \theta(\tau = 0)$ is *not* present in Eq. (A3.22), so the sum over $l_{n=0} = l_0$ *remains* in I_0. Changing the notation (from $\bar{\theta}_n$ to θ_n) and using Eq. (A3.25) in Eqs. (A3.23) and (A3.22), we obtain

$$\rho(\theta, \theta') = (2\pi\alpha\epsilon)^{-\frac{\bar{N}}{2}} \int_{-\infty}^{\infty} d\theta_1 \int_{-\infty}^{\infty} d\theta_2 \cdots \int_{-\infty}^{\infty} d\theta_{\bar{N}-1}$$

$$\times \sum_{l_0=-\infty}^{\infty} \exp\left(-\frac{1}{2\alpha\epsilon}(\theta_{\bar{N}} - \theta_{\bar{N}-1})^2 - \epsilon V(\theta_{\bar{N}-1})\right) \cdots$$

$$\times \exp\left(-\frac{1}{2\alpha\epsilon}(\theta_1 - \theta_0 + 2\pi l_0)^2 - \epsilon V(\theta_0)\right)$$

$$= (2\pi\alpha\epsilon)^{-\frac{\bar{N}}{2}} \sum_{l=-\infty}^{\infty} \prod_{n=1}^{\bar{N}-1} \int_{-\infty}^{\infty} d\theta_n$$

$$\times \exp\left[-\sum_{n=0}^{\bar{N}-1}\left(\frac{1}{2\alpha\epsilon}(\theta_{n+1} - \theta_n + 2\pi l \delta_{n,0})^2 \right.\right.$$

$$\left.\left. + \epsilon V(\theta_n)\right)\right] \qquad \text{(A3.26)}$$

where a change of notation (from l_0 to l) has been made on the RHS of (A3.26). The $l = 0$ contribution to Eq. (A3.26) is given by

$$\rho_{l=0}(\theta, \theta') = (2\pi\alpha\epsilon)^{-\frac{\bar{N}}{2}} \prod_{n=1}^{\bar{N}} \int_{-\infty}^{\infty} d\theta_n \, \exp\left(-\epsilon \sum_{n=0}^{\bar{N}-1} L_n\right) \qquad \text{(A3.27)}$$

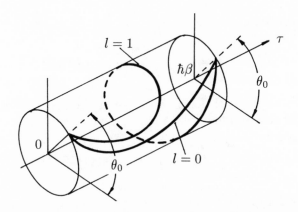

FIG. A3.1. Paths with winding numbers $l = 0$ and 1, satisfying the condition $\theta(\tau = 0) = \theta_0$ and $\theta(\tau = \hbar\beta) = \theta_0 + 2\pi l$.

where

$$L_n = \frac{\hbar^2}{2\alpha}\left(\frac{\theta_{n+1} - \theta_n}{\hbar\epsilon}\right)^2 + V(\theta_n) \qquad \text{(A3.28)}$$

The expression (A3.27) has the same form as the density matrix for the *unconstrained* matrix given in Eq. (A1.19). Taking the limit $\epsilon \to 0$, the steps (A1.20)–(A1.23) can be repeated, yielding

$$\rho_{l=0}(\theta, \theta') = \int_{\theta(0)=\theta'}^{\theta(\hbar\beta)=\theta} \mathcal{D}\theta(\tau)\, e^{-\frac{1}{\hbar}S[\theta]} \qquad \text{(A3.29)}$$

where the Euclidean action $S[\theta]$ is given by

$$S[\theta] = \int_0^{\hbar\beta} d\tau\, L[\theta(\tau), \dot{\theta}(\tau)] = \int_0^{\hbar\beta} d\tau\left(\frac{I}{2}\dot{\theta}^2 + V(\theta)\right) \qquad \text{(A3.30)}$$

For $l \neq 0$, the term involving l in Eq. (A3.26) can be written

$$\theta_{n+1} - \theta_n + 2\pi l\delta_{n,0} = \theta_1 - (\theta_0 - 2\pi l), \qquad \text{for } n = 0 \qquad \text{(A3.31)}$$

Equation (A3.31) shows that the $l \neq 0$ contribution to Eq. (A3.26) can be obtained by simply *changing the boundary condition* at $\tau = 0$ from $\theta(0) = \theta'$ to

$$\theta(0) = \theta' - 2\pi l \qquad \text{(A3.32)}$$

Then the $l \neq 0$ contribution to Eq. (A3.26) can be written as

$$\rho_l(\theta, \theta') = \int_{\theta(0)=\theta'-2\pi l}^{\theta(\hbar\beta)=\theta} \mathcal{D}\theta(\tau)\, e^{-\frac{1}{\hbar}S[\theta]} \qquad \text{(A3.33)}$$

Introducing this result into Eq. (A3.26), we have for $\theta = \theta'$

$$\rho(\theta, \theta) = \sum_{l=-\infty}^{\infty} \rho_l(\theta, \theta) = \sum_{l=-\infty}^{\infty} \int_{\theta(0)=\theta-2\pi l}^{\theta(\hbar\beta)=\theta} \mathcal{D}\theta(\tau)\, e^{-\frac{1}{\hbar} S[\theta]} \qquad (A3.34)$$

The partition function for a particle on a circle is then obtained by introducing Eq. (A3.34) into Eq. (A3.4)

$$Z = \sum_{l=-\infty}^{\infty} \int_0^{2\pi} d\theta_0 \int_{\theta_0}^{\theta_0+2\pi l} \mathcal{D}\theta(\tau)\, e^{-\frac{1}{\hbar} S[\theta]} \qquad (A3.35)$$

We have rewritten Eq. (A3.35) in a more standard but equivalent form. The path integral for $\rho(\theta, \theta)$ and Z is a sum over the path integrals involving paths with various winding numbers l. A pictorial representation of such paths is given in Fig. A3.1. A path with winding number l is shown to wind around the cylinder l times.

A4

FREE PARTICLE ON A CIRCLE

As a simple example of the evaluation of the partition function (A3.35), we consider a free particle for which $V(\theta) = 0$. First, we present a calculation of $\rho(\theta, \theta')$ directly from quantum mechanics. By inserting a complete set of the energy eigenstates $|n\rangle$ and $|n'\rangle$, we have

$$
\begin{aligned}
\rho^{\text{free}}(\theta, \theta') &= \langle\theta|e^{-\beta\widehat{H}}|\theta'\rangle \\
&= \sum_{n,n'} \langle\theta|n\rangle\langle n|e^{-\beta\widehat{H}}|n'\rangle\langle n'|\theta'\rangle \\
&= \sum_{n} \langle\theta|n\rangle\langle n|\theta'\rangle e^{-\beta E_n} \qquad\qquad (A4.1)
\end{aligned}
$$

Using Eqs. (A3.2) on the RHS of Eq. (A4.1), we obtain for a free rotator

$$
\rho^{\text{free}}(\theta, \theta') = \rho^{(0)}(\theta, \theta') = \sum_{n} \psi_n(\theta)\psi_n^*(\theta')e^{-\beta E_n} \qquad (A4.2)
$$

The energy eigenstates $\psi_n(\theta)$ satisfy the Schrödinger equation of a free rotator

$$
-\left(\frac{\alpha}{2}\right)\frac{\partial^2}{\partial\theta^2}\psi_n(\theta) = E_n\psi_n(\theta) \qquad (A4.3)
$$

The eigenfunctions $\psi_n(\theta)$ and the eigenvalues E_n of this equation are

$$
\psi_n(\theta) = (2\pi)^{-\frac{1}{2}}e^{in\theta} \qquad (A4.4)
$$

and

$$
E_n = \frac{\alpha}{2}n^2 \qquad (A4.5)
$$

respectively. Introducing Eqs. (A4.4) and (A4.5) into Eq. (A4.1), we find

$$
\rho^{(0)}(\theta, \theta') = \frac{1}{2\pi}\sum_{n=-\infty}^{\infty} \exp\left(in(\theta - \theta') - \frac{\alpha\beta n^2}{2}\right) \qquad (A4.6)
$$

Next we turn to the path-integral evaluation of $\rho(\theta, \theta')$, which turns out to be much more laborious than the straightforward quantum-mechanical method. Let us consider a single l contribution to Eq. (A3.26), which we can write explicitly as

$$\rho_l^{(0)}(\theta, \theta') = (2\pi\alpha\epsilon)^{-\frac{\bar{N}}{2}} \int_{-\infty}^{\infty} d\theta_1 \, \exp\left(-\frac{1}{2\alpha\epsilon}(\theta_1 - \theta_{0,l})^2\right)$$

$$\times \int_{-\infty}^{\infty} d\theta_2 \, \exp\left(-\frac{1}{2\alpha\epsilon}(\theta_2 - \theta_1)^2\right) \cdots$$

$$\times \int_{-\infty}^{\infty} d\theta_{\bar{N}-1} \, \exp\left(-\frac{1}{2\alpha\epsilon}(\theta_{\bar{N}} - \theta_{\bar{N}-1})^2\right) \qquad (A4.7)$$

where $\theta_{0,l} = \theta_0 - 2\pi l$. The Gaussian integrals on the RHS of Eq. (A4.7) can be evaluated with the use of formula (A3.20). Consider first the integral over θ_1, defined by

$$I_1(\theta_{0,l}, \theta_2) = \int_{-\infty}^{\infty} d\theta_1 \, \exp\left(-\frac{1}{2\alpha\epsilon}\left[(\theta_1 - \theta_{0,l})^2 + (\theta_2 - \theta_1)^2\right]\right)$$

$$= \exp\left(-\frac{1}{2\alpha\epsilon}(\theta_{0,l}^2 + \theta_2^2)\right)$$

$$\times \int_{-\infty}^{\infty} d\theta_1 \, \exp\left(-\frac{1}{\alpha\epsilon}\left[\theta_1^2 - \theta_1(\theta_{0,l} + \theta_2)\right]\right)$$

$$= \exp\left(-\frac{1}{2\alpha\epsilon}(\theta_{0,l}^2 + \theta_2^2)\right)(\pi\alpha\epsilon)^{\frac{1}{2}} \exp\left(\frac{1}{4\alpha\epsilon}(\theta_{0,l} + \theta_2)^2\right)$$

$$= \frac{(2\pi\alpha\epsilon)^{\frac{1}{2}}}{2^{\frac{1}{2}}} \exp\left(-\frac{1}{4\alpha\epsilon}(\theta_{0,l} - \theta_2)^2\right) \qquad (A4.8)$$

The integral over θ_2 in Eq. (A4.7) is obtained, using Eq. (A4.8), in the form

$$I_2 = \int_{-\infty}^{\infty} d\theta_2 \, I_1(\theta_{0,l}, \theta_2) \exp\left(-\frac{(\theta_3 - \theta_2)^2}{2\alpha\epsilon}\right)$$

$$= (\pi\alpha\epsilon)^{\frac{1}{2}} \int_{-\infty}^{\infty} d\theta_2 \, \exp\left(-\frac{(\theta_{0,l} - \theta_2)^2}{4\alpha\epsilon} - \frac{(\theta_3 - \theta_2)^2}{2\alpha\epsilon}\right) \qquad (A4.9)$$

The exponent on the RHS of Eq. (A4.9) is given by

$$-\frac{1}{4\alpha\epsilon}\left[\theta_{0,l}^2 + \theta_2^2 - 2\theta_{0,l}\theta_2 + 2(\theta_3^2 + \theta_2^2) - 4\theta_2\theta_3\right]$$

$$= -\frac{1}{4\alpha\epsilon}\left[\theta_{0,l}^2 + 2\theta_3^2 + 3\theta_2^2 - 2\theta_2(\theta_{0,l} + 2\theta_3)\right] \qquad (A4.10)$$

The integral I_2 is evaluated, using Eq. (A4.10) and the formula (A3.20), as follows

$$I_2 = (\pi\alpha\epsilon)^{\frac{1}{2}} \exp\left(-\frac{(\theta_{0,l}^2 + 2\theta_3^2)}{4\alpha\epsilon}\right)$$

$$\times \int_{-\infty}^{\infty} d\theta_2 \exp\left(-\frac{[3\theta_2^2 - 2\theta_2(\theta_{0,l} + 2\theta_3)]}{4\alpha\epsilon}\right)$$

$$= (\pi\alpha\epsilon)^{\frac{1}{2}} \exp\left(-\frac{(\theta_{0,l}^2 + 2\theta_3^2)}{4\alpha\epsilon}\right)\left(\frac{4\pi\alpha\epsilon}{3}\right)^{\frac{1}{2}} \exp\left(\frac{(\theta_{0,l} + 2\theta_3)^2}{12\alpha\epsilon}\right)$$

$$= \frac{2\pi\alpha\epsilon}{3^{\frac{1}{2}}} \exp\left(-\frac{(\theta_{0,l} - \theta_3)^2}{6\alpha\epsilon}\right) \tag{A4.11}$$

Equations (A4.8) and (A4.11) imply that the last integration (over $\theta_{\bar{N}-1}$) yields

$$I_{\bar{N}-1} = \frac{(2\pi\alpha\epsilon)^{\frac{\bar{N}-1}{2}}}{(\bar{N})^{\frac{1}{2}}} \exp\left(-\frac{(\theta_{0,l} - \theta_{\bar{N}})^2}{2\alpha\epsilon\bar{N}}\right) \tag{A4.12}$$

Introducing this result into Eq. (A4.7), we have, using the fact that $\epsilon\bar{N} = \beta$, $\theta_{0,l} = \theta(0) - 2\pi l = \theta' - 2\pi l$, and $\theta_{\bar{N}} = \theta(\hbar\beta) = \theta$

$$\rho_l^{(0)}(\theta, \theta') = (2\pi\alpha\epsilon)^{-\frac{\bar{N}}{2}} I_{\bar{N}-1}$$

$$= \frac{1}{(2\pi\alpha\epsilon\bar{N})^{\frac{1}{2}}} \exp\left(-\frac{(\theta - \theta' + 2\pi l)^2}{2\alpha\epsilon\bar{N}}\right)$$

$$= \frac{1}{(2\pi\alpha\beta)^{\frac{1}{2}}} \exp\left(-\frac{(\theta - \theta' + 2\pi l)^2}{2\alpha\beta}\right) \tag{A4.13}$$

From Eqs. (A3.34) and (A4.13), we obtain the density matrix of a free planar rotator

$$\rho^{(0)}(\theta, \theta') = \frac{1}{(2\pi\alpha\beta)^{\frac{1}{2}}} \sum_{l=-\infty}^{\infty} \exp\left(-\frac{(\theta - \theta' - 2\pi l)^2}{2\alpha\beta}\right) \tag{A4.14}$$

where we have made a change ($l \to -l$) in the dummy summation variable to bring Eq. (A4.14) to a more convenient form.

Though the expression (A4.14) looks quite different from the quantum-mechanical result (A4.6), it can be actually derived from it with the use of the Poisson formula

$$\sum_{n=-\infty}^{\infty} \delta(p - n) = \sum_{l=-\infty}^{\infty} e^{2\pi ilp} \tag{A4.15}$$

We note that the LHS of Eq. (A4.15) is a periodic train of δ functions and the RHS is its Fourier-series representation. Let us rewrite the RHS of Eq. (A4.6) so that it involves the LHS of Eq. (A4.15)

$$\rho^{(0)}(\theta, \theta') = \frac{1}{2\pi} \sum_{n=-\infty}^{\infty} \exp\left(in(\theta - \theta') - \frac{\alpha\beta}{2}n^2\right)$$

$$= \frac{1}{2\pi} \int_{-\infty}^{\infty} dp \sum_{n=-\infty}^{\infty} \delta(p-n) \exp\left(ip(\theta - \theta') - \frac{\alpha\beta}{2}p^2\right)$$

$$= \frac{1}{2\pi} \sum_{l=-\infty}^{\infty} \int_{-\infty}^{\infty} \frac{dp}{2\pi}$$

$$\times \exp\left(ip(\theta - \theta' + 2\pi l) - \frac{\alpha\beta}{2}p^2\right) \tag{A4.16}$$

where the Poisson formula (A4.15) has been used. The p integral in Eq. (A4.16) can be performed using the formula (A3.20), and the expression (A4.14) is obtained.

Let us now consider the correlator for a free rotator. Applying the equation of motion method to this problem, we obtain [see Eq. (3.106)]

$$R(\tau) = \left\langle e^{i\theta[\tau]} e^{-i\theta[0]} \right\rangle_0$$

$$= \frac{e^{-\alpha\tau/2\hbar}}{\mathrm{Tr}\left(e^{-\beta\widehat{H}_0}\right)} \sum_{n=-\infty}^{\infty} \exp\left(-\frac{1}{2}\alpha\beta n^2 + \frac{1}{\hbar}\alpha n\tau\right) \tag{A4.17}$$

where $\widehat{H}_0$ is the Hamiltonian (A3.5) with V set equal to zero. Letting $n \to n+1$ in the n summation on the RHS of Eq. (A4.17), the following identity is obtained

$$R(\tau) = R(\hbar\beta - \tau) \tag{A4.18}$$

Thus, the correlator is symmetric about the midpoint of the interval $(0, \hbar\beta)$. This property is in accord with the general correlator property

$$\langle A[\tau]B[0]\rangle = \langle B[0]A[\tau - \hbar\beta]\rangle = \langle B[\hbar\beta - \tau]A[0]\rangle \tag{A4.19}$$

We note that the first equality follows from the cyclic invariance of the trace, while the second is a result of the stationarity of the correlator. An example of a correlator satisfying Eq. (A4.19) is seen on the RHS of Eq. (3.126). The latter equation is the zero winding number part of the phase correlator (3.117). To make a transition to the winding number representation, we use the Poisson formula (A4.15) and rewrite the correlator (A4.17) as

$$R(\tau) = \frac{e^{-\alpha\tau/2\hbar}}{Z_0} \sum_{n=-\infty}^{\infty} \int_{-\infty}^{\infty} dp \, \delta(p-n) \exp\left(-\frac{1}{2}\alpha\beta p^2 + \frac{1}{\hbar}\alpha p\tau\right)$$

$$= \frac{e^{-\alpha\tau/2\hbar}}{Z_0} \int_{-\infty}^{\infty} dp \sum_{l=-\infty}^{\infty}$$

$$\times \exp\left[-\frac{1}{2}\alpha\beta p^2 + p\left(\frac{\alpha}{\hbar}\tau + 2\pi i l\right)\right] \tag{A4.20}$$

Performing the p integrations, we obtain from this equation

$$R(\tau) = \frac{1}{Z_0} \left(\frac{2\pi}{\alpha\beta} \right)^{\frac{1}{2}} \exp \left(-\frac{\alpha\tau}{2\hbar^2\beta}(\hbar\beta - \tau) \right)$$
$$\times \sum_{l=-\infty}^{\infty} \exp \left(-\frac{2\pi^2}{\alpha\beta}l^2 + \frac{2\pi i\tau}{\hbar\beta}l \right) \qquad \text{(A4.21)}$$

where

$$Z_0 = \left(\frac{2\pi}{\alpha\beta} \right)^{\frac{1}{2}} \sum_{l=-\infty}^{\infty} \exp \left(-\frac{2\pi^2}{\alpha\beta}l^2 \right) \qquad \text{(A4.22)}$$

Taking the $l = 0$ component of equation (A4.21), we have

$$R_{l=0}(\tau) = \exp \left(-\frac{\alpha\tau}{2\hbar^2\beta}(\hbar\beta - \tau) \right) \qquad \text{(A4.23)}$$

which agrees with the correlator (3.126) derived by the Feynman path-integral method. Numerical estimates show that $R_{l=0}(\tau)$ does not differ significantly from the complete correlator (A4.21) when the parameter $\alpha\beta$ is either much larger than one or below one. The difference is, in fact, responsible for the reentrant shape of the phase diagram calculated with the use of the correlator (3.126).

A5

PARTICLE ON A CIRCLE—HIGH-TEMPERATURE LIMIT

In this Appendix, we derive the partition function in the classical (high-temperature) limit characterized by the condition

$$\beta = \frac{1}{k_B T} \ll \frac{2}{\alpha} \tag{A5.1}$$

We consider the planar rotator described by the Hamiltonian (A3.5). The partition function is, according to Eq. (A3.4),

$$Z = \int_0^{2\pi} d\theta \, \rho(\theta, \theta) \tag{A5.2}$$

where the density matrix is, from Eq. (A3.7), given by the "time-sliced" expression

$$\rho(\theta, \theta) = \langle \theta | e^{-\epsilon \widehat{H}} \cdots e^{-\epsilon \widehat{H}} | \theta \rangle \tag{A5.3}$$

According to the condition (A5.1), in the high-T limit, the entire time interval $\hbar\beta = \hbar\bar{N}\epsilon$ is small. Thus, we need only the *first* slice, of width $\epsilon = \beta$, to evaluate the expression (A5.3). Thus we write

$$\rho(\theta, \theta) \simeq \langle \theta_0 | e^{-\epsilon \widehat{H}(\theta_0)} | \theta_0 \rangle \delta_{\theta, \theta_0} \tag{A5.4}$$

where $\epsilon = \beta$. To evaluate the matrix element on the RHS of Eq. (A5.4), we use the expression

$$\langle \theta_0 | e^{-\epsilon \widehat{H}} | \theta_0 \rangle = (2\pi\alpha\epsilon)^{-\frac{1}{2}} \sum_{l=-\infty}^{\infty} \exp\left(-\frac{1}{2\alpha\epsilon}(2\pi l)^2 - \epsilon V(\theta_0) \right) \tag{A5.5}$$

This expression can be obtained by applying the same steps that have led us from the expression (A3.8) to the formula (A3.21). For $\epsilon = \beta$ such that the condition (A5.1) holds, the series in Eq. (A5.5) converges rapidly, so it is legitimate to replace it by the leading $l = 0$ term, yielding

$$\rho(\theta, \theta) = (2\pi\alpha\beta)^{-\frac{1}{2}} e^{-\beta V(\theta_0)} \delta_{\theta, \theta_0} \tag{A5.6}$$

The prefactor $(2\pi\beta\alpha)^{-\frac{1}{2}}$ is the classical limit of the partition function for a free rotator. In fact, putting $V(\theta_0) = 0$ in Eq. (A5.5) and keeping only the $l = 0$ term, we have

$$\rho^{(0)}(\theta_0, \theta_0) = (2\pi\alpha\epsilon)^{-\frac{1}{2}} \tag{A5.7}$$

Introducing the result (A5.6) into Eq. (A5.2), we obtain the partition function in the high-temperature approximation

$$Z^{\text{class}} = (2\pi\alpha\beta)^{-\frac{1}{2}} \int_0^{2\pi} d\theta\, e^{-\beta V(\theta)} \tag{A5.8}$$

A6

PERTURBATION EXPANSION FOR THERMAL AVERAGES

Consider the thermal average of an operator $\widehat{O}$ defined as

$$\left\langle \widehat{O} \right\rangle = \frac{\mathrm{Tr}\left(e^{-\beta\widehat{H}}\widehat{O}\right)}{\mathrm{Tr}\left(e^{-\beta\widehat{H}}\right)} \tag{A6.1}$$

where $\widehat{H}$ is the Hamiltonian of a system that may correspond to a single particle or a many-body system, including quantum fields. We assume that $\widehat{H}$ can be separated as follows

$$\widehat{H} = \widehat{H}_0 + \widehat{H}_1 \tag{A6.2}$$

where $\widehat{H}_0$ is the Hamiltonian of the unperturbed system (leading to an exactly soluble problem) and $\widehat{H}_1$ is the perturbation. It is often desirable to have an expansion formula for $\left\langle \widehat{O} \right\rangle$ in terms of a small perturbation $\widehat{H}_1$. This can be derived by introducing the imaginary-time evolution operator $\widehat{\sigma}(\tau)$ (see Fetter and Walecka, 1971)

$$e^{-\widehat{H}\tau/\hbar} = e^{-(\widehat{H}_0+\widehat{H}_1)\tau/\hbar} = e^{-\widehat{H}_0\tau/\hbar}\widehat{\sigma}(\tau) \tag{A6.3}$$

where τ is the imaginary time. Differentiating Eq. (A6.3) with respect to τ and multiplying the resulting equation on the left by $e^{\widehat{H}_0\tau/\hbar}$, we obtain

$$\frac{d\widehat{\sigma}(\tau)}{d\tau} = -\frac{1}{\hbar}\widehat{H}_1[\tau]\widehat{\sigma}(\tau) \tag{A6.4}$$

where

$$\widehat{H}_1[\tau] = e^{\widehat{H}_0\tau/\hbar}\widehat{H}_1 e^{-\widehat{H}_0\tau/\hbar} \tag{A6.5}$$

Equation (A6.5) defines the "imaginary-time" operator in the interaction picture. Equation (A6.4) is to be solved with the initial condition

$$\widehat{\sigma}(\tau = 0) = 1 \tag{A6.6}$$

which follows from Eq. (A6.3) by setting $\tau = 0$. To first order in $\widehat{H}_1$, we obtain

$$\widehat{\sigma}_1(\tau) = 1 - \frac{1}{\hbar}\int_0^\tau d\tau'\, \widehat{H}_1[\tau'] \tag{A6.7}$$

The solution can be carried out, formally, to infinite order in $\widehat{H}_1$ by applying the method of iteration and introducing the time-ordered products of operators. The resulting expansion for $\widehat{\sigma}(\tau)$ becomes

$$
\widehat{\sigma}(\tau) = \sum_{n=0}^{\infty} (-1)^n \frac{1}{n!} \frac{1}{\hbar} \int_0^{\tau} d\tau_1 \cdots \frac{1}{\hbar} \int_0^{\tau} d\tau_n \ \mathrm{T}_\tau \left(\widehat{H}_1[\tau_1] \cdots \widehat{H}_1[\tau_n] \right)
$$
$$
= \mathrm{T}_\tau \exp \left(-\frac{1}{\hbar} \int_0^{\tau} d\tau' \ \widehat{H}_1[\tau'] \right) \tag{A6.8}
$$

where T_τ is the time-ordering operator (see Fetter and Walecka, 1971). Using Eq. (A6.3), we can express Eq. (A6.1) in terms of $\widehat{\sigma}(\hbar\beta)$

$$
\left\langle \widehat{O} \right\rangle = \frac{\mathrm{Tr} \left[e^{-\beta \widehat{H}_0} \widehat{\sigma}(\hbar\beta) \widehat{O} \right]}{\mathrm{Tr} \left[e^{-\beta \widehat{H}_0} \widehat{\sigma}(\hbar\beta) \right]} \tag{A6.9}
$$

Equation (A6.9) can be written in terms of the thermodynamic averages of operators over the unperturbed ensemble. If we define, for some operator $\widehat{B}$, such an average by

$$
\left\langle \widehat{B} \right\rangle_0 = \frac{\mathrm{Tr} \left(e^{-\beta \widehat{H}_0} \widehat{B} \right)}{\mathrm{Tr} \left(e^{-\beta \widehat{H}_0} \right)} \tag{A6.10}
$$

then the expression (A6.9) becomes

$$
\left\langle \widehat{O} \right\rangle = \frac{\left\langle \widehat{\sigma}(\hbar\beta) \widehat{O} \right\rangle_0}{\left\langle \widehat{\sigma}(\hbar\beta) \right\rangle_0} \tag{A6.11}
$$

where the operator $\widehat{\sigma}(\hbar\beta)$ is, according to Eq. (A6.8), given by

$$
\widehat{\sigma}(\hbar\beta) = \mathrm{T}_\tau \exp \left(-\frac{1}{\hbar} \int_0^{\hbar\beta} d\tau \ \widehat{H}_1[\tau] \right) \tag{A6.12}
$$

It is often adequate to calculate the quantity $\left\langle \widehat{O} \right\rangle$ to first order in $\widehat{H}_1$. Using the expression (A6.7) in Eq. (A6.11), we obtain

$$
\left\langle \widehat{O} \right\rangle \simeq \left\langle \left(1 - \frac{1}{\hbar} \int_0^{\hbar\beta} d\tau \ \widehat{H}_1 \right) \widehat{O} \right\rangle_0
$$
$$
= \frac{\mathrm{Tr} \left[e^{-\beta \widehat{H}_0} \left(1 - \frac{1}{\hbar} \int_0^{\hbar\beta} d\tau \ \widehat{H}_1 \right) \widehat{O}[0] \right]}{\mathrm{Tr} \left(e^{-\beta \widehat{H}_0} \right)} \tag{A6.13}
$$

where we have used the definition (A6.10) for the unperturbed average. The expression on the RHS of Eq. (A6.13) is reminiscent of the Kubo formula for the linear response to a static perturbation $\widehat{H}_1$. This relationship can be made more apparent by assuming that the operator $\widehat{H}_1$ is proportional to the operator $\widehat{O}$

$$\widehat{H}_1 = -\zeta\widehat{O} \tag{A6.14}$$

where ζ represents a time-independent external field. Introducing the expression (A6.14) into Eq. (A6.13), we obtain to order $\widehat{H}_1$ a Kubo-like formula

$$\delta\left\langle\widehat{O}\right\rangle = \left\langle\widehat{O}\right\rangle - \left\langle\widehat{O}\right\rangle_0 = \frac{\zeta}{\hbar}\int_0^{\hbar\beta} d\tau \left\langle\widehat{O}[\tau]\widehat{O}[0]\right\rangle_0 \tag{A6.15}$$

which expresses the deviation of the average of $\widehat{O}$, from its unperturbed value, in terms of the correlation function of the operator $\widehat{O}$. The formulas (A6.13) and (A6.15) provide a useful tool for the determination of the transition temperature of interacting systems in the mean-field approximation (see Chapter 3).

A7

PATH INTEGRAL FOR THE CORRELATION FUNCTION

The correlation function in Eq. (A6.15) represents a special case of the two-point time-ordered correlation function

$$G(\tau_A, \tau_B) = Z^{-1} \operatorname{Tr} \left(\mathrm{T}_\tau \, \widehat{O}[\tau_A] \widehat{O}[\tau_B] e^{-\beta \widehat{H}} \right) \tag{A7.1}$$

where Z is the partition function

$$Z = \operatorname{Tr} \left(e^{-\beta \widehat{H}} \right) \tag{A7.2}$$

We note that Eq. (A7.1) agrees with the definition of the finite-temperature Green's function, in the Matsubara (imaginary-time) formalism (see Fetter and Walecka, 1971). To express this Green's function as a Feynman path integral, we recall the time-slicing method, applied in Appendix A1 to the partition function for one-body motion. For the sake of simplicity, we start again with a system described by the one-body Hamiltonian

$$\widehat{H} = -\frac{\hbar^2}{2m} \frac{d^2}{dx^2} + V(x) \tag{A7.3}$$

The operator $\widehat{O} = \widehat{O}(x)$ is a function of the coordinate x. The case of $\widehat{O}(x) = \cos x$ is of special interest for Josephson systems. Inserting the complete set of coordinate eigenstates $|x\rangle$ into the trace in Eq. (A7.1), we obtain

$$\operatorname{Tr} \left\{ \mathrm{T}_\tau \, \widehat{O}[\tau_A] \widehat{O}[\tau_B] e^{-\beta \widehat{H}} \right\}$$

$$= \sum_n \langle n | \, \mathrm{T}_\tau \, \widehat{O}[\tau_A] \widehat{O}[\tau_B] e^{-\beta \widehat{H}} | n \rangle$$

$$= \sum_n \int_{-\infty}^{\infty} dx \int_{-\infty}^{\infty} dx'$$

$$\times \langle n | x \rangle \langle x | \, \mathrm{T}_\tau \, \widehat{O}[\tau_A] \widehat{O}[\tau_B] e^{-\beta \widehat{H}} | x' \rangle \langle x' | n \rangle \tag{A7.4}$$

Using Eq. (A1.5), the summation over n in Eq. (A7.4) can be performed, yielding for the Green's function (A7.1)

$$G(\tau_A, \tau_B) = Z^{-1} \int_{-\infty}^{\infty} dx \int_{-\infty}^{\infty} dx'$$

$$\times \langle x| \, \mathrm{T}_\tau \, \widehat{O}[\tau_A]\widehat{O}[\tau_B]e^{-\beta\widehat{H}}|x'\rangle \, \delta(x - x') \qquad (A7.5)$$

We next concentrate on the path-integral formulation for the matrix element $\langle x| \, \mathrm{T}_\tau \, \widehat{O}[\tau_A]\widehat{O}[\tau_B]e^{-\beta\widehat{H}}|x'\rangle$. For $0 < \tau_A < \tau_B < \hbar\beta$, we have using the cyclic invariance of the trace

$$
\begin{aligned}
&\langle x| \, \mathrm{T}_\tau \, \widehat{O}[\tau_A]\widehat{O}[\tau_B]e^{-\beta\widehat{H}}|x'\rangle \\
&= \langle x|e^{-\beta\widehat{H}}\widehat{O}[x, \tau_A]\widehat{O}[x, \tau_B]|x'\rangle \\
&= \langle x|e^{-(\hbar\beta-\tau_A)\widehat{H}/\hbar}\widehat{O}(x)e^{-(\tau_A-\tau_B)\widehat{H}/\hbar}\widehat{O}(x)e^{-\tau_B\widehat{H}/\hbar}|x'\rangle \\
&= \int_{-\infty}^{\infty} dx_A \int_{-\infty}^{\infty} dx_B \, \langle x|e^{-(\hbar\beta-\tau_A)\widehat{H}/\hbar}|x_A\rangle \\
&\quad \times \langle x_A|\widehat{O}(x)e^{-(\tau_A-\tau_B)\widehat{H}/\hbar}|x_B\rangle\langle x_B|\widehat{O}(x)e^{-\tau_B\widehat{H}/\hbar}|x'\rangle \quad (A7.6)
\end{aligned}
$$

where we have inserted the complete coordinate eigenstates $|x_A\rangle$ and $|x_B\rangle$ at the times τ_A and τ_B, respectively. The coordinate eigenstates satisfy the eigenvalue equations

$$\langle x_i|\widehat{O}(x) = O(x_i)\langle x_i|, \qquad i = A, B \qquad (A7.7)$$

where $O(x_i)$ is the c number eigenvalue of the operator $\widehat{O}(x)$ for $x = x_i = x(\tau_i)$. Using Eqs. (A7.7), Eq. (A7.6) can be written as

$$
\begin{aligned}
&\langle x| \, \mathrm{T}_\tau \, \widehat{O}[\tau_A]\widehat{O}[\tau_B]e^{-\beta\widehat{H}}|x'\rangle \\
&= \int_{-\infty}^{\infty} dx_A \int_{-\infty}^{\infty} dx_B \, \langle x|e^{-(\hbar\beta-\tau_A)\widehat{H}/\hbar}|x_A\rangle O(x_A) \\
&\quad \times \langle x_A|e^{-(\tau_A-\tau_B)\widehat{H}/\hbar}|x_B\rangle O(x_B)\langle x_B|e^{-\tau_B\widehat{H}/\hbar}|x'\rangle \quad (A7.8)
\end{aligned}
$$

Performing the slicing of the time intervals $[0, \tau_B]$, $[\tau_B, \tau_A]$, $[\tau_A, \hbar\beta]$ and using the expressions (A1.17)–(A1.23), we obtain from Eq. (A7.8)

$$
\begin{aligned}
&\langle x| \, \mathrm{T}_\tau \, \widehat{O}[\tau_A]\widehat{O}[\tau_B]e^{-\beta\widehat{H}}|x'\rangle \\
&= \int_{-\infty}^{\infty} dx_1 \cdots \int_{-\infty}^{\infty} dx_{\bar{N}-1} \left(\frac{m}{2\pi\epsilon\hbar^2}\right)^{\frac{\bar{N}}{2}} \\
&\quad \times O(x_A)O(x_B) \exp\left(-\epsilon \sum_{n=0}^{\bar{N}-1} L_n\right) \\
&= \int_{x(0)=x'}^{x(\hbar\beta)=x} \mathcal{D}x(\tau) \, O(\tau_A)O(\tau_B)e^{-\frac{1}{\hbar}S[x]} \qquad (A7.9)
\end{aligned}
$$

where

$$S[x] = \int_0^{\hbar\beta} d\tau \, L[x(\tau), \dot{x}(\tau)] \qquad (A7.10)$$

The Lagrangian L in Eq. (A7.10) is given by Eq. (A1.21). Introducing the RHS of Eq. (A7.9) into the expression (A7.5), the Green's function is, for $\tau_A > \tau_B$,

$$G(\tau_A, \tau_B) = Z^{-1} \int_{-\infty}^{\infty} dx \int_{x(0)=x}^{x(\hbar\beta)=x} \mathcal{D}x(\tau)\, O(\tau_A)O(\tau_B)e^{-\frac{1}{\hbar}S[x]}$$

$$= Z^{-1} \int \mathcal{D}x(\tau)\, O(\tau_A)O(\tau_B)e^{-\frac{1}{\hbar}S[x]} \tag{A7.11}$$

where the path integral on the RHS of Eq. (A7.11) implies an integration over *all* the paths $x(\tau)$, *regardless* of the boundary conditions on the paths. It can be shown that the same path-integral formulation holds also when $\tau_A < \tau_B$. Thus Eq. (A7.11) is a general result.

The computation of the correlation function $\left\langle \hat{O}[\tau_A]\hat{O}[\tau_B] \right\rangle$ can be reduced to a functional differentiation of the partition function $Z[\zeta]$ in the presence of an external τ-dependent field $\zeta(\tau)$. We define $Z[\zeta(\tau)]$ by a path integral

$$Z[\zeta(\tau)] = \int_{-\infty}^{\infty} dx \int_{x(0)=x}^{x(\hbar\beta)=x} \mathcal{D}x(\tau)\, e^{-\frac{1}{\hbar}S_\zeta[x]} \tag{A7.12}$$

where the action $S_\zeta[x]$ is given by

$$S_\zeta[x] = \int_0^{\hbar\beta} d\tau \left[\frac{m}{2}\left(\frac{dx}{d\tau}\right)^2 + V(x) - \zeta(\tau)O(\tau) \right] \tag{A7.13}$$

We note that Eqs. (A7.12) and (A7.13) follow from Eqs. (A1.27) and (A1.23), respectively. The Lagrangian in Eq. (A7.13) is obtained by augmenting Eq. (A1.21) by a time-dependent perturbation $-\zeta(\tau)O(\tau)$. A simple choice of $\zeta(\tau)$ of the form

$$\zeta(\tau) = \zeta_A\,\delta(\tau - \tau_A) + \zeta_B\,\delta(\tau - \tau_B) \tag{A7.14}$$

leads to the following contribution of the external field to the action (A7.13)

$$-\int_0^{\hbar\beta} d\tau\, \zeta(\tau)O(\tau) = -[\zeta_A O(\tau_A) + \zeta_B O(\tau_B)] \tag{A7.15}$$

Inserting Eq. (A7.15) into Eq. (A7.13), we obtain from Eq. (A7.12)

$$Z^{-1}[0]\left(\frac{\partial^2 Z[\zeta(\tau)]}{\partial\zeta_A\partial\zeta_B} \right)_{\zeta=0}$$

$$= Z^{-1}[0]\int_{-\infty}^{\infty} dx$$

$$\times \int_{x(0)=x}^{x(\hbar\beta)=x} \mathcal{D}x(\tau)\, O(\tau_A)O(\tau_B)e^{-\frac{1}{\hbar}S[x]} \tag{A7.16}$$

which reproduces the path-integral formula for the Green's function [see Eq. (A7.11)]. The expression on the LHS of Eq. (A7.16) defines the *functional derivative* of the generating functional $Z[\zeta]$. Hence,

$$G(\tau_A, \tau_B) = \left(\hbar^2 Z^{-1}[\zeta] \frac{\delta}{\delta\zeta(\tau_A)} \frac{\delta}{\delta\zeta(\tau_B)} Z[\zeta(\tau)] \right)_{\zeta=0} \qquad (A7.17)$$

The method of the generating functional can be extended to many-body systems and quantum fields. Consider, for example, the one-dimensional scalar field $\psi(x, \tau)$ representing the transverse displacement of an elastic string at point $x \in [0, L]$ and time $\tau \in [0, \hbar\beta]$ (see Appendix A2). The corresponding generating functional $Z[\zeta(x, \tau)]$ is the partition function of the string in the presence of an external field $\zeta(x, \tau)$. Recalling Eqs. (A2.21)–(A2.24), we have

$$Z[\zeta(x, \tau)] = \int_{\psi(x,0)=\psi(x,\hbar\beta)} \mathcal{D}\psi(x, \tau) \, \exp\left(-\frac{1}{\hbar} S_\zeta[\psi] \right) \qquad (A7.18)$$

where the action functional is given by

$$S_\zeta[\psi] = \int_0^{\hbar\beta} d\tau \int_0^L dx \left\{ \frac{1}{2}\rho \left[\left(\frac{\partial\psi(x, \tau)}{\partial\tau} \right)^2 \right. \right.$$
$$\left. \left. +c^2 \left(\frac{\partial\psi(x, \tau)}{\partial x} \right)^2 \right] - \zeta(x, \tau)\psi(x, \tau) \right\} \qquad (A7.19)$$

The two-point Green's function $G(x, \tau; x', \tau')$ is defined by

$$G(x, \tau; x', \tau') = Z^{-1} \int \mathcal{D}\psi(x, \tau) \, \psi(x, \tau)\psi(x', \tau')e^{-\frac{1}{\hbar}S[\psi]} \qquad (A7.20)$$

where

$$S[\psi] = S_{\zeta=0}[\psi] \qquad (A7.21)$$

Equation (A7.20) can be regarded, on one hand, as a generalization of Eq. (A7.11), resulting from the mapping given in Eq. (A2.19). On the other hand, generalizing the functional formula (A7.17) to space- and time-dependent $\zeta(x, \tau)$, the *same* expression (A7.20) is obtained from the formula

$$G(x, \tau; x', \tau')$$
$$= \left(\hbar^2 Z^{-1}[\zeta(x, \tau)] \frac{\delta}{\delta\zeta(x, \tau)} \frac{\delta}{\delta\zeta(x', \tau')} Z[\zeta(x, \tau)] \right)_{\zeta=0} \qquad (A7.22)$$

where $Z[\zeta(x, \tau)]$ is given by the generating functional (A7.18).

A8

CORRELATION FUNCTION IN THE CLASSICAL LIMIT

To establish a background for the derivation of Eq. (2.3), we first derive the high-temperature limit of the correlation function. We start with the simplest case of the one-body motion and confine ourselves to the equal-time correlation function of two different operators $\widehat{M}$ and $\widehat{N}$

$$\left\langle \widehat{M}[\tau_A]\widehat{N}[\tau_A] \right\rangle = Z^{-1} \int_{-\infty}^{\infty} dx \, \langle x|\widehat{M}[\tau_A]\widehat{N}[\tau_A]e^{-\beta\widehat{H}}|x\rangle \tag{A8.1}$$

where $\widehat{M}$ and $\widehat{N}$ are both functions of the coordinate x. The matrix element on the RHS of Eq. (A8.1) can be written, with the use of Eq. (A7.8), as

$$\langle x|\widehat{M}[\tau_A]\widehat{N}[\tau_A]e^{-\beta\widehat{H}}|x\rangle$$
$$= \int_{-\infty}^{\infty} dx_A \int_{-\infty}^{\infty} dx_B \, \langle x|e^{-(\hbar\beta-\tau_A)\widehat{H}/\hbar}|x_A\rangle$$
$$\times M(x_A)\langle x_A|x_B\rangle N(x_B)\langle x_B|e^{-\tau_A\widehat{H}/\hbar}|x\rangle$$
$$= \int_{-\infty}^{\infty} dx_A \, \langle x|e^{-(\hbar\beta-\tau_A)\widehat{H}/\hbar}|x_A\rangle$$
$$\times M(x_A)N(x_A)\langle x_A|e^{-\tau_A\widehat{H}/\hbar}|x\rangle \tag{A8.2}$$

where we have used the orthogonality of the coordinate eigenstates, implying: $\langle x_A|x_B\rangle = \delta(x_A - x_B)$. The matrix elements on the RHS of Eq. (A8.2) are again evaluated by the time-slicing method [see Eq. (A7.9)]. In the high-temperature limit, the entire interval $[0, \hbar\beta]$ is small. Thus, it is sufficient to keep only the first slice extending over the interval $\tau \in [0, \tau_A]$, where $\tau_A = \hbar\epsilon = \hbar\beta$. This implies for the matrix elements

$$\langle x_A|e^{-(\hbar\beta-\tau_A)\widehat{H}/\hbar}|x\rangle = \langle x|x_A\rangle$$
$$= \delta(x - x_A) \tag{A8.3}$$

and

$$\langle x_A|e^{-\tau_A\widehat{H}/\hbar}|x\rangle = \langle x_A|e^{-\beta\widehat{H}}|x\rangle$$

$$= \left(\frac{m}{2\pi\beta\hbar^2}\right)^{\frac{1}{2}}$$

$$\times \exp\left(-\frac{m}{2\beta\hbar^2}(x_A - x)^2 - \beta V(x)\right) \qquad (A8.4)$$

where Eq. (A1.17) has been used. Inserting Eqs. (A8.3) and (A8.4) into Eq. (A8.2), we obtain from Eq. (A8.1) the equal-time correlation function (in the classical limit)

$$\left\langle \widehat{M}[\tau_A]\widehat{N}[\tau_A]\right\rangle_{\text{class}}$$

$$= Z^{-1}\left(\frac{m}{2\pi\beta\hbar^2}\right)^{\frac{1}{2}} \int_{-\infty}^{\infty} dx \int_{-\infty}^{\infty} dx_A \, \delta(x - x_A)M(x_A)N(x_A)$$

$$\times \exp\left(-\frac{m}{2\beta\hbar^2}(x_A - x)^2 - \beta V(x)\right)$$

$$= Z^{-1}\left(\frac{m}{2\pi\beta\hbar^2}\right)^{\frac{1}{2}} \int_{-\infty}^{\infty} dx \, M(x)N(x)e^{-\beta V(x)} \qquad (A8.5)$$

The partition function Z, in the high-temperature limit, can be derived from Eq. (A1.9), using a similar time-slicing procedure, with the result

$$Z = \left(\frac{m}{2\pi\beta\hbar^2}\right)^{\frac{1}{2}} \int_{-\infty}^{\infty} dx \, e^{-\beta V(x)} \qquad (A8.6)$$

Using expression (A8.6) in Eq. (A8.5), we obtain

$$\left\langle \widehat{M}[\tau_A]\widehat{N}[\tau_A]\right\rangle = \frac{\int_{-\infty}^{\infty} dx \, M(x)N(x)e^{-\beta V(x)}}{\int_{-\infty}^{\infty} dx \, e^{-\beta V(x)}} \qquad (A8.7)$$

which coincides with the standard expression for the equal-time correlation function in classical statistical mechanics of a one-body system. This expression can be generalized to many-body systems and quantum fields by invoking the mapping (A2.19) in the form

$$\{x\} \rightarrow \{\psi(x)\} \qquad (A8.8)$$

We note that the τ dependence of $x(\tau)$ and $\psi(x,\tau)$ is irrelevant for the equal-time correlation function, because of the stationarity property [see Eq. (A8.5)]. As a concrete example, consider the real scalar field $\psi(x,\tau)$ describing the displacement of the string. For each position x we have a whole set of all possible values of the function $\psi(x)$. Then the equal-time correlation function becomes (in the classical limit)

$$\left\langle \widehat{\psi}[x,\tau_A]\widehat{\psi}[x',\tau_A]\right\rangle_{\text{class}} = \frac{\int \mathcal{D}\psi(x) \, \psi(x)\psi(x')e^{-\beta V[\psi,\frac{\partial\psi}{\partial x}]}}{\int \mathcal{D}\psi(x) \, e^{-\beta V[\psi,\frac{\partial\psi}{\partial x}]}} \qquad (A8.9)$$

where $V[\psi, \partial\psi/\partial x]$ is the functional of the potential energy given by

$$V\left[\psi, \frac{\partial\psi}{\partial x}\right] = \frac{\rho c^2}{L} \int_0^L dx \left(\frac{\partial\psi}{\partial x}\right)^2 \qquad \text{(A8.10)}$$

A9

DERIVATION OF EQ. (2.3)

Let us consider the classical correlation function

$$\langle \psi(\mathbf{r}) \psi^*(\mathbf{r}') \rangle = \left\langle \widehat{\psi}[\mathbf{r}, \tau] \widehat{\psi}^\dagger[\mathbf{r}', \tau] \right\rangle_{\text{class}} \tag{A9.1}$$

where $\psi(\mathbf{r})$ is the complex classical field describing the superconducting order parameter

$$\psi(\mathbf{r}) = |\psi(\mathbf{r})| e^{i\theta(\mathbf{r})} \tag{A9.2}$$

and $\widehat{\psi}[\mathbf{r}, \tau]$ is the operator for this field in the Heisenberg representation. Expressing the RHS of Eq. (A9.1) with the use of Eq. (A8.9), we have

$$\langle \psi(\mathbf{r}) \psi^*(\mathbf{r}') \rangle = \frac{\int \mathcal{D}\psi(\mathbf{r}) \, e^{-\beta F[\psi, \boldsymbol{\nabla}\psi]} \psi(\mathbf{r}) \psi^*(\mathbf{r}')}{\int \mathcal{D}\psi(\mathbf{r}) \, e^{-\beta F[\psi, \boldsymbol{\nabla}\psi]}} \tag{A9.3}$$

where $F[\psi, \boldsymbol{\nabla}\psi]$ is the "potential" part of the Lagrangian, which is, for a d-dimensional neutral superconductor, given by

$$F[\psi, \boldsymbol{\nabla}\psi] = \int d^d r \, f(\psi, \boldsymbol{\nabla}\psi)$$

$$= \int d^d r \left(\frac{\hbar^2}{2m^*} |\boldsymbol{\nabla}\psi|^2 + \alpha |\psi|^2 + \frac{\beta}{2} |\psi|^4 \right) \tag{A9.4}$$

On the RHS of Eq. (A9.4), we recognize the superconducting part of the free energy of the time-independent Ginzburg–Landau model [see Eq. (2.1)]. In performing the path integration over the complex field $\psi(\mathbf{r})$, we follow Rice (1965) and introduce small fluctuations ψ_1 and θ_1 (of the magnitude and angle of ψ, respectively) about the broken-symmetry state (see Fig. 2.1). Thus we make the ansatz

$$\psi(\mathbf{r}) = [\psi_0 + \psi_1(\mathbf{r})] e^{i\theta_1(\mathbf{r})} \tag{A9.5}$$

where ψ_0 and $\psi_1(\mathbf{r})$ are real. Introducing Eq. (A9.5) into Eq. (A9.4), we have

$$F[\psi, \boldsymbol{\nabla}\psi] = \int d^d r \left(\alpha(\psi_0^2 + \psi_1^2 + 2\psi_0\psi_1) + \frac{\beta}{2}(\psi_0^4 + 4\psi_0^3\psi_1 + 4\psi_0\psi_1^3 \right.$$

$$\left. + 6\psi_0^2\psi_1^2 + \psi_1^4) + \frac{\hbar^2}{2m^*} |[\boldsymbol{\nabla}\psi_1 + i(\psi_0 + \psi_1)\boldsymbol{\nabla}\theta_1]|^2 \right) \tag{A9.6}$$

Using Eq. (2.2), we have $\beta\psi_0^2 = -\alpha$ (for $T < T_c$), and we see that the third and the fifth terms on the RHS of Eq. (A9.6) mutually cancel. In the quadratic approximation, we keep only terms up to second order in ψ_1 and θ_1, and obtain from Eq. (A9.6)

$$F[\psi, \boldsymbol{\nabla}\psi] = \int d^d r \left(\alpha\psi_0^2 + \frac{\beta}{2}\psi_0^4 \right)$$
$$+ \int d^d r \left(-2\alpha\psi_1^2 + \frac{\hbar^2}{2m^*}(\boldsymbol{\nabla}\psi_1)^2 + \frac{\hbar^2}{2m^*}\psi_0^2(\boldsymbol{\nabla}\theta_1)^2 \right)$$
$$= F_0 + F_1 \tag{A9.7}$$

Introducing this expression into Eq. (A9.3), the $e^{-\beta F_0}$ terms in the denominator and numerator cancel, and we obtain with use of Eq. (A9.5)

$$\langle\psi(\mathbf{r})\psi^*(\mathbf{r}')\rangle = \left(\int \mathcal{D}\psi_1\,\mathcal{D}\theta_1\, e^{-\beta F_1}[\psi_0 + \psi_1(\mathbf{r})][\psi_0 + \psi_1(\mathbf{r}')]e^{i[\theta_1(\mathbf{r})-\theta_1(\mathbf{r}')]} \right)$$
$$\Big/ \int \int \mathcal{D}\psi_1\,\mathcal{D}\theta_1\, e^{-\beta F_1} \tag{A9.8}$$

Since the amplitude and phase fluctuations contribute separately to the free-energy functional F_1 of Eq. (A9.7), the path integration in Eq. (A9.8) can be factorized as follows

$$\langle\psi(\mathbf{r})\psi^*(\mathbf{r}')\rangle = \left\langle e^{i[\theta_1(\mathbf{r})-\theta_1(\mathbf{r}')]} \right\rangle \left[\psi_0^2 + \langle\psi_1(\mathbf{r})\psi_1(\mathbf{r}')\rangle \right] \tag{A9.9}$$

where

$$\left\langle e^{i[\theta_1(\mathbf{r})-\theta_1(\mathbf{r}')]} \right\rangle$$
$$= \int \mathcal{D}\theta_1 \exp\left(-\frac{\beta\hbar^2\psi_0^2}{2m^*} \int d^d r\,(\boldsymbol{\nabla}\theta_1)^2 \right) e^{i[\theta_1(\mathbf{r})-\theta_1(\mathbf{r}')]}$$
$$\Big/ \int \int \mathcal{D}\theta_1 \exp\left(-\frac{\beta\hbar^2\psi_0^2}{2m^*} \int d^d r\,(\boldsymbol{\nabla}\theta_1)^2 \right) \tag{A9.10}$$

and

$$\langle\psi_1(\mathbf{r})\psi_1(\mathbf{r}')\rangle$$
$$= \int \mathcal{D}\psi_1 \exp\left(-\beta \int d^d r \left[\frac{\hbar^2}{2m^*}(\boldsymbol{\nabla}\psi_1)^2 - 2\alpha\psi_1^2 \right] \right) \psi_1(\mathbf{r})\psi_1(\mathbf{r}')$$
$$\Big/ \int \int \mathcal{D}\psi_1 \exp\left(-\beta \int d^d r \left[\frac{\hbar^2}{2m^*}(\boldsymbol{\nabla}\psi_1)^2 - 2\alpha\psi_1^2 \right] \right) \tag{A9.11}$$

Well below the superconducting transition temperature, we have $\psi_0 \gg \langle\psi_1(\mathbf{r})\psi_1(\mathbf{r}')\rangle$ for all $|\mathbf{r} - \mathbf{r}'|$, and the correlation function (A9.9) is dominated by the first term, yielding

$$\langle\psi(\mathbf{r})\psi^*(\mathbf{r}')\rangle \simeq \psi_0^2 \left\langle e^{i[\theta_1(\mathbf{r})-\theta_1(\mathbf{r}')]} \right\rangle \tag{A9.12}$$

We use the method of the Fourier series to evaluate the path integrals in Eq. (A9.10). Thus we expand the real function $\theta_1(\mathbf{r})$ into a complex Fourier series

$$\theta_1(\mathbf{r}) = \sum_{\mathbf{k}=-\infty}^{\infty} \theta_k e^{i\mathbf{k}\cdot\mathbf{r}} \tag{A9.13}$$

where the complex coefficients $\theta_k = \theta_k^r + i\theta_k^i$ satisfy the reality conditions

$$\begin{aligned} \theta_k^r &= \theta_{-k}^r \\ \theta_k^i &= -\theta_{-k}^i \end{aligned} \tag{A9.14}$$

Using Eqs. (A9.13) and (A9.14), we have

$$\begin{aligned} \theta_1(\mathbf{r}) - \theta_1(\mathbf{r}') &= \sum_k (\theta_k^r + i\theta_k^i)[(\cos\mathbf{k}\cdot\mathbf{r} - \cos\mathbf{k}\cdot\mathbf{r}') \\ &\quad + i(\sin\mathbf{k}\cdot\mathbf{r} - \sin\mathbf{k}\cdot\mathbf{r}')] \\ &= 2\sum_{k>0} [\theta_k^r(\cos\mathbf{k}\cdot\mathbf{r} - \cos\mathbf{k}\cdot\mathbf{r}') \\ &\quad - \theta_k^i(\sin\mathbf{k}\cdot\mathbf{r} - \sin\mathbf{k}\cdot\mathbf{r}')] \end{aligned} \tag{A9.15}$$

Fourier transforming the θ_1 part of the free energy, we have

$$\begin{aligned} \int d^d r \, (\boldsymbol{\nabla}\theta_1)^2 &= -\int d^d r \sum_{k,k'} e^{i(\mathbf{k}+\mathbf{k}')\cdot\mathbf{r}} \mathbf{k}\cdot\mathbf{k}'\theta_k\theta_{k'} \\ &= 2\Omega \sum_{k>0} k^2 [(\theta_k^r)^2 + (\theta_k^i)^2] \end{aligned} \tag{A9.16}$$

where Ω is the volume of the superconductor. The path integral (A9.10) can be expressed as a multiple Gaussian integral over θ_k^r and θ_k^i. Using Eqs. (A9.15) and (A9.16), the phase correlator becomes

$$\begin{aligned} &\left\langle e^{i[\theta_1(\mathbf{r}) - \theta_1(\mathbf{r}')]} \right\rangle \\ &= Z_\theta^{-1} \prod_{k>0} \int_{-\infty}^{\infty} d\theta_k^r \int_{-\infty}^{\infty} d\theta_k^i \, \exp\Bigg(-\frac{\beta\hbar^2\psi_0^2\Omega}{m^*}[(\theta_k^r)^2 + (\theta_k^i)^2]k^2 \\ &\quad + 2i\,[\theta_k^r(\cos\mathbf{k}\cdot\mathbf{r} - \cos\mathbf{k}\cdot\mathbf{r}') \\ &\quad - \theta_k^i(\sin\mathbf{k}\cdot\mathbf{r} - \sin\mathbf{k}\cdot\mathbf{r}')]\Bigg) \end{aligned} \tag{A9.17}$$

where

$$Z_\theta = \prod_{k>0} \int_{-\infty}^{\infty} d\theta_k^r \int_{-\infty}^{\infty} d\theta_k^i \, \exp\left(-\frac{\beta\hbar^2\psi_0^2\Omega}{m^*}[(\theta_k^r)^2 + (\theta_k^i)^2]k^2\right) \tag{A9.18}$$

The product $\prod_{k>0}$ shows that $\mathbf{k}$ and $-\mathbf{k}$ vectors are to be counted as a *single mode* [see Eq. (A9.14)]. Using the formula (A3.20), the θ_k^r integration in Eq. (A9.17) can be performed with the result

$$\frac{\int_{-\infty}^{\infty} d\theta_k^r \, \exp\left(-\frac{\beta\hbar^2\psi_0^2\Omega}{m^*}(\theta_k^r)^2 k^2 + 2i\theta_k^r(\cos \mathbf{k}\cdot\mathbf{r} - \cos \mathbf{k}\cdot\mathbf{r}')\right)}{\int_{-\infty}^{\infty} d\theta_k^r \, \exp\left(-\frac{\beta\hbar^2\psi_0^2\Omega}{m^*}(\theta_k^r)^2 k^2\right)}$$

$$= \exp\left(-\frac{m^*(\cos \mathbf{k}\cdot\mathbf{r} - \cos \mathbf{k}\cdot\mathbf{r}')^2}{\beta\hbar^2\psi_0^2\Omega k^2}\right) \tag{A9.19}$$

The integration over θ_k^i can be done in a similar way, and the expression (A9.18) then yields

$$\left\langle e^{i[\theta_1(\mathbf{r}) - \theta_1(\mathbf{r}')]} \right\rangle$$

$$= \exp\left(-\frac{m^*}{\beta\hbar^2\psi_0^2\Omega} \sum_{k>0} \frac{1}{k^2}[(\cos \mathbf{k}\cdot\mathbf{r} - \cos \mathbf{k}\cdot\mathbf{r}')^2\right.$$

$$\left. + (\sin \mathbf{k}\cdot\mathbf{r} - \sin \mathbf{k}\cdot\mathbf{r}')^2]\right) \tag{A9.20}$$

A simple manipulation of the exponent in Eq. (A9.16) then yields the desired expression (2.3). We note that the latter expression is written with the k summation over *all* vectors $\mathbf{k}$. A special case of Eq. (A9.20) is the thermal average

$$\langle \cos \theta_1(\mathbf{r}) \rangle = \left\langle e^{i\theta_1(\mathbf{r})} \right\rangle$$

$$= \exp\left(-\frac{m^*}{\beta\hbar^2\psi_0^2\Omega} \sum_{k>0} \frac{1}{k^2}\right) \tag{A9.21}$$

The expression on the RHS of Eq. (A9.21) is independent of $\mathbf{r}$, as expected for a translationally invariant system. The k sum in the exponent diverges for one- and two-dimensional systems in the long-wavelength limit, confirming the absence of broken symmetry in such systems (at any finite temperature).

A10

PATH INTEGRAL FOR THE TRANSITION AMPLITUDE

The purpose of this Appendix is to show that the (real-time) quantum-mechanical transition amplitude $K(q, t; q', t')$ defined by [see Eq. (4.65)]

$$\psi(q, t) = \int_{-\infty}^{\infty} dq' \, K(q, t; q', t') \psi(q', t') \tag{A10.1}$$

can be expressed as a Feynman path integral [see Eq. (4.67)]

$$K(q, t; q', t') = \int_{q(t')=q'}^{q(t)=q} \mathcal{D}q(t) \, \exp\left(\frac{i}{\hbar} \int_{t'}^{t} dt'' \, L^{(r)}[q, \dot{q}]\right) \tag{A10.2}$$

where $L^{(r)}[q, q']$ is the Lagrangian of the particle, given by

$$L^{(r)}[q, \dot{q}] = M\left(\frac{dq}{dt}\right)^2 - V(q) \tag{A10.3}$$

First, we show that $K(q, t; q', t')$ can be written as

$$K(q, t; q', t') = \langle q | e^{-i(t-t')\widehat{H}/\hbar} | q' \rangle \tag{A10.4}$$

where $\widehat{H}$ is the (real-time) Hamiltonian of the particle. We start by noting that Eq. (A10.1) can be written as

$$\langle q | t \rangle = \int_{-\infty}^{\infty} dq' \, K(q, t; q', t') \langle q' | t' \rangle \tag{A10.5}$$

Equation (A10.5) is obtained by introducing into Eq. (A10.1) the inner products [see Eq. (A1.3)]

$$\langle q | t \rangle = \int_{-\infty}^{\infty} dq' \, \delta(q - q') \psi(q', t) = \psi(q, t) \tag{A10.6}$$

Next we define the time-evolution operator $\widehat{U}(t, t')$ by

$$|t\rangle = \widehat{U}(t, t') | t' \rangle = \int_{-\infty}^{\infty} dq' \, \widehat{U}(t, t') | q' \rangle \langle q' | t' \rangle \tag{A10.7}$$

Forming an inner product of Eq. (A10.7) with $\langle q|$, we obtain

$$\langle q|t\rangle = \langle q|\widehat{U}(t,t')|t'\rangle = \int_{-\infty}^{\infty} dq'\, \langle q|\widehat{U}(t,t')|q'\rangle\langle q'|t'\rangle \tag{A10.8}$$

Comparing Eqs. (A10.8) and (A10.5), we obtain

$$K(q,t;q',t') = \langle q|\widehat{U}(t,t')|q'\rangle \tag{A10.9}$$

To complete the proof of Eq. (A10.4), we need the expression for the time-evolution operator. Inserting the relation [see Eq. (A10.7)]

$$|t\rangle = \widehat{U}(t,t')|t'\rangle \tag{A10.10}$$

into the Schrödinger equation

$$i\hbar\frac{\partial}{\partial t}|t\rangle = \widehat{H}|t\rangle \tag{A10.11}$$

we obtain an operator equation

$$i\hbar\frac{\partial}{\partial t}\widehat{U}(t,t') = \widehat{H}\widehat{U}(t,t') \tag{A10.12}$$

For a conservative system, described by Eq. (A10.3), the Hamiltonian $\widehat{H}$ is time-independent and Eq. (A10.12) is solved by

$$\widehat{U}(t,t') = Ce^{-i\widehat{H}t/\hbar} \tag{A10.13}$$

The integration constant C is found from the initial condition

$$\widehat{U}(t',t') = \widehat{1} = Ce^{-i\widehat{H}t'/\hbar} \tag{A10.14}$$

From Eqs. (A10.3) and (A10.4), we find

$$\widehat{U}(t,t') = e^{-i\widehat{H}(t-t')/\hbar} \tag{A10.15}$$

Substituting this result into Eq. (A10.9), Eq. (A10.4) is verified.

The path-integral formulation (A10.2) is obtained by realizing that the RHS of (A10.4) is of the same form as the expression for the density matrix. In fact, if we put $\beta = i(t-t')/\hbar$ in Eq. (A1.7), we have

$$\rho^{(r)}(q,q') = \langle q|e^{-i\widehat{H}(t-t')/\hbar}|q'\rangle = K(q,t;q',t') \tag{A10.16}$$

where the superscript (r) indicates the "transition" to real times in the density matrix. Using this correspondence, Eq. (A1.22) gives us

$$\rho^{(r)}(q,q') = \int_{q(t')=q'}^{q(t)=q} \mathcal{D}q(t)\, e^{-\frac{1}{\hbar}S^{(r)}[q]} \tag{A10.17}$$

where the "real-time" action $S^{(r)}[q]$ is obtained from Eqs. (A1.21) and (A1.23) by making the replacement $\tau = it''$, yielding

$$S^{(r)}[q] = i \int_{t'}^{t} dt'' \left[-\frac{1}{2} M \left(\frac{dq}{dt''} \right)^2 + V[q(t'')] \right]$$

$$= -i \int_{t'}^{t} dt'' \, L^{(r)}[q, \dot{q}] \tag{A10.18}$$

Introducing the RHS of Eq. (A10.18) in Eq. (A10.17) and using relation (A10.16), the expression (A10.2) is obtained.

A11

EXTREMAL PATH (BOUNCE) AND FLUCTUATIONS ABOUT IT

In this Appendix we discuss the contribution to the transition amplitude $K(0, 0; \beta)$, coming from the fluctuations of the paths $q(\tau)$ about the extremal path $q_B(\tau)$. This path satisfies the classical equation of motion of a particle in an inverted potential [see Eq. (4.70) with $\bar{q} = q_B$]

$$- M \frac{d^2 q_B}{d\tau^2} + \frac{\partial V}{\partial q_B} = 0 \qquad (A11.1)$$

We use the cubic potential (4.100), which has been found relevant for tunneling in superconducting interference devices (Caldeira and Leggett, 1983). Introducing this potential into Eq. (A11.1), we obtain

$$\frac{d^2 q_B}{d\tau^2} = \omega_0^2 \left(q_B - \frac{3 q_B^2}{2 q_1} \right) \qquad (A11.2)$$

Introducing the dimensionless variables

$$z = q_B(\tau)/q_1 \qquad (A11.3)$$

and

$$\omega_0 \tau = u \qquad (A11.4)$$

Eq. (A11.2) can be written

$$\frac{d^2 z}{du^2} = z - \frac{3}{2} z^2 \qquad (A11.5)$$

This equation is to be solved with the boundary conditions [see eq.(4.92)]

$$z(0) = z(\infty) = 0 \qquad (A11.6)$$

and

$$\frac{dz(0)}{du} = \frac{dz(\infty)}{du} = 0 \qquad (A11.7)$$

Multiplying Eq. (A11.5) by dz/du and integrating over the variable u, we have

$$\int \frac{d^2 z}{du^2} \frac{dz}{du} \, du = \frac{1}{2} \int \frac{d}{du} \left[\left(\frac{dz}{du} \right)^2 \right] du = \int \left(z - \frac{3}{2} z^2 \right) dz \qquad (A11.8)$$

which implies

$$\left(\frac{dz}{du}\right)^2 = z^2 - z^3 + C \tag{A11.9}$$

The integration constant C is obtained by applying the boundary conditions (A11.6) and (A11.7) to Eq. (A11.9), yielding $C = 0$. We thus get from Eq. (A11.9) integrating over u from $\hbar\beta\omega_0/2$ to u

$$\int_{z=1}^{z(u)} \frac{dz}{(z^2 - z^3)^{\frac{1}{2}}} = \int_{\hbar\beta\omega_0/2}^{u} du \tag{A11.10}$$

where the lower limits of integration correspond to the condition that the particle rebounds at $q = q_1$ at time $\tau = \hbar\beta/2$. The integral on the LHS of Eq. (A11.10) is given by

$$\int_1^{z(u)} \frac{dz}{z(1-z)^{\frac{1}{2}}} = -2\operatorname{arctanh}[1 - z(u)]^{\frac{1}{2}} \tag{A11.11}$$

Introducing this result into Eq. (A11.10), we have with use of Eqs. (A11.4) and (A11.5)

$$-\left\{\tanh\left[\frac{1}{2}\omega_0\left(\tau - \frac{1}{2}\hbar\beta\right)\right]\right\} = \left(1 - \frac{q_B(\tau)}{q_1}\right)^{\frac{1}{2}} \tag{A11.12}$$

which implies the following formula for the bounce

$$q_B(\tau) = q_1 \operatorname{sech}^2\left[\frac{1}{2}\omega_0\left(\tau - \frac{1}{2}\hbar\beta\right)\right] \tag{A11.13}$$

Let us now consider Eq. (4.88) for this *specific* form of $q_B(\tau)$. From Eq. (4.100), we have with use of Eq. (A11.13)

$$V''(q_B) = M\omega_0^2\left(1 - \frac{3q_B}{q_1}\right)$$
$$= M\omega_0^2\left\{1 - 3\operatorname{sech}^2\left[\frac{1}{2}\omega_0\left(\tau - \frac{1}{2}\hbar\beta\right)\right]\right\} \tag{A11.14}$$

Substituting this expression into Eq. (4.88), we obtain

$$\left(-\frac{d^2}{d\tau^2} + \omega_0^2\left\{1 - 3\operatorname{sech}^2\left[\frac{1}{2}\omega_0\left(\tau - \frac{1}{2}\hbar\beta\right)\right]\right\}\right)\xi_s(\tau) = \lambda_s\xi_s(\tau) \tag{A11.15}$$

It is convenient to make the following change of variables

$$y = \frac{1}{2}\omega_0\left(\tau - \frac{1}{2}\hbar\beta\right), \qquad \xi_s(\tau) = \bar{\xi}_s(y) \tag{A11.16}$$

which brings (A11.15) to the form

$$\left(-\frac{d^2}{dy^2} - 12\operatorname{sech}^2 y\right)\bar{\xi}_s = \bar{\omega}_s\bar{\xi}_s \tag{A11.17}$$

where $\bar{\omega}_s$ is a dimensionless eigenvalue given by

$$\bar{\omega}_s = \frac{4(\lambda_s - \omega_0^2)}{\omega_0^2} \tag{A11.18}$$

Equation (A11.17) can be reduced, by a further substitution, to a hypergeometric equation (Landau and Lifshitz, 1958). The eigenvalues $\bar{\omega}_s$ follow from the condition

$$\bar{\omega}_s = -(t - s)^2, \qquad s = 0, 1, 2, \ldots \tag{A11.19}$$

where the parameter t must satisfy the condition $t(t + 1) = 12$, implying $t = 3$. For $s = 0, 1$, and 2, Eq. (A11.19) yields $\bar{\omega}_s = -9, -4$, and -1, respectively. These are the discrete eigenvalues of the three bound states of Eq. (A11.17). Introducing these eigenvalues into Eq. (A11.18), we find

$$\frac{\lambda_s}{\omega_0^2} = \begin{cases} -5/4, & \text{for } s = 0 \\ 0, & \text{for } s = 1 \\ 3/4, & \text{for } s = 2 \end{cases} \tag{A11.20}$$

The eigenvalues with $s \geq 3$ belong to the continuous spectrum of Eq. (A11.17). We note that the $\xi_{s=0}$ solution, with negative eigenvalue, is an *unstable* mode, which is directly responsible for the quantum decay. *Any* potential $V(q)$ that allows quantum tunneling (such as that of Fig. 4.4) must yield such a mode. The $\xi_{s=1}$ solution, with zero eigenvalue, is a *Goldstone-like* mode, since it corresponds to a *rigid displacement* of the bounce function $q_B(\tau)$. This can be shown as follows. If we put $\lambda_1 = 0$ into Eq. (4.88), we get

$$\hat{L}_B\xi_1(\tau) = 0 \tag{A11.21}$$

where $\hat{L}_B$ is a linear differential operator given by

$$\hat{L}_B = -\frac{d^2}{d\tau^2} + \frac{1}{M}V''[q_B(\tau)] \tag{A11.22}$$

Equations (A11.21) and (A11.22) imply

$$\hat{L}_B\left(\frac{d\xi_1}{d\tau}\right) = -\frac{d^3\xi_1}{d\tau^3} + \frac{1}{M}V''[q_B(\tau)]\frac{d\xi_1}{d\tau} = 0 \tag{A11.23}$$

It is easy to see that $dq_B/d\tau$ satisfies a differential equation of the form (A11.23). The bounce solution $q_B(\tau)$ satisfies the equation [see Eq. (4.70)]

$$-M\frac{d^2q_B}{d\tau^2} + \frac{\partial V}{\partial q_B} = 0 \tag{A11.24}$$

Differentiating this equation with respect to τ, we have

$$-\frac{d^3 q_B}{d\tau^3} + \frac{1}{M} V''[q_B(\tau)]\frac{dq_B}{d\tau} = 0 \tag{A11.25}$$

where we have used

$$\frac{\partial}{\partial \tau}\left(\frac{\partial V}{\partial q_B}\right) = V''[q_B(\tau)]\frac{dq_B}{d\tau} \tag{A11.26}$$

Comparison of Eqs. (A11.23) and (A11.25) implies that $dq_B/d\tau$ is an eigenfunction of operator $\widehat{L}_B$ with *zero eigenvalue*. For a small translation in imaginary time, we have

$$q_B(\tau + \tau_0) = q_B(\tau) + \tau_0 \frac{dq_B}{d\tau} \tag{A11.27}$$

This implies that the mode $\xi_1 = dq_B/d\tau$ generates *translations* in time, which cost *no energy*. Hence, $\xi_1 = \xi_{s=1}$ is a Goldstone-like mode. Finally, the ξ_2 solution, with positive eigenvalue, is an ordinary harmonic oscillator mode.

A12

BOUNCE CONTRIBUTION TO THE TRANSITION AMPLITUDE

According to Eq. (4.92), the extremal path $q_B(\tau)$ and fluctuations about it contribute to the transition amplitude K by

$$K_B = \exp\left(-\frac{1}{\hbar}S_B\right)Z_\xi \tag{A12.1}$$

where

$$Z_\xi = \int_{\xi(0)=\xi(\hbar\beta)=0} \mathcal{D}\xi(\tau)\, \exp\left(-\frac{1}{\hbar}S_\xi\right) \tag{A12.2}$$

In this Appendix, we consider the evaluation of K_B with the goal of deriving Eq. (4.101). In Eq. (4.89) we have obtained the action S_ξ in a diagonalized form

$$S_\xi = \frac{1}{2}M\hbar\beta \sum_s \lambda_s\eta_s^2 \tag{A12.3}$$

If all λ_s were positive, then the path integral Z_ξ would be given by a multiple Gaussian integral

$$Z_\xi = J\prod_n \left(\int_{-\infty}^{\infty} d\eta_n\right) \exp\left(-\frac{1}{2}M\beta \sum_s \lambda_s\eta_s^2\right) \tag{A12.4}$$

where J is the Jacobian of the transformation (4.78). However, as shown in Eq. (A11.20), the modes ξ_0 and ξ_1 have negative and zero eigenvalues, respectively, so their contribution to Z_ξ must be treated by special methods, which will be discussed in the following.

A12.1. Contribution of Goldstone mode ξ_1

The Goldstone mode ξ_1 has eigenvalue $\lambda_1 = 0$. If we proceed to calculate the contribution of this mode to Z_ξ by formally introducing this eigenvalue into (A12.4), then we end up with a divergent η_1 integral. This problem can be resolved by using the method of the *time-translation collective coordinate* (Polyakov, 1977). In this method, one replaces the expansion given in Eqs. (4.77) and (4.78) by

$$q(\tau) = q_B(\tau - \eta_1) + \sum_{s\neq 1}\eta_s\xi_s(\tau - \eta_1) \tag{A12.5}$$

In Eq. (A12.5) the $\xi_1(\tau)$ mode is replaced by the *position*, $\tau_B = \eta_1$, of the bounce. This formulation is based on the observation that the Goldstone mode corresponds to translations of the bounce in time, so that there exists an infinite number of equivalent bounces located at different times $\tau_B \in [0, \hbar\beta]$. The path integral over ξ_1 is then replaced by an integral over $\eta_1 = \tau_B$. Using Eq. (A12.5), we now consider the evaluation of the expression for K_B. We start from the general formula (4.92)

$$K_B = \int_{q(0)=q(\hbar\beta)=0} \mathcal{D}q(\tau) \exp\left(-\frac{1}{\hbar}(S_B + S_\xi)\right) \qquad (A12.6)$$

Applying the time-slicing method [see Eq. (A1.19)], the integration measure in Eq. (A12.6) is

$$\int \mathcal{D}q(\tau) = \int_{-\infty}^{\infty} dq(\tau_1) \int_{-\infty}^{\infty} dq(\tau_2) \cdots \int_{-\infty}^{\infty} dq(\tau_{\bar{N}-1})$$

$$= J_1 \int_0^{\hbar\beta} d\eta_1 \prod_{s \neq 1} \int d\eta_s \qquad (A12.7)$$

where J_1 is the Jacobian of the transformation (A12.5) taken at a discrete set of times $\tau = \tau_i$. We use the identity

$$J_1 = \left(\det \widehat{M}\right)^{\frac{1}{2}} \qquad (A12.8)$$

where $\widehat{M}$ is a matrix with the elements

$$M_{\alpha\beta} = \sum_{i=1}^{\bar{N}-1} \frac{\partial q(\tau_i)}{\partial \eta_\alpha} \frac{\partial q(\tau_i)}{\partial \eta_\beta}$$

$$\xrightarrow[\bar{N}\to\infty]{} \frac{\bar{N}}{\hbar\beta} \int_0^{\hbar\beta} d\tau \frac{\partial q(\tau)}{\partial \eta_\alpha} \frac{\partial q(\tau)}{\partial \eta_\beta} \qquad (A12.9)$$

On the RHS of Eq. (A12.9) we have made a transition from the discrete set τ_i to a continuum time τ. The integrand on the RHS of Eq. (A12.9) can be evaluated by differentiation of Eq. (A12.5), yielding

$$dq(\tau) = -\dot{q}_B(\tau)\, d\eta_1 + \sum_{s \neq 1} \xi_s(\tau - \eta_1)\, d\eta_s \qquad (A12.10)$$

Let us consider the first three diagonal matrix elements of $\widehat{M}$. With the use of Eq. (A12.9) and (A12.10) we have

$$M_{11} = \frac{\bar{N}}{\hbar\beta} \int_0^{\hbar\beta} d\tau\, [\dot{q}_B(\tau)]^2 \qquad (A12.11)$$

$$M_{00} = \frac{\bar{N}}{\hbar\beta} \int_0^{\hbar\beta} d\tau\, \xi_0^2(\tau - \eta_1) = \bar{N} \qquad (A12.12)$$

$$M_{22} = \frac{\bar{N}}{\hbar\beta} \int_0^{\hbar\beta} d\tau\, \xi_2^2(\tau - \eta_1) = \bar{N} \qquad (A12.13)$$

where the ortho-normalization condition (4.79) has been used in Eqs. (A12.12) and (A12.13). The same condition implies that the off-diagonal elements of $\widehat{M}$ vanish. For $\alpha = 1$, $\beta \neq 1$, this can be shown by using the relation [see Eqs. (A11.23)–(A11.27)]

$$\frac{\partial q_B}{\partial \eta_1} = -\frac{\partial q_B}{\partial \tau} = -\xi_1(\tau) \tag{A12.14}$$

From Eqs. (A12.5), (A12.9) and (A12.14), we have

$$
\begin{aligned}
M_{1\beta} &= \frac{\bar{N}}{\hbar\beta} \int_0^{\hbar\beta} d\tau \, \frac{\partial q_B(\tau - \eta_1)}{\partial \eta_1} \xi_\beta(\tau - \eta_1) \\
&= -\int_0^{\hbar\beta} d\tau \, \xi_1(\tau)\xi_\beta(\tau) = \hbar\delta_{1,\beta}
\end{aligned} \tag{A12.15}
$$

Thus the determinant of $\widehat{M}$ is given by the product of the diagonal matrix elements, so that the Jacobian is, from Eqs. (A12.8) and (A12.11)–(A12.13),

$$J_1 = \text{const} \times \left(\frac{1}{\hbar\beta} \int_0^{\hbar\beta} d\tau \, (\dot{q}_B)^2 \right)^{\frac{1}{2}} = \text{const} \times \left(\frac{1}{\hbar\beta M} S_B \right)^{\frac{1}{2}} \tag{A12.16}$$

where we have used Eq. (4.116). The constant factor results from the product of the $\bar{N}$ factors that appear in Eqs. (A12.11)–(A12.13). This constant will be discarded, as it is expected to cancel with a similar factor in K_0 [see Eq. (4.104)]. Note that multiplication of the partition function by a constant, which is independent of the physical parameters, does not lead to measurable effects.

A12.2. Contribution of unstable mode ξ_0

From Eq. (A12.4), the contribution of the mode η_0 can be formally written as

$$Z_{\eta_0} = \int_{-\infty}^{\infty} d\eta_0 \, \exp\left(\frac{1}{2}M\beta|\lambda_0|\eta_0^2 \right) \tag{A12.17}$$

This integral is manifestly divergent. Langer (1967) proposed a method of analytic continuation of the η_0 integration, which shows that the unstable mode produces a *finite* contribution to Z_ξ that has an *imaginary* part. We now present the main arguments of this approach following Shulman (1981).

Let $F(\mu)$ be a function of the complex variable μ, defined by the real axis integral in the complex z plane

$$F(\mu) = \pi^{-\frac{1}{2}} \int_{-\infty}^{\infty} dz \, e^{-\mu z^2} \tag{A12.18}$$

Equation (A12.18) implies

$$F(\mu) = \mu^{-\frac{1}{2}}, \qquad \text{for } \mathrm{Re}\,\mu > 0 \tag{A12.19}$$

We shall now analytically continue $F(\mu)$ to $\mathrm{Re}\,\mu \leq 0$ by letting $F(\mu) = \mu^{-\frac{1}{2}}$ in the *complex* μ plane. The integral (A12.18), as it stands, however, does not define $F(\mu)$ for $\mathrm{Re}\,\mu \leq 0$. To proceed, we make a change of the integration contour from the real axis to a contour C' on which

$$z = e^{-i\theta} x \tag{A12.20}$$

where $-\infty < x < \infty$ and θ is an angle of rotation, to be specified shortly. After this change is made, Eq. (A12.18) becomes

$$F(\mu) = \pi^{-\frac{1}{2}} \int_{C'} dz\, e^{-\mu z^2} = \pi^{-\frac{1}{2}} \int_{\infty}^{-\infty} dx\, e^{-i\theta} \exp\left(-\mu e^{-2i\theta} x^2\right) \tag{A12.21}$$

Note that the limits of the x integration correspond to the counter-clockwise rotation of the real axis. The integral (A12.21) converges for $\mathrm{Re}\left(\mu e^{-2i\theta}\right) > 0$, implying

$$-\frac{\pi}{2} < \arg \mu - 2\theta < \frac{\pi}{2} \tag{A12.22}$$

If we set $\theta = \pi/2$, Eq. (A12.22) yields the following range for $\arg \mu$

$$\frac{\pi}{2} < \arg \mu < \frac{3\pi}{2} \tag{A12.23}$$

From this equation we see that for $\theta = \pi/2$ the function $F(\mu)$ is defined for $\mathrm{Re}\,\mu \leq 0$. Putting $\theta = \pi/2$ in Eq. (A12.21), we obtain

$$F(\mu) = i\pi^{-\frac{1}{2}} \int_{-\infty}^{\infty} dx\, \exp(-|\mu|x^2) = i|\mu|^{-\frac{1}{2}} \tag{A12.24}$$

We note that the RHS of Eq. (A12.21) agrees with the analytically continued function $\mu^{-\frac{1}{2}}$.

As pointed out by Langer (1967) and discussed thoroughly by Shulman (1981) and Kleinert (1990), the expression (A12.24) cannot be used directly to evaluate the contribution of the ξ_0 mode to Z_ξ. The reason is that the imaginary part of Z_{ξ_0}, shown in Eq. (A12.24), is produced by *positive* values of η_0 only. We note that an increase of η_0 in the positive direction corresponds to the increase of the amplitude of the bounce, representing the decay channel of the metastable state. On the other hand, a decrease of η to negative values represents the return of the particle to the potential well at $q_0 = 0$. Thus, the range $\eta \in [-\infty, 0]$ is expected to contribute a real part of K_B, representing the renormalization of the well energy. In view of this, the correct contribution of the ξ_0 mode is obtained by analytic continuation of Eq. (A12.17), where the integration range is $[0, \infty]$. This amounts to replacing the lower limit in Eq. (A12.21) by zero, resulting in a multiplication of the RHS of Eq. (A12.24) by a factor of $\frac{1}{2}$. Hence, the correct expression for the ξ_0 contribution to Z_ξ is

$$Z_0 = \frac{i}{2}\left(\frac{\pi}{|\mu|}\right)^{\frac{1}{2}} = \frac{i}{2}\left(\frac{2\pi}{\beta M|\lambda_0|}\right)^{\frac{1}{2}} \qquad (A12.25)$$

The remaining discrete mode (contributing to Z_ξ) is the stable η_2 mode with positive eigenvalue λ_2. Its contribution is given by the (convergent) integral

$$Z_2 = \int_{-\infty}^{\infty} d\eta_2 \, \exp\left(-\frac{1}{2}M\beta\lambda_2\eta_2^2\right) = \left(\frac{2\pi}{M\beta\lambda_2}\right)^{\frac{1}{2}} \qquad (A12.26)$$

From Eqs. (A12.6) and (A12.7), we obtain with the use of Eqs. (A12.16), (A12.25), and (A12.26)

$$K_B = \hbar\beta J_1 Z_0 Z_2 Z_c \exp\left(-\frac{1}{\hbar}S_B\right)$$

$$= i\pi\left(\frac{\hbar S_B}{\beta M^3|\lambda_0|\lambda_2}\right)^{\frac{1}{2}} Z_c \exp\left(-\frac{1}{\hbar}S_B\right) \qquad (A12.27)$$

where Z_c represents the contribution of the continuum modes ξ_s with $s > 3$. Equation (A12.27) is the desired result for the contribution of the bounce to the transition amplitude, quoted in Eq. (4.101).

A13

EXPRESSION FOR BOUNCE ACTION

For the potential (4.100), the bounce action S_B is given by the expression

$$S_B = \frac{36V_0}{5\omega_0} \tag{A13.1}$$

where V_0 is the barrier height. This expression has been used in Eq. (4.109) to obtain the final formula for the decay rate Γ. Equation (A13.1) is derived by substituting the potential (4.100) into Eq. (4.111)

$$S_B = 2 \int_0^{q_1} dq\, [2MV(q)]^{\frac{1}{2}} = 2M\omega_0 \int_0^{q_1} dq\, q \left(1 - \frac{q}{q_1}\right)^{\frac{1}{2}} \tag{A13.2}$$

The integral on the RHS of Eq. (A13.2) is evaluated by making a substitution $u = q/q_1$

$$S_B = 2M\omega_0 q_1^2 \int_0^1 du\, u(1-u)^{\frac{1}{2}} = 2M\omega_0 q_1^2 \frac{\Gamma(2)\Gamma(3/2)}{\Gamma(7/2)} = \frac{8M\omega_0 q_1^2}{15} \tag{A13.3}$$

where the definite integral over u is found in standard tables (Dwight, 1961). The barrier height is given by the value of $V(q)$ at $q = q_m$, where q_m is the position of the maximum of the function (4.100), given by

$$q_m = \frac{2}{3} q_1 \tag{A13.4}$$

Using this value, Eq. (4.100) yields for the barrier height

$$V_0 = \frac{2}{27} M\omega_0^2 q_1^2 \tag{A13.5}$$

Calculating q_1^2 from this equation and substituting the result into Eq. (A13.3), we obtain Eq. (A13.1).

A14

DISSIPATIVE ACTION FOR WEAK DAMPING

For weak damping, the correction ΔS_B to the bounce action due to dissipation obtained in Eq. (4.135) is of the form

$$\Delta S_B = S_D[q_B] = \frac{\eta}{2\pi} \int_0^{\hbar\beta} d\tau \int_0^{\hbar\beta} d\tau' \left(\frac{q_B(\tau) - q_B(\tau')}{\tau - \tau'} \right)^2 \tag{A14.1}$$

where

$$q_B(\tau) = q_1 \operatorname{sech}^2(\omega_0 \tau/2) \tag{A14.2}$$

It is convenient to first Fourier transform the expression (A14.1) and then perform the integrations in the real-frequency domain. Starting from expression (4.40) and letting $\beta \to \infty$, we have

$$S_D[q_B] = \hbar\beta\eta \sum_n |\omega_n| q_n^{(B)} q_{-n}^{(B)} \xrightarrow[\beta \to \infty]{} \frac{\eta}{\pi} \int_0^\infty d\omega \, \omega |q_B(\omega)|^2 \tag{A14.3}$$

where $q_n^{(B)}$ are the coefficients of the Fourier series for $q_B(\tau)$, and $q_B(\omega)$ is the Fourier integral transform of $q_B(\tau)$

$$q_B(\omega) = \int_{-\infty}^\infty d\tau \, e^{i\omega\tau} q_B(\tau)$$

$$= 2q_1 \int_0^\infty d\tau \, \cos(\omega\tau) \operatorname{sech}^2(\omega_0\tau/2)$$

$$= \left(\frac{4\pi q_1}{\omega_0^2} \right) \frac{\omega}{\sinh(\pi\omega/\omega_0)} \tag{A14.4}$$

where tables of Fourier transforms are used on the RHS of (A14.4) (Robinson, 1968). Inserting this expression into Eq. (A14.3), we obtain in the limit of $\beta \to \infty$

$$S_D[q_B] = \frac{\eta}{\pi} \left(\frac{4\pi q_1}{\omega_0^2} \right)^2 \int_0^\infty d\omega \, \frac{\omega^3}{\sinh^2(\pi\omega/\omega_0)} \tag{A14.5}$$

To evaluate this integral, we use the formula (Ryzhik and Gradshteyn, 1965)

$$\int_0^\infty dx \, \frac{x^{\mu-1}}{\sinh^2 ax} = \frac{4}{(2a)^\mu} \Gamma(\mu)\zeta(\mu - 1) \tag{A14.6}$$

where ζ is the Riemann theta function. Thus we obtain

$$\Delta S_B = S_D[q_B] = \frac{24}{\pi^3}\zeta(3)\eta q_1^2 \qquad (A14.7)$$

Taking $\zeta(3) = 1.2$, Eq. (A14.7) yields for the dissipative correction a value

$$S_D[q_B] \simeq 0.9\eta q_1^2 \qquad (A14.8)$$

A15

BOUNCE TRAJECTORY FOR STRONG DAMPING

Following Larkin and Ovchinnikov (1983), we consider the solution of the algebraic equation [see Eq. (4.141)]

$$- M\omega_0^2 \left(\widetilde{q}_n - \frac{3}{2q_1} \sum_m \widetilde{q}_m \widetilde{q}_{n-m} \right) - 2\eta |\omega_n| \widetilde{q}_n = 0 \qquad (\text{A15.1})$$

This equation is solved by the ansatz

$$\widetilde{q}_n = \alpha e^{-b|n|} \qquad (\text{A15.2})$$

First, we establish a relationship between the constants α and b. By letting $n = 0$ in Eq. (A15.1), we obtain

$$- M\omega_0^2 \left(\widetilde{q}_0 - \frac{3}{2q_1} \sum_m \widetilde{q}_m \widetilde{q}_{-m} \right) = 0 \qquad (\text{A15.3})$$

Introducing the ansatz (A15.2) into Eq. (A15.3) and performing the m summation with the use of geometric series, we have

$$- M\omega_0^2 \left(\alpha - \frac{3\alpha^2}{2q_1} \coth b \right) = 0 \qquad (\text{A15.4})$$

This equation implies the following relation between α and b

$$\alpha = \frac{2q_1}{3} \tanh b \qquad (\text{A15.5})$$

Next, let us consider Eq. (A15.1) for $n = 1$

$$- M\omega_0^2 \left(\widetilde{q}_1 - \frac{3}{2q_1} \sum_m \widetilde{q}_m \widetilde{q}_{1-m} \right) - 2\eta \omega_1 \widetilde{q}_1 = 0 \qquad (\text{A15.6})$$

The m summation is performed by regrouping the terms in the following manner

$$\sum_m \widetilde{q}_m \widetilde{q}_{1-m} = (\widetilde{q}_0 \widetilde{q}_{1-0} + \widetilde{q}_1 \widetilde{q}_{1-1}) + (\widetilde{q}_2 \widetilde{q}_{1-2} + \widetilde{q}_{-1} \widetilde{q}_{1+1})$$

$$+ (\widetilde{q}_3 \widetilde{q}_{1-3} + \widetilde{q}_{-2} \widetilde{q}_{1+2}) + \cdots$$
$$= 2\widetilde{q}_0 \widetilde{q}_1 + 2(\widetilde{q}_1 \widetilde{q}_2 + \widetilde{q}_2 \widetilde{q}_3 + \cdots) \qquad (\text{A15.7})$$

where we have used the property $\widetilde{q}_n = \widetilde{q}_{-n}$. Using the ansatz (A15.2), the RHS of Eq. (A15.7) becomes

$$\sum_m \widetilde{q}_m \widetilde{q}_{1-m} = 2\alpha^2 \left[e^{-b} + e^{-3b} \left(1 + e^{-2b} + e^{-4b} + \cdots \right) \right]$$

$$= 2\alpha^2 e^{-b} \left(1 + \frac{e^{-2b}}{1 - e^{-2b}} \right) = \frac{\alpha^2}{\sinh b} \qquad (A15.8)$$

Introducing this result in Eq. (A15.6), we obtain

$$-M\omega_0^2 \left[1 - \frac{3\alpha}{2q_1} \left(\frac{e^b}{\sinh b} \right) \right] - 2\eta\omega_1$$

$$= M\omega_0^2 \left(\frac{e^b}{\cosh b} - 1 \right) - 2\eta\omega_1 = 0 \qquad (A15.9)$$

where we have applied the relationship (A15.5) to eliminate the constant α/q_1. Using the identity

$$\frac{e^b}{\cosh b} - 1 = \tanh b \qquad (A15.10)$$

Eq. (A15.9) yields (setting $\omega_1 = 2\pi/\hbar\beta$)

$$b = \operatorname{arctanh}\left(\frac{2\eta\omega_1}{M\omega_0^2} \right) = \operatorname{arctanh}\left(\frac{4\pi\eta}{\hbar\beta M\omega_0^2} \right) \qquad (A15.11)$$

Using this result, Eq. (A15.5) yields

$$\alpha = \frac{8\pi\eta q_1}{3\hbar\beta M\omega_0^2} \qquad (A15.12)$$

When the argument of the arctanh in Eq. (A15.11) is equal to one, the constant b diverges. Thus we may define a critical temperature T_0 by

$$T_0 = \frac{\hbar M\omega_0^2}{4\pi\eta k_B} \qquad (A15.13)$$

Equation (A15.11) has a finite and real solution only when $T < T_0$. At $T = T_0$, the bounce trajectory actually collapses to a constant (Riseborough et al., 1985). This can be verified by considering $q_B(\tau)$ as a function of τ. Using Eq. (A15.2), the Fourier series (4.131) can be summed as a geometric series, yielding

$$q_B(\tau) = \alpha \sum_n e^{-b|n| - i\omega_n \tau} = \alpha \left[\left(\frac{1}{1 - e^{-b - i\omega_1 \tau}} + \text{c.c.} \right) - 1 \right]$$

$$= \alpha \frac{\coth(b/2)}{1 + \left(\frac{\sin(\pi\tau/\hbar\beta)}{\sinh(b/2)} \right)^2} \xrightarrow[\substack{T \to T_0 \\ b \to \infty}]{} \alpha_0 \qquad (A15.14)$$

where α_0 is given by Eq. (A15.2) with $\beta = \beta_0 = 1/k_B T_0$. Using Eq. (A15.13) for T_0, Eq. (A15.2) yields

$$\alpha_0 = \frac{8\pi\eta q_1 k_B T_0}{3\hbar M\omega_0^2} = \frac{2q_1}{3} = q_m \qquad (A15.15)$$

where q_m is the position of the top of the barrier [see Eq. (A13.4)]. Hence, we see that at $T = T_0$ the bounce trajectory *collapses* to a constant equal to q_m. For $T \geq T_0$, there is no quantum tunneling, only thermal activation determines the decay rate.

A16

CUMULANT EXPANSION THEOREM FOR $Z_G[\Delta]$

The purpose of this Appendix is to present a derivation of the cumulant expansion theorem that is used to bring the functional $Z_G[\Delta]$ to the form given in Eq. (5.36). Dividing and multiplying Eq. (5.35) by Z_{G_0}, we have

$$Z_G[\Delta(\mathbf{r}, \tau)] = Z_{G\lambda=1}[\Delta] \tag{A16.1}$$

where

$$Z_{G\lambda}[\Delta] = Z_{G_0} \left\langle T_\tau \exp\left[-\frac{1}{\hbar} \int_0^{\hbar\beta} d\tau \, V_\lambda(\tau)\right] \right\rangle_0$$

$$= \mathrm{Tr}\left\{ e^{-\beta K_0} T_\tau \exp\left[-\frac{1}{\hbar} \int_0^{\hbar\beta} d\tau \, V_\lambda(\tau)\right] \right\} \tag{A16.2}$$

The operator $V_\lambda(\tau)$ is equal to $\lambda V(\tau)$, where $V(\tau)$ is a general perturbation ($D(\tau)$ of Eq. (5.26) is a particular realization of $V(\tau)$).

The functional $Z_{G\lambda}$ satisfies the following differential equation

$$\frac{1}{Z_{G\lambda}} \frac{\partial Z_{G\lambda}}{\partial \lambda} = -\mathrm{Tr}\left[e^{-\beta K_0} T_\tau \exp\left(-\frac{1}{\hbar} \int_0^{\hbar\beta} d\tau \, V_\lambda(\tau)\right) \frac{1}{\hbar} \int_0^{\hbar\beta} d\tau \, V(\tau) \right]$$

$$\Big/ \mathrm{Tr}\left[e^{-\beta K_0} T_\tau \exp\left(-\frac{1}{\hbar} \int_0^{\hbar\beta} d\tau \, V_\lambda(\tau)\right) \right] \tag{A16.3}$$

A16.1. Proof of Eq. (A16.3)

Differentiating Eq. (A16.2) with respect to λ, we obtain

$$\frac{\partial Z_{G\lambda}}{\partial \lambda} = Z_{G_0} \frac{\partial}{\partial \lambda} \left\langle T_\tau \sum_{n=0}^{\infty} (-1)^n (n!)^{-1} \left(\frac{1}{\hbar} \int_0^{\hbar\beta} d\tau \, V_\lambda(\tau)\right)^n \right\rangle_0$$

$$= Z_{G_0} \sum_{n=0}^{\infty} (-1)^n (n!)^{-1} \frac{1}{\hbar} \int_0^{\hbar\beta} d\tau_1 \cdots \frac{1}{\hbar} \int_0^{\hbar\beta} d\tau_n$$

$$\times \frac{\partial}{\partial \lambda} \langle T_\tau V_\lambda(\tau_1) \cdots V_\lambda(\tau_n) \rangle_0 \tag{A16.4}$$

Since $V_\lambda(\tau) = \lambda V(\tau)$, we have

$$\frac{\partial}{\partial \lambda} \langle T_\tau V_\lambda(\tau_1) \cdots V_\lambda(\tau_n) \rangle_0 = n\lambda^{n-1} \langle T_\tau V_\lambda(\tau_1) \cdots V_\lambda(\tau_n) \rangle_0 \qquad (A16.5)$$

Introducing this result into Eq. (A16.3), we obtain

$$\begin{aligned}
\frac{\partial Z_{G\lambda}}{\partial \lambda} &= -Z_{G_0} \sum_{n=1}^{\infty} \frac{(-\lambda)^{n-1}}{(n-1)!} \frac{1}{\hbar} \int_0^{\hbar\beta} d\tau_1 \cdots \frac{1}{\hbar} \int_0^{\hbar\beta} d\tau_n \langle T_\tau V(\tau_1) \cdots V(\tau_n) \rangle_0 \\
&= -Z_{G_0} \sum_{n=1}^{\infty} \frac{(-\lambda)^{n-1}}{(n-1)!} \frac{1}{\hbar} \int_0^{\hbar\beta} d\tau_1 \cdots \frac{1}{\hbar} \int_0^{\hbar\beta} d\tau_{n-1} \frac{1}{\hbar} \int_0^{\hbar\beta} d\tau \\
&\quad \times \langle T_\tau V(\tau_1) \cdots V(\tau_{n-1}) V(\tau) \rangle_0 \\
&= -Z_{G_0} \left\langle T_\tau \left(1 - \frac{\lambda}{1!\hbar} \int_0^{\hbar\beta} d\tau_1 V(\tau_1) \right. \right. \\
&\quad \left. \left. + \frac{\lambda^2}{2!\hbar^2} \int_0^{\hbar\beta} d\tau_1 \int_0^{\hbar\beta} d\tau_2 V(\tau_1) V(\tau_2) + \cdots \right) \frac{1}{\hbar} \int_0^{\hbar\beta} d\tau V(\tau) \right\rangle_0 \\
&= -Z_{G_0} \left\langle T_\tau \exp\left(-\frac{1}{\hbar} \int_0^{\hbar\beta} d\tau V_\lambda(\tau) \right) \frac{1}{\hbar} \int_0^{\hbar\beta} d\tau V(\tau) \right\rangle_0 \qquad (A16.6)
\end{aligned}$$

The τ_n integration has been factored out so that the remaining integration can be summed as an exponential series. Using the definition

$$\langle \cdots \rangle_0 = \frac{\text{Tr}\left(e^{-\beta K_0} \cdots\right)}{\text{Tr}\left(e^{-\beta K_0}\right)} \qquad (A16.7)$$

on the RHS of Eq. (A16.6), we have

$$\begin{aligned}
\frac{\partial Z_{G\lambda}}{\partial \lambda} = -\text{Tr}\Bigg[e^{-\beta K_0} T_\tau \exp\left(-\frac{1}{\hbar} \int_0^{\hbar\beta} d\tau V_\lambda(\tau) \right) \\
\times \frac{1}{\hbar} \int_0^{\hbar\beta} d\tau V(\tau) \Bigg] \qquad (A16.8)
\end{aligned}$$

Dividing this equation by Eq. (A16.2), Eq. (A16.3) is proven.

Noting the disentangling theorem (5.7), we have with use of Eq. (5.39)

$$e^{-\beta K_0} T_\tau \exp\left(-\frac{1}{\hbar} \int_0^{\hbar\beta} d\tau V_\lambda(\tau) \right) = e^{-\beta K_\lambda} \qquad (A16.9)$$

Using this equation on the RHS of Eq. (A16.3), we have

$$\begin{aligned}
\frac{1}{Z_{G\lambda}} \frac{\partial Z_{G\lambda}}{\partial \lambda} &= -\frac{\text{Tr}\left(e^{-\beta K_\lambda} \frac{1}{\hbar} \int_0^{\hbar\beta} d\tau V(\tau)\right)}{\text{Tr}\left(e^{-\beta K_\lambda}\right)} \\
&= -\left\langle \frac{1}{\hbar} \int_0^{\hbar\beta} d\tau V(\tau) \right\rangle_\lambda \qquad (A16.10)
\end{aligned}$$

where we have used the definition of $\langle \cdots \rangle_\lambda$ given in Eq. (5.38). We now define a function

$$F_\lambda = -\ln Z_{G\lambda} \qquad (A16.11)$$

in terms of which Eq. (A16.10) is written as

$$\frac{dF_\lambda}{d\lambda} = \left\langle \frac{1}{\hbar} \int_0^{\hbar\beta} d\tau\, V(\tau) \right\rangle_\lambda = \left\langle \frac{1}{\lambda\hbar} \int_0^{\hbar\beta} d\tau\, V_\lambda(\tau) \right\rangle_\lambda \qquad (A16.12)$$

Integrating this equation over λ, we obtain

$$F_1 - F_0 = \int_0^1 \frac{d\lambda}{\lambda} \left\langle \frac{1}{\hbar} \int_0^{\hbar\beta} d\tau\, V_\lambda(\tau) \right\rangle_\lambda \qquad (A16.13)$$

Equations (A16.1) and (A16.11) imply

$$Z_G[\Delta] = Z_{G,\lambda=1} = \exp(-F_1) \qquad (A16.14)$$

Substituting for F_1 from Eq. (A16.13), Eq. (A16.4) yields

$$Z_G[\Delta] = Z_{G_0} \exp\left(-\int_0^1 \frac{d\lambda}{\lambda} \left\langle \frac{1}{\hbar} \int_0^{\hbar\beta} d\tau\, V_\lambda(\tau) \right\rangle_\lambda \right) \qquad (A16.15)$$

This completes the proof of the cumulant expansion theorem used in Eq. (5.36).

A17

GREEN'S FUNCTION IN THE PRESENCE OF $V(\tau)$

The exponent on the RHS of Eq. (A16.15) can be expressed as follows [see Eqs. (5.40)–(5.52) for a proof]

$$-\int_0^1 \frac{d\lambda}{\lambda} \left\langle \frac{1}{\hbar} \int_0^{\hbar\beta} d\tau\, V_\lambda(\tau) \right\rangle_\lambda = \mathrm{Tr}\left(\ln \widehat{G}_0 + \ln \widehat{G}^{-1} \right) \qquad (A17.1)$$

where $\widehat{G}_0$ and $\widehat{G}$ are Matsubara Green's functions (in Nambu matrix notation), defined as

$$\widehat{G}_0 = \widehat{G}_{\lambda=0}$$
$$\widehat{G} = \widehat{G}_{\lambda=1} \qquad (A17.2)$$

On the RHS of Eqs. (A17.2), we have a Green's function $\widehat{G}_\lambda$ for a variable coupling constant λ

$$\widehat{G}_\lambda(\mathbf{r},\tau;\mathbf{r}',\tau') = -\left\langle \mathrm{T}_\tau\, \widehat{\psi}(\mathbf{r},\tau)\widehat{\psi}^\dagger(\mathbf{r},\tau) \right\rangle_\lambda \qquad (A17.3)$$

where the average $\langle \cdots \rangle_\lambda$ excludes the trace over the Nambu space [as in Eq. (5.32)], and the τ dependence of the spinor-field operators is given by

$$\widehat{\psi}(\mathbf{r},\tau) = e^{\bar{K}_\lambda \tau/\hbar}\widehat{\psi}(\mathbf{r})e^{-\bar{K}_\lambda \tau/\hbar}$$
$$\widehat{\psi}^\dagger(\mathbf{r},\tau) = e^{\bar{K}_\lambda \tau/\hbar}\widehat{\psi}^\dagger(\mathbf{r})e^{-\bar{K}_\lambda \tau/\hbar} \qquad (A17.4)$$

where $\bar{K}_\lambda$ is (for the superconducting pairing interaction)

$$\bar{K}_\lambda = K_0 + \bar{D}_\lambda(\tau) = K_0 + \lambda \int d^3r\, \widehat{\psi}^\dagger(\mathbf{r},\tau)\widehat{D}(\mathbf{r},\tau)\widehat{\psi}(\mathbf{r},\tau) \qquad (A17.5)$$

A17.1. Integral equation for $\widehat{G}_\lambda$

The proof of Eq. (A17.1) given in Section 5.2 is based on expanding $\widehat{G}_\lambda$ in the perturbation $\bar{D}_\lambda$. We now derive an equation of motion for $\widehat{G}_\lambda$, and then solve it as a series expansion in parameter λ using Fredholm's method. Differentiating Eq. (A17.3) with respect to τ, we have

$$\hbar \frac{\partial \widehat{G}_\lambda}{\partial \tau} = -\left\langle \mathrm{T}_\tau \left[K_0 + \bar{D}_\lambda(\tau), \widehat{\psi}(\mathbf{r},\tau) \right] \widehat{\psi}^\dagger(\mathbf{r}',\tau') \right\rangle - \hbar\, \delta(\mathbf{r}-\mathbf{r}')\, \delta(\tau-\tau')$$
$$= -\left[\widehat{k}_0 + \lambda\widehat{D}(\mathbf{r},\tau) \right] \widehat{G}_\lambda - \hbar\, \delta(\mathbf{r}-\mathbf{r}')\, \delta(\tau-\tau') \qquad (A17.6)$$

where $\widehat{k}_0$ and $\widehat{D}$ are Nambu matrices defined by Eqs. (5.25) and (5.26), respectively. Equation (A17.6) can be transformed to an integral equation by introducing the free-particle Green's function $\widehat{G}_0$ [see Eq. (A17.2)], which satisfies

$$\left(-\hbar\frac{\partial}{\partial\tau} - \widehat{k}_0\right)\widehat{G}_0(\mathbf{r} - \mathbf{r}', \tau - \tau') = \hbar\,\delta(\mathbf{r} - \mathbf{r}')\,\delta(\tau - \tau') \qquad (A17.7)$$

Subtracting this equation from Eq. (A17.6), we obtain

$$\left(-\hbar\frac{\partial}{\partial\tau} - \widehat{k}_0\right)\left[\widehat{G}_\lambda(\mathbf{r}, \tau; \mathbf{r}', \tau') - \widehat{G}_0(\mathbf{r} - \mathbf{r}', \tau - \tau')\right]$$

$$= \lambda\widehat{D}(\mathbf{r}, \tau)\widehat{G}_\lambda(\mathbf{r}, \tau; \mathbf{r}', \tau')$$

$$= \lambda\int_0^{\hbar\beta} d\tau'' \int d^3r''\,\delta(\mathbf{r} - \mathbf{r}'')\,\delta(\tau - \tau'')$$

$$\times \widehat{D}(\mathbf{r}'', \tau'')\widehat{G}_\lambda(\mathbf{r}'', \tau''; \mathbf{r}', \tau') \qquad (A17.8)$$

Having expressed the RHS of this equation as a superposition of δ functions, we find, using Eq. (A17.7), the solution for $\widehat{G}_\lambda - \widehat{G}_0$ as a superposition of $\widehat{G}_0$ functions. Thus we obtain the desired integral equation for $\widehat{G}_\lambda$ in the form

$$\widehat{G}_\lambda(\mathbf{r}, \tau; \mathbf{r}', \tau') = \widehat{G}_0(\mathbf{r}, \tau; \mathbf{r}', \tau') + \frac{1}{\hbar}\int_0^{\hbar\beta} d\tau'' \int d^3r''$$

$$\times \widehat{G}_0(\mathbf{r}, \tau; \mathbf{r}'', \tau'')\widehat{D}_\lambda(\mathbf{r}'', \tau'')$$

$$\times \widehat{G}_\lambda(\mathbf{r}'', \tau''; \mathbf{r}', \tau') \qquad (A17.9)$$

where we introduced the matrix $\widehat{D}_\lambda = \lambda\widehat{D}$.

Next we Fourier transform Eq. (A17.9) to bring it to a canonical form. This allows us to apply the Fredholm method of solution. We use the expansions

$$\widehat{G}_\lambda(\mathbf{r}, \tau; \mathbf{r}', \tau') = \frac{1}{\hbar\beta\Omega}\sum_{m,m'}\sum_{k,k'}\widehat{G}_\lambda(\mathbf{k}, m; \mathbf{k}', m')$$

$$\times \exp\left[-i(\omega_m\tau - \omega_{m'}\tau') + i(\mathbf{k}\cdot\mathbf{r} - \mathbf{k}'\cdot\mathbf{r}')\right] \quad (A17.10)$$

$$\widehat{G}_0(\mathbf{r}, \tau; \mathbf{r}', \tau') = \frac{1}{\hbar\beta\Omega}\sum_n\sum_q\widehat{G}_0(\mathbf{q}, n)$$

$$\times \exp\left[-i\omega_n(\tau - \tau') + i\mathbf{q}\cdot(\mathbf{r} - \mathbf{r}')\right] \quad (A17.11)$$

$$\frac{1}{\hbar}\widehat{D}_\lambda(\mathbf{r}, \tau) = \sum_p\sum_l\widehat{D}_\lambda(\mathbf{l}, p)\exp\left(-i\omega_p\tau + i\mathbf{l}\cdot\mathbf{r}\right) \qquad (A17.12)$$

where Ω is the sample volume. Inserting Eqs. (A17.10)–(A17.12) into (A17.9), we obtain a set of algebraic equations

$$\widehat{G}_\lambda(\mathbf{k}, n; \mathbf{k}', n') = \delta_{\mathbf{k},\mathbf{k}'}\delta_{n,n'}\widehat{G}_0(\mathbf{k}, n) + \sum_{n'',\mathbf{k}''} \widehat{G}_0(\mathbf{k}, n)\widehat{D}_\lambda(\mathbf{k} - \mathbf{k}'', n - n'')$$
$$\times \widehat{G}_\lambda(\mathbf{k}'', n''; \mathbf{k}', n') \tag{A17.13}$$

A17.2. Fredholm solution of Eq. (A17.13)

Using a "condensed" notation $\mathbf{n} \equiv (\mathbf{k}, n, \sigma)$, where σ indicates the components of the Nambu spinors, Eq. (A17.13) can be written as a matrix equation

$$\left(\widehat{G}_\lambda - \lambda\widehat{T}\widehat{G}_\lambda\right)_{\mathbf{n},\mathbf{n}'} = \left(\widehat{G}_0\right)_{\mathbf{n},\mathbf{n}'} \tag{A17.14}$$

where the matrix $\widehat{T}$ represents the *kernel* of the integral equation

$$\widehat{T} = \widehat{G}_0\widehat{D} \tag{A17.15}$$

Equation (A17.4) is now in a canonical Fredholm form, and the solution for the diagonal matrix element $(\widehat{G}_\lambda)_{\mathbf{n},\mathbf{n}}$ is

$$\left(\widehat{G}_\lambda\right)_{\mathbf{n},\mathbf{n}} = \widehat{G}_0(\mathbf{n}) + \frac{1}{\Delta(\lambda)}\sum_{\mathbf{n}',\mathbf{n}} \widehat{\Delta}_{\mathbf{n},\mathbf{n}'}(\lambda)\widehat{G}_0(\mathbf{n}')\delta_{\mathbf{n},\mathbf{n}'} \tag{A17.16}$$

which can be written as

$$\widehat{G}_\lambda(\mathbf{n}, \mathbf{n}) = \widehat{G}_0(\mathbf{n}) + \frac{1}{\Delta(\lambda)}\widehat{\Delta}_{\mathbf{n},\mathbf{n}}(\lambda)\widehat{G}_0(\mathbf{n}) \tag{A17.17}$$

where $\widehat{\Delta}(\lambda)$ and $\Delta(\lambda)$ are the Fredholm matrix and determinant, respectively, defined by series expansions in the coupling constant λ (Mathews and Walker, 1970). They satisfy the following relations

$$\text{Tr}\,\widehat{\Delta}(\lambda) = -\lambda\frac{d\Delta(\lambda)}{d\lambda} \tag{A17.18}$$

and

$$\Delta(\lambda) = \exp\left[\text{Tr}\ln(\widehat{I} - \lambda\widehat{T})\right] \tag{A17.19}$$

where $\widehat{I}$ is the unit matrix.

REFERENCES

Abeles, B. (1977). *Phys. Rev.* **B15**, 2828.

Abeles, B., and J. J. Hanak (1971). *Phys. Lett.* **34A**, 165.

Abeles, B., P. Sheng, M. C. Coutts, and Y. Arie (1975). *Adv. in Phys.* **24**, 407.

Abrahams, E., and T. Tsuneto (1966). *Phys. Rev.* **152**, 416.

Abramowitz, M., and T. A. Stegun (1964). *Handbook of Mathematical Functions.* Dover, New York.

Abrikosov, A. A., L. P. Gorkov, and I. Yu. Dzyaloshinskii (1965). *Quantum Field Theoretical Methods in Statistical Physics.* Pergamon Press, Oxford.

Aharonov, Y., and D. Bohm (1959). *Phys. Rev.* **115**, 485.

Ambegaokar, V., and A. Baratoff (1963). *Phys. Rev. Lett.* **10**, 486; erratum *Phys. Rev. Lett.* **11**, 104.

Ambegaokar, V., U. Eckern, and G. Schön (1982). *Phys. Rev. Lett.* **46**, 211.

Anderson, P. W. (1959). *J. Phys. Chem. Solids* **11**, 26.

Anderson, P. W. (1964). *Lectures on the Many Body Problems*, Vol. 2. E. R. Caianello, ed., Academic Press, New York.

Anderson, P. W. (1984). *Basic Notions of Condensed Matter Physics.* Frontiers in Physics, Benjamin, New York.

Anderson, P. W., K. A. Muttalib, and T. V. Ramakrishnan (1983). *Phys. Rev.* **28**, 117.

Anderson, P. W., and J. M. Rowell (1963). *Phys. Rev. Lett.* **10**, 230.

Anderson, P. W., and G. Yuval (1971). *J. Phys.* **C4**, 607.

Anderson, P. W., G. Yuval, and D. R. Hamann (1970). *Phys. Rev.* **B1**, 4464.

Bardeen, J., and M. J. Stephen (1965). *Phys. Rev.* **140**, A 1197.

Beasley, R. M., J. E. Mooij, and T. D. Orlando (1979). *Phys. Rev. Lett.* **42**, 1165.

Bednorz, J., and K. A. Müller (1986). *Z. Phys.* **B66**, 189.

Belitz, D. (1987). *Phys. Rev.* **B35**, 1636, 1651.

Bellman, R. (1961). *A Brief Introduction to Theta Functions.* Holt, Rinehart, and Winston, Inc., New York.

Berezinskii, V. L. (1971). *Sov. Phys. JETP* **32**, 493.

Berry, M. V. (1984). *Proc. R. Soc.* **A392**, 45.

Blatter, G., V. B. Geshkenbein, and V. M. Vinokur (1991). *Phys. Rev. Lett.* **66**, 3297.

Bogoliubov, N. N., V. V. Tolmachev, and D. V. Shirkov (1959). *A New Method in the Theory of Superconductivity.* Consultants Bureau, New York.

Bradley, R. M., and S. Doniach (1984). *Phys. Rev.* **B30**, 1138.

Bray, A. J., and M. A. Moore (1982). *Phys. Rev. Lett.* **49**, 1545.

Brown, R., and E. Šimánek (1986). *Phys. Rev.* **B34**, 2957.

Brown, R., and E. Šimánek (1992). *Phys. Rev.* **B45**, 6069.

Caldeira, A. O., and A. J. Leggett (1981). *Phys. Rev. Lett.* **46**, 211.

Caldeira, A. O., and A. J. Leggett (1983). *Ann. Phys.* (New York) **149**, 374.

Callan, C., and S. Coleman (1977). *Phys. Rev.* **D16**, 1762.

Caroli, C., P. G. de Gennes, and J. Matricon (1964). *Phys. Lett.* **9**, 307.

Cha, M., M. P. A. Fisher, S. M. Girvin, M. Wallin, and A. P. Young (1991). *Phys. Rev.* **B44**, 6883.

Chakravarty, S. (1982). *Phys. Rev. Lett.* **49**, 681.

Chakravarty, S., G. L. Ingold, S. Kivelson, and A. Luther (1986). *Phys. Rev. Lett.* **56**, 2303.

Chakravarty, S., S. Kivelson, G. Zimanyi, and B. I. Halperin (1987). *Phys. Rev.* **B35**, 7256.

Coffey, M. (1993). Preprint.

Coleman, S. (1973). *Comm. Math. Phys.* **31**, 259.

de Gennes, P. G. (1966). *Superconductivity in Metals and Alloys.* Benjamin, New York.

de Gennes, P. G., private communication cited by Worthington et al. (1978).

de Gennes, P. G., and J. Matricon (1964). *Rev. Mod. Phy.* **36**, 45.

Deutscher, G., B. Bandyopadhyay, T. Chui, P. Lindenfeld, W. L. McLean, and T. Worthington (1980). *Phys. Rev. Lett.* **44**, 1150

Deutscher, G., M. Gershenson, E. Grünbaum, and Y. Imry (1973). *J. Vac. Sci. Technol.* **10**, 697.

Deutscher, G., Y. Imry, and L. Gunther (1974). *Phys. Rev.* **B10**, 10.

Deutscher, G., and M. L. Rapaport (1978). *J. Phys.* (Paris), Colloq. **39**, C6-581.

Dolan, G. J., G. V. Chandrasekhar, J. R. Dinger, C. Field, and F. Holtzberg (1989). *Phys. Rev. Lett.* **62**, 827.

Doniach, S. (1981). *Phys. Rev.* **B28**, 5063.

Doniach, S. (1984). in *Percolation, Localization and Superconductivity, NATO Advanced Studies Institute Series B.* New York.

Duan, J. M. (1993). *Phys. Rev.* **B48**, 333.

Duan, J. M., and A. J. Leggett (1992). *Phys. Rev. Lett.* **68**, 1216.

Duan, J. M., and E. Šimánek (1993). Preprint.

Dwight, H. B. (1961). *Tables of Integrals and Other Mathematical Data.* Macmillan Company, New York.

Ebisawa, H., S. Maekawa, and H. Fukuyama (1985). *J. Phys. Soc. Jpn.* **54**, 2257.

Eckern, U., and A. Schmid (1989). *Phys. Rev.* **B39**, 6441.

Eckern, U., G. Schön, and V. Ambegaokar (1984). *Phys. Rev.* **B30**, 6419.

Efetof, K. B. (1980). *Zh. Eksp. Theor. Fiz.* **78**, 1017.

Efetof, K. B., and M. M. Salomaa (1981). *J. Low Temp. Phys.* **42**, 35.

Eliashberg, G. M. (1988). *JETP Lett.* **46**, 581.

Fazekas, P. (1982). *Z. Phys.* **B45**, 215.

Fazio, R., A. van Otterlo, G. Schön, H. S. J. van der Zant, and J. E. Mooij (1992). *Helvetica Physica Acta* **65**, 228.

Feigel'man, M. V., V. B. Geshkenbein, A. I. Larkin, and V. M. Vinokur (1989). *Phys. Rev. Lett.* **63**, 2303.

Ferrell, R. A. (1964). *Phys. Rev. Lett.* **13**, 330.

Ferrell, R. A., and B. Mirhashem (1988). *Phys. Rev.* **B37**, 648.

Ferrell, R. A., and R. Prange (1963). *Phys. Rev. Lett.* **10**, 479.

Fetter, A. L. (1967). *Phys. Rev.* **162**, 143.

Fetter, A. L. (1971). *Phys. Rev. Lett.* **27**, 986.

Fetter, A. L., P. C. Hohenberg, and P. Pincus (1966). *Phys. Rev.* **147**, 140.

Fetter, A. L., and J. D. Walecka (1971). *Quantum Theory of Many-Particle Systems*. McGraw–Hill, New York.

Feynman, R. P. (1955). *Progress in Low Temperature Physics*, Vol. I. C. J. Gorter, ed., North Holland, Amsterdam.

Feynman, R. P. (1972). *Statistical Mechanics*. Benjamin, New York.

Feynman, R. P., and A. R. Hibbs (1965). *Quantum Mechanics and Path Integrals*. McGraw–Hill, New York.

Fisher, M. P. A. (1990). *Phys. Rev. Lett.* **65**, 923.

Fisher, M. P. A. (1991). *Physica* **A177**, 553.

Fisher, M. P. A., and G. Grinstein (1988). *Phys. Rev. Lett.* **60**, 208.

Fisher, M. P. A., G. Grinstein, and S. M. Girvin (1990). Phys. Rev. Lett. **64**, 587.

Fisher, M. P. A., and D. H. Lee (1989). *Phys. Rev.* **B39**, 2756.

Fisher, M. P. A., T. A. Tokuyasu, and A. P. Young (1991). *Phys. Rev. Lett.* **66**, 2931.

Fisher, M. P. A., and W. Zwerger (1985). *Phys. Rev.* **B32**, 6190.

Fishman, R. S. (1989). *Phys. Rev.* **B40**, 11014

Friedman, B. (1964). *Principles and Techniques of Applied Mathematics*. Wiley, New York.

Fruchter, L., A. P. Malozemoff, I. A. Campbell, J. Sanchez, M. Konczykowski, R. Griessen, and F. Holtzberg (1991). *Phys. Rev.* **B43**, 8709.

Fukuyama, H., H. Ebisawa, and S. Maekawa (1984). *J. Phys. Soc. Jpn.* **53**, 3560.

Ginzburg, V. L. (1960). *Sov. Phys. Solid State* **2**, 1824.

Giovannini, B., and L. Weiss (1978). *Solid State Commun.* **28**, 1005.

Glazman, L. I., and N. Ya. Fogel (1984). *Sov. J. Low. Temp. Phys.* **10**, 51.

Grabert, H., U. Weiss, and P. Talkner (1984). *Z. Phys. B—Condensed Matter* **55**, 87.

Gross, E. P. (1961). *Nuovo Cimento* **20**, 454.

Hagen, S. J., C. J. Lobb, R. L. Green, M. G. Forrester, and J. H. Kang (1990). *Phys. Rev.* **B41**, 11630.

Haldane, F. D. M., and Y. S. Wu (1985). *Phys. Rev. Lett* **55**, 2887.

Hall, H. E., and W. F. Vinen (1956). *Proc. R. Soc.* **A238**, 215.

Halperin, B. I. (1984). *Phys. Rev. Lett.* **52**, 1583.

Halperin, B. I., and D. R. Nelson (1979). *J. Low Temp. Phys.* **36**, 1165.

Hauge, E. H., and P. C. Hemmer (1971). *Physica Norwegica* **5**, 209.

Hebard, A. F. (1979). in *Inhomogeneous Superconductors, AIP Conf. Proc. No. 58.* AIP, New York.

Hebard, A. F., and M. A. Paalanen (1990). *Phys. Rev. Lett.* **65**, 927.

Hebard, A. F., and J. M. Vandenberg (1980). *Phys. Rev. Lett.* **44**, 50.

Henley, E. M., and W. Thirring (1962). *Elementary Quantum Field Theory.* McGraw–Hill, New York.

Hertz, J. A. (1976). *Phys. Rev.* **B13**, 1165.

Hubbard, J. (1959). *Phys. Rev. Lett.* **3**, 77.

Hubbard, J. (1964). *Proc. R. Soc.* **A277**, 237.

Hurault, J. P. (1968). *J. Phys. Chem. Solids* **29**, 1765.

Ivanchenko, Yu. M., and L. A. Zil'berman (1968). *JETP Lett.* **8**, 113.

Jackel, L. D., J. P. Gordon, E. L. Hu, R. E. Howard, A. L. Fetter, D. M. Tennants, R. W. Epworth, and J. Kurkijärwi (1981). *Phys. Rev. Lett.* **47**, 697.

Jackson, J. D. (1975). *Classical Electrodynamics.* John Wiley, New York.

Jacobs, L., J. V. José, and M. A. Novotný (1984). *Phys. Rev. Lett.* **53**, 2177.

Jacobs, L., J. V. José, M. A. Novotný, and A. Goldman (1987). *Europhys. Lett.* **3**, 1295.

Jaeger, H. M., D. B. Haviland, and A. M. Goldman (1986). *Phys. Rev.* **B34**, 4920.

Jaeger, H. M., D. B. Haviland, B. G. Orr, and A. M. Goldman (1989). *Phys. Rev.* **B40**, 182.

José, J. V. (1984). *Phys. Rev.* **B29**, 2836.

José, J. V., L. P. Kadanoff, S. Kirkpatrick, and D. Nelson (1977). *Phys. Rev.* **B16**, 1217.

Josephson, B. D. (1962). *Phys. Lett.* **1**, 251.

Kampf, A., and G. Schön (1987). *Phys. Rev.* **B36**, 3651.

Kampf, A., and G. Schön (1988). *Physica* **B152**, 239.

Kawabata, A. (1977). *J. Phys. Soc. Japan* **43**, 1491.

Kirchhoff, G. (1883). *Vorlesungen über Mathematische Physik: Mechanik.* P. G. Teubner, Leipzig.

Kleinert, H. (1990). *Path Integrals in Quantum Mechanics, Statistics, and Polymer Physics.* World Scientific Publishing Co., Singapore.

Kobayashi, S. (1990). *Phase Transitions*, Vols. 24–26. Gordon and Breach Science Publ., New York.

Kobayashi, S. (1992). *Surface Science Reports* **16**, 1.

Kobayashi, S., Y. Tada, and W. Sasaki (1980). *J. Phys. Soc. Jpn.* **49**, 2075.

Kosterlitz, J. M. (1974). *J. Phys.* **C7**, 1047.

Kosterlitz, J. M., and D. J. Thouless (1973). *J. Phys.* **C6**, 1181.

Kramers, H. A. (1940). *Physics* **7**, 284.

Kubo, R. (1962). *J. Phys. Soc. Jpn.* **17**, 975.

Lamb, H. (1945). *Hydrodynamics.* Dover Publications, Inc., New York.

Landau, L. D., V. L. reported by Ginzburg (1950). *Usp. Fiz. Nauk.* **42**, 333.

Landau, L. D., and E. M. Lifshitz (1958). *Quantum Mechanics.* Pergamon Press, London.

Landau, L. D., and E. M. Lifshitz (1975). *Fluid Dynamics.* Pergamon Press, Oxford.

Landau, L. D., and E. M. Lifshitz (1980). *Statistical Physics.* Pergamon Press, Oxford.

Langer, J. S. (1967). *Ann. Phys.* **41**, 108.

Langer, J. S., and V. Ambegaokar (1967). *Phys. Rev.* **164**, 498.

Larkin, A. I., and Yu. N. Ovchinnikov (1983). *JETP Lett.* **37**, 382.

Larkin, A. I., Yu. N. Ovchinnikov, and A. Schmid (1988). *Physica* **B152**, 266.

Lee, D. H., and M. P. A. Fisher (1991). *Int. J. Mod. Phys.* **5**, 2675.

Lee, P. A., and T. V. Ramakrishnan (1985). *Rev. Mod. Phy.* **57**, 287.

Leggett, A. J. (1975). *Rev. Mod. Phys.* **47**, 331.

Likharev, K. K., and A. B. Zorin (1985). *J. Low Temp. Phys.* **59**, 347.

Lin, T. H., X. Y. Shao, M. K. Wu, P. H. Hor, C. H. Jin, C. W. Chu, N. Evans, and R. Bayuzick (1984). *Phys. Rev.* **B29**, 1493.

Liu, Y., D. B. Haviland, L. I. Glazman, and A. M. Goldman (1992). *Phys. Rev. Lett.* **68**, 2224.

Lobb, C. J., D. W. Abraham, and M. Tinkham (1983). *Phys. Rev.* **B27**, 150.

Ma, Shang-Keng., and R. Rajaraman (1975). *Phys. Rev.* **D11**, 1701.

Maekawa, S., and H. Fukuyama (1982). *J. Phys. Soc. Jpn.* **51**, 1380.

Maekawa, S., H. Fukuyama, and S. Kobayashi (1981). *Solid State Commun.* **37**, 45.

Mathews, J., and R. L. Walker (1970). *Mathematical Methods of Physics*. Benjamin, New York.

McCauley, J. L. (1979). *J. Phys. A: Math. Gen.* **12**, 1999.

McCumber, D. E., and B. I. Halperin (1970). *Phys. Rev.* **B1**, 1054.

McLean, W. L., and M. J. Stephen (1979). *Phys. Rev.* **B19**, 5925.

Mermin, N. D. (1967). *J. Math. Phys.* **8**, 1061.

Messiah, A. (1961). *Quantum Mechanics*, Vol. 1. Interscience Publishers, New York.

Mooij, J. E. (1984). in *Percolation, Localization and Superconductivity*. A. M. Goldman and S. Wolf, eds., Plenum, New York.

Mooij, J. E., and G. Schön (1992). in *Single Charge Tunneling*. H. Grabert and M. H. Devoret, eds., Plenum Press, New York.

Mooij, J. E., B. J. van Wees, L. J. Geerligs, M. Peters, R. Fazio, and G. Schön (1990). *Phys. Rev. Lett.* **65**, 645.

Mota, A. C., A. Pollini, P. Visani, K. A. Müller, and J. G. Bednorz (1987). *Phys. Rev.* **B36**, 4011.

Mota, A. C., A. Pollini, P. Visani, K. A. Müller, and J. G. Bednorz (1988). *Physica Scripta* **37**, 823.

Mott, N. F. (1974). *Metal Insulator Transitions*. Taylor and Francis, London.

Mott, N. F., and E. A. Davis (1971). *Electronic Processes in Noncrystalline Materials*. Clarendon Press, Oxford.

Mühlschlegel, B. (1962). *J. Math. Phys.* **3**, 522.

Mühlschlegel, B., D. J. Scalapino, and R. Denton (1972). *Phys. Rev.* **B6**, 1767.

Nambu, Y. (1960). *Phys. Rev.* **117**, 648.

Nelson, D., and J. M. Kosterlitz (1977). *Phys. Rev. Lett.* **39**, 1201.

Newbower, R. S., M. R. Beasley, and M. Tinkham (1972). *Phys. Rev.* **B5**, 864.

Nozieres, P., and W. F. Vinen (1966). *Philos. Mag.* **14**, 667.

Onsager, L. (1949). *Suppl. Nuovo Cimento* **6**, 279.

Orr, B. G., H. M. Jaeger, and A. M. Goldman (1985). *Phys. Rev.* **B32**, 4920.

Orr, B. G., H. M. Jaeger, A. M. Goldman, and C. G. Kuper (1986). *Phys. Rev. Lett.* **56**, 378.

Patton, S. R., W. Lamb, and D. Stroud (1979). in *Inhomogeneous Superconductors, AIP Conf. Proc. No. 58.* AIP, New York.

Pearl, J. (1964). *Appl. Phys. Lett.* **5**, 65.

Pincus, P., P. Chaikin, and C. F. Coll (1973). *Solid State Commun.* **12**, 1265.

Pitaevskii, L. P. (1961). *Sov. Phys. JETP* **13**, 451

Pokrovskii, V. L., and G. V. Uimin (1973). *Phys. Lett.* **45A**, 467.

Polyakov, A. M. (1977). *Nucl. Phys.* **B120**, 429.

Popov, V. N. (1983). *Functional Integrals in Quantum Field Theory and Statistical Physics.* Reidel, Boston.

Ramond, P. (1981). *Field Theory: A Modern Primer.* Benjamin, New York.

Rapaport, D. C. (1972). *J. Phys.* **C5**, 1830.

Rice, T. M. (1965). *Phys. Rev.* **140**, A1889.

Rice, T. M. (1967). *J. Math. Phys.* **8**, 158.

Riseborough, P. S., P. Hänggi, and E. Freidkin (1985). *Phys. Rev.* **A32**, 489.

Robinson, P. D. (1968). *Fourier and Laplace Transforms.* Routledge and Kegan Paul Ltd., London.

Rogers, C. T., and R. A. Buhrman (1984). *Phys. Rev. Lett.* **53**, 1272.

Rogers, C. T., and R. A. Buhrman (1985). *Phys. Rev. Lett.* **55**, 859.

Rosenblatt, J. (1974). *Phys. Rev. Sppl.* **9**, 217.

Rosenblatt, J., H. Cortés, and P. Pellan (1970). *Phys. Lett.* **A33**, 143.

Rosenblatt, J., A. Raboutou, and P. Pellan (1975). *Proc. 13th Conf. Low Temp. Phys.* Plenum Press, New York.

Ryzhik, I. M., and I. S. Gradshteyn (1965). *Tables of Integrals, Series and Products.* Academic Press, New York.

Scalapino, D. J. (1969). in *Tunneling Phenomena in Solids.* E. Burstein and S. Lindquist, eds., Plenum, New York.

Schafroth, M. R. (1951). *Helv. Phys. Acta* **24**, 645.

Schiff, L. (1968). *Quantum Mechanics.* McGraw–Hill, New York.

Schmid, A. (1983). *Phys. Rev. Lett.* **51**, 1506.

Schmidt, V. V. (1967). *Proc. 9th Conf. Low Temp. Phys.*, Vol. 2B. Moscow.

Schön, G., and A. D. Zaikin (1990). *Phys. Reports* **198**, 237.

Shulman, L. (1981). *Techniques and Applications of Path Integration.* Wiley, New York.

Šimánek, E. (1979). *Solid State Commun.* **31**, 419.

Šimánek, E. (1980a). *Phys. Rev.* **B22**, 459.

Šimánek, E. (1980b). *Phys. Rev. Lett.* **45**, 1442.

Šimánek, E. (1982). *Phys. Rev.* **B23**, 5762.

Šimánek, E. (1983). *Solid State Commun.* **48**, 1023.

Šimánek, E. (1985). *Phys. Rev.* **B32**, 500.

Šimánek, E. (1989a). *Phys. Rev.* **B39**, 11384.

Šimánek, E. (1989b). *Phys. Lett.* **A139**, 183.

Šimánek, E. (1990). *Physica* **A164**, 147.

Šimánek, E. (1991). *Phys. Lett.* **154A**, 309.

Šimánek, E. (1992). *Phys. Rev.* **B46**, 14054.

Šimánek, E. (1993). *Phys. Lett.* **177A**, 367.

Šimánek, E., and R. Brown (1986). *Phys. Rev.* **B34**, 3495.

Šimánek, E., and K. Stein (1984). *Physica* **129A**, 40.

Simkin, M. V. (1991). *Phys. Rev.* **B44**, 7074.

Stanley, H. E., and T. A. Kaplan (1966). *Phys. Rev. Lett.* **17**, 913.

Stephen, M., and H. Suhl (1964). *Phys. Rev. Lett.* **13** 797.

Stone, M. (1977). *Phys. Lett.* **67B**, 186.

Stratonovich, R. L. (1958). *Soviet Phys. Dokl.* **2**, 416.

Strongin, M., R. S. Thompson, O. F. Kammerer, and J. E. Crow (1970). *Phys. Rev.* **B1**, 1078.

Suhl, H. (1965). *Phys. Rev. Lett.* **14**, 226.

Susskind, L., and J. Glogower (1964). *Physics* **1**, 49.

Testardi, L. R., W. G. Moulton, H. Mathias, H. K. Ng, and C. M. Rey (1988). *Phys. Rev.* **B37**, 2324.

t'Hooft, G. (1976). *Phys. Rev. Lett.* **37**, 8.

Tinkham, M. (1975). *Introduction to Superconductivity.* McGraw–Hill, New York.

Tucker, J., and B. I. Halperin (1971). *Phys. Rev.* **B3**, 378.

van der Zant, H. S. J., F. C. Fritschy, T. P. Orlando, and J. E. Mooij (1991). *Phys. Rev. Lett.* **66**, 2531.

Villain, J. (1975). *J. Phys. (Paris)* **36**, 581.

von Klitzing, K., G. Dorda, and M. Pepper (1980). *Phys. Rev. Lett.* **45**, 494.

Voss, R. F., and R. A. Webb (1981). *Phys. Rev. Lett.* **47**, 265.

Ward, R. C., S. Cremer, and E. Šimánek (1978). *Phys. Rev.* **B4**, 2162.

Wegner, F. (1967). *Z. Phys.* **206**, 465.

Weinkauf, A., and J. Zittartz (1974). *Solid State Commun.* **14**, 365.

Widom, A., and S. Badjou (1988). *Phys. Rev.* **B37**, 7915.

Wiegel, F. W. (1973). *Physica* **65**, 321.

Wilczek, F. (1982). *Phys. Rev. Lett.* **49**, 957.

Wilks, J. (1967). *The Properties of Liquid and Solid Helium*. Oxford University Press, Oxford.

Wood, D. M., and D. Stroud (1982). *Phys. Rev.* **B25**, 1600.

Worthington, T., P. Lindenfeld, and G. Deutscher (1978). *Phys. Rev. Lett.* **41**, 316.

Yamada, R., S. Katsumoto, and S. Kobayashi (1992). *J. Phys. Soc. Jpn.* **61**, 2656.

Yang, C. N. (1962). *Rev. Mod. Phys.* **34**, 694.

Yoshikawa, N., T. Akeoshi, M. Kojima, and M. Sugahara (1987). *Japanese J. Appl. Phys.* **26**, 949.

Young, A. P. (1978). *J. Phys.* **C11**, L453.

INDEX